Lecture Notes in Computer Science　　10440

Commenced Publication in 1973
Founding and Former Series Editors:
Gerhard Goos, Juris Hartmanis, and Jan van Leeuwen

More information about this series at http://www.springer.com/series/7409

Ladjel Bellatreche · Sharma Chakravarthy (Eds.)

Big Data Analytics and Knowledge Discovery

19th International Conference, DaWaK 2017
Lyon, France, August 28–31, 2017
Proceedings

 Springer

Editors
Ladjel Bellatreche
LIAS/ISAE-ENSMA
Chasseneuil
France

Sharma Chakravarthy
University of Texas at Arlington
Arlington, TX
USA

ISSN 0302-9743   ISSN 1611-3349   (electronic)
Lecture Notes in Computer Science
ISBN 978-3-319-64282-6   ISBN 978-3-319-64283-3   (eBook)
DOI 10.1007/978-3-319-64283-3

Library of Congress Control Number: 2017947764

LNCS Sublibrary: SL3 – Information Systems and Applications, incl. Internet/Web, and HCI

Printed on acid-free paper

This Springer imprint is published by Springer Nature
The registered company is Springer International Publishing AG
The registered company address is: Gewerbestrasse 11, 6330 Cham, Switzerland

Preface

Big data analytics and knowledge discovery technologies have been growing over time and they are now a practical need in every major small or large company. The adoption of these technologies by researchers and industries encourages these companies to integrate heterogeneous, distributed, autonomous, and evolving data with high variety coming from traditional (databases) and advanced sources—such as sensors and social networks, IoT— into a single large database to enable advanced querying, analysis, and recommendation. With the explosion of the diversity of deployment platforms motivated by high-performance computing (HPC) and advanced programming paradigms, offering the process of retrieval, knowledge discovery from this huge amount of heterogeneous complex data and sentiment analysis forms the litmus test for research in this area. Faced with this volume of data managed by the new generation of data warehouses, companies are very sensitive to non-functional requirement satisfaction, by proposing mathematical cost models estimating different metrics such as query performance, elasticity, etc.

During the past few years, the International Conference on Big Data Analytics and Knowledge Discovery (DaWaK) has become one of the most important international scientific events bringing together researchers, developers, and practitioners to discuss the latest research issues and experiences in developing and deploying a new generation of data warehouses and knowledge discovery systems, applications, and solutions. This year's conference (DaWaK 2017) built on this tradition of facilitating the cross-disciplinary exchange of ideas, experience, and potential research directions. DaWaK 2017 sought to introduce innovative principles, methods, models, algorithms and solutions, industrial products, and experiences to challenging problems faced in the development of a new generation of data warehouses in the big data era, knowledge discovery, data-mining applications, and the emerging area of HPC.

This year we received 97 papers and the Program Committee finally selected 24 full papers and 11 short papers, making an acceptance rate of 25%. The accepted papers cover a number of broad research areas on both theoretical and practical aspects of new generations of data warehouses and knowledge discovery. In the area of big data, the topics covered included the modeling and designing of a new generation of data warehouses by considering the big data aspects, data flows, NoSQL databases, cloud computing, cost models, data streaming, advanced programming paradigms (map-reduce, Hadoop), query optimization, data privacy, multidimensional analysis of text documents, and data warehousing and big data for real-world applications such as agricultural trade data, air quality measurement, electrical vehicle, etc. In the areas of data mining and knowledge discovery, the topics included traditional data-mining topics such as frequent item sets, association, etc., and machine-learning techniques. It was especially notable that some papers covered emerging real-world applications with a special focus on social networks.

Due to the maturity of DaWaK, several editors of well-known journals support our conference. This year, we had three special issues of well-known journals: *Distributed and Parallel Databases*, Springer, *Journal of Concurrency and Computation: Practice and Experience*, Wiley, and *Transactions on Large-Scale Data- and Knowledge-Centered Systems*, TLDKS, Springer.

We would like to thank all authors for submitting their research papers to DaWaK 2017. We express our gratitude to all the Program Committee members and to the external reviewers, who reviewed the papers thoroughly and in a timely manner. Finally, we would like to thank Gabriela Wagner for her endless help and support.

Hope you enjoy the proceedings.

June 2017

Ladjel Bellatreche
Sharma Chakravarthy

Organization

Program Committee Co-chairs

Ladjel Bellatreche   LIAS/ISAE-ENSMA, Poitiers, France
Sharma Chakravarthy   The University of Texas at Arlington, USA

Program Committee

Alberto Abelló   Universitat Politecnica de Catalunya, Spain
Sonali Agarwal   Indian Institute of information Technology, Allahabad,
                              India
Mohammed Al-Kateb   Teradata Labs, USA
Toshiyuki Amagasa   University of Tsukuba, Japan
Torben Bach Pedersen   Aalborg University, Denmark
Elena Baralis   Politecnico di Torino, Italy
Ladjel Bellatreche   ENSMA, France
Sadok Ben Yahia   Faculty of Sciences of Tunis, Tunisia
Jorge Bernardino   ISEC - Polytechnic Institute of Coimbra, Portugal
Mikael Berndtsson   University of Skovde, Sweden
Vasudha Bhatnagar   Delhi University, India
Omar Boussaid   University of Lyon, France
Stephane Bressan   National University of Singapore, Singapore
Sharma Chakravarthy   The University of Texas at Arlington, USA
Isabelle Comyn-Wattiau   ESSEC Business School, Paris, France
Bruno Cremilleux   Université de Caen, France
Alfredo Cuzzocrea   University of Trieste, Italy
Laurent d'Orazio   University of Rennes 1, France
Karen Davis   University of Cincinnati, USA
Claudia Diamantini   Università Politecnica delle Marche, Italy
Alin Dobra   University of Florida, USA
Josep Domingo-Ferrer   Universitat Rovira i Virgili, Spain
Dejing Dou   University of Oregon, USA
Curtis Dyreson   Utah State University, USA
Markus Endres   University of Augsburg, Germany
Leonidas Fegaras   The University of Texas at Arlington, USA
Filippo Furfaro   DIMES - University of Calabria, Italy
Pedro Furtado   Universidade de Coimbra, Portugal, Portugal
Carlos Garcia-Alvarado   Amazon
Kazuo Goda   University of Tokyo, Japan
Matteo Golfarelli   University of Bologna, Italy
Sergio Greco   University of Calabria, Italy
Takahiro Hara   Osaka University, Japan

Frank Hoppner	Ostfalia University of Applied Sciences, Germany
Yoshiharu Ishikawa	Nagoya University, Japan
Stéphane Jean	LIAS/ISAE-ENSMA and University of Poitiers, France
Lili Jiang	Umeå University, Sweden
Vana Kalogeraki	Athens University of Economics and Business, Greece
Selma Khouri	LCSI/ESI, Algeria and LIAS/ISAE-ENSMA, France
Uday Kiran	University of Tokyo, Japan
Nhan Le-Thanh	Nice Sophia Antipolis University, France
Jens Lechtenboerger	Westfalische Wilhelms - Universität Münster, Germany
Wookey Lee	Inha University, South Korea
Carson K. Leung	University of Manitoba, Canada
Sofian Maabout	University of Bordeaux, France
Sanjay Kumar Madria	Missouri University of Science and Technology, USA
Yannis Manolopoulos	Aristotle University, Greece
Patrick Marcel	Université François Rabelais Tours, France
Amin Mesmoudi	Poitiers University, France
Jun Miyazaki	Tokyo Institute of Technology, Japan
Anirban Mondal	Shiv Nadar University, India
Yasuhiko Morimoto	Hiroshima University, Japan
Makoto Onizuka	Osaka University, Japan
Carlos Ordonez	University of Houston, USA
Alex Poulovassilis	Birkbek, University of London, UK
Praveen Rao	University of Missouri-Kansas City, USA
Goce Ristanoski	Data61, CSIRO, Australia
Laura Rusu	IBM, Australia
Alkis Simitsis	HP Labs, USA
David Taniar	Monash University, Australia
Olivier Teste	IRIT, University of Toulouse, France
Dimitri Theodoratos	New Jersey Institute of Technology, USA
Predrag Tosic	Washington State University, USA
Panos Vassiliadis	University of Ioannina, Greece
Guangtao Wang	NTU, Singapore
Robert Wrembel	Poznan University of Technology, Poland
Haruo Yokota	Tokyo Institute of Technology, Japan
Osmar Zaiane	University of Alberta, Canada

Additional Reviewers

Bijay Neupane	Aalborg University, Denmark
Christian Thomsen	Aalborg University, Denmark
Muhammad Aamir Saleem	Aalborg University, Denmark
Yuya Sasaki	Osaka University, Japan
Hieu Hanh Le	Tokyo Institute of Technology, Japan
Dominique Li	University of Tours, France
Yuto Hayamizu	University of Tokyo, Japan

Hiroyuki Yamda	University of Tokyo, Japan
Sharanjit Kaur	University of Delhi, India
Rakhi Saxena	University of Delhi, India
Swagata Duari	University of Delhi, India
Fan Jiang	University of Manitoba, Canada
Adam Pazdor	University of Manitoba, Canada
Syed Tanbeer	University of Manitoba, Canada
Xiaoying Wu	Wuhan University, China
Aggeliki Dimitriou	National Technical University of Athens, Greece
Souvik Sinha	New Jersey Institute of Technology, USA
Antonio Corral	University of Almeria, Spain
Anastasios Gounaris	Aristotle University of Thessaloniki, Greece
Johannes Kastner	University of Augsburg, Germany
Lena Rudenko	University of Augsburg, Germany
Rohit Kumar	Université Libre de Bruxelles, Belgium
Rana Faisal Munir	Universitat Politècnica de Catalunya, Spain
Luca Cagliero	Politecnico di Torino, Italy
Evelina Di Corso	Politecnico di Torino, Italy
Paolo Garza	Politecnico di Torino, Italy
Trung Dung LE	Université de Rennes 1, France
Anas Katib	University of Missouri-Kansas City, USA
Monica Senapati	University of Missouri-Kansas City, USA
Dig Vijay Kumar Yarlagadda	University of Missouri-Kansas City, USA
Rodrigo Rocha Silva	São Paulo State Technological College, FATEC-MC, Brazil
Emanuele Storti	Università Politecnica delle Marche, Italy
Alex Mircoli	Università Politecnica delle Marche, Italy

Contents

Non-functional Requirements Satisfaction

Machine Learning

Data Flow Management and Optimization

New Generation Data Warehouses Design

Evaluation of Data Warehouse Design Methodologies in the Context of Big Data

Francesco Di Tria[✉], Ezio Lefons, and Filippo Tangorra

Dipartimento di Informatica, Università degli Studi di Bari Aldo Moro,
via Orabona 4, 70125 Bari, Italy
{francesco.ditria,ezio.lefons,filippo.tangorra}@uniba.it

Abstract. The data warehouse design methodologies require a novel approach in the Big Data context, because the methodologies have to provide solutions to face the issues related to the 5 Vs (Volume, Velocity, Variety, Veracity, and Value). So it is mandatory to support the designer through automatic techniques able to quickly produce a multidimensional schema using and integrating several data sources, which can be also unstructured and, therefore, need an ontology-based reasoning. Accordingly, the methodologies have to adopt agile techniques, in order to change the multidimensional schema as the business requirements change, without a complete design process. Furthermore, hybrid approaches must be used instead of the traditional data-driven or requirement-driven approaches, in order to avoid missing the adhesion to user requirements and to produce a valuable multidimensional schema compliant with data sources. In the paper, we perform a metric comparison among different methodologies, in order to demonstrate that methodologies classified as hybrid, ontology-based, automatic, and agile are tailored for the Big Data context.

Keywords: Metrics · Automation · Agility · Ontology

1 Introduction

The design of data warehouses in the context of Big Data deals with new data sources, such as social networks and sensor networks, that daily generate massive data (*Volume* issue) [1]. Data generated from the Web are sometimes unstructured (*Variety* issue), such as tweets or blog posts [2, 3]. However, just these data are of great importance, since they are able to provide valuable feedbacks (*Value* issue) to companies about user preferences [4]. New data sources arise very frequently (*Velocity* issue) and, then, a key factor of success depends on the ability to exploit new available data of good quality (*Veracity* issue) as soon as possible in order to be competitive. We discuss each issue in what follows.

- *Variety* issue. Recent papers have introduced a semantic level in multidimensional design, on the basis of an ontological approach [5–9]. Since an ontology is a machine-processable conceptual representation of a domain of interest, it is used for solving in automatic way syntactical and semantic inconsistencies in the schemas integration process, even in presence of unstructured data.

© Springer International Publishing AG 2017
L. Bellatreche and S. Chakravarthy (Eds.): DaWaK 2017, LNCS 10440, pp. 3–18, 2017.
DOI: 10.1007/978-3-319-64283-3_1

- *Volume* issue. Using a data warehouse for analyzing big data means applying the multidimensional modeling and OLAP operators without importing and storing tens of terabytes or more into the data warehouse (ETL process). The strategy is based on a virtual data warehouse, where the data to be used in the analytical phase are immediately available, and so, in the underlying architecture, the movement of data among systems is avoided and the delays of the importing phase for feeding the data warehouse are discarded [10]. As an alternative to the virtual data warehouse approach, emergent non relational models adopted in NoSQL databases provide more flexibility, for they allow denormalized and join-less schemas that can be exploited for analysing data according to novel paradigms, besides the traditional OLAP operators [11]. So, *non relational* models are actually replacing traditional logical models (*viz* ROLAP and MOLAP) [12].
- *Veracity* issue. A secondary aim of the ETL process is to feed the data warehouse using cleaned data in order to ensure data quality. On the other hand, in a virtual environment ensuring quality of the data means running a batch process that marks dirty data and, therefore, presents only a selected subset of data, in order to discard information unnecessary for decision making [13].
- *Velocity* issue. The velocity is related to the necessity of integrating a new data source and accepting new business requirements as soon as possible, in order to timely provide updated information. This aim can be reached using automatic and agile techniques. The former simulates the reasoning of an expert designer, by avoiding repetitive tasks and human error [14, 15]. The latter produces changes in a data warehouse schema without performing a complete design process, letting the data warehouse evolve as business requirements change [16].
- *Value* issue. This problem is faced using hybrid methodologies, which take into account the best features of traditional methodologies. Applying these methodologies, the designer is sure to produce a multidimensional schema that not only agrees with data source but also does not miss any requirement and does not discard any data source. As a counterpart, these methodologies result quite complex because they integrate and reconcile both the requirement and the data oriented approaches [17–20].

In the paper, we discuss the criteria that characterize the design of data warehouses in the context of Big Data and present a set of metrics for the evaluation of these criteria. In detail, the metrics evaluate both the level of automation and agility of a methodology in reference to the quality of the multidimensional schema produced by that methodology. The contribution is a demonstration that, in the Big Data context, methodologies classified as *ontology-based*, *automatic*, and *agile* require less effort than the others, while *hybrid* methodologies are able to produce multidimensional schemas of a better quality, especially when novel logical models are used. To do so, we present a case study applied to several methodologies, falling in different categories, and we perform a metric comparison of the design process in order to highlight the features of each methodology.

The paper is organized as follow. Section 2 overviews the classification of the methodologies based on suitable criteria. Section 3 illustrates the metrics we used for each classification criterion. Section 4 presents the experimental results of the application of the metrics to different methodologies. Section 5 concludes the paper with an outline of the experiments and future research issues.

2   Methodology Classification

Design methodologies can be classified according to several general criteria, that take into account how a methodology conceives and perceives the design problem and the methods and strategies it provides for supporting the designer. The criteria surveyed in [21] are, to name a few, the *paradigm*, which may be data-driven, requirement-driven, or hybrid, and the *application*, which may be automatic, semiautomatic, or manual. Among all the criteria present in literature, we are only interested in those relevant for the Big Data context, namely the *paradigm* and the *application*. Then, we add both the *agility* and the *ontological approach* criteria, for they affect the *Velocity*, *Value*, and *Variety* issues.

According to the *paradigm*, methodologies can be

- *Data-driven*. A data-driven approach, also known as *supply-driven*, aims at defining multidimensional schemas through a *made-by-expert* reengineering process of the data sources, minimizing the participation of end users and, consequently, going towards a possible failure of their expectations, due to a missing of adhesion to requirements. Furthermore, these approaches need well-structured data, since functional dependencies are taken into account in the remodeling phase [22]. Therefore, when dealing with unstructured data, they need a manual pre-processing phase devoted to normalize data.
- *Requirement-driven*. A requirement-driven approach, also known as *demand-driven* or *goal-oriented*, aims at defining multidimensional schemas using business goals resulting from the needs of the decision makers. The data sources are considered later, when the ETL process must be designed. In this step, the multidimensional concepts have to be mapped on the data sources in order to program the feeding plan of the data warehouse. At this point, it may happen that the designer discovers that the needed data are not available or, in case of unstructured data, these mappings result an hard task to be completed. On the other hand, some data sources containing interesting information, albeit available, may be omitted and not exploited [23].
- *Hybrid*. This class can be further specialized in *pure* hybrid methodologies and *integration-derived* methodologies. The former group includes all the methodologies that perform the design process considering simultaneously the data sources and the business goals. The latter comprises methodologies that combine and integrate a data-driven approach with a requirement-driven one, which, in turn, can be divided into *sequential* and *parallel* hybrid methodologies. In sequential hybrid methodologies, the two stages are executed according to a prefixed order, and the output of the first stage is used as input of the second stage. In parallel hybrid methodologies, the two stages are executed independently and, at the end, the comparison and integration of the schemas coming from the different stages are performed [20]. Since hybrid methodologies capture the best features of both the opposite paradigms, the designer can produce a multidimensional schema compliant with data sources, without missing any requirement.

According to the *application*, methodologies can be

- *Automatic/Semiautomatic.* In a (semi)automatic methodology, (a number of) the steps of the design process are executed in a supervised way using algorithms able to simulate an expert designer's reasoning. This reduces the design efforts and supports the designer by avoiding errors and repetitive tasks. On the basis of automatic phases, we are also able to include and to integrate new data sources on the fly, as required by the business goals. So, the data warehouse evolves rapidly in a manner consistent with business needs and uses the novel data sources as they become available on the market [24].
- *Manual.* A manual methodology does not provide any form of automation supporting the design process. This means that the whole design process in based on both the designer's experience and choices.

According to the *agility*, methodologies can be

- *Incremental.* A methodology including incremental steps allows to dynamically modify a previous schema without performing a design process starting from zero and to adapt it to novel business requirements. So, the data warehouse evolves as business goals change. In this case, the methodology provides solutions to quickly consider further requirements emerged in the lifecycle, on the basis of agile methods that address the frequent changes in user requirements, in order to have the minimum impact on the design process [25].
- *Non incremental.* Non incremental methodologies require to follow all the steps of the design process every time new data sources must be considered or new business requirements emerge. It is worth noting that new requirements may emerge as a consequence of the actual business trend and, then, they affect the current data warehouse that may become obsolete. Therefore, traditional methodologies that present a slow reaction to changes in business requirements and do not allow a fast adjustment to a data warehouse are not suitable for the data warehouse's consistent evolution.

According to the *ontological-approach*, methodologies can be

- *Ontology-based.* An ontology is a formal and shared definition of a domain at a very high level of abstraction, where concepts along their relationships are represented. These can be used to add a knowledge layer for solving syntactic and semantic inconsistencies when integrating data sources, even if these are unstructured. Furthermore, these are machine-processable and, then, reasoning can be automated in order to apply a design process and directly derive a multidimensional schema from the ontological concepts [8].
- *Human-based.* These methodologies always rely on the designer experience and knowledge in order to solve inconsistencies arising in the integration process. We observe that the integration involves two levels: (i) schema integration, usually applied in data-driven methodologies for creating a global and reconciled schema; and (ii) data integration, usually performed in the ETL process when data are first extracted and then loaded into the data warehouse in a standardized form. (As an example, the format of dates can be different in the various data sources).

According to the *logical modeling*, methodologies can be

- *ROLAP*, based on the relational model, produces popular star or snow-flake schemas;
- *MOLAP*, based on multidimensional structures;
- *HOLAP*, based on hybrid ROLAP and MOLAP solutions; and
- *Non relational*. Emergent logical models, widely known as non-relational, are utilized in the Big Data context for NoSQL databases: the key-value, column-oriented, graph-oriented, and document-oriented models. They allow a more flexible design, due to their schema-less nature, and present a strong denormalization of data. Furthermore, NoSQL databases are characterized by low answering times and their ability to exploit horizontal scalability.

3 Metrics for Design Evaluation of Methodologies

For the criteria established in Sect. 2, we defined a set of metrics for evaluating the data warehouse design process in the context of Big Data. The metrics, along with their relationships with the criteria and the Big Data issues, are summarized in Table 1. In detail, we evaluate the effort to be spent by adopting a methodology (*i.e.*, the *costs*) and the quality of the multidimensional schema produced by that methodology (*i.e.*, the *benefits*) [26].

Table 1. Criteria and metrics for evaluating Big Data Warehouses.

Criterion	Issue	Metrics
Automation	Velocity	Metrics for methodology evaluation
Agility	Velocity	Metrics for methodology evaluation
Ontological approach	Variety	Metrics for methodology evaluation
Hybrid approach	Value	Metrics for schema quality evaluation

3.1 Metrics for Methodology Evaluation

The metrics we defined for evaluating the design effort of a methodology M are grouped into five classes: *artifacts*, *phases*, *automation*, *ontological-approach*, and *agility*.

The *artifacts* metrics aim at evaluating the number of artifacts necessary to start/complete the data warehouse design. The underlying assumption is that the higher the number of artifacts, the greater the design effort.

- $nia(M)$, number of artifacts to be given in input to M;
- $noa(M)$, number of artifacts produced in output by M;
- $ninta(M)$, number of intermediate artifacts produced by M; and
- $na(M)$, number of artifacts involved in M or $nia(M) + noa(M) + ninta(M)$.

The *phases* metrics check the number of phases to be executed in the methodology. So, we consider a methodology M presenting a high number of phases quite difficult to be adopted and, then, we penalize methodologies requiring a complex workflow.

- $np(M)$, number of phases in M.

The *automation* metrics aim at evaluating the automation degree of methodology M.

- $npa(M)$, number of phases automatically executed in M; and
- $naa(M)$, number of artifacts automatically produced by M.

The *ontological-approach* metrics aim at evaluating whether the methodology uses an ontology.

- $no(M)$, number of ontologies used by M.

The *agility* metrics aim at evaluating the agility degree of the methodology. Practically, the agile phases are those that can be avoided when performing an adjustment to the schema.

- $npag(M)$, number of agile phases in M.

Indeed, phases that can be automatically executed do not affect the design effort. Accordingly, also agile phases reduce the design effort. On the basis of this assumption, we minimize the computation of these metrics in presence of a high number of automatic and/or agile phases or artifacts. Similarly, the ontology is an artifact that reduces the effort in the design process. So, the final design effort d using a methodology M is evaluated as follows

$$d(M) = (np(M) - npa(M)) + (np(M) - npag(M)) + (na(M) - naa(M) - no(M)).$$

3.2 Metrics for Schema Quality Evaluation

The metrics we defined for evaluating the quality of a schema S produced by M are:

- $c(S_M)$, the *complexity* of the schema S produced by adopting M;
- $mr(S_M)$, the number of missing requirements;
- $v(S_M)$, the number of violations of the schema S produced by adopting M; and
- $peq(S_M)$, the percentage of queries executable against S.

The metrics for evaluating the *complexity* of a star schema S (*viz* $c(S_M)$) are those proposed by Serrano *et al.* —*cf.* [27] for a complete dissertation.

- $nft(S)$, the number of fact tables of S;
- $ndt(S)$, the number of dimension tables of S; and
- $nfk(S)$, the number of foreign keys of all fact tables of S.

In [28], the authors introduce a new metrics, namely $nmft(S)$, which measures the number of facts in the fact tables or the number of attributes in the fact tables that are not foreign keys. Moreover, they state that nft, ndt, nfk, and $nmft$ are good indicators for the data warehouse quality, since they affect understandability and cognitive complexity. So, we can use only these metrics for measuring the complexity of a schema S. Therefore, we define the *complexity* c of schema S as the sum of the single metrics, that is,

$$c(S_M) = nft(S) + ndt(S) + nfk(S) + nmft(S).$$

So, the quality q of a schema S_M is evaluated as follows

$$q(S_M) = \frac{1}{c(S_M)+mr(S_M)+v(S_M)} + peq(S_M).$$

4 Experimental Results

The case study refers to the analysis of scientific activities. The aim is counting the number of publications and the number of awards won per author.

Two are the data sources we utilized: (a) DBLP database; and (b) BIOS document. The DBLP database includes the scientific publications, available in xml format at http:// dblp.uni-trier.de/xml/, and stores a set of records, wherein each record is *article, inproceedings, proceedings, book, incollection, phdthesis, mastersthesis,* or *www.* Records present almost the same structure. Some of the common fields are *author, editor, title, booktitle, pages, year.* Several elements of the xml file are not normalized, such as *pages* in *article* (`<pages>58–65</pages>`), where the fields *from* and *to* are not specified, though the elements present a regular expression. The same happens for *author,* where the *first name* and the *last name* are not structured. Other elements present a strong redundancy. As an example, the elements *author* and *editor* are repeated for each record, for a person *identifier* is not used. This may leads to problems of homonymy and inconsistencies among records (see [29] for an example). The BIOS document includes the bibliographies of scientists, available in JSON format at https://docs.mongodb.com/ manual/reference/bios-example-collection for MongoDB databases. The information in the document includes: the author identifier, his/her first name, his/her last name, his/her title, his/her birth date, his/her death date, the list of contributions, the list of awards, containing the name of the award, the date he/she won the award, and the name of the Institution that released the award. We note that the document is equivalent to unstructured data, for it is schema-less. Indeed, some fields (*Death* and *Title,* for example) are absent for some scientists. Furthermore, the list of contributions is an array and the list of awards is an array of records. So, the document cannot be represented as a relation, for it is not in 1NF.

We evaluated the methodologies: DFM [22], UML [23], GRAHAM [14], and OBDWSD [9]. Their classification is reported in Table 2. In order to show the generality of the metrics, each methodology falls in a different class.

Table 2. Classification of case study methodologies.

Methodology	Approach	Automation	Agility	Ontological-approach	Logical model
DFM	Data-driven	Semi-automatic	–	–	ROLAP
UML	Requirement-driven	–	–	–	ROLAP
GRAHAM	Hybrid	Semi-automatic	✓	✓	Key-value
OBDWSD	Hybrid	Semi-automatic	–	✓	ROLAP

4.1 Methodology Evaluation

The complete workflows of the methodologies are depicted in Fig. 1. This figure shows phases as rectangles and artefacts as ovals. The dotted-line rectangles and ovals represent, respectively, automatically executed phases and automatically produced artefacts. Accordingly, a dotted-line represents an incremental step, whereas a methodology provides an agile approach.

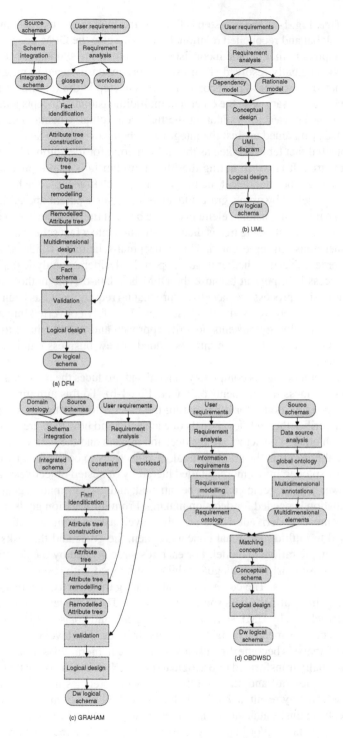

Fig. 1. Workflow of case study methodologies.

In DFM (*see,* Fig. 1-a), the first step is the integration of the source schemas in order to produce a global and reconciled relational schema. Since the DFM does not adopt an ontological approach, then all the inconsistencies, both at the syntactical and semantic levels, must be manually resolved. To do so, we preliminarily need to materialize and normalize each data source. The second independent step is the requirement analysis, which produces a glossary with the emergent multidimensional concepts and a preliminary workload to be used for validating the final conceptual schema. These concepts must be manually matched against the integrated schema in order to label potential fact tables. A potential fact table is taken as the starting point for the automatic construction of an attribute tree. It is worth noting that the algorithm fails in presence of complex schemas because it does not detect the many-to-many relationships. Each attribute tree is manually remodelled by prune, graft, add node, and change parent operations, in order to bring out the multidimensional elements on the basis of the designer's experience and choices. A remodelled attribute tree is then transformed into a fact schema, where measures and dimensions are represented. This conceptual schema is validated against the workload in order to verify whether it can support the desiderate analytical queries. The validation process is important because the DFM is a data-driven methodology and at the end of the design process we have to ensure that no requirement has been discarded. If so, the schema must be revised; otherwise, it can be safely converted into a relational schema. This methodology presents no agile approach and then we have to repeat the whole process if a new data source must be added or new business requirements must be implemented.

The UML methodology is completely manual and produces the final data warehouse schema using business requirements only (*see,* Fig. 1-b). To this aim, first we create a strategic dependency model for representing the main actors using the $i*$ framework. This model describes the high-level goals of each actor and how they are correlated and depend on each other. The actors are the decision makers and the data warehouse. This high-level goal represents the strategic goal, or objectives to be reached by the organization, and is exploded into a more detailed hierarchy of nested objectives: (a) decision goals, to answer how strategic goals can be satisfied; and (b) information goals, to define which information is needed for decision making. From information goals, information requirements must be derived. These are high-level tasks aiming at defining queries strictly related to multidimensional concepts. Then, the goals and the tasks are represented in a strategic rationale model. For each resource needed by the decision maker, the data warehouse must reach its goal, which is equipped with resources as measures and a context of analysis. The context aims at creating the dimensions, by defining the level of aggregations and the data sources to be used. These models are used to design the conceptual schema by UML stereotypes. Each goals and goals' resources of the data warehouse become, respectively, facts and fact attributes. The level of aggregations of the contexts of analysis become dimensions, structured in hierarchies. Finally, the UML diagram is manually transformed into a logical schema. It is worth noting that no schema integration is carried out and, then, all the data inconsistencies must be solved in the ETL phase, which may result an hard task due to the difficulty of mapping multidimensional concepts against data sources not in early stages. Also in this case, changes in business requirements and/or in data sources cause a complete redesign process.

In GRAHAM (*see,* Fig. 1-c), the requirement analysis is identical to that of UML, but from $i*$ diagrams we have to produce (i) a set of constraints that will be used for both fact identification and data remodeling, and (ii) a workload that will be user for validation. The schema integration phase is based on an ontological approach. The ontology used in this case study is that of OpenCyc, the open source version of Cyc [30]. On the basis of the similarity degree of concepts [7], we automatically obtain a conceptual schema that can be transformed into a logical schema. Now, we consider the constraints for the fact identification, for the construction of the attribute trees—as it happens in DFM. The only difference is that the construction algorithm detects the many-to-many relationships [24]. Constraints are also used for the next automatic phase, which is the remodeling of the attribute trees in a supervised way instead of relying on the designer's choices. The next automatic phase is the validation of the remodelled attribute trees against the queries of the workload [31]. If validated, the attribute trees can be transformed into a logical schema on the basis of the Key-Value logical model used in NoSQL systems. This methodology presents an incremental step thanks to which a constraint can be used directly in the remodeling phase, for modifying a logical schema given a pre-existing attribute tree.

In OBDWSD (*see,* Fig. 1-d), the first step is the design of a requirement ontology. It comes from a requirement analysis phase, devoted to discovering goals of decision makers, from which information requirements are defined. Finally, information requirements are related to multidimensional concepts, *i.e.* measures and context of analysis. The methodology is based on a GUI where the designer can enter the goals of the decision makers, along with the multidimensional concepts identified in the requirement analysis phase. From these, a requirement ontology, based on the $i*$ framework, is produced. Similarly, each data source is represented using an ontology, which is then merged for defining a global ontology. At this point, an algorithm is used to automatically identify interesting multidimensional concepts present in the ontology. This phase is the multidimensional annotation, which produces multidimensional elements, such as facts and measures. In order to identify facts, the numerical property of a class is examined. If it is greater than a threshold fixed by the designer, the class is marked as a fact and the numerical properties as measures. Finally, each class having a many-to-one relationship with the fact is marked as a dimension. It is worth noting that the methodology does not address how the threshold is fixed and how to handle classes that do not present numerical properties (*i.e.*, factless fact table).

The application of the metrics to these methodologies is reported in Table 3. The hardest methodology is the DFM, while the UML presents low values due to its simplicity. On the other hand, GRAHAM and OBDWSD, despite their hybrid approach, require less effort than others, because they provide automatic, ontology-based, and agile phases.

Table 3. Design effort *d* per methodology.

	DFM	UML	GRAHAM	OBDWSD
np	8	3	7	6
na	9	5	9	8
npa	2	0	6	4
naa	2	0	4	4
no	0	0	1	2
npag	0	0	3	0
d	**21**	**11**	**9**	**10**

4.2 Schema Evaluation

The resulting schemas derived by the application of the methodologies are shown in Fig. 2. We omit the intermediate steps of the design for preserving space.

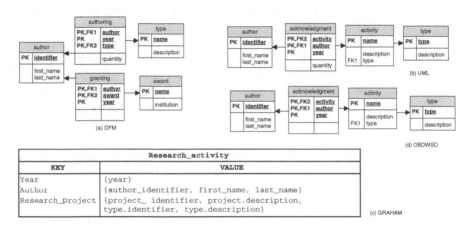

Fig. 2. Schema per methodology.

In the DFM (*see,* Fig. 2-a), the materialization of data sources produces two different schemas, which must be manually integrated. To this end, we observe that the concept of *author* in DBLP overlaps to that of *scientist* in BIOS. On the basis of the requirements, we are interested in analyzing publications and granting. Matching the requirements against the integrated schema, we derive that potential fact tables are *authoring* for publications' analyses and *granting* for awards' analyses. Each of them generates a different fact schema, having *author* as a shared dimension.

In the UML (*see,* Fig. 2-b), the only decision maker represented in the *i** diagrams is a *business analyst* who works for an independent company involved in making a survey about scientific activities and who depends on the data warehouse to reach his/her goal, which consists in "*knowing the most active scientists*". For each resource needed by the decision maker—*information about public acknowledgments*, in this case—the data warehouse must reach its goal, which is equipped with resources as measures and a context of analysis. The context aims at creating the *scientists' profiles*. Finally, the conceptual diagram derived from the $i*$ diagrams is manually transformed into a logical schema. Here, without knowing data sources, the designer assumes that possible instances for *activity* are *proceedings*, *journal*, *book*, while possible occurrences for *type* are *paper* for scientific publications and *ACM*, *IEEE*, … for awards. So, the measure is the yearly *quantity* of activities per *author*.

In GRAHAM (*see*, Fig. 2-c), the integrated schema differs from that obtained in the DFM, because, on the basis of the ontological representation of the domain, the algorithm detected a generalization among *paper*, *merit*, and *award* and this leads to the introduction of the *research project* concept, whose occurrences are specialized according to the *type* attribute. For the fact identification, the constraints derived from the $i*$ diagrams are used. Now, the algorithm for matching concepts takes over that *acknowledgments* and *granting* are synonyms[1]. So, *granting* is tagged as a potential fact table and an attribute tree is constructed having *granting* as a root. The attribute tree is remodelled in a supervised way and, then, transformed into a logical schema on the basis of the Key-Value logical model. We observe that this document presents no measures, for no numerical properties are present in the data sources.

In OBDWSD (*see,* Fig. 2-d), using both the requirement ontology, derived from the $i*$ diagrams, and the global ontology, derived from data sources, the matched concept is *acknowledgment*. Indeed, in the data sources, this is a non-numerical fact, where the number of acknowledgments must be derived by aggregation functions. The entities marked as dimensions are *year*, *scientist*, and *activity*, whose specializations are on turn *paper* and *award*. So, a further level, namely *type*, is added, for it is ontologically subsumed by *activity*.

The application of the metrics to the schemas is reported in Table 4.

Table 4. Schema quality per methodology.

	DFM	UML	GRAHAM	OBDWSD
nft	2	1	1	1
ndt	4	4	4	4
nfk	4	3	0	3
nmft	1	1	0	0
mr	0	0	0	0
v	0	2	0	0
peq	1	1	1	1
quality	1.090909	1.090909	1.2	1.125

None of the schemas presents missing requirements and the queries of the analytic workload can be executed against all these schemas. However, the information power of the schemas is not equal, for in UML an aggregation level has been lost and, conversely, an unnecessary aggregation level has been added. For this reason, we computed two violations. The final result shows that GRAHAM present the highest schema quality and OBDWSD presents a very close value.

---

[1] http://www.thesaurus.com/browse/granting.

5   Conclusion

In the paper, we discussed the following criteria for designing a data warehouse in the context of Big Data: automation, ontological and hybrid approaches, and agility. For each of these, we presented the set of metrics we used for suitably comparing several methodologies.

On the basis of the experimentation, we conclude that the design paradigm affects the schema quality, while the degree of automation, the ontological approach, and the agility affect the design effort. In the context of Big Data, the aim is to increase the schema quality (*i.e.*, the *Value*), while reducing the design effort by automatic and agile phases (*i.e.*, *Velocity*), when dealing also with unstructured data (*i.e.*, *Variety*).

Future research is devoted to the evaluation of both the *Volume* and *Veracity* issues. Since these are mainly addressed in the ETL phase, further criteria and metrics will be introduced for evaluating the physical properties of the system architecture and the effort for designing the ETL phase in reference to a given paradigm.

References

1. Chen, M., Mao, S., Liu, Y.: Big data: a survey. Mob. Netw. Appl. **19**(2), 171–209 (2014)
2. Buneman, P., Davidson, S., Fernandez, M., Suciu, D.: Adding structure to unstructured data. In: Afrati, F., Kolaitis, P. (eds.) ICDT 1997. LNCS, vol. 1186, pp. 336–350. Springer, Heidelberg (1997). doi:10.1007/3-540-62222-5_55
3. Rehman, N.U., Mansmann, S., Weiler, A., Scholl, M.H.: Building a data warehouse for twitter stream exploration. In: International Conference on Advances in Social Networks Analysis and Mining, pp. 1341–1348. IEEE Computer Society (2012)
4. Waters, R.D., Jamal, J.Y.: Tweet, tweet, tweet: a content analysis of nonprofit organizations' twitter updates. Public Relat. Rev. **37**(3), 321–324 (2011)
5. He, L., Chen, Y., Meng, N., Liu, L.Y.: An ontology-based conceptual modeling method for data warehouse. In: International Conference on Information Technology, Computer Engineering and Management Sciences, vol. 4, pp. 130–133. IEEE (2011)
6. Vranesic, H., Rovan, L.: Ontology-based data warehouse development process. In: International Conference on Information Technology Interfaces, pp. 205–210. IEEE Computer Society (2009)
7. Di Tria, F., Lefons, E., Tangorra, F.: Ontological approach to data warehouse source integration. In: Gelenbe, E., Lent, R. (eds.) Information Sciences and Systems. Lecture Notes in Electrical Engineering, vol. 264, pp. 251–259. Springer, Heidelberg (2013). doi: 10.1007/978-3-319-01604-7_25
8. Khouri, S., Bellatreche, L.: DWOBS: data warehouse design from ontology-based sources. In: Yu, J.X., Kim, M.H., Unland, R. (eds.) DASFAA 2011. LNCS, vol. 6588, pp. 438–441. Springer, Heidelberg (2011). doi:10.1007/978-3-642-20152-3_34
9. Thenmozhi, M., Vivekanandan, K.: A tool for data warehouse multidimensional schema design using ontology. Int. J. Comput. Sci. Issues **10**(2), 161–168 (2013)
10. Farooq, F., Sarwar, S.M.: Real-time data warehousing for business intelligence. In: Proceedings of the 8th International Conference on Frontiers of Information Technology, pp. 38:1–38:7. ACM, New York (2010)

11. Dehdouh, K., Bentayeb, F., Boussaid, O., Kabachi, N.: Columnar NoSQL CUBE: aggregation operator for columnar NoSQL data warehouse. In: 2014 IEEE International Conference on Systems, Man and Cybernetics, pp. 3828–3833. IEEE (2014)
12. Chevalier, M., El Malki, M., Kopliku, A., Teste, O., Tournier, R.: How can we implement a multidimensional data warehouse using NoSQL? In: Hammoudi, S., Maciaszek, L., Teniente, E., Camp, O., Cordeiro, J. (eds.) ICEIS 2015. LNBIP, vol. 241, pp. 108–130. Springer, Cham (2015). doi:10.1007/978-3-319-29133-8_6
13. Labrinidis, A., Jagadish, H.V.: Challenges and opportunities with big data. Proc. VLDB Endow. 5(12), 2032–2033 (2012). VLDB Endowment
14. Di Tria, F., Lefons, E., Tangorra, F.: Data warehouse automatic design methodology. In: Hu, W., Kaabouch, N. (eds.) Big Data Management, Technologies, and Applications, pp. 115–149. IGI Global, Hershey (2014)
15. Phipps, C., Davis, K.C.: Automating data warehouse conceptual schema design and evaluation. In: Lakshmanan, L.V.S. (ed.) Design and Management of Data Warehouses, vol. 58, pp. 23–32. CEUR-WS.org, Toronto (2002)
16. Corr, L., Stagnitto, J.: Agile data warehouse design: collaborative dimensional modeling, from whiteboard to star schema. DecisionOne Consulting (2011)
17. Mazón, J.N., Trujillo, J.: A hybrid model driven development framework for the multidimensional modeling of data warehouses! ACM SIGMOD Rec. 38(2), 12–17 (2009)
18. Mazón, J.N., Trujillo, J., Lechtenbörger, J.: Reconciling requirement-driven data warehouses with data sources via multidimensional normal forms. Data Knowl. Eng. 63, 725–751 (2007)
19. Di Tria, F., Lefons, E., Tangorra, F.: Academic data warehouse design using a hybrid methodology. Comput. Sci. Inf. Syst. 12(1), 135–160 (2015)
20. Di Tria, F., Lefons, E., Tangorra, F.: Hybrid methodology for data warehouse conceptual design by UML schemas. Inf. Softw. Technol. 54(4), 360–379 (2012)
21. Romero, O., Abelló, A.: A survey of multidimensional modeling methodologies. Int. J. Data Warehous. Min. 5, 1–23 (2009)
22. Golfarelli, M., Maio, D., Rizzi, S.: The dimensional fact model: a conceptual model for data warehouses. Int. J. Coop. Inf. Syst. 7(2), 215–247 (1998)
23. Mazón, J.N., Trujillo, J., Serrano, M., Piattini, M.: Designing data warehouses: from business requirement analysis to multidimensional modeling. In: REBNITA, vol. 5, pp. 44–53 (2005)
24. dell'Aquila, C., Di Tria, F., Lefons, E., Tangorra, F.: Dimensional fact model extension via predicate calculus. In: 24th International Symposium on Computer and Information Sciences, pp. 211–217. IEEE (2009)
25. Cohen, J., Dolan, B., Dunlap, M., Hellerstein, J.M., Welton, C.: MAD skills: new analysis practices for big data. Proc. VLDB Endow. 2(2), 1481–1492 (2009). VLDB Endowment
26. Di Tria, F., Lefons, E., Tangorra, F.: Cost-benefit analysis of data warehouse design methodologies. Inf. Syst. 63, 47–62 (2017)
27. Serrano, M.A., Calero, C., Piattini, M.: Metrics for data warehouse quality. In: Effective Databases for Text & Document Management, pp. 156–173. IGI Global (2003)
28. Serrano, M., Calero, C., Sahraoui, H.A., Piattini, M.: Empirical studies to assess the understandability of data warehouse schemas using structural metrics. Softw. Qual. J. 16(1), 79–106 (2008)
29. Ley, M.: DBLP: some lessons learned. Proc. VLDB Endow. 2(2), 1493–1500 (2009). VLDB Endowment

30. Foxvog, D.: Cyc. In: Poli, R., Healy, M., Kameas, A. (eds.) Theory and Applications of Ontology: Computer Applications, pp. 259–278. Springer, Dordrecht (2010). doi:10.1007/978-90-481-8847-5_12
31. dell'Aquila, C., Di Tria, F., Lefons, E., Tangorra, F.: Logic programming for data warehouse conceptual schema validation. In: Bach Pedersen, T., Mohania, M.K., Tjoa, A.M. (eds.) DaWaK 2010. LNCS, vol. 6263, pp. 1–12. Springer, Heidelberg (2010). doi:10.1007/978-3-642-15105-7_1

Optimal Task Ordering in Chain Data Flows: Exploring the Practicality of Non-scalable Solutions

Georgia Kougka and Anastasios Gounaris[✉]

Department of Informatics, Aristotle University of Thessaloniki,
Thessaloniki, Greece
{georkoug,gounaria}@csd.auth.gr

Abstract. Modern data flows generalize traditional Extract-Transform-Load and data integration workflows in order to enable end-to-end data processing and analytics. The more complex they become, the more pressing the need for automated optimization solutions. Optimizing data flows comes in several forms, among which, optimal task ordering is one of the most challenging ones. We take a practical approach; motivated by real-world examples, such as those captured by the TPC-DI benchmark, we argue that exhaustive non-scalable solutions are indeed a valid choice for chain flows. Our contribution is that we thoroughly discuss the three main directions for exhaustive enumeration of task ordering alternatives, namely backtracking, dynamic programming and topological sorting, and we provide concrete evidence up to which size and level of flexibility of chain flows they can be applied.

1 Introduction

Data analysis in a highly dynamic environment becomes more and more critical in order to extract high-quality information from raw data and derive actionable information in a timely manner. To this end, we typically employ fully automated *data-centric flows* (or *data flows*) both for business intelligence [4,10] and scientific purposes [13], which typically execute under demanding performance requirements, e.g., to complete in a few seconds. These flows generalize traditional Extract-Transform-Load (ETL) and data integration flows through the incorporation of data analytics [8,17]. Meeting the demanding performance requirements, combined with the volatile nature of the environment and the data, gives rise to the need for efficient data flow optimization techniques.

Data flow optimization techniques cover a wide spectrum from deciding on the order of the constituent tasks to detailed low-level configuration of the underlying execution engine [12]. In this work, we focus on the former aspect, namely the specification of the execution order of the constituent tasks. In practice, this is usually the result of a manual procedure, which, in many cases results in non-optimal flow execution plans. Furthermore, even if a data flow is optimal for a specific input data set, it may prove significantly suboptimal for another data set

© Springer International Publishing AG 2017
L. Bellatreche and S. Chakravarthy (Eds.): DaWaK 2017, LNCS 10440, pp. 19–32, 2017.
DOI: 10.1007/978-3-319-64283-3_2

with different characteristics [7]. We tackle this problem through the proposal
of optimization algorithms that can provide the optimal execution order of the
tasks in a *chain* (or *linear*) data flow in an efficient manner and relieve the flow
designers from the burden of selecting the task ordering on their own. We consider a single optimization objective, namely the minimization of the sum of the
task execution costs; we assume that the execution cost of each task depends
on the volume of data to be processed, which in turn depends on the relative
position of the task in the execution flow.

The main challenges in flow optimization that need to be addressed and differentiate the problem from that of traditional query optimization, discussed in
[3,9], are that the tasks need not belong to a set of well-defined algebra, such as
the relational one, there exist arbitrary precedence constraints among operators,
and flows can consist of dozens of tasks, whereas, typically, operators in query
plans are fewer. The main implication is that query optimization techniques,
which operate on plans with up to a few tens of operators that belong to the
relational algebra (according to which operator reordering is typically permitted), are not applicable. Nevertheless, they are successful in their domain and
this is the reason the data flow solutions proposed in this work are partially
inspired by query optimization, as we explain later. Overall, to date, there are
very few proposals that deal with (or are applicable to) task reordering in data
flows [8,16,20]. A common characteristic of these proposals is that they are too
slow to find an exact solution in small flows [8], or they can find significantly
suboptimal (approximate) solutions for bigger flows [16,20].

In this work, we go beyond the state-of-the-art with regards to exact solutions
in chain data flows, i.e., data flows where the tasks form a sequence. Chain data
flows are a main building block in generic data flows. Optimization of chain
flows is a big step towards optimization of more generic ones. Exact solutions
cannot scale in general, but, partially inspired by real worlds-like data flows,
such as those in the TPC-DI benchmark [14], we show that they can be applied
in several cases in practice. The main contribution of this work is the proposal
of two additional exact solutions that significantly improve upon the technique
in [8] in terms of time overhead and the size of data flows they can handle.[1]

The remainder of this paper is structured as follows. In Sect. 2, we present the
notation, the problem statement and the motivation from TPC-DI. Our exact
solutions are explained and evaluated in Sects. 3 and 4, respectively. We mention
related work in Sect. 5, and we conclude in the next section, which discusses the
issues in optimizing more generic flows than chain ones.

2   Preliminaries

In this paper, we deal with the problem of re-ordering the tasks of a chain
data flow without violating existing precedence constraints between tasks, while
the performance of the flow is maximized. The data flow is represented as a

---

[1] An abstract of these ideas, without considering TPC-DI, have appeared in [11] in
less than a page.

directed acyclic graph (DAG), denoted as $G = (T, E)$, where each task $t_i \in T$, $i = 1, \ldots, n$ corresponds to a node in the graph and the edges between nodes represent intermediate data shipping among tasks; i.e., in data flows, the exchange of data between tasks is explicitly represented through edges. The tasks that have no incoming edges are termed as *sources*, and those without outgoing edges as *sinks*.

The precedence constraints are captured by another directed acyclic graph $PC = (T, D)$, such that each ordered pair of vertices (t_i, t_j) that either belongs to D or the transitive closure of PC corresponds to the requirement that, in any valid G, there must exist a directed path from t_i to t_j. In other words, the PC graph corresponds to a higher-level, non-executable flow representation, where the exact task ordering is not defined; only a partial ordering is defined instead.

We focus on linear or chain data flows, where the flow contains only one sink and one source and each task in between has exactly one incoming and outgoing edge. Chain data flows can be regarded as sub-flows within generic data flows. For example, each path from a source to a sink in a generic flow forms a chain.

Further, we assume that each task t_i has a processing cost per input record (or tuple) c_i, and selectivity sel_i. The selectivity denotes the average number of returned tuples per input tuple. For filtering tasks, $sel_i < 1$; for data sources and operators that just manipulate the input $sel = 1$, whereas, for operators that may produce more output records for each input record, $sel_i > 1$. Figure 1(left) shows an example of a chain data flow with 5 tasks, their precedence constraints, and one possible valid G that respects such constraints. In the middle, example task metadata are presented. As will be shown later (e.g., Fig. 5), there are multiple other task orderings that respect the constraints as well, e.g., placing t_3 after t_4.

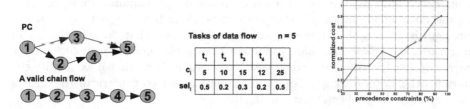

Fig. 1. An example of PC and G (left), task metadata (middle) and indicative performance improvements (right).

Problem Statement: Given a set of tasks T with known cost and selectivity values, and a corresponding precedence constraint graph PC, we aim to find a valid task ordering to form a chain G that minimizes the *sum cost metric (SCM)* per source tuple. SCM is defined as follows:

$$SCM(G) = \sum_{i=1}^{n} (\prod_{j \in Pred(t_i)} sel_j) \, c_i,$$

where $Pred(t_i)$ is the set of tasks that precede t_i in G; if the set is null, the product of selectivities is set to 1. In the example of Fig. 1, $Pred(t_4) = \{t_1, t_2, t_3\}$. The optimal plan is denoted as P.

Figure 1(right) shows indicative performance improvements. We examine 100 randomly generated data flows consisting of $n = 15$ tasks with $c_i \in [1, 100]$, $sel_i \in (0, 2]$ and 20%–95% precedence constraints. A chain flow has 100% precedence constraints, when the transitive closure of its PC has $\frac{n(n-1)}{2}$ edges. The case of 100% precedence constraints is when there is a single valid ordering with no ordering alternatives. We use the percentage of precedence constraints as an efficient way to quantify the flexibility in reordering operators in a flow. The initial random ordering has normalized cost 1. As can be observed from Fig. 1(right), the improvements can be of a factor of 3 and more.

Note that the input set of tuples are processed by all the tasks of the chain data flow until they are filtered out, but typically, some of the input tuple attributes may not be required by every flow activity. According to [21], the unnecessary tuple attributes just run through the flow, resembling an assembly-line model. The execution of a flow activity is not affected by the unnecessary attributes. This implies that the tasks of a flow have the ability to be reordered as long as the precedence constraints between the tasks are preserved.

2.1   Problem Complexity

In [2], Burge et al. proved that finding the optimal ordering of tasks is an NP-hard problem when (i) each flow task is characterized by its cost per input record and selectivity; (ii) the cost of each task is a linear function of the number of records processed and that number of records depends on the product of the selectivities of all preceding tasks (assuming independence of selectivities for simplicity); and (iii) the optimization criterion is the minimization of the sum of the costs of all tasks. All the above conditions hold for our case, so our problem is intractable. Moreover, in [2] it is discussed that "*it is unlikely that any polynomial time algorithm can approximate the optimal plan to within a factor of $O(n^\theta)$*", where θ is some positive constant. Note that if we modify the optimization criterion, e.g., to optimize the bottleneck cost metric or the critical path renders the problem tractable [1, 18].

2.2   Chains in TPC-DI

TPC-DI [14] is the TPC standard for data integration flows. Although it focuses on extensions to ETLs rather than on both advanced ETLs and analytics, it is the closest standard to our scenarios. TPC-DI aims to model the data integration processes of a retail brokerage firm. More importantly, it has been implemented for an open-source data flow engine, namely Pentaho Kettle.[2] Figure 2 shows an example of part of the implementation, which is responsible for building one of the seven dimension tables in the underlying data warehouse defined by the standard.

---

[2] http://www.essi.upc.edu/dtim/blog/post/tpc-di-etls-using-pdi-aka-kettle.

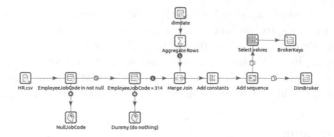

Fig. 2. A TPC-DI flow implemented in Kettle.

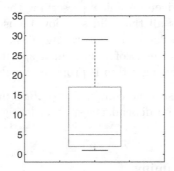

Fig. 3. Boxplot diagram for the length of chains in TPC-DI flows.

Next, we proceed to a simple analysis of the flows referring to historical loading, as these are implemented in Kettle. First, we extract chain sub-flows according to the following procedure: for each sink, we take the longest path from any of the connected sources. The boxplot diagram in Fig. 3 reveals significant information regarding the size of such chains. More specifically, most of the chains considered contain less than 20 tasks. Also, only 2 chains have size larger than 30 tasks; these chains are removed from the boxplot as outliers.

Analyzing the percentage of precedence constraints, it is found that there is a small degree of flexibility in re-ordering tasks. For example, in the largest chain in the boxplot, there are 97% precedence constraints. This pattern appears in all flows with size more than the median. For smaller chains, such as those contained in the flow in Fig. 2, the constraints drop to 87%, which is still high. However, according to the evidence in Fig. 1(right), significant improvements of 10–15% can still be achieved.

3   Accurate Algorithms for Linear Execution Plans

In this section, we present three accurate algorithms for reordering chain data flows in order to generate an optimal execution plan. The algorithms are based on backtracking, dynamic programming and generation of all topological sortings, respectively. Our main novelty here is that we examine a topological sorting-based algorithm, despite its worst-case complexity. Counter-intuitively, as we show in the evaluation, the algorithm is practical not only for the type of flow chains appearing in TPC-DI, but also for much larger n, when there are many precedence constraints and, in general, can scale better than the two other options. However, still, it cannot be applied to arbitrary flows of very large size.

3.1 Backtracking

The *Backtracking* algorithm finds all the possible execution plans generated after reordering the tasks of a given data flow preserving the precedence constraints. The algorithm enumerates all the valid sub-flow plans after applying a set of recursive calls on these sub-flows until generating all the possible data flow plans. It backtracks when a placement of a task in a specific position violates the precedence constraints. The algorithm is proposed for flow optimization in [8].

Complexity: The worst case time complexity of *Backtracking* is factorial (i.e., $O(n!)$), since, if there are no dependencies, all orderings will be examined in a brute force manner.

3.2 Dynamic Programming

This algorithm is extensively used as part of the System R-type of query optimization to produce (linear) join orderings [15]. The rationale of the dynamic programming algorithm (termed as *DP* henceforth) for data flows remains the same, that is to calculate the cost of task subsets of size n based on subsets of size $n - 1$. For each of these subsets, we keep only the optimal solutions, which are valid with regards to the precedence constraints. Specifically, the *DP* algorithm considers each flow of size n as a flow of $(n - 1)$ tasks followed by the nth task; the key point is that the former part is the optimal subset of size $n-1$, which has been found from previous step; then the algorithm exhaustively examines which of the n flow tasks is the one that, when added at the end, yields an optimal subplan of size n. For example, the algorithm starts by calculating subsets that consist of only one task $\{t_1\}$, then $\{t_2\}$, $\{t_3\}$ and so on. In a similar way, in the second step, it examines subsets containing two tasks, i.e., $\{t_1, t_2\}$, $\{t_1, t_3\}$ and so on, until it examines the complete flow $\{t_1, t_2, ..., t_n\}$. The number of the optimal (non-empty) subsets of a flow is equal to $2^n - 1$.

Complexity: The time complexity is $O(n^2 2^n)$. This is because we examine all subsets of n tasks, which are $O(2^n)$. For each subset, which is up to size $O(n)$, we examine whether each element can be placed at the end of the subplan. Each such check involves testing whether any of the rest $n - 1$ tasks violate a precedence constraint, when placed before the n-th task. Overall, for each element, we make $O(n)$ comparisons. So, the overall time complexity is $O(2^n)O(n)O(n) = O(n^2 2^n)$. The space complexity is derived by the size of the auxiliary data structures employed. We use three vectors of size $2^n - 1$ as

---

Algorithm 1. Dynamic Programming

Require: 1. A set of n tasks, $T=\{t_1, ..., t_n\}$. 2. A directed acyclic graph PC with precedence constraints.

Ensure: A directed acyclic graph P representing the optimal plan
   {Initialize *PartialPlan*, *Costs* and *Sel* of size $2^n - 1$}
1: **for all** $i \in \{2, ..., n\}$ **do**
2:     $\text{PartialPlan}(2^{i-1}) = t_i$
3:     $\text{Costs}(2^{i-1}) = c_i$
4:     $\text{Sel}(2^{i-1}) = sel_i$;
5: **end for**
6: **for all** $s \in \{2, ..., n\}$ **do**
7:     $R \leftarrow Subsets(T, s)$ {X is a set with all subsets of T of size s}
       {r is a specific subset of size s}
8:     $tempBest \leftarrow \infty$
9:     **for each** $r \in R$ **do**
10:        **for all** $i \in \{1, ..., r.length()\}$ **do**
11:            $tempSet \leftarrow r - r(i)$
12:            $pos1 \leftarrow findIndex(tempSet)$
13:            $pos2 \leftarrow findIndex(r(i))$
14:            **if** r(i) has all predecessors in *tempSet* **then**
15:                $TempPlan \leftarrow tempSet, r(i)$
16:                $costTempPlan \leftarrow Costs(pos1) + Sel(pos1)Costs(pos2)$
17:                **if** $costTempPlan < tempBest$ **then**
18:                    $tempBest \leftarrow costTempPlan$
19:                    $k \leftarrow pos1 + pos2$
20:                    update(*PartialPlan(k)*, *Costs(k)*, *Sel(k)*)
21:                **end if**
22:            **end if**
23:        **end for**
24:    **end for**
25: **end for**
26: $P \leftarrow PartialPlan(2^n - 1)$

---

explained in the implementation details, one of which stores elements of size $O(n)$. So the space complexity is $O(n2^n)$.

Implementation Issues: In order to implement the algorithm, we use three vectors of size $2^n - 1$, namely *PartialPlan*, *Costs* and *Sel*. According to the algorithm implementation, the i-th cell corresponds to the combination of tasks for which the bit is 1 in its binary representation. For example, if $i = 13$, then the binary representation of this position is $(1101)_2$. Specifically, this means that $partialPlan[13]$ corresponds to the optimal ordering of the 1st, 2nd and 4th tasks. Analogously, the partial plan $\{1, 3, 4, 5\}$ is stored in position $2^{1-1} + 2^{3-1} + 2^{4-1} + 2^{5-1} = 29$ of the *partialPlan* matrix. The *Costs* and *Sel* vectors hold the aggregate cost and selectivity of the subplans, respectively. The last cell of *PartialPlan* and *Costs* contain the optimal plan and its total cost, respectively. A complete pseudocode is shown in Algorithm 1. For the sake of simplicity of presentation, the algorithm is not fully optimized; e.g., in line 18, the update of vertices may occur only once after the final best plan is found.

An example: We give an example of the algorithm with a flow with $n = 5$; the task metadata are shown in Fig. 1. The *DP* example is in Fig. 4. First of all, all the subsets R of T of length $s = \{1, 2, ..., n\}$ are found. For single task subsets, such as $\{t_1\}, \{t_2\}, ..., \{t_n\}$, *DP* estimates their position in the *partialPlan* matrix, e.g. $\{2\}$ subset is positioned in $partialPlan(2^{2-1})$. For subsets with length greater than 1, e.g., the subset $\{1, 3, 4\}$, we examine the case that each element of that subset is placed at the end of the subset. If the precedence constraints are violated, *DP* continues to the next placement. If the precedence constrains are not violated, the algorithm estimates the cost of the valid partial plan with that element positioned at the end of the subset, reusing the results of the orderings of smaller subsets. Similarly, the cost of all orderings in the subset is estimated and the algorithm finds the ordering of the subset with the minimum cost. The optimal partial plan, its cost and the product of task selectivities are stored in the corresponding position in the *partialPlan* and *Costs* and *Sel* vertices, respectively.

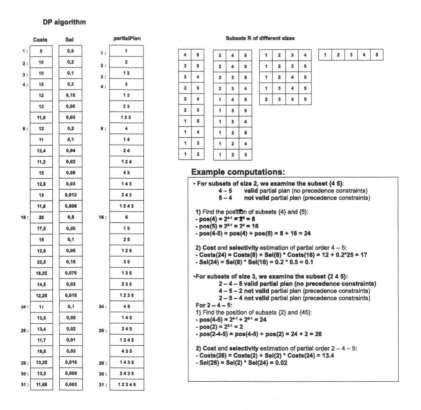

Fig. 4. Example of the DP algorithm.

Algorithm 2. TopSort

Require: 1. A set of n tasks, T={t_1, ..., t_n}. 2. A directed acyclic graph PC with precedence constraints.

Ensure: An ordering of the tasks P representing the optimal plan.

1: P={t_1, t_2, ..., t_n} {P is initialized with a valid topological ordering of PC.}
2: i=1
3: minCost $\leftarrow$ computeSCM(P)
4: **while** $i < n$ {n is the total number of tasks} **do**
5:    k $\leftarrow$ location of t_i in P
6:    k1 $\leftarrow$ k + 1
7:    **if** $P(k1)$ task has prerequisite t_i **then**
8:       // **Rotation stage**
9:       Rotate the elements of P from positions i to k
10:      cost $\leftarrow$ computeSCM(P)
11:      i$\leftarrow$ i+1
12:    **else**
13:       // **Swapping stage**
14:       Swap the k and $k1$ elements of P
15:      cost $\leftarrow$ computeSCM(P)
16:      i $\leftarrow$ 1
17:    **end if**
18:    **if** cost < minCost **then**
19:      bestP $\leftarrow$ P
20:      minCost = cost
21:    **end if**
22: **end while**
23: P $\leftarrow$ bestP

Correctness. If *PartialPlan* is of size $n = 1$, the optimal solution is trivial and is found by the algorithm during initialization in lines 1–3 of Algorithm 1. We assume that a *PartialPlan* of size $n - 1$ is optimal and we need to prove that *PartialPlan* of size n is also optimal. The sketch of the proof will be based on contradiction. Let us assume that the *DP* does not produce the optimal solution. Any linear solution of size n consists of a *PartialPlan* of size $n - 1$ followed by the n-th task; *DP* checks all the alternatives for the n-th task. So, there is a different optimal solution, where the *PartialPlan* of size $n - 1$ is different of *DP*'s *PartialPlan* of the same size. According to the *SCM*, the cost of the subplan of size n is computed as the sum of two components: the cost of subplan of size $n - 1$ and the cost of the n-th task times the selectivity of the first $n - 1$ tasks. The costs of the solutions of size n, which end with the same task, differ only in the first component. According to our assumptions, the cost of *DP*'s *PartialPlan* of size $n - 1$ cannot be higher than any other subplan solution of size $n - 1$ by definition. Consequently, there is no other solution different from *DP*'s solution that can yield lower cost. This completes the proof.

3.3   Topological Sorting

The *TopSort* algorithm is a topological sorting algorithm based on [19], which finds all the possible topological sortings given a partial ordering of a finite set; in our case the partial ordering is due to the precedence constraints. The reason behind using this algorithm is that it (implicitly) prunes invalid plans very efficiently and it generates a new plan based on a previous plan after performing a minimal change. For the purposes of this work, we adapted the topological sorting algorithm in order to generate all the possible execution plans of a data flow and detect the execution plan with the minimum cost. The algorithm assumes that it can receive as input a valid task permutation $t_1 \to t_2 \to t_3 \to ... \to t_n$, which is trivial since it can be done in linear time. We generate all other valid execution plans by applying two main operations, namely, cyclic rotations and swapping adjacent tasks.

Firstly, the process of generating all the valid flow execution plans begins with the topological sorting of the $n - 1$ tasks $t_2 \to t_3 \to ... \to t_n$ of the flow. Based on this partial sorting, we generate all the valid orderings of the $t_1 \to t_2 \to t_3 \to ... \to t_n$ plan. Specifically, in the first stage of the algorithm the task t_1 is placed on the left part of the partial plan $t_2 \to t_3 \to ... \to t_n$ and in the next steps of this stage, we swap it with the tasks on its right, while the tasks of the partial plan maintain their relative position. The t_1 task stops moving when such a swap violates a precedence constraint. Then, as the task t_1 cannot be further transposed, the second stage of algorithm begins with a right-cyclic rotation of another partial plan consisted of t_1 and all the tasks that precede it, which means all the tasks which are positioned to its left. In this way, t_1 is placed to its initial position. Similarly, we generate all the topological sortings of $t_2 \to t_3 \to ... \to t_n$, $t_3 \to t_4 \to ... \to t_n$ and so on. For each generated plan, we estimate the total execution cost. A pseudocode is presented in Algorithm 2.

Complexity: Since the algorithm checks all the permutations the time complexity is $O(n!)$ in the worst case. However, compared to other algorithms that produce all topological sortings, it is more efficient [19]. The space complexity is $O(n)$ because only one plan is stored in main memory at any point of execution.

Implementation Issues: The algorithm exhaustively checks all the permutations that satisfy the precedence constraints, and as such, it always finds the optimal solution for linear flows. No specific data structures are required. As shown in Algorithm 2, we employ a *computeSCM* function needs to be constructed in a way that does not compute the cost of each ordering from scratch, which is too naive, but leverages the computations of the previous plans taking into account the local changes in the new plan. Note that we can implement *TopSort* in a different way, where the tasks are checked from right to left. Although in [19] this flavour is claimed to be capable of yielding better performance, this has not been verified in our flows.

Example: In Fig. 5, an example of finding the optimal plan of a flow using *Top-Sort* is presented. In this example, the running steps of *topSort* algorithm are depicted, given as input a valid flow execution plan (*Initial plan order* plan label)

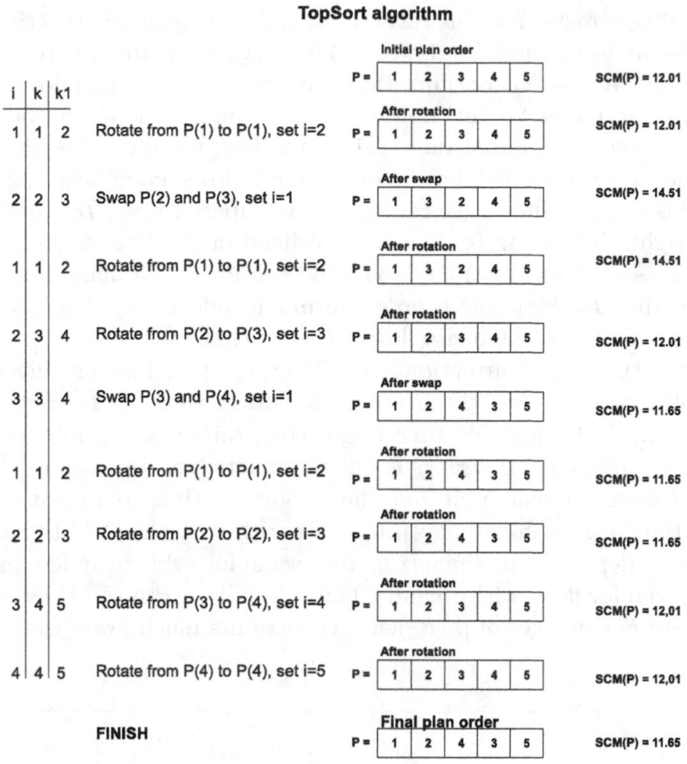

TopSort algorithm

i	k	k1		Initial plan order	
				P = [1][2][3][4][5]	SCM(P) = 12.01
1	1	2	Rotate from P(1) to P(1), set i=2	After rotation P = [1][2][3][4][5]	SCM(P) = 12.01
2	2	3	Swap P(2) and P(3), set i=1	After swap P = [1][3][2][4][5]	SCM(P) = 14.51
1	1	2	Rotate from P(1) to P(1), set i=2	After rotation P = [1][3][2][4][5]	SCM(P) = 14.51
2	3	4	Rotate from P(2) to P(3), set i=3	After rotation P = [1][2][3][4][5]	SCM(P) = 12.01
3	3	4	Swap P(3) and P(4), set i=1	After swap P = [1][2][4][3][5]	SCM(P) = 11.65
1	1	2	Rotate from P(1) to P(1), set i=2	After rotation P = [1][2][4][3][5]	SCM(P) = 11.65
2	2	3	Rotate from P(2) to P(2), set i=3	After rotation P = [1][2][4][3][5]	SCM(P) = 11.65
3	4	5	Rotate from P(3) to P(4), set i=4	After rotation P = [1][2][3][4][5]	SCM(P) = 12,01
4	4	5	Rotate from P(4) to P(4), set i=5	After rotation P = [1][2][3][4][5]	SCM(P) = 12,01
			FINISH	Final plan order P = [1][2][4][3][5]	SCM(P) = 11.65

Fig. 5. Example of the TopSort algorithm.

and assuming the metadata of Fig. 1. Each of the given plans describe a plan generated after either a rotation or a swap action.

4 Evaluation of the Time Overhead

In this section, we conduct a thorough evaluation of the time overhead of the accurate optimization algorithms. We use a machine with an Intel Core i5 660 CPU and 6 GB of RAM. We construct synthetic flows so that we can evaluate the algorithms in a wide range of parameter combinations. More specifically, we produce the optimal ordering of tasks to form a chain flow after we have (i) created random PC graphs with a configurable ratio of precedence constraints, and (ii) chosen task metadata randomly. The time overhead does not actually depend on the task metadata.

Backtracking cannot scale in the percentage of PCs and *DP* cannot scale in the number of tasks. So, we compare them against *TopSort* separately.

Figure 6(top-left) presents the average execution time of the *DP* algorithm compared to the *TopSort* solution for 50% precedence constraints, and

$n = 15, ..., 20$ flow tasks. For this ratio of PCs, *Backtracking* is practically inapplicable. The main conclusions that can be drawn from this figure is that *DP* algorithm is not a practical optimization solution even for small flows that consist of 19 flow activities for this number of constraints; the execution of a flow with 20 tasks requires over 3 days using our test machine. *TopSort* runs at least 50 times faster than *DP*, but its execution follows a similar pattern to *DP*. However, this figure is indicative regarding the superiority of *TopSort* over *DP*. In the top-right part of Fig. 6, the time overhead of *Backtracking* compared to *TopSort* is presented for $n = 15, ..., 34$ and 90% PCs. The main observation of this figure is that *Backtracking* is orders of magnitude slower than *TopSort* and cannot scale even for medium-size flows of more than 30 tasks.

We now turn our attention to *TopSort* for further investigation. Figure 6(bottom-left) shows the average execution time of *TopSort* for flows with $n = 10, ..., 70$ having 98% precedence constraints, which implies that the number of the possible re-orderings is quite restricted, as also observed in TPC-DI. *TopSort* does not scale well, but the important thing is that it can run in acceptable time, e.g., a minute, even in flows with 60 tasks. Additionally, Fig. 6 (bottom-right) depicts that *TopSort* cannot scale for arbitrarily few precedence constraints even for flows with 15 and 20 flow activities. But, for these flow sizes, it can tolerate percentages of precedence constraints much lower than 90%.

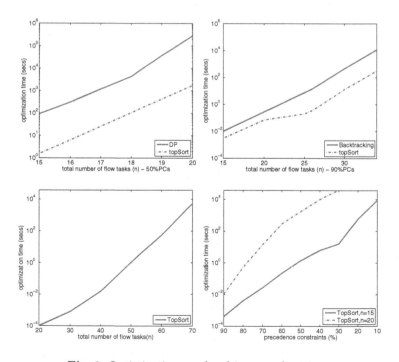

Fig. 6. Optimization overhead in several settings.

5   Related Work

Further to the discussion of related work in the introduction, task ordering for data flows is an area that has been significantly influenced by query processing techniques. In [5], an optimization algorithm for query plans with dependency constraints between algebraic operators is presented. In [20], a proposal for data integration that is also applicable to data flow task ordering is discussed. The corresponding techniques are approximate rather than exact solutions for chain flows, and they are either similar or inferior to those considered in [11]. ETL flows are analyzed in [16], where ETL execution plans are considered as states and transitions, such as swap, merge, split, distribute and so on, are employed to generate new states in order to navigate through the state space, which corresponds to the execution plan alternatives. As shown in [11], these techniques can deviate from the optimal solutions in chain dataflows by several factors.

In addition, there is a significant portion of proposals on flow optimization that proceed to flow structure optimizations but do not perform task reordering, as we do; an overview of the complete spectrum of data flow optimization techniques appears in [12]. Note that deciding the task ordering before execution in order to determine the flow structure is orthogonal to deciding the scheduling order of tasks at runtime, which is another well investigated area, e.g., see [6].

6   Conclusions

In this work, we deal with the problem of optimally ordering the constituent tasks of a chain data flow in order to minimize the sum of the task execution costs. We are motivated by the significant limitations of fully-automated optimization solutions for data flows, as, nowadays, the optimization of the complex data flows is left to the flow designers and is a manual procedure. We are also motivated by the fact that real-world flows, such as those in TPC-DI, are not very flexible, in terms of the alternative valid task orderings. As such, carefully crafted exhaustive solutions become applicable. We propose two such solutions, a dynamic programming one and one that efficiently generates all topological orderings. We explain the technical details involved and we show that the latter approach is both dominant and practical under realistic assumptions.

Our work is a big step towards optimal task ordering in generic flows instead of simple chain flows, e.g., the complete flows assumed by the TPC-DI benchmark. For this goal to be met, three further issues need to be resolved: (i) to devise solutions that, using chain optimization as a building block, apply to complete data flows; in the way we defined chains hereby, these chains are overlapping and their isolated optimizations may not be compatible among overlapping chains and do not guarantee optimality when combined; (ii) to develop efficient ways to collect the required statistical metadata and detect precedence constraints; and (iii) to take into account other metrics than the minimization of the sum of the costs with a view to considering realistic issues, such as pipelined execution and parallel execution on multi-core engines, which better reflect the running time of a flow.

References

1. Agrawal, K., Benoit, A., Dufossé, F., Robert, Y.: Mapping filtering streaming applications. Algorithmica **62**(1–2), 258–308 (2012)
2. Burge, J., Munagala, K., Srivastava, U.: Ordering pipelined query operators with precedence constraints. Technical report 2005–40, Stanford InfoLab (2005)
3. Chaudhuri, S.: An overview of query optimization in relational systems. In: PODS, pp. 34–43 (1998)
4. Chaudhuri, S., Dayal, U., Narasayya, V.: An overview of business intelligence technology. Commun. ACM **54**, 88–98 (2011)
5. Florescu, D., Levy, A., Manolescu, I., Suciu, D.: Query optimization in the presence of limited access patterns. In: SIGMOD, pp. 311–322. ACM (1999)
6. Grehant, X., Demeure, I., Jarp, S.: A survey of task mapping on production grids. ACM Comput. Surv. **45**(3), 37:1–37:25 (2013)
7. Halasipuram, R., Deshpande, P.M., Padmanabhan, S.: Determining essential statistics for cost based optimization of an ETL workflow. In: EDBT, pp. 307–318 (2014)
8. Hueske, F., Peters, M., Sax, M., Rheinländer, A., Bergmann, R., Krettek, A., Tzoumas, K.: Opening the black boxes in data flow optimization. PVLDB **5**(11), 1256–1267 (2012)
9. Ioannidis, Y.E.: Query optimization. ACM Comput. Surv. **28**(1), 121–123 (1996)
10. Jovanovic, P., Romero, O., Abelló, A.: A unified view of data-intensive flows in business intelligence systems: a survey. T. Large-Scale Data Knowl. Centered Syst. **29**, 66–107 (2016)
11. Kougka, G., Gounaris, A.: Optimization of data-intensive flows: is it needed? Is it solved? In: DOLAP, pp. 95–98 (2014)
12. Kougka, G., Gounaris, A., Simitsis, A.: The many faces of data-centric workflow optimization: a survey. CoRR, abs/1701.07723 (2017)
13. Ogasawara, E.S., de Oliveira, D., Valduriez, P., Dias, J., Porto, F., Mattoso, M.: An algebraic approach for data-centric scientific workflows. PVLDB **4**, 1328–1339 (2011)
14. Poess, M., Rabl, T., Caufield, B.: TPC-DI: the first industry benchmark for data integration. PVLDB **7**(13), 1367–1378 (2014)
15. Selinger, P.G., Astrahan, M.M., Chamberlin, D.D., Lorie, R.A., Price, T.G.: Access path selection in a relational database management system. In: SIGMOD, pp. 23–34 (1979)
16. Simitsis, A., Vassiliadis, P., Sellis, T.K.: State-space optimization of ETL workflows. IEEE Trans. Knowl. Data Eng. **17**(10), 1404–1419 (2005)
17. Simitsis, A., Wilkinson, K., Castellanos, M., Dayal, U.: Optimizing analytic data flows for multiple execution engines. In: SIGMOD, pp. 829–840 (2012)
18. Srivastava, U., Munagala, K., Widom, J., Motwani, R.: Query optimization over web services. In: Proceedings of the 32nd International Conference on Very Large Data Bases VLDB, pp. 355–366 (2006)
19. Varol, Y.L., Rotem, D.: An algorithm to generate all topological sorting arrangements. Comput. J. **24**(1), 83–84 (1981)
20. Yerneni, R., Li, C., Ullman, J., Garcia-Molina, H.: Optimizing large join queries in mediation systems. In: Beeri, C., Buneman, P. (eds.) ICDT 1999. LNCS, vol. 1540, pp. 348–364. Springer, Heidelberg (1999). doi:10.1007/3-540-49257-7_22
21. Zinn, D., Bowers, S., McPhillips, T., Ludäscher, B.: Scientific workflow design with data assembly lines. In: Proceedings of the 4th Workshop on Workflows in Support of Large-Scale Science, WORKS 2009, pp. 14:1–14:10. ACM (2009)

Exploiting Mathematical Structures of Statistical Measures for Comparison of RDF Data Cubes

Claudia Diamantini, Domenico Potena, and Emanuele Storti[✉]

Dipartimento di Ingegneria dell'Informazione, Universita Politecnica delle Marche,
via Brecce Bianche, 60131 Ancona, Italy
{c.diamantini,d.potena,e.storti}@univpm.it

Abstract. A growing number of public institutions all over the world has recently started to publish statistical data according to the RDF Data Cube vocabulary, as open and machine-readable Linked Data. Although this approach allows easier data access and consumption, appropriate mechanisms are still needed to perform proper comparisons of statistical data. Indeed, the lack of an explicit representation of how statistical measures are calculated still hinders their interpretation and use. In this work, we discuss an approach for the analysis and schema-level comparison of distributed data cubes, which is based on the formal and mathematical representation of measures. Relying on a knowledge model, we present and evaluate a set of logic-based functionalities able to support novel typologies of comparison of different data cubes.

Keywords: Statistical datasets · RDF Data Cube · Multidimensional modeling · Logic reasoning

1 Introduction

Performance analysis, which is traditionally related to project and process assessment in the business domain, has recently become an important tool also for evaluating efficiency and effectiveness of organizations, infrastructures and public services. Comparison of performances between different public administrations are able to provide better guidance to decision makers, foster civic participation and also support investors in choosing the best location for investments.

Several public administrations have already paved the way for the publication of thousands of statistical datasets in open and machine-readable formats. In order to make these datasets more easily accessed, recently the RDF Data Cube vocabulary[1] has been proposed by W3C for publishing statistical data on the Web, according to the Linked Data approach. By resorting to the multidimensional model, this format allows to represent observation values for a given statistical measure/indicator (e.g., the pollution in terms of CO_2), along a set of

---

[1] https://www.w3.org/TR/vocab-data-cube/.

© Springer International Publishing AG 2017
L. Bellatreche and S. Chakravarthy (Eds.): DaWaK 2017, LNCS 10440, pp. 33–41, 2017.
DOI: 10.1007/978-3-319-64283-3_3

dimensions (e.g., the time and the place of the measurement). Benefits include easier access to data (which are publicly available and can be retrieved through a query language) and integration of multiple distributed data sources. Much work has been done in the Literature to investigate approaches for comparison of datasets published in the Data Cube format, but besides the concrete steps taken towards the integration of dimensions and corresponding members, appropriate mechanisms to evaluate and compare measures are yet to come. From one side, a multitude of indicator frameworks have been proposed in the Literature for various domains, but only a few are generally accepted. Furthermore, measures not only have an aggregate [1,2] but also a compound nature, and there is still a lack of shared, explicit and unambiguous ways to define their semantics. As a consequence, no meaningful comparisons of measures can be made without the awareness of how they are calculated.

As a motivating example, let us consider a community of users willing to compare statistics on the main pollutants responsible for acidification, eutrophication and ground-level ozone pollution, namely sulphur dioxide (SOx), nitrogen oxides (NOx), ammonia (NH3) together with particulate matter up to $10\,\mu m$ (PM10). Let us suppose two different municipalities, $City_A$ and $City_B$, provide statistical data in RDF Data Cube through the following measures and dimensions:

- $City_A$ provides datasets measuring the *total emissions* (which sums up NH3 and PM10 values) and the *total emissions per person*, with respect to dimensions time and place;
- $City_B$ includes datasets measuring values of pollutants *SOx*, *NOx* and *PM10*, with respect to time and place, the *total emissions* (as a sum of SOx + NOx + NH3), and the *total population* of the city with respect to time.

If users were interested in comparing the total emission, they would firstly require to understand how this indicator is calculated by the two municipalities. Determining whether two statistical datasets are comparable or not, at to what extent, requires a considerable effort for users, which should be aware of how measures are calculated, which possible transformation are to be performed, which conditions must satisfy the dimensional schemas.

With the purpose to address the above-mentioned issues, in this paper we present a logic-based approach to enable the comparison of statistical datasets published as Linked Open Data according to the RDF Data Cube vocabulary. The approach relies on the formal representation of indicators through an ontology, which allows the possibility to express also their calculation formula. Mapping between measures of the data cubes and these formal definitions enable to declaratively express their semantics. On the top of this model, a set of services are developed to evaluate (1) if an indicator can be calculated over a given data cube and (2) if two data cubes are comparable, by exploiting reasoning functionalities over the model.

The rest of this work is organised as follows: in Sect. 2 we discuss the model used to formally represent statistical indicators with their calculation formulas and the RDF Data Cube vocabulary, that are exploited in Sect. 3 to provide functionalities capable to support the comparisons of Linked datasets, for which

we provide an experimental evaluation. Finally, in Sect. 4 we conclude the paper
and outline future work.

2 Model and Data Representation

According to the multidimensional model, given a cube C we refer to its schema
$ds(C)$ as a tuple $\langle \mathcal{M}_c, \mathcal{D}_c \rangle$, where $\mathcal{M}_c$ is a set of measures (or indicators) and
$\mathcal{D}_c$ is a set of dimensions. An indicator can be either atomic or compound, built
by combining other indicators. Dependencies of a compound indicator ind on its
building elements are defined through a mathematical expression $f(s_1, \ldots, s_n)$,
i.e. a *Formula* expressing how the indicator is computed in terms of s_i, which
are in turn indicators or constants [3].

For the case study presented in Sect. 1, the following atomic indicators are defined: *SOXEmission, NOXEmission, NH3Emission, PM10Emission,
TotalPopulation*. As the two cities measure the total population according to different definitions, the two corresponding compound indicators are
TotalEmission with a formula equal to *NH3Emission + PM10Emission*
and *TotalEmission2* with formula equal to *SOXEmission + NOXEmission
+ NH3Emission*. Finally, indicator *TotEmissionPerCapita* is defined as
$\frac{TotalEmission}{TotalPopulation}$.

In this work, we refer to a scenario where data cubes are distributed, and
a knowledge base of indicator definitions is shared among the cubes. In order
to properly state the link between a measure $ind_c \in \mathcal{M}_c$ of a cube C and a
corresponding indicator definition ind in the knowledge base, a mapping $ind_c \rightarrow
ind$ must be defined between them.

Representation of indicators. In the context of this work, indicators and their
formulas are formally represented in KPIOnto (http://w3id.org/kpionto). For
the purpose of this work we focus on classes kpi:Indicator and kpi:Formula.
The former represents a quantitative metric (or measure) together with a set of
properties, e.g. one or more compatible dimensions, a formula, a unit of measurement, a business objective and an aggregation function. The latter formally
represents an indicator as a function of other indicators. Operators are represented as defined by OpenMath [4], an extensible XML-based standard for representing the semantics of mathematical objects. On the other hand, operands
can be defined as indicators, constants or, recursively, as other formulas.

Representation of data cubes. To address the limits of early approaches for
representation of statistical data on the web (e.g., the capability to properly
represent dimensions, attributes and measures or to group together data values
sharing the same structure), the Data Cube vocabulary (QB) [5], was proposed
by W3C to publish statistical data as RDF following the Linked Data principles.
According to the multidimensional model, the QB language defines the schema of
a cube as a set of dimensions, attributes and measures through the corresponding
classes qb:DimensionProperty, qb:AttributeProperty and qb:MeasureProperty.
Data instances are represented in QB as a set of qb:Observations. To make an

example about the case study in the Introduction, the data structure of one of the datasets for City$_B$ includes the following components:

- `cityB:PM10`, a qb:MeasureProperty for the PM10 pollutant;
- `sdmx-dimension:timePeriod`, a qb:DimensionProperty for the time of the observation;
- `cityB:place`, a qb:DimensionProperty for the location of the PM10 sensor.

Please note that the prefix *"qb:"* stands for the specification of the Data Cube vocabulary[2], while *"cityB:"* is a custom namespace for describing measures, dimensions and members of the dataset for City$_B$, and *"sdmx-dimension:"* points to the SDMX vocabulary for standard dimensions[3]. In order to make datasets comparable, the approach we take in this work is to rely on KPIOnto as reference vocabulary to define indicators. As such, instances of MeasureProperty as defined in Data Cube datasets have to be semantically aligned with instances of `kpi:Indicator`, through a RDF property as follows: `cityB:PM10 rdfs:isDefinedBy kpi:PM10Emission`.

3 Structural Comparison of RDF Data Cubes

In this Section we discuss a set of services aimed to support analysis and comparisons of statistical datasets. The architecture includes the data layer defined in the previous Section, containing the definitions of indicators according to KPI-Onto vocabulary. On its top, a set of services are built, relying on logic based functionalities in Prolog for manipulation of formulas (e.g., for equation solving) [6]. Access to a dataset is performed through SPARQL queries over the corresponding endpoint, which may serve a library of different data cubes belonging to the same organization. Mappings between local MeasureProperties and indicator definitions (through `rdfs:isDefinedBy` properties) can be defined both at dataset level (at publication time, by the dataset publisher) or at central level (later on, by final users). Hereafter, we assume the following mappings have been defined between dataset measures and KPIOnto indicators:

```
cityA : TotEmission          →  kpi : TotalEmission
cityA : TotEmissionPerPerson →  kpi : TotEmissionPerCapita
cityB : SOX  →  kpi : SOXEmission
cityB : NOX  →  kpi : NOXEmission
cityB : PM10 →  kpi : PM10Emission
cityB : TotEmission   →  kpi : TotalEmission2
cityB : TotPopulation →  kpi : TotalPopulation
```

For what concerns dimensions, for simplicity we assume that the time dimension is defined as *sdmx-dimension:timePeriod* in all datasets[4].

---

[2] https://www.w3.org/TR/vocab-data-cube/.

[3] http://purl.org/linked-data/sdmx/2009/dimension.

[4] Please note that *owl:sameAs* links can be defined between different definitions of the same dimension for interoperability purposes.

3.1   Computability and Comparability

We provide here the notion of comparability between data cubes, which is based on the capability to compute an indicator for a cube.

Definition 1 *(Computability of an indicator for a cube)*. Given an indicator $ind_x \in \mathcal{KB}$ with a valid formula $f(ind_{x_1}, \ldots, ind_{x_n})$ and a cube $C = \langle \mathcal{M}_c, \mathcal{D}_c \rangle$, ind_x is computable for C iif $(\exists ind_c \in \mathcal{M}_C$ such that $ind_c \to ind_x) \vee (\forall ind_{x_i} \in \{ind_{x_1}, \ldots, ind_{x_n}\} \exists ind_{c_i} \in \mathcal{M}_c$ such that $ind_{c_i} \to ind_{x_i})$.

In the first case, if $ind_c \to ind_x$ then the indicator is fully computable in C, with no restriction on the dimensional schema. In the second case, ind_x is computable from $ind_{c_1}, \ldots, ind_{c_n}$ only for what concerns their common dimensional schema cds, i.e. the set of dimensions in common among their corresponding cubes. An indicator is actually computable from other indicators in a cube if their cds is non-empty. The following definition allows to assess the comparability of two cubes.

Definition 2 *(Comparability of two cubes)*. Given two cubes C_x, C_y and an indicator $ind \in \mathcal{KB}$ such that ind is computable for C_x from $\{ind_{x_1}, \ldots, ind_{x_n}\}$ and for C_y from $\{ind_{y_1}, \ldots, ind_{y_m}\}$, the cubes C_1 and C_2 are comparable through ind iif $cds(ind_{x_1}, \ldots, ind_{x_n}) \cap cds(ind_{y_1}, \ldots, ind_{y_m}) \neq \emptyset$.

In other terms, two cubes are comparable with respect to a specific indicator that is either provided by them or can be calculated from them. The indicator can be compared only along the dimensions that are in common between the cubes. If $cds(ind_{x_1}, \ldots, ind_{x_n}) = cds(ind_{y_1}, \ldots, ind_{y_m})$ then the cubes are *fully* comparable for ind. Otherwise, the comparability is limited and the reliability of any analysis performed over different dimensional schemas must be properly addressed: in order to obtain a valid answer, the query for each data cube must aggregate at the most abstract level (i.e., ALL) any dimension that is not in the common dimensional schema.

3.2   Comparison Functionalities

The services discussed in this subsection are devised to assess computability and comparability at schema-level according to the given definitions.

Calculation of computable indicators for data cubes. Given an indicator ind, a dataset library and its corresponding endpoint, the service returns those datasets in which the indicator at hand is computable. The approach relies on the exploitation of KPIOnto definitions of indicator formulas, and Logic Programming functions capable to manipulate them. Firstly, if ind is not directly available in the library, the service (1) derives all alternative ways to calculate ind. Then (2) it searches into the library for combinations of datasets including those measures. While this last step is performed through SPARQL queries, in order to fulfill step 1 a Logic Programming function named get_formulas is called. This function is capable to manipulate the whole set of defined formulas and find alternative rewritings by recursively applying mathematical axioms (e.g., commutativity, associativity, distributivity and properties of equality). In case the indicator is compound and is provided with a formula f, the function returns the *class of equivalent formulas*, i.e. the set of formulas that are mathematically equivalent to f, and hence valid for the indicator. If the indicator is atomic, the function firstly tries to derive a formula for *ind* by reverting other existing formulas.

Let us suppose the user searches for indicator *TotalEmission* in datasets of City$_A$ and City$_B$. Given that such an indicator is directly available only in City$_A$, the service calls the `get_formulas` predicate, which returns three solutions, i.e. $s_1 = TotEmissionPerCapita * TotalPopulation$, $s_2 = NH3Emission + PM10Emission$ and $s_3 = TotalEmission2 - SOxEmission - NOxEmission + PM10Emission$. At step 2, each solution is tested against the libraries. Checking a solution means to verify, through SPARQL queries, that every operand of the solution is measured by a dataset in the library at hand. For City$_B$, solutions s_1 and s_2 are not valid, as they lack both $TotEmissionPerCapita$ (needed by s_1) and $NH3Emission$ (needed by s_2). However, solution s_3 can be used, as the cube provides all the 4 needed measures. Hence, the indicator at hand is computable for both cities. As for City$_B$, the common dimensional schema is given (applying the derived formula) by $cds(TotalEmission2, SOxEmission, NOxEmission, PM10Emission) = \{timePeriod\} \cap \{timePeriod, place\} \cap \{timePeriod, place\} \cap \{timePeriod, place\}$ and hence it includes only the *timePeriod* dimension.

Comparison of data cubes. This service is aimed to extract measures provided by each library of datasets through function `get_all_indicators` and returns the indicators in common, by implementing the notion of comparability between data cubes.

```
get_all_indicators(endpoint):
    measures←get_measures(endpoint)
    ∀ m ∈ measures:
        indicators←get_ind_from_mea(m,endpoint)
    availableIndicators←derive_all_indicators(indicators)
    return availableIndicators
```

In detail, the function `get_all_indicators` firstly retrieves through SPARQL all the MeasureProperties from each library of datasets, and their corresponding KPIOnto indicator. Then, the service calls the logic function `derive_all_indicators`, which is capable to derive all indicators that can be calculated from those provided in input. The function exploits `get_formulas` to decompose all the available indicators in any possible way, and each of these rewriting is checked against the list in input. If there is a match, the solution is selected. Once computable measures in common between sources are found, their common dimensional schema is retrieved.

Let us consider libraries $City_A$ and $City_B$. By using `derive_all_indicators`, the reasoner infers the full set of computable indicators:
I_A = {$kpi{:}TotalEmission$, $kpi{:}TotEmissionPerCapita$, $kpi{:}TotalPopulation$};
I_B = {$kpi{:}TotalEmission2$, $kpi{:}NOXEmission$, $kpi{:}SOXEmission$, $kpi{:}PM10$ $Emission$, $kpi{:}TotalPopulation$, $kpi{:}NH3Emission$, $kpi{:}TotalEmission$, $kpi{:}Tot$ $EmissionPerCapita$}. Indeed, the third indicator for I_A is derived from the $TotEmissionsPerCapita$ formula. On the other hand, for I_B, indicator $NH3Emission$ was derived from the $TotalEmission2$ formula, given that the other operands were available, while $TotalEmission$ is calculated from this last and $PM10Emission$. In turn, $TotEmissionsPerCapita$ is calculated from this last and $TotalPopulation$. As a conclusion, the two libraries share the indicator set $I_A \cap I_B$ = {$kpi{:}TotalEmission$, $kpi{:}TotEmissionPerCapita$, $kpi{:}TotalPopulation$}. Please note that without the explicit representation of formulas and logic reasoning on their structure, no result would have been obtained. Finally, the three indicators are comparable only through dimension $sdmx{-}dimension{:}timePeriod$.

3.3 Experimentation

In this Section we evaluate running times of `get_formula` and `derive_all_indicators`, the most computationally complex steps in the procedures, on synthetically generated knowledge bases containing a set of formulas. We fixed the number of operands per formula to 2, while formulas are generated as summation of two randomly chosen indicators. As a result, each knowledge base is a totally connected graph of formulas. Please note that, in real scenarios, several independent graphs of formulas of this kind may co-exist in the same knowledge base. Given that logic predicates perform manipulations within each connected graph, the existence of other graphs does not have any impact on the running times. In the tests, given that these predicates involve a number of recursive execution increasing combinatorially with the input size, we bounded the maximum

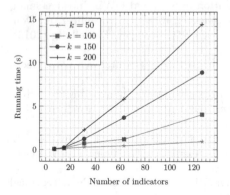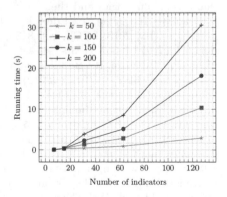

Fig. 1. Running times for predicates `get_formulas` (left, values are averaged) and `derive_all_indicators` (right).

number of formulas that can be inferred at each iteration to a value specified by a parameter k, to keep the complexity of the experimental procedure under control. In the following, each test has been executed with k = 50, 100, 150, 200. As results show, with larger input files, smaller values of parameter k should be considered to keep running times at a reasonable level (Fig. 1).

4   Conclusion

In this work, we proposed a logic-based approach for schema-level comparison of statistical datasets published following the RDF Data Cube vocabulary. Advantages of the approach include the possibility for users to define reusable and shareable formula definitions and hence define custom indicators to more easily integrate several data cubes. The ontology providing the vocabulary for the definition of indicators has been already exploited in various applications, ranging from performance monitoring in the context of collaborative organizations or Ambient Assisted Living to ontology-based data exploration of indicators [7].

More fine-grained typologies of schema comparisons can be performed by considering together measures, dimensions and their hierarchies in levels, which however are not entirely supported by RDF Data Cube. To this aim, to better guide integration, a richer representation for dimension hierarchies is provided by QB4OLAP [8], an extension to RDF Data Cube that will be taken into account for future work. Furthermore, we are investigating an approach to evaluate similarities between mathematical expressions representing formulas. This will extend the notion of comparability given in this paper in order to assess not only perfect matches between cubes but also approximate matches, by calculating a degree of similarity between mathematical structures.

References

1. Neumayr, B., Schütz, C., Schrefl, M.: Semantic enrichment of OLAP cubes: multidimensional ontologies and their representation in SQL and OWL. In: Meersman, R., Panetto, H., Dillon, T., Eder, J., Bellahsene, Z., Ritter, N., Leenheer, P., Dou, D. (eds.) OTM 2013. LNCS, vol. 8185, pp. 624–641. Springer, Heidelberg (2013). doi:10.1007/978-3-642-41030-7_46
2. Nebot, V., Berlanga, R.: Building data warehouses with semantic web data. Decis. Support Syst. **52**(4), 853–868 (2012)
3. Diamantini, C., Potena, D., Storti, E.: ScmPI: a semantic framework for the collaborative construction and maintenance of a shared dictionary of performance indicators. Future Gener. Comput. Syst. **54**, 352–365 (2015)
4. Buswell, S., Caprotti, O., Carlisle, D.P., Dewar, M.C., Gaetano, M., Kohlhase, M.: The open math standard. Technical report, version 2.0, The Open Math Society (2004). http://www.openmath.org/standard/om20
5. Cyganiak, R., Reynolds, D., Tennison, J.: The RDF data cube vocabulary. Technical report, World Wide Web Consortium (2014)
6. Sterling, L., Bundy, A., Byrd, L., O'Keefe, R., Silver, B.: Solving symbolic equations with PRESS. J. Symb. Comput. **7**(1), 71–84 (1989)

7. Diamantini, C., Potena, D., Storti, E.: Extended drill-down operator: digging into the structure of performance indicators. Concurr. Comput. Practice Exp. **28**(15), 3948–3968 (2016)
8. Etcheverry, L., Vaisman, A., Zimányi, E.: Modeling and querying data warehouses on the semantic web using QB4OLAP. In: Bellatreche, L., Mohania, M.K. (eds.) DaWaK 2014. LNCS, vol. 8646, pp. 45–56. Springer, Cham (2014). doi:10.1007/978-3-319-10160-6_5

S2D: Shared Distributed Datasets, Storing Shared Data for Multiple and Massive Queries Optimization in a Distributed Data Warehouse

Rado Ratsimbazafy$^{(\boxtimes)}$, Omar Boussaid, and Fadila Bentayeb

Université de Lyon, Lyon 2, ERIC EA 3083,
5 Avenue Pierre Mendés France, F69676 Bron Cedex, France
{rado.ratsimbazafy,omar.boussaid,fadila.bentayeb}@univ-lyon2.fr
https://eric.ish-lyon.cnrs.fr/

Abstract. Nowadays, with the constantly increasing amount of data, we are facing a growing number of users, who are characterized by a frequent and a massively concurrent data access. The large number of users pose multiple query optimization problems. In a distributed data warehousing system such as Hadoop/Hive, queries are evaluated one at a time and processed with the MapReduce paradigm. The massive query execution usually overloads and slows down the entire distributed environment mainly due to multiple data scan tasks. In this paper we aim to optimize the multiple query execution performance on Hive. We propose Shared Distributed Datasets (*S2D*), a method that dynamically looks for and shares common data among queries. The evaluation shows that, compared to Hive, *S2D* consumes on average 20% less memory in the *Map*-scan task and it is 12% faster regarding the execution time of interactive and reporting queries from *TPC-DS*.

Keywords: Big data warehousing · Query optimization · Distributed environment · Data sharing

1 Introduction

Nowadays, big data systems are facing an intensive demand to leverage and discover new insights inside big data. This intensive demand concerns thousands of queries arriving at the same time. This problem has been well known as Multiple Query Optimization (MQO) [7] in databases or data warehousing field. Many solutions and approaches have been proposed in order to optimize the multiple query problem, such as view materialization, caching, horizontal or vertical fragmentation and so on.

However in Hadoop/Hive environment those solutions need to be rethought or scaled up, as they are not suitable in big data context. We look for a responsive approach with big data warehouse properties by proposing a distributed solution.

Inside the Hadoop/Hive ecosystem the SQL-like queries are executed into MapReduce, Tez, Impala or Spark workflows. The efficiency of this multiple

© Springer International Publishing AG 2017
L. Bellatreche and S. Chakravarthy (Eds.): DaWaK 2017, LNCS 10440, pp. 42–50, 2017.
DOI: 10.1007/978-3-319-64283-3_4

analytical process is strongly affected by the performance of those workflows. However in presence of intensive queries, the system evaluates the query one by one, and ignores common subexpressions or shared data.

Processing intensive query, in big data environment, has recently attracted tremendous attention, in particular with regard to the system tuning [2,3,5], job scheduling [4,9] and tasks and/or data sharing [1,6,8]. These works discovered that reusing intermediate results improve MapReduce efficiency during job execution. Nonetheless the outputs produced by MapReduce execution model [10] are still stored in the distributed file system (disk). This is less efficient when the outputs are stored in memory.

The main focus of our research is to investigate adequate solutions to reduce query response time of typical OLAP queries and to improve scalability using a distributed computation environment, which takes advantage of characteristics specific to the distributed context.

In this paper we present *S2D*, a methodology that allows us to reduce computation time in workload processing: first, *S2D* is able to find the shared data among workloads, it means that the common data that can be shared between a group of queries, is saved in a logical representation; Secondly, depending on the shared data we construct a block files which is the physical representation on memory or on disk if it is recurrently asked; Lastly, instead of block files from HDFS, we give to the Hive execution engine, as input, the needed physical representation of the shared data.

To illustrate the working principle of *S2D*, we examine all the queries that are written in SQL-like language. We transform them into relational algebra, and we search for the common *Projection* (π). In HDFS, each similar π between queries is from the same block files. We regroup the data that is queried the most frequently from several block files into a smaller block file. This reduces the amount of data lines and accelerates the scan task. The first results we obtained from *S2D* method are encouraging. They show that for multiple query optimization, *S2D* reduces Hive execution time and memory usage.

The outline of the paper is organized as follows: Sect. 2 overviews the main research issues related to *S2D*. Section 3 details and explains our method. Section 4 points out the efficiency of our solution, and Sect. 5 concludes the paper and suggests future work.

2   Related Work

As far as we know, in MapReduce or MapReduce-like execution engines, queries are transformed into a workflow of jobs. Every job in the workflow may depend on the other jobs. A given job may not be started if all of his dependencies are not finished.

We outline, in this section, the research studies that arise in multiple query optimization on Hadoop environment. We classify the following research into three groups:

(1) **System tuning approaches:** Herodotou H. et al. have proposed *Starfish* to improve MapReduce efficiency. *Starfish* configures the Hadoop system,

automatically, to meet the best performance to process the workload [2,3]. With *Wisedb* Marcus R. et al. have introduced a resource provisioning and query placement system to handle multiple query processing [5]. The Cloud computing elasticity plays an important role on these approaches.

(2) **Job scheduling techniques:** Zaharia M. et al. have investigated the cluster resource division, by allocating to each user's jobs a small private cluster [9]. Isard M. et al. have introduced *Quincy*, a fair scheduler on Hadoop for concurrent jobs [4], cluster nodes are divided into slots, among which all jobs are distributed.

(3) **Task and Data sharing methods:** Wang et al. [8] have proposed to combine multiple jobs into a single one and to share the materialized map outputs between them. However, their solution did not provide any system level optimizations for recurrent queries. With *MRShare*, Nykiel T. et al. [6] have studied to compute queries with similar jobs to process them as a single execution. However, the solution did not work on queries with "join" function. Elghandour I. et al. have developed *Restore* [1] to reuse "stored" Map and/or Reduce intermediate results. The system attempt to find the sharing opportunities to avoid redundant work for similar jobs. For recurrent queries *ReStore* has materialized MapReduce intermediate outputs for future reuse.

Comparing multiple workflows, to find all similar jobs, is a complicated task. Our research work focuses mainly on data sharing issues like *Restore* [1]. We differ from those listed approaches in several ways: (1) we pre-compute all the possible shared data before compiling SQL-like queries into MapReduce or Spark jobs, (2) we re-use that information to speed up the computation by colocating all shared data, (3) we only store reusable shared data to reduce *"the scan jobs"* from the distributed file systems (HDFS).

3   Overview of Shared Distributed Datasets

Our proposal leveraged the similarities between queries to find all the shared data. Our approach is based on two main phases: **(1) finding the data that can be shared:** from an input group of similar queries, we identified which data can be shared among them and we output the *logical representation* of those shared data (Subsect. 3.1), and **(2) the construction of shared distributed datasets:** once the shared data is described by their *logical representation*, we built the *physical representation*. We gathered all the needed records in identified and collocated block files (Subsect. 3.2).

3.1   Phase 1: The Logical Representation

To achieve this phase we evaluated the Select-Projection-Join (*SPJ*) expressions for each query from the workload. We used the *SPJ* form, because in *SPJ* we had all relational algebra operations to process a query.

1. *First of all,* from the massive queries we clustered all the queries in order to regroup queries by similarities. To this end we used a machine learning algorithm, *proximus*[1]. This part was detailed and presented in another research paper [13]. In this work we focused only on a single group of similar queries.
2. *Secondly,* for each cluster of similar queries, we compared all the projections (π) to find queries that share the same block files. Since we expected to perform all aggregation function in the Reduce task, at this point we did not need to consider them as a feature for the sharing opportunity research.
3. *Lastly,* the number of entries that we got from the block files can be considerable so we redefined the size of our shared dataset, by using the selection operation (σ) from the *SPJ*. σ helped us to select the subset of records from the relation R (all joined tables) that satisfied the predicate (p) condition. Our algorithm took in comparison all σ related to the attributes in the π.

Algorithm 1 was designed to formalize this phase. This algorithm details the working principle of our approach at this step.

Algorithm 1. The sharing opportunity research

    input : Queries workload
    output: Sharing opportunities SO
1: **for** *all queries q_i in query workload* **do**
    // Translate each query into a list of SPJ operations
2:     $Sq_i \leftarrow SPJ(q_i)$
    // Cluster queries by similarity
3:     Cluster Sq_i into c_i cluster
4:     $C \leftarrow c_i$
5: **end**
6: **for** c_i *in* C **do**
7:     Find inside all projection operations all shared attiributes
    // looking for the best set of shared attibutes
8:     $SO \leftarrow \pi_{<a_1,a_2,...,a_*>}(R)$
9:     **if** $\exists$ *selection operations applied on attributes* **then**
        // reduce the shared data record with the help of selection operations
10:         $SO \leftarrow \pi_{<a_1,a_2,...,a_*>}(R)\sigma_{<p(a_1),p(a_2),...,p(a_*)>}$
11:     **end**
12: **end**
13: **return** SO

Running Example: To illustrate our approach, in this phase we will use an example of two queries Q_1 and Q_2. Those queries are built from templates query 3 and query 52 from TPC-DS benchmark.

Code 1.1. Sharing opportunities: Logical representation

```
1- date_dim : {d_date_sk, d_year} sharpen by {d_moy=12}
2- item : {i_item_sk, i_brand_id}
3- store_sales : {ss_ext_sales_price, ss_item_sk, ss_sold_date_sk}
```

---

[1] http://compbio.case.edu/koyuturk/software/proximus/.

(A) After a comparison we noticed that Q_1 and Q_2 have a common part and they use almost the same data.
(B) We transform Q_1 and Q_2 in their SPJ form. At this point we use the SPJ query representation to identify all the shared data.
(C) The output from our algorithm $Code$ 1.1 is the logical representations of identified shared datas between Q_1 and Q_2.

3.2   Phase 2: The Physical Representation

This phase is the physical construction of the identified shared data. In this phase the program performed multiple tasks:

1. For a given cluster of queries, we took as input all the logical representations of the shared data from the previous phase.
2. We computed the most asked shared data and give it a high priority to be constructed.
3. We built and stored the shared data related to each cluster or the most used sharing opportunities depending on the Hadoop cluster resources.
4. We kept in memory, or saved on defined nodes, all block files on HDFS for future reuse. This gave us the opportunity to scan and to find quickly the data that we needed.

Figure 1 illustrates the process of $S2D$ for our given examples Q_1 and Q_2. Let's suppose that we need from blk_1 data from lines (100 to 200) and (500 to 600) and from blk_3 data from lines (50 to 75) and (250 to 500). In MapReduce a Map task will be needed to scan all the blocks from HDFS. The intermediate results from this Map will contain all the needed data lines. This Map task will be performed each time when a query needs those same data lines, this is due to query at a time evaluation policy. With $S2D$ we pre-process all the needed lines and we save them in identified and collocated block files. This aims to speed up the Map-scan task and reduces the number of needed scans because the number of new block files is much lower than the original.

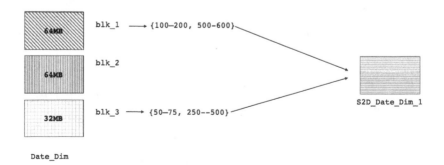

Fig. 1. $S2D$ construction

In *S2D*, we had the logical representation of shared data and a defined number of physical shared data. The physical representation could be deleted anytime and rebuilt on demand since we kept all the logical representation. Moreover saving the logic form may reduce the storage unit consumption.

4 Experimental Evaluation

4.1 Experimental Setup

Environment and resources: we experimented our approach on Hortonworks data cloud (HDC) services[2], a platform for analyzing and processing data. HDC provides an Hadoop preinstalled and preconfigured cluster on Amazon web service. We deployed a cluster composed by five m4.xlarge (4 vCPU - 16 Gb RAM) computer. **Data**: the implementation of our algorithm presented in this paper has been tested with TPC-DS benchmark [11]. TPC-DS provides data and query generator in order to evaluate the performance of a decision support system [12]. Due to lack of space we limit our experiments with 2TB (TPC-DS scale factor 2000). For all generated tables, the number of records average $\sim$ 13 billion lines. **Queries**: TPC-DS utilizes 99 query templates, which are used by the TPC-DS query generator to produce a fully qualified and valid SQL queries. Those query templates assess the interactive nature of OLAP queries, the data extraction queries, ad-doc queries and reporting queries. In this evaluation, we intentionally defined our workload to take only interactive and reporting query templates. We included templates with a variety of joins and aggregations, as well as complex multi-level aggregations. Bellow the TPC-DS query templates (QT) that we choose as:

– Interactive queries: QT_{19}, QT_{42}, QT_{55}, QT_{52}
– Report queries: QT_3, QT_7, QT_{27}, QT_{43}

We varied the number of generated queries per template from 10 to 100, we got 80 to 800 queries for this evaluation.

Query similarity. The query similarity computation gave us a view of which generated queries share data. Figure 2 summarizes the data sharing opportunities per query templates (QT). In this figure we can see that queries based on QT_3 shared data with QT_{52} such as QT_7 with QT_{27} and QT_{19} with QT_{55}. Queries based on QT_{42} and QT_{43} don't share data with other query templates. In this experimentation we used QT_{42} and QT_{43} to test recurrent executions.

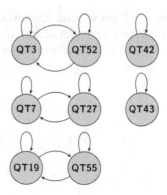

Fig. 2. Shared data research

---

[2] http://hortonworks.com/products/cloud/aws/.

4.2   Experimental Results and Discussion

We are aware that building all logical representations of shared data is not suitable. With the high number of queries and a wide variety of selection predicate, we may have a large amount of possible shared data. In this experimentation, we focused on two groups of queries: queries from QT_{19} with QT_{55} and queries from QT_{43}. We chose to present how $S2D$ works with a small group of queries as mentioned in Subsect. 3.1.

Storage occupation. Each time a MapReduce performs a scan task, all block files may be loaded in memory. We measured all block files from the needed dimension. We also measured the size of all built $S2D$. In Fig. 3 we compared the dimension's size and $S2D$'s size. Figure 3 shows that $S2D$ data size is lower than the dimension's size. This is due to data preparation, we only kept in $S2D$ the needed entries.

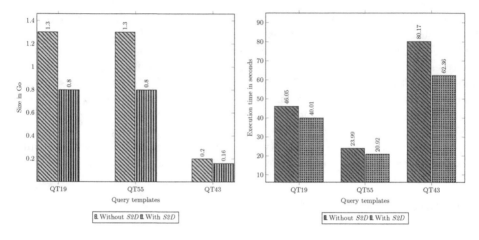

Fig. 3. Data size to scan   **Fig. 4.** Execution time in seconds

Query Execution. Execution time in Hive is highly relied on the system's and cluster's configuration. We measured in our experimentation the number of needed tasks for our set of queries. Table 1 shows that the number of MapReduce tasks does not change. We had this result since the number of block files to scan

Table 1. Number of MapReduce task required per query

Query	Map (without $S2D$)	Reduce (without $S2D$)	Map (with $S2D$)	Reduce (with $S2D$)
QT_{19}	10	2	10	2
QT_{55}	4	6	4	6
QT_{43}	5	34	5	34

is the same. We did not merge any block files. Nonetheless Fig. 4, shows better execution time with our approach. This is due to the data size and the number of processed record lines inside the block files.

Discussion One of the main goals of this evaluation was to attempt to find a way to optimize multiple query execution by sharing data. These results support and negate some of our hypotheses. It was predicted that collocating all shared data, in order to reduce disk scan tasks, would result in a better query response time, and storing the logical representation of shared data helped to reduce the disk space usage but computing the query similarity may took a considerable preparation time, when unknown new queries are executed.

Previous studies in tasks or data sharing approach differ from our method: it was common for most of them to focus on the MapReduce by storing intermediate results or computing once all similar jobs [1,6]. Due to the Hadoop version and Spark usage, we decided to not compare our results with theirs, since it would be, in our opinion, a biased comparison.

In our implementations and uses, *S2D* may resemble to materialized views, but there are technological differences. In definition materialized views contain a copy of query results from a single point of time; in *S2D* we only store the most used part of the needed records, from the logical representation of the shared data. Materialized views needs a data refresh operations. With *S2D*, when the massive query execution is finished, we keep the most needed data entries and all the logical representation will be saved. This allows us to accelerate the construction of shared distributed datasets when its needed.

Our research only focuses on data sharing, whereas it might be important to include the task sharing as well. In fact, the inclusion of the task sharing would enable us to reduce the number of similar executed task. These results are a step forward in OLAP query execution, for example to store the data CUBE as *S2D* and perform analysis by using distributed memory.

5   Conclusion and Future Work

This paper stresses the importance of data sharing in optimization for multiple query execution. With our implementation of *S2D*, we obtained expected results that share the proper subset of data providing efficient workload processing. We demonstrate that our approach is able to optimize workload execution. Even though our work has some limitations, we believe that our contribution could be the starting point for optimization in multiple complex query processing like OLAP. We are considering to experiment the data CUBE creation and storage by using *S2D* were the data warehouse's dimensions are distributed over multiple cloud providers or geographically separated.

We also need to test the fault tolerance of our algorithm. We are aware that we cannot create all the logical representation of shared data. We are working on an algorithm to improve the physical representation selection. Future work will mainly cover the development of additional features of the method, such as pre-computed join inside *S2D*.

References

1. Elghandour, I., Aboulnaga, A.: ReStore: reusing results of MapReduce jobs. Proc. VLDB Endowment **5**(6), 586–597 (2012)
2. Herodotou, H.: Automatic tuning of data-intensive analytical workloads. Ph.D. thesis, Duke University (2012)
3. Herodotou, H., Lim, H., Luo, G., Borisov, N., Dong, L., Cetin, F.B., Babu, S.: Starfish: a self-tuning system for big data analytics. CIDR **11**, 261–272 (2011)
4. Isard, M., Prabhakaran, V., Currey, J., Wieder, U., Talwar, K., Goldberg, A.: Quincy: fair scheduling for distributed computing clusters. In: Proceedings of the ACM SIGOPS 22nd symposium on Operating systems principles, pp. 261–276 (2009)
5. Marcus, R., Papaemmanouil, O.: WiSeDB: a learning-based workload management advisor for cloud databases. CoRR abs/1601.08221 (2016)
6. Nykiel, T., Potamias, M., Mishra, C., Kollios, G., Koudas, N.: MRShare: sharing across multiple queries in mapreduce. Proc. VLDB Endowment **3**(1–2), 494–505 (2010)
7. Sellis, T.K.: Multiple-query optimization. ACM Trans. Database Syst. **13**, 23–52 (1988)
8. Wang, G., Chan, C.-Y.: Multi-query optimization in MapReduce framework. Proc. VLDB Endowment **7**, 145–156 (2013)
9. Zaharia, M., Borthakur, D., Sarma, J.S., Elmeleegy, K., Shenker, S., Stoica, I.: Job scheduling for multi-user mapreduce clusters. EECS Department, University of California, Berkeley, Technical report. UCB/EECS-2009-55 (2009)
10. Dean, J., Ghemawat, S.: MapReduce: simplified data processing on large clusters. In: Proceedings 6th Symposium on Operating System Design and Implementation OSDI, pp. 137–150 (2004)
11. Nambiar, R.O., Poess, M.: The making of TPC-DS. In: Proceedings of the 32nd International Conference on Very Large Data Bases, pp. 1049–1058. VLDB Endowment (2006)
12. Poess, M., Nambiar, R.O., Walrath, D.: Why you should run TPC-DS: a workload analysis. In: Proceedings of the 33rd International Conference on Very Large Data Bases. VLDB Endowment (2007)
13. Ratsimbazafy R., Boussaid O., Bentayeb F.: Stratégie pour le traitement des processus décisionnels massifs dans un big data warehouse. In: Proceedings of the EDA 2016, Aix-en-Provence (2016)

Cloud and NoSQL Databases

Enforcing Privacy in Cloud Databases

Somayeh Sobati Moghadam[✉], Jérôme Darmont[✉], and Gérald Gavin

Université de Lyon, Lyon 2, Lyon 1, ERIC EA 3083,
5 avenue Pierre Mendès France, 69676 Bron Cedex, France
ssobati@eric.univ-lyon2.fr, jerome.darmont@univ-lyon2.fr,
gerald.gavin@univ-lyon1.fr

Abstract. Outsourcing databases, i.e., resorting to Database-as-a-Service (DBaaS), is nowadays a popular choice due to the elasticity, availability, scalability and pay-as-you-go features of cloud computing. However, most data are sensitive to some extent, and data privacy remains one of the top concerns to DBaaS users, for obvious legal and competitive reasons. In this paper, we survey the mechanisms that aim at making databases secure in a cloud environment, and discuss current pitfalls and related research challenges.

Keywords: Databases · Cloud computing · DBaaS · Data privacy · Data encryption

1 Introduction

Cloud computing offers a variety of services via a pay-per-use model on the Internet. The flexibility that cloud computing offers is very appealing for many organizations, especially mid-sized and small ones, because it provides reduced start-up costs and means to financially cope with variations in system usage. Outsourcing data to the cloud is particularly interesting [75]. However, some data are especially sensitive, e.g., personal data, health-related data, business data, and generally data used in decision-support processes. Outsourcing a database in the cloud raises security issues, some related to cloud architectures (e.g., untrusted service providers, curious cloud employees...), and others related to such concerns as data privacy, integrity and availability. With increasingly sophisticated internal and external cloud attacks, traditional security mechanisms are no longer sufficient to protect cloud databases [77].

Let us consider a Database-as-a-Service (DBaaS) scenario (Fig. 1) where a user outsources a database at one or more Cloud Service Providers' (CSPs). The objective is to eliminate storage and minimize computation at the user's to take full advantage of cloud benefits. Yet, anything beyond the user is considered untrusted. CSPs might indeed be honest but curious, i.e., read the user's data, or even be malicious or maliciously hacked, i.e., alter data or provide fake query results. Network transactions are also considered unsafe. The user must thus protect sensitive data and queries before sending them to CSPs, and safely reconstruct query results "at home".

© Springer International Publishing AG 2017
L. Bellatreche and S. Chakravarthy (Eds.): DaWaK 2017, LNCS 10440, pp. 53–73, 2017.
DOI: 10.1007/978-3-319-64283-3_5

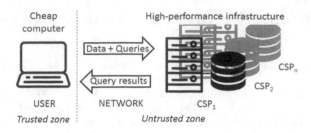

Fig. 1. Database outsourcing scenario

In this paper, we survey the security mechanisms that may be exploited in our cloud database scenario, particularly in terms of privacy. Another recent survey intersects ours [58], but more deeply focuses on cryptography, while we adopt a broader scope on security, put more emphasis on querying efficiency and survey non-cryptographic methods (Sect. 2). We also mostly target database practitioners and researchers with no background in cryptography. Moreover, we review cryptographic methods that are not covered in [58], i.e., secret sharing, Private Information Retrieval (PIR) and oblivious RAM (ORAM) schemes. We classify cryptographic tools that can be exploited within cloud database scenarios into secret sharing schemes (Sect. 3), index-based methods (Sect. 4) and secure databases (Sect. 5). Finally, we conclude this paper by a global discussion (Sect. 6).

2   Non-cryptographic Methods

2.1   Differential Privacy

Differential privacy aims at protecting data privacy when performing statistical queries [24]. While global statistics are public, individual data must remain private. To achieve this goal, a noise term is added to statistical query results, e.g., to an average salary, thus preventing the computation of individual salaries. It would indeed be easy to compute a new salary that has just been added in the database, knowing averages avg and the number of records n in the dataset: $avg_{n+1} \times (n+1) - avg_n \times (n)$.

A randomized algorithm A enforces ϵ-differential privacy if and only if, for any two databases DB_1 and DB_2 that differ on exactly one record, the ratio between the probability that A outputs O on DB_1 and DB_2 is bounded by a constant: $P(A(DB_1) = O) \div P(A(DB_2) = O) \leq e^\epsilon$ [76]. The tradeoff is that the smaller ϵ is, the better privacy is enforced, but the worse accuracy is. Thus, one major challenge is reducing the amount of noise while still satisfying differential privacy. Yet, in a DBaaS context, a curious CSP would still have access to fine-grained data, which is incompatible with our data outsourcing scenario, where we aim at protecting data event from the CSP.

2.2   Data Anonymization

Data anonymization irreversibly modifies data in a way that prevents the identi-fication of sensitive information, while allowing querying data for releasing use-ful statistical information [25]. Some database management systems (DBMSs) natively include anonymization schemes. For example, Oracle Data Masking Pack provides data masking for various types of data, replacing real data with realistic-looking values [51]. Obscuring query processing and results may also be achieved either by rejecting queries leading to privacy disclosure or through methods such as k-anonymity and its variants t-closeness and l-diversity, which transform k distinguishable records into k indistinguishable records [67].

In a DBaaS context, privacy-preserving queries in a distributed environment can be achieved by table perturbation and reconstruction [5]. Perturbation ran-domly replaces an element in a table by another with probability p. Then, recon-struction can estimate COUNT queries over the perturbed table. Unfortunately, other aggregation functions are not supported. Similarly, random data distortion techniques may be used, e.g., the zero-sum method provides an accurate esti-mation of summation for range queries [66], but it induces a trade-off between privacy and accuracy. Finally, adding different amounts of noise to query answers can be used, e.g., with iReduct [74]. iReduct initially estimates query answers and iteratively refines its estimates to minimize relative errors.

Data anonymization is less complex than encryption and straightforwardly allows querying anonymized data. However, it typically reduces data granularity, which may cause a loss of effectiveness and correctness in computation [1]. More-over, the various anonymization methods have different impacts on data utility. As a result, the same anonymized table may provide accurate answers to some queries and inaccurate results to others. More importantly, data anonymization cannot provide an adequate level of security since private personal data can be re-identified by an adversary who has some knowledge about data [23].

2.3   Data Fragmentation

In data fragmentation [2], data are assumed not to be sensitive *per se*. What is sensitive is their association with other data. Privacy is thus guaranteed by con-cealing such associations with respect to a predefined set of security constraints that express restrictions on one or more attributes in a table [20,37]. For instance, given a *Patient* table, constraint $C = \{Name, Illness\}$ indicates that associa-tions between patient names and illnesses should not be disclosed. Then, the table is split into fragments such that attributes listed in a constraint belong to different fragments. For instance, table *Patient* would be split in two frag-ments $Patient_1$ and $Patient_2$, with $Name \in Patient_1$ and $Illness \in Patient_2$. Yet, most data fragmentation approaches apply to numerical data and specific methods most be used to handle categorical data [56].

In a cloud context, data can be partitioned at independent CSPs' with respect to security constraints. When a query is issued, an appropriate subquery is trans-mitted to each CSP, then the results are pieced together at the user's. Moreover, intrinsically sensitive attributes such as social security numbers are stored locally at the user's. Eventually, fragmentation-based approaches yield little overhead on query computation, but are vulnerable in cases of CSP collusion and CSP inference on data updates [35]. Additionally, retaining sensitive data at the user's requires local storage capacities, which is incompatible with our data outsourcing scenario.

3   Secret Sharing-Based Methods

Secret sharing is a particular cryptographic method introduced by Shamir in which a secret piece of data is mathematically divided into so-called shares that are stored at n participants' [60]. One single participant has no means to reconstruct the secret. A subset of $k \leq n$ participants is indeed required to reconstruct the secret, providing perfect theoretical privacy when at most $k - 1$ participants collude, i.e., exchange shares. Moreover, computations can run directly on shares, outputting unintelligible results that can only be put together through k participants. Finally, up to $n - k$ participants may disappear without compromising data availability; and message authentication code or signature can be applied to guarantee data integrity. Thus, secret sharing is a promising solution to security challenges in cloud data outsourcing [35] since one can easily imagine participants being CSPs. In the following subsections, we review the secret sharing-based approaches that aim at outsourcing databases, and then discuss them globally.

3.1   Verifiable Secret Sharing

Thompson et al.'s scheme allows participants to collaboratively compute aggregation queries without gaining knowledge of intermediate results [68]. A lightweight cryptographic scheme is introduced for privacy-preserving computation and verification of SUM and AVG aggregation queries. Moreover, users can verify query results with the help of signatures, while the data values contributing to the results are kept secret from both users and the CSPs. The query issuer indeed interacts with a single CSP to obtain aggregation results and can verify whether the CSP returns correct results. However, aggregation queries other than SUM and AVG cannot be computed with this scheme.

Attasena et al. specifically target cloud data warehouses through a flexible, verifiable secret sharing scheme named fVSS [11]. fVSS minimizes shared data volume, provides inner and outer data verification to check data correctness and the honesty of CSPs, improves the ability to update shares in case of CSP failure, and adjusts share volume with respect to CSP pricing policies. Moreover, in addition to queries explicitly handled by previous schemes, fVSS also allow grouping queries that are ubiquitous in On-Line Analysis Processing (OLAP). However, although some queries can be computed directly over shares (e.g., exact match queries), others require that some or all data are decrypted first (e.g., range queries).

Wang et al. propose a framework for secure and efficient query processing of relational data in the cloud that allows exact match and range queries, as well as updates [71]. B+-tree indexes are also built to optimize query response. Both data and indexes are organized into matrices, encrypted and stored at CSPs'. Additionally, data integrity is achieved by using checksum and an index structure. This framework is robust against statistical attacks.

Statistical attacks refer to an adversary obtaining some information about ciphertexts, i.e., encrypted data, through prior knowledge about plaintexts, i.e., clear data [35]. For instance, the adversary may know plaintext distribution or frequency. Then, by extracting statistics from ciphertexts, the adversary can infer ranges containing dense data or highlight ciphertexts bearing the same frequency as plaintexts.

3.2 Order-Preserving Secret Sharing

Agrawal et al. propose a complete approach to execute exact match, range, and aggregation queries over shares outsourced at multiple CSPs [3].

Original data are divided using order-preserving polynomials such that the order of shares is the same as that of original data. However, while this solution allows efficiently processing any kind of queries, including updates, it is susceptible to statistical attacks [35].

Hadavi et al. introduce a framework to provide data privacy based on threshold secret sharing [36]. First, secret values are encrypted by an Order Preserving Encryption (OPE) scheme (Sect. 5.1). Then, a B+-tree is built over ciphertexts and sent to an index server. The user receives query responses, including record numbers, from the index server and can then request these records from the CSPs. As Agrawal et al.'s scheme, this approach supports different kind of queries over shares, including exact match, range and aggregation queries, as well as updates. Moreover, it provides stronger security than Agrawal et al.'s scheme. It is indeed secure against frequency attacks, and an extension uses distribution perturbation to improve its robustness against statistical attacks in general [35]; but at the price of a higher computational overhead.

3.3 Discussion

Table 1 provides a comparison of secret sharing-based database outsourcing methods with respect to the queries that can run directly on shares and security features beyond privacy and availability, i.e., integrity checks and robustness against statistical attacks.

Despite secret sharing's benefits, it is not trivial to process some queries directly over shares, especially queries requiring data ordering. Obviously, it is not efficient to send back all shares in response to a query, and execute the query over reconstructed values at the user's. However, the techniques that allow directly processing such queries as range queries over shares reveal some information about plaintexts, e.g., duplicates [3]. Moreover, since every secret is shared n times, global share size can be quite large. Communication between the user

Table 1. Comparison of secret sharing-based methods

	Allowed queries				Additional security features	
	Exact match	Range	Aggregate	Update	Integrity	Statistical attacks
Thompson et al. [68]	No	No	SUM/AVG	No	Signature	Not robust
Attasena et al. [11]	Yes	Yes	Yes	Yes	Signature	Not robust
Wang et al. [71]	Yes	Yes	Yes	Yes	Checksum	Robust
Agrawal et al. [3]	Yes	Yes	Yes	Yes	None	Not robust
Hadavi et al. [36]	Yes	Yes	Yes	Yes	None	Robust

and multiple CSPs is not optimal in terms of bandwidth resources either. As a result, storage and communication overhead of secret sharing-based approaches is remarkably high for moderately large databases [44].

4   Index-Based Methods

In databases, data encryption is usually managed at the tuple level [22], which prevents any computation over ciphertexts. Thus, indexes based on plaintexts are stored together with the encrypted database to help return ciphertexts in response to queries. We distinguish three types of index-based methods, namely bucketization, order preserving indexing and indexes used in Searchable Encryption (SE) schemes. We review them in the following subsections before providing a global discussion.

4.1   Bucketization-Based Indexing

Bucketization-based indexing involves dividing data into buckets and providing explicit labels for each bucket [49]. The domain of an attribute is partitioned into a set of non-overlapping buckets. Labels may preserve the order of values in the original domain or not. They are stored along with encrypted tuples. Such indexing allows exact match, range (if data order is preserved) and join queries, but also induce false positives in query answers. Thus, query post-processing is needed at the user's to filter out false positives [59].

Hacigümüs et al. partition data as in histogram construction, e.g., by equi-depth and equi-width partitioning [33]. Then, it assigns a random tag to each bucket. Any table $T(A_1, A_2, ..., A_n)$ from a database is stored at the CSP's as $T^S(etuple, A_1^S, A_2^S, ..., A_n^S)$, where $etuple$ is the encrypted tuple and each A_i^S is the index of attribute A_i. Each query is rewritten into server-side and user-side subqueries Q^S and Q^C, respectively. Q^S is executed by the CSP over ciphertexts using indexes A_i^S. The result of Q^S is then sent back to the user, who decrypts it and executes Q^C to filter out false positives. Query rewriting requires maintaining metadata, including bucket labels.

With the help of an homomorphic function, this approach is extended to support aggregation queries over ciphertexts [34]. The homomorphic encryption function is based on the Privacy Homomorphism (PH) scheme [57], which relies on the difficulty of factoring large composite integers, just like the famous Rivest-Shamir-Adleman (RSA) public-key cryptosystem. Unfortunately, Mykletun and Tsudik demonstrate that the CSP can obtain plaintexts with access to ciphertexts only [49].

Based on their rebuttal of Hacigümüs et al., Mykletun and Tsudik propose a simple alternative for supporting aggregation queries [49]. The user precomputes aggregation values (e.g., SUM and COUNT) for each bucket, encrypts and stores them at the CSP's. Moreover, instead of using the PH scheme, Mykletun and Tsudik use provably secure additive homomorphic encryption schemes such as Paillier's [53] and El Gamal's [27]. Precomputing aggregations decreases security risks, but requires extra storage and makes updates more complex. Updates must indeed be executed in two steps: (1) actual data update and (2) update of related precomputed aggregates in a bucket.

Hore et al. also address shortcomings of Hacigümüs et al.'s method. They notably optimize the accuracy of range queries to minimize false positives in query results [40]. They also introduce a re-bucketization technique, in which the user can fine-tune bucketization to achieve a desired level of privacy. Moreover, they propose a new method for answering range queries on multidimensional data [38]. Range queries over multiple attributes, e.g., *age* < 20 and *salary* > 25k, are allowed, while minimizing the cost of multidimensional bucketization. Yet, a threshold is defined to help the user control the trade-off between risk of data disclosure and cost.

4.2   Order-Preserving Indexing

Order Preserving Encryption Scheme (OPES). Agrawal et al.'s OPES is an OPE indexing scheme that supports range and equality queries over integers [4]. OPES transforms plaintexts with an order preserving function so that transformed values (e.g., index values) follow a target distribution. Comparison operations can be directly applied at the CSP's without inducing spurious tuples nor false positives. However, this scheme has been demonstrated to be vulnerable to statistical attacks [45].

OPE with Splitting and Scaling (OPESS). The OPESS scheme encrypts XML databases [72]. Wang et al. adopt splitting and scaling techniques to create index values following a uniform distribution. Plaintext order is preserved over indexes. Moreover, identical clear values are transformed into different indexes so that this scheme is robust against statistical attacks. However, this scheme flattens the frequency distribution of index values.

B+tree Indexing. Shmueli et al. [61] and Damiani et al. [21] use B+-trees built on database plaintext attribute values to preserve order in secure environments. B+-trees must either be stored in a trusted machine [36] or be encrypted at the CSP's, where each B+-tree is stored in a table with two attributes: node identifier and node content. In addition to ordering, B+-tree indexes support exact match, range and grouping queries, as well as predicates such as LIKE. For example, to execute a range query, the user sends a sequence of queries until reaching the leaf corresponding to the range's lower bound. Then, the node identifier helps retrieve all the tuples belonging to the range. The advantage of such indexing is that index content is not visible to the CSP and reveals no information about underlying plaintexts [21].

4.3 Searchable Encryption

SE allows the CSP to run keyword-based searches on encrypted data [63] that are particularly suitable to data outsourcing [16]. Considering a set of documents $\{D_i\}_{i=1,n}$ and an index of keywords $\{w_j\}_{j=1,m}$ describing the documents, users encrypt both documents D_i with any secure encryption scheme using a key $\mathcal{K}_{Enc}$ and keywords w_j with a searchable scheme using a key $\mathcal{K}_{Index}$. The encrypted documents and index are then outsourced. When searching for documents containing some keywords, the user sends a so-called trapdoor encapsulating the keywords to the CSP [65]. Then, the CSP can search the encrypted index and the trapdoor to find the corresponding documents and send them back to the user. Both symmetric (private) and asymmetric (public) key encryption can be used to build symmetric SE (SSE) and asymetric SE (ASE) schemes, respectively [65]. ASE schemes support various query types such as range and subset queries, but are computationally intensive. SSE is more efficient than ASE, but supports fewer query types. SE induces a trade-off between security, efficiency and query expressiveness. SE schemes with higher levels of security induce higher complexity, while SE schemes supporting more query types are either less secure and/or less efficient [16]. Moreover, most SE schemes reveal access patterns, i.e., which documents contain a keyword. Only techniques based on PIR or ORAM do not.

PIR [17,18,47] enables a user to retrieve data from an outsourced database while preventing the CSP from learning any information about retrieved data [19], i.e., PIR enforces query privacy. Unfortunately, in a single-server setting, the only thing a user can do is retrieving the whole database, which induces communication overhead and annihilates the benefits of outsourcing. However, in a multiple-server setting where copies of the database are stored at k non-communicating/colluding CSPs, a user can hide queries by querying each server for a part of data, so that no server knows the whole query.

ORAM allows reading and writing to memory without revealing access patterns to the CSP [32]. In ORAM schemes, a user stores encrypted data at the CSP's and continuously shuffles and re-encrypts data as they are accessed [64]. Let $P = (q_1, \ldots, q_n)$ be an access pattern. The shuffling process induces the transformation of each query q_i into multiple queries, producing a new access pattern P'. An ORAM protocol is secure if two access patterns $\text{ORAM}(P)$ and $\text{ORAM}(P')$ are computationally indistinguishable. ORAM can be implemented using symmetric or fully homomorphic encryption (Sect. 5.1). An alternative solution for hiding access patterns is to frequently send fake queries to CSPsto prevent any adversary from inferring correlations between frequently queried data [48]. Yet, generating fake but realistic-looking queries is a challenge.

Unfortunately, a common limitation of PIR and ORAM schemes is a prohibitive query execution time [73].

4.4 Discussion

Table 2 provides a comparison of index-based methods with respect to the query types they allow and whether they require a post-processing step to eliminate false positives.

Table 2. Comparison of index-based methods

	Allowed queries			Post-processing
	Exact match	Range	Aggregation	
Hacigümüs et al. [33]	Yes	Yes	Yes	Yes
Mykletun & Tsudik [49]	No	No	Yes	No
Hore et al. [40]	Yes	Yes	No	Yes
Agrawal et al. [4]	Yes	Yes	No	No
Wang et al. [72]	Yes	Yes	No	No
Shmueli et al. [61], Damiani et al. [21]	Yes	Yes	No	No
Searchable Encryption [17,18,32,47,63,65]	Yes	No	No	No

When defining an indexing method, it is important to consider two conflicting requirements. On one hand, the index should be related to the data well enough to allow efficient query execution. On the other hand, this relationship between plaintexts and the index should minimize the risk of any disclosure or loss of privacy [28,59]. For example, in bucketization-based indexing, decreasing the number of buckets impairs performance, while a larger number of buckets increases the risk of data disclosure. In our database outsourcing scenario, a critical drawback of bucketization-based indexing is the loss of data granularity, which prevents grouping operations. The CSP can indeed not distinguish between tuples in buckets and the user has to filter intermediate results sent by the CSP to reconstruct the global result. Hence, bucketization-based methods induce computational overhead at the user's, too. Such an overhead can be high, especially for queries that return a large number of encrypted tuples [70], e.g., grouping queries running on fine-grained data.

Finally, although index-based approaches are quite popular in cloud data outsourcing due to their efficiency [35,59], their main limitation lies in data update. Typically, such methods but SE exploit the distribution of plaintexts, while update operations may change it, making index regeneration unavoidable [35]. As a result, index-based solutions but SE are suitable for read-only data. Yet, SE schemes are either too costly or too limited in query expressiveness to be used in practice.

5   Secure Databases

5.1   CryptDB

CryptDB is a pioneer system that allows efficient SQL query processing over ciphertexts into a DBMS [54]. The properties of a cryptographic scheme determine the kinds of queries that can be directly executed over ciphertexts. Thus, CryptDB implements several schemes with respect to different user-determined security requirements and query needs. Thus, we first describe these cryptographic schemes, and then we detail CryptDB's architecture.

Query-Aware Encryption Schemes

Random Encryption (RND). RND schemes are the strongest security schemes. They indeed guarantee semantic security, i.e., it is computationally impossible to distinguish two ciphertexts. For instance, let x be a plaintext value and E an RND encrypting function. If, using the same encryption key, $e_1 = E(x)$ and $e_2 = E(x)$, then with high probability, $e_1 \neq e_2$. However, RND schemes do not allow any computations nor queries over ciphertexts. They are only designed for safe storage.

Homomorphic Encryption (HE). HE allows performing arbitrary arithmetic operations over ciphertexts without decryption [31] while still providing semantic security. For instance, with an additive HE scheme, for any two encryptions $E(x)$ and $E(y)$, there exists a function f such that $f(E(x), E(y)) = E(x + y)$. Fully homomorphic encryption (FHE) is prohibitively slow and requires so much computing power that it cannot be used in practice as of today. However, partially homomorphic encryption (PHE) is efficient for specific operations and can be used in practice. PHE allows either addition or multiplication over ciphertexts and guarantees semantic security. Paillier's [53] and El Gamal's [27] are examples of PHE schemes. For instance, with Paillier's PHE, the product of two encryptions encrypts the sum of the encrypted values, i.e., $E(x) \times E(y) = E(x + y)$.

Deterministic Encryption (DET). DET encrypts identical data values into identical encryptions when using the same key, i.e., $\forall x, y$: $x = y \Leftrightarrow E(x) = E(y)$. Thus, DET allows queries with equality predicates, equi-joins, as well as `GROUP BY`, `COUNT` and `DISTINCT` queries [55]. DET is secure only when there is no redundancy in data. It is not robust against statistical attacks. Although some public key encryption schemes allow exact match queries with stronger security guarantees [15], search takes linear time with the size of the database, while DET operates in logarithmic time [13], thus explaining its adoption in CryptDB.

Order Preserving Encryption (OPE). OPE is a deterministic encryption scheme that preserves plaintext order in ciphertexts. Let x and y be two plaintext values and E an OPE scheme. If $x \leq y$, then $E(x) \leq E(y)$. This feature allows range queries, MIN and MAX aggregations, and ordering over ciphertexts. In terms of security, OPE is weaker than DET because it reveals data order. Yet, it can provide sufficient security for some applications, e.g., when the adversary does not possess any prior knowledge, while increasing the efficiency of query processing [52].

Table 3 summarizes the features of the cryptographic schemes used in CryptDB.

Table 3. Features of CryptDB's encryption schemes

Allowed queries	RND	HE	DET	OPE
DISTINCT	No	No	Yes	Yes
WHERE (=, ≠)	No	No	Yes	Yes
Range queries	No	No	No	Yes
ORDER BY	No	No	No	Yes
JOIN	No	No	Yes	Yes
SUM, AVG	No	Yes	No	No
MIN, MAX	No	No	No	Yes
GROUP BY	No	No	Yes	Yes
Information leakage	None	None	Duplicates	Data order

CryptDB's Architecture. CryptDB follows three principles to solve the problem of querying encrypted databases: (1) *SQL-aware encryption* that uses cryptographic schemes within SQL queries; (2) *adjustable query-based encryption* to minimize data leakage; and (3) *chain cryptographic keys in user passwords* to enable data decryption only for authorized users with access privileges.

In CryptDB's core, encryption is structured in multiple embedded levels akin to onion layers. Each onion layer helps process given classes of queries. The outermost layers are RND and HE, HE actually being Paillier's PHE scheme. RND and HE provide the highest level of security, whereas inner layers, OPE and DET, provide more functionality. The OPE layer is an enhancement of [14]. Eventually, two new cryptographic schemes enable join operations.

Ciphertext access is achieved through a trusted proxy server that encrypts data, rewrites queries (by anonymizing table and attribute names and encrypting constants) and decrypts query results. The proxy server stores encryption keys, the database schema and the onion layers of all attributes in the database. When a query is issued, the proxy dynamically peels off onion layers downs to a layer corresponding to the given computation. For instance, consider the query SELECT * FROM employee WHERE name = 'Alice'. First, the proxy issues a query to peel off the RND layer for attribute name down to the DET layer. Then, the proxy rewrites the query as SELECT * FROM T1 WHERE A2 = '0xac18f', where T1 and A2 denote the anonymization of table employee and attribute name, respectively, and 0xac18f = E_{DET}('Alice'). Similary, aggregation query

SELECT SUM(salary) FROM employee would translate as SELECT SUM$_{HE}$(A3) FROM T1, where SUM$_{HE}$ is a user-defined function implementing Paillier's PHE and A3 is the anonymization of attribute salary.

5.2 MONOMI

While CryptDB offers one of the first practical solutions for secure DBMSs, there are still a lot of queries that are not supported, especially OLAP-like queries. As an illustration, CryptDB supports only 2 queries out of 22 from the TPC-H decision support benchmark [69]. Thence, MONOMI builds upon CryptDB to allow the execution of analytical workloads [70].

To this aim, MONOMI adds in a designer that optimizes the physical database layout at the CSP's and a query planner that splits query execution between the CSP and the user. The optimal plan for executing some queries may indeed involve sending intermediate results between the user and the CSP several times to execute different parts of a query [70]. For instance, to run a SUM/GROUP BY/HAVING query, MONOMI computes the SUM and GROUP BY at the CSP's through the HE and DET encryption schemes, respectively. Then, since HE does not preserve data order, the HAVING statement is executed at the user's after decryption. This strategy helps MONOMI allow 19 out of the 22 queries of TPC-H.

5.3 Multi-valued Order Preserving Encryption (MV-OPE)

Lopes et al. rightly claim that "little attention has been devoted to determine how a data warehouse hosted in a cloud should be encrypted to enable analytical queries processing" [46]. Thence, they propose the MV-OPE scheme that allows GROUP BY queries over ciphertexts. Such a scheme could replace CryptDB's and MONOMI's OPE without having to compute anything at the user's.

Generally speaking, MV-OPE extends OPE by encrypting the same plaintext into different ciphertexts while preserving the order of the plaintexts [41]. Thus, given two clear values x and y and an MV-OPE function E, if $x < y$ then $E(x) < E(y)$. MV-OPE can be used to compute operations such as equality, difference, inequalities, minimum, maximum and count [46]. MV-OPE improves robustness against statistical attacks and only leaks the order of data. Lopes et al.'s scheme combines MV-OPE with FHE (Sect. 5.1). Moreover, as CryptDB and MONOMI, it involves a secure host, e.g., a trusted proxy server. Despite using FHE, Lopes et al. experimentally show that computing queries over ciphertexts at the CSP's is significantly faster than computing them at the user's after decryption.

5.4 Secure Trusted Hardware

Trusted hardware devices are widely used for security, e.g., smart cards for secure authentication and secure coprocessors in automated teller machines (ATMs).

Quite naturally, the idea of processing queries inside tamper-proof enclosures of trusted hardware, such as a secure coprocessor or Field Programmable Gate Array (FPGA)-based secure programmable hardware [26], came up. Such components are physically hosted at the CSP's. They have access to encryption keys and allow performing a limited set of queries over ciphertexts.

TrustedDB. TrustedDB is an SQL database processing engine that makes use of IBM 4764/5 cryptographic coprocessors [10] to run custom queries securely [12]. Coprocessors offer several cryptographic schemes such as the Advanced Encryption Standard (AES), the Triple Data Encryption Standard (3DES), RSA, pseu-do-random number generation and cryptographic hash functions. Yet, cryptographic coprocessors are significantly constrained in both computation ability and memory capacity. Thus, a trade-off must be considered between cheap query processing on untrusted main processors (at the CSP's) and expensive computation inside secure coprocessors.

Sensitive data can only be decrypted and processed by the user or a secure coprocessor. Only non-sensitive data are stored unencrypted at the CSP's. When a query is issued, it is encrypted at the user's, rewritten as a set of subqueries and executed at the CSP's or in the secure coprocessor database engine, with respect to data sensitivity. The final result is assembled, encrypted by the secure coprocessor and sent back to the user.

Cipherbase. Cipherbase aims at deploying trusted hardware for secure data processing in the cloud [9]. Cipherbase actually extends Microsoft SQL Server with in-server, customized FPGA-based trusted hardware. The FPGA is a trusted black box for computing operations over ciphertexts, which are encrypted with a non-homomorphic encryption scheme such as AES. The FPGA decrypts data internally, processes the operations and encrypts the result back. As in TrustedDB, query processing on non-sensitive data is handled by the CSP.

5.5   Discussion

CryptDB is much cited, but is quite insecure and introduces some loopholes. Its onion adjustable encryption architecture is indeed unidirectional, i.e., once an attribute is set down to a weak scheme such as DET, it never returns to a higher encryption level [43]. Moreover, attributes targeted by exact match and range queries are encrypted with DET and OPE, respectively, and are vulnerable to statistical attacks. As a result, once an exact match or range query is issued, the system becomes vulnerable ever after. DET and OPE have even been shown to be much more insecure than previously expected [50]. Additionally, peeling down onion layers induces an overhead, especially in the case of big tables.

Moreover, although CryptDB does support many types of queries, there are still many unsupported types of queries, e.g., predicate evaluation on more than one attribute. MONOMI addresses this shortcoming, but retains the same security mechanisms as CryptDB. MONOMI also induces a heavy communication overhead between the user and the CSP, since intermediate results may be exchanged several times to execute different parts of a query [70].

Despite a distributed architecture, Lopes et al.'s solution requires a trusted server to securely execute `GROUP BY` queries. In our database outsourcing scenario, all service providers that are external to the user's are considered untrusted. Thus, Lopes et al.'s trusted server would be located at the user's, inducing costs that do not fit our scenario. Additionally, this solution does not support `MIN` and `MAX` aggregation operators directly over ciphertexts.

Finally, beside computation ability and memory capacity limitations, trusted hardware is still very expensive, which is again contrary to our scenario that aims at using cheap commodity machines in the cloud. Moreover, leaving unencrypted attributes jeopardizes ciphertext, because relationships between ciphertexts and plaintexts may reveal information about ciphertexts [8].

6   Conclusion

Although encryption methods enforce privacy, in some cases, the impact on performance makes them inapplicable to cloud databases. It is indeed currently impossible to develop a system that meets both state-of-the-art cryptographic security standards and query performance requirements. In this final section, we provide a global discussion on security, performance and storage requirements for secure databases, before concluding the paper.

6.1   Security

The DET and OPE schemes, which are notably used in CryptDB, allow efficiently performing queries over ciphertexts. Database optimization techniques, e.g., usual indexing methods, can also be used to enhance query performance. However, DET and OPE leak a non-negligible amount of information and are vulnerable to statistical attacks [42]. For example, a large fraction of tuples from DET encrypted attributes can be decrypted by statistical attacks [50]. The vulnerability of DET is extremely detrimental to DBs with high redundancy, e.g., data warehouses. The weak security of OPE makes it inappropriate, too. It is indeed even worse than DET in terms of security [29,42]. Eventually, a recent class of generic attacks against private range query schemes invalidates much of the existing literature [42].

Thus, FHE looks like a more appropriate choice for encryption. In particular, PHE encryption can be used to sum ciphertexts, but the cost of decryption at the client's can remain high. As of today, it is indeed usually more efficient to decrypt data at the client's and then perform the aggregation, rather than processing aggregation queries over ciphertexts at the CSP's [70]. Yet, FHE is likely to become a viable alternative in the upcoming decade, with both new FHE schemes and improvements in hardware performance. However, since preserving the order of data is necessary when running queries such as sorting, grouping and range operations, the issue of designing order preserving FHE schemes will have to be addressed.

6.2   Query Post-processing

Tuple and table-level encryption are casually considered preferable to attribute-level encryption, because of lower startup costs at the user's and minimal storage costs at the CSP's [39]. However, the loss of data granularity is an important deficiency in scenarios such as OLAP. Thus, some solutions that use tuple-level encryption (Sect. 4) handle query processing by means of auxiliary indexes at the CSP's (e.g., bucketization-based indexing) and perform final query processing at the user's. Similarly, MONOMI splits the execution of queries between the user and CSP. In such solutions, it is essential to cut down the bandwidth required to transfer intermediate results and user computational resources for user side query processing [70], which is quite an open issue. CPU and storage usage at the user's must indeed be minimum for maintaining the benefits of outsourcing.

6.3   Storage Overhead

CryptDB, MONOMI and Cipherbase use attribute-level encryption, i.e., each attribute value is encrypted independently [9], at the cost of storage overhead. For instance, using classical AES in Cipher-Block Chaining (CBC) mode, a 32-bit integer is encrypted on 256 bits [9]. Worse, Paillier's PHE scheme, which is used in CryptDB, operates over 2048-bit ciphertext [70]. MONOMI addresses this issue by packing multiple values from a single tuple into one PHE encryption, using Ge and Zdonik's scheme [30]. This optimization works properly for a table with many PHE-encrypted attributes, but would complicate partial updates that reset some but not all attribute values packed into a PHE tuple encryption [55]. Thus, although security vs. performance is necessarily a tradeoff, there is still some room for improving the storage overhead of cryptographic schemes, especially for secret sharing schemes.

6.4   Computational Overhead

Operations at the CSP's should not involve any expensive arithmetic operations such as modular multiplication or exponentiation [62]. However, for instance in Paillier's scheme, encrypting the sum of two clear values x and y requires multiplying ciphertexts $E(x)$ and $E(y)$ modulo a 2048-bit public key, i.e., $E(x + y) = E(x) \times E(y)$. Such modular multiplications are computationally expensive, especially on big tables.

MONOMI implements a grouped homomorphic addition optimization. All to-be-aggregated attributes are packed in such a way that aggregation queries can be computed with a single modular multiplication. This implies that all queries must be declared ahead of time, which it is not possible for all applications, e.g., OLAP ad-hoc navigation. Yet, performance optimization techniques, such as indexing, partitioning or view materialization, can apply onto ciphertexts. However, although they speed up some queries, they also slow down others [70]. As a result, it is crucial to select a cryptographic method that meets all usage constrains. Again, a tradeoff must be defined to meet the intended level of privacy while minimizing the impact on performance.

6.5   Wrap-up

In this paper, we review the security mechanisms that can nowadays be used in the deployment of cloud databases. We particularly focus on the cryptographic schemes and the (would-be) secure systems that enable executing queries over ciphertexts without decryption. This survey highlights the potential benefits of existing solutions in a cloud computing context, but also that one must take great care about security guarantees before selecting one such solution.

Moreover, cryptography cannot prevent all attacks by malicious adversaries, e.g., Distributed Denial of Service (DDoS) attacks. It is thus essential to clearly specify the objectives of cloud database deployment, to adopt security mechanisms that are adapted to these objectives. Such preliminary work shall determine the initialization of secure protocols, the choice of cryptographic schemes, the need for a trusted third party, etc.

Finally, since computational performance is currently still a bottleneck, resorting to data distribution and query parallelization must be a priority. Thus, cloud frameworks such as Hadoop [7] and Spark [6] should be exploited in future secure cloud DBMSs.

References

1. Aggarwal, C.C., Yu, P.S.: A general survey of privacy-preserving data mining models and algorithms. In: Aggarwal, C.C., Yu, P.S. (eds.) Privacy-Preserving Data Mining: Models and Algorithms, pp. 11–52. Springer, Boston (2008)
2. Aggarwal, G., Bawa, M., Ganesan, P., Garcia-Molina, H., Kenthapadi, K., Motwani, R., Srivastava, U., Thomas, D., Ying, X.: Two can keep a secret: a distributed architecture for secure database services. In: 2nd Biennial Conference on Innovative Data Systems Research (CIDR), Asilomar, CA, USA, pp. 186–199 (2005)
3. Agrawal, D., El Abbadi, A., Emekçi, F., Metwally, A.: Database management as a service: challenges and opportunities. In: 25th International Conference on Data Engineering (ICDE), Shanghai, China, pp. 1709–1716 (2009)

4. Agrawal, R., Kiernan, J., Srikant, R., Xu, Y.: Order-preserving encryption for numeric data. In: ACM SIGMOD International Conference on Management of Data (SIGMOD), Paris, France, pp. 563–574 (2004)
5. Agrawal, R., Srikant, R., Thomas, D.: Privacy preserving OLAP. In: ACM SIG-MOD International Conference on Management of Data (SIGMOD), Baltimore, MD, USA, pp. 251–262 (2005)
6. Apache Software Foundation. Apache Spark - Lightning-fast cluster computing (2016). https://spark.apache.org
7. Apache Software Foundation. Hadoop (2016). http://hadoop.apache.org
8. Arasu, A., Blanas, S., Eguro, K., Kaushik, R., Kossmann, D., Ramamurthy, R., Venkatesan, R.: Orthogonal security with cipherbase. In: 6th Biennial Conference on Innovative Data Systems Research (CIDR), Asilomar, CA, USA (2013)
9. Arasu, A., Eguro, K., Joglekar, M., Kaushik, R., Kossmann, D., Ramamurthy, R.: Transaction processing on confidential data using Cipherbase. In: 31st IEEE International Conference on Data Engineering (ICDE), Seoul, Korea, pp. 435–446 (2015)
10. Arnold, T.W., Buscaglia, C.U., Chan, F., Condorelli, V., Dayka, J.C., Santiago-Fernandez, W., Hadzic, N., Hocker, M.D., Jordan, M., Morris, T.E., Werner, K.: IBM 4765 cryptographic coprocessor. IBM J. Res. Dev. **56**(1), 10 (2012)
11. Attasena, V., Harbi, N., Darmont, J.: fVSS: A new secure and cost-efficient scheme for cloud data warehouses. In: 7th International Workshop on Data Warehousing and OLAP (DOLAP), Shanghai, China, pp. 81–90 (2014)
12. Bajaj, S., Sion, R.: TrustedDB: a trusted hardware based database with privacy and data confidentiality. In: ACM SIGMOD International Conference on Management of Data (SIGMOD), Athens, Greece, pp. 205–216 (2011)
13. Bellare, M., Boldyreva, A., O'Neill, A.: Deterministic and efficiently searchable encryption. In: Menezes, A. (ed.) CRYPTO 2007. LNCS, vol. 4622, pp. 535–552. Springer, Heidelberg (2007). doi:10.1007/978-3-540-74143-5_30
14. Boldyreva, A., Chenette, N., Lee, Y., O'Neill, A.: Order-preserving symmetric encryption. In: Joux, A. (ed.) EUROCRYPT 2009. LNCS, vol. 5479, pp. 224–241. Springer, Heidelberg (2009). doi:10.1007/978-3-642-01001-9_13
15. Boneh, D., Di Crescenzo, G., Ostrovsky, R., Persiano, G.: Public key encryption with keyword search. In: Cachin, C., Camenisch, J.L. (eds.) EUROCRYPT 2004. LNCS, vol. 3027, pp. 506–522. Springer, Heidelberg (2004). doi:10.1007/978-3-540-24676-3_30
16. Bösch, C., Hartel, P.H., Jonker, W., Peter, A.: A survey of provably secure searchable encryption. ACM Comput. Surv. **47**(2), 18:1–18:51 (2014)
17. Cachin, C., Micali, S., Stadler, M.: Computationally private information retrieval with polylogarithmic communication. In: Stern, J. (ed.) EUROCRYPT 1999. LNCS, vol. 1592, pp. 402–414. Springer, Heidelberg (1999). doi:10.1007/3-540-48910-X_28
18. Chang, Y.-C.: Single database private information retrieval with logarithmic communication. In: Wang, H., Pieprzyk, J., Varadharajan, V. (eds.) ACISP 2004. LNCS, vol. 3108, pp. 50–61. Springer, Heidelberg (2004). doi:10.1007/978-3-540-27800-9_5
19. Chor, B., Kushilevitz, E., Goldreich, O., Sudan, M.: Private Information Retrieval. Journal of the ACM **45**(6), 965–981 (1998)
20. Ciriani, V., De Capitani, S., di Vimercati, S., Foresti, S.J., Paraboschi, S., Samarati, P.: Selective data outsourcing for enforcing privacy. J. Comput. Secur. **19**(3), 531–566 (2011)

21. Damiani, E., De Capitani di Vimercati, S., Jajodia, S., Paraboschi, S., Samarati, P.: Balancing confidentiality and efficiency in untrusted relational DBMSs. In: 10th ACM Conference on Computer and Communications Security (CCS), Washington, DC, USA, pp. 93–102 (2003)

22. Davida, G.I., Wells, D.L., Kam, J.B.: A database encryption system with subkeys. ACM Trans. Database Syst. **6**(2), 312–328 (1981)

23. de Montjoye, Y.-A., Hidalgo, C.A., Verleysen, M., Blondel, V.D.: Unique in the crowd: the privacy bounds of human mobility. Nature Scientific Reports 3, Article number: 1376 (2013). http://www.nature.com/articles/srep01376

24. Dwork, C.: Differential privacy. In: Bugliesi, M., Preneel, B., Sassone, V., Wegener, I. (eds.) ICALP 2006. LNCS, vol. 4052, pp. 1–12. Springer, Heidelberg (2006). doi:10.1007/11787006_1

25. Dwork, C.: Differential privacy. In: van Tilborg, H.C.A., Jajodia, S. (eds.) Encyclopedia of Cryptography and Security, pp. 338–340. Springer, New York (2011)

26. Eguro, K., Venkatesan, R.: FPGAs for trusted cloud computing. In: 22nd International Conference on Field Programmable Logic and Applications (FPL), Oslo, Norway, pp. 63–70 (2012)

27. El Gamal, T.: A public key cryptosystem and a signature scheme based on discrete logarithms. IEEE Trans. Inf. Theory **31**(4), 469–472 (1985)

28. Elovici, Y., Waisenberg, R., Shmueli, E., Gudes, E.: A structure preserving database encryption scheme. In: Jonker, W., Petković, M. (eds.) SDM 2004. LNCS, vol. 3178, pp. 28–40. Springer, Heidelberg (2004). doi:10.1007/978-3-540-30073-1_3

29. Furukawa, J.: Short comparable encryption. In: Gritzalis, D., Kiayias, A., Askoxylakis, I. (eds.) CANS 2014. LNCS, vol. 8813, pp. 337–352. Springer, Cham (2014). doi:10.1007/978-3-319-12280-9_22

30. Ge, T., Zdonik, S.B.: Answering aggregation queries in a secure system model. In: 33rd International Conference on Very Large Data Bases (VLDB), Vienna, Austria, pp. 519–530 (2007)

31. Gentry, C.: A fully homomorphic encryption scheme. Ph.D. thesis, Stanford University (2009)

32. Goldreich, O., Ostrovsky, R.: Software protection and simulation on oblivious RAMs. J. ACM **43**(3), 431–473 (1996)

33. Hacıgümüş, H., Iyer, B.R., Li, C., Mehrotra, S.: Executing SQL over encrypted data in the database-service-provider model. In: ACM SIGMOD International Conference on Management of Data (SIGMOD), Madison, WI, USA, pp. 216–227 (2002)

34. Hacıgümüş, H., Iyer, B.R., Mehrotra, S.: Efficient execution of aggregation queries over encrypted relational databases. In: Lee, Y.J., Li, J., Whang, K.-Y., Lee, D. (eds.) DASFAA 2004. LNCS, vol. 2973, pp. 125–136. Springer, Heidelberg (2004). doi:10.1007/978-3-540-24571-1_10

35. Hadavi, M.A., Damiani, E., Jalili, R., Cimato, S., Ganjei, Z.: AS5: a secure searchable secret sharing scheme for privacy preserving database outsourcing. In: Pietro, R., Herranz, J., Damiani, E., State, R. (eds.) DPM/SETOP -2012. LNCS, vol. 7731, pp. 201–216. Springer, Heidelberg (2013). doi:10.1007/978-3-642-35890-6_15

36. Hadavi, M.A., Jalili, R.: Secure data outsourcing based on threshold secret sharing; towards a more practical solution. In: 36th International Conference on Very Large Data Bases (VLDB) PhD Workshop, Singapore, pp. 54–59 (2010)

37. Hadavi, M.A., Noferesti, M., Jalili, R., Damiani, E.: Database as a service: towards a unified solution for security requirements. In: 36th Annual IEEE Computer Software and Applications Conference (COMPSAC) Workshops, Izmir, Turkey, pp. 415–420 (2012)

38. Hore, B., Mehrotra, S., Canim, M., Kantarcioglu, M.: Secure multidimensional range queries over outsourced data. VLDB J. **21**(3), 333–358 (2012)
39. Hore, B., Mehrotra, S., Hacigümüç, H.: Managing and querying encrypted data. In: Gertz, M., Jajodia, S. (eds.) Handbook of Database Security, pp. 163–190. Springer, Boston (2008)
40. Hore, B., Mehrotra, S., Tsudik, G.: A privacy-preserving index for range queries. In: 30th International Conference on Very Large Data Bases (VLDB), Toronto, Canada, pp. 720–731 (2004)
41. Kadhem, H., Amagasa, T., Hiroyuki Kitagawa, M.-O.: Multivalued-order preserving encryption scheme: a novel scheme for encrypting integer value to many different values. IEICE Trans. Inf. Syst. **93–D**(9), 2520–2533 (2010)
42. Kellaris, G., Kollios, G., Nissim, K., O'Neill, A.: Generic attacks on secure outsourced databases. In: 23rd ACM Conference on Computer and Communications Security (CCS), Vienna, Austria, pp. 1329–1340 (2016)
43. Kerschbaum, F., Grofig, P., Hang, I., Härterich, M., Kohler, M., Schaad, A., Schröpfer, A., Tighzert, W.: Adjustably encrypted in-memory column-store. In: ACM SIGSAC Conference on Computer and Communications Security (CCS), Berlin, Germany, pp. 1325–1328 (2013)
44. Krawczyk, H.: Secret sharing made short. In: Stinson, D.R. (ed.) CRYPTO 1993. LNCS, vol. 773, pp. 136–146. Springer, Heidelberg (1994). doi:10.1007/3-540-48329-2_12
45. Liu, Z., Chen, X., Yang, J., Jia, C., You, I.: New order preserving encryption model for outsourced databases in cloud environments. J. Netw. Comput. Appl. **59**, 198–207 (2016)
46. Lopes, C.C., Times, V.C., Matwin, S., Ciferri, R.R., Ciferri, C.D.A.: Processing OLAP queries over an encrypted data warehouse stored in the cloud. In: Bellatreche, L., Mohania, M.K. (eds.) DaWaK 2014. LNCS, vol. 8646, pp. 195–207. Springer, Cham (2014). doi:10.1007/978-3-319-10160-6_18
47. Lueks, W., Goldberg, I.: Sublinear scaling for multi-client private information retrieval. In: Böhme, R., Okamoto, T. (eds.) FC 2015. LNCS, vol. 8975, pp. 168–186. Springer, Heidelberg (2015). doi:10.1007/978-3-662-47854-7_10
48. Mavroforakis, C., Chenette, N., O'Neill, A., Kollios, G., Canetti, R.: Modular order-preserving encryption, Revisited. In: ACM SIGMOD International Conference on Management of Data, Melbourne, Australia, pp. 763–777 (2015)
49. Mykletun, E., Tsudik, G.: Aggregation queries in the database-as-a-service model. In: Damiani, E., Liu, P. (eds.) DBSec 2006. LNCS, vol. 4127, pp. 89–103. Springer, Heidelberg (2006). doi:10.1007/11805588_7
50. Naveed, M., Kamara, S., Wright, C.V.: Inference attacks on property-preserving encrypted databases. In: 22nd ACM SIGSAC Conference on Computer and Communications Security (CCS), Denver, CO, USA, pp. 644–655 (2015)
51. Oracle Corporation. Data Masking Best Practices. White paper (2013)
52. Özsoyoglu, G., Singer, D.A., Chung, S.S.: Anti-tamper databases: querying encrypted databases. In: 17th Annual IFIP WG 11.3 Working Conference on Data and Application Security (DBSec), Estes Park, CO, USA, pp. 133–146 (2003)
53. Paillier, P.: Public-key cryptosystems based on composite degree residuosity classes. In: Stern, J. (ed.) EUROCRYPT 1999. LNCS, vol. 1592, pp. 223–238. Springer, Heidelberg (1999). doi:10.1007/3-540-48910-X_16
54. Popa, R.A., Redfield, C.M.S., Zeldovich, N., Balakrishnan, H.: CryptDB: protecting confidentiality with encrypted query processing. In: 23rd ACM Symposium on Operating Systems Principles (SOSP), Cascais, Portugal, pp. 85–100 (2011)

55. Popa, R.A.: Building practical systems that compute on encrypted data. Ph.D. thesis, Massachusetts Institute of Technology (2014)
56. Ricci, S., Domingo-Ferrer, J., Sánchez, D.: Privacy-preserving cloud-based statistical analyses on sensitive categorical data. In: Torra, V., Narukawa, Y., Navarro-Arribas, G., Yañez, C. (eds.) MDAI 2016. LNCS (LNAI), vol. 9880, pp. 227–238. Springer, Cham (2016). doi:10.1007/978-3-319-45656-0_19
57. Rivest, R.L., Adleman, L., Dertouzos, M.L.: On data banks and privacy homomorphisms. Found. Secure Comput. **4**(11), 169–180 (1978)
58. Saleh, E., Alsa'deh, A., Kayed, A., Meinel, C.: Processing over encrypted data: between theory and practice. SIGMOD Rec. **45**(3), 5–16 (2016)
59. Samarati, P., De Capitani di Vimercati, S.: Data protection in outsourcing scenarios: issues and directions. In: 5th ACM Symposium on Information, Computer and Communications Security (ASIACCS), Beijing, China, pp. 1–14 (2010)
60. Shamir, A.: How to share a secret. Commun. ACM **22**(11), 612–613 (1979)
61. Shmueli, E., Waisenberg, R., Elovici, Y., Gudes, E.: Designing secure indexes for encrypted databases. In: Jajodia, S., Wijesekera, D. (eds.) DBSec 2005. LNCS, vol. 3654, pp. 54–68. Springer, Heidelberg (2005). doi:10.1007/11535706_5
62. Sion, R.: Towards secure data outsourcing. In: Gertz, M., Jajodia, S. (eds.) Handbook of Database Security - Applications and Trends, pp. 137–161. Springer, Boston (2008)
63. Song, D.X., Wagner, D., Perrig, A.: Practical techniques for searches on encrypted data. In: IEEE Symposium on Security and Privacy (SP), Berkeley, CA, USA, pp. 44–55 (2000)
64. Stefanov, E., van Dijk, M., Shi, E., Fletcher, C.W., Ren, L., Xiangyao, Y., Devadas, S.: Path ORAM: an extremely simple oblivious RAM protocol. In: ACM SIGSAC Conference on Computer and Communications Security (CCS), Berlin, Germany, pp. 299–310 (2013)
65. Sun, W., Lou, W., Hou, Y.T., Li, H.: Privacy-preserving keyword search over encrypted data in cloud computing. In: Jajodia, S., Kant, K., Samarati, P., Singhal, A., Swarup, V., Wang, C. (eds.) Secure Cloud Computing, pp. 189–212. Springer, New York (2014). doi:10.1007/978-1-4614-9278-8_9
66. Sung, S.Y., Liu, Y., Xiong, H., Ng, P.A.: Privacy preservation for data cubes. Knowl. Inf. Syst. **9**(1), 38–61 (2006)
67. Sweeney, L.: k-Anonymity: a model for protecting privacy. Int. J. Uncertain. Fuzziness Knowl. Based Syst. **10**(5), 557–570 (2002)
68. Thompson, B., Haber, S., Horne, W.G., Sander, T., Yao, D.: Privacy-preserving computation and verification of aggregate queries on outsourced databases. In: Goldberg, I., Atallah, M.J. (eds.) PETS 2009. LNCS, vol. 5672, pp. 185–201. Springer, Heidelberg (2009). doi:10.1007/978-3-642-03168-7_11
69. Transaction Performance Processing Council. TPC Benchmark H (Decision Support) Standard Specification Revision 2.1 (2014). http://www.tpc.org
70. Tu, S., Kaashoek, M.F., Madden, S., Zeldovich, N.: Processing analytical queries over encrypted data. Proc. VLDB Endowment **6**(5), 289–300 (2013)
71. Wang, S., Agrawal, D., El Abbadi, A.: A comprehensive framework for secure query processing on relational data in the cloud. In: Jonker, W., Petković, M. (eds.) SDM 2011. LNCS, vol. 6933, pp. 52–69. Springer, Heidelberg (2011). doi:10.1007/978-3-642-23556-6_4
72. Wang, W.H., Lakshmanan, L.V.S.: Efficient secure query evaluation over encrypted XML databases. In: 32nd International Conference on Very Large Data Bases, Seoul, Korea, pp. 127–138 (2006)

73. Williams, P., Sion, R.: Access privacy and correctness on untrusted storage. ACM Trans. Inf. Syst. Secur. **16**(3), 12 (2013)
74. Xiao, X., Bender, G., Hay, M., Gehrke, J.: iReduct: differential privacy with reduced relative errors. In: ACM SIGMOD International Conference on Management of Data (SIGMOD), Athens, Greece, pp. 229–240 (2011)
75. Xiong, L., Chitti, S., Liu, L.: Preserving data privacy in outsourcing data aggregation services. ACM Trans. Internet Technol. **7**(3), 17 (2007)
76. Yang, Y., Zhang, Z., Miklau, G., Winslett, M., Xiao, X.: Differential privacy in data publication and analysis. In: ACM SIGMOD International Conference on Management of Data, Scottsdale, AZ, USA, pp. 601–606 (2012)
77. Yuhanna, N., Gilpin, M., Knoll, A.: Your Enterprise Database Security Strategy 2010 (2009). Forrester - http://www.oracle.com/us/ciocentral/forrester-database-security-396253.pdf

TARDIS: Optimal Execution of Scientific Workflows in Apache Spark

Daniel Gaspar[1]([⊠]), Fabio Porto[1], Reza Akbarinia[2], and Esther Pacitti[3]

[1] LNCC - National Laboratory for Scientific Computing,
Av. Getúlio Vargas, 333, Petrópolis, RJ 25651-075, Brazil
gaspar@lncc.br
[2] INRIA - National Institute for Research in Computer Science
and Control, 161 Rue Ada, 34095 Montpellier, France
[3] LIRMM - Montpellier Laboratory of Informatics, Robotics and Microelectronics,
860 Rue de St Priest, 34095 Montpellier, France

Abstract. The success of using workflows for modeling large-scale scientific applications has fostered the research on parallel execution of scientific workflows in shared-nothing clusters, in which large volumes of scientific data may be stored and processed in parallel using ordinary machines. However, most of the current scientific workflow management systems do not handle the memory and data locality appropriately. *Apache Spark* deals with these issues by chaining activities that should be executed in a specific node, among other optimizations such as the in-memory storage of intermediate data in *RDDs (Resilient Distributed Datasets)*. However, to take advantage of the RDDs, *Spark* requires existing workflows to be described using its own API, which forces the activities to be reimplemented in Python, Java, Scala or R, and this demands a big effort from the workflow programmers.

In this paper, we propose a parallel scientific workflow engine called *TARDIS*, whose objective is to run existing workflows inside a *Spark* cluster, using RDDs and smart caching, in a completely transparent way for the user, i.e., without needing to reimplement the workflows in the *Spark* API. We evaluated our system through experiments and compared its performance with *Swift/K*. The results show that *TARDIS* performs better (up to 138% improvement) than *Swift/K* for parallel scientific workflow execution.

1 Introduction

Over the last years, the volume of data produced by scientific simulations and experiments has been increasing in a astronomical rate. This increase is mainly a consequence of advances in sensors and the thriving of the Internet of Things, which amplify the quantities that are analysed and stored during an experiment. This leads to a field of study, called *big data*, that is interested in studying how to collect, store and process these enormous volumes of data.

Fortunately, there has also been plenty of development in the high-performance computing area. Usually, big data is stored and processed in parallel

© Springer International Publishing AG 2017
L. Bellatreche and S. Chakravarthy (Eds.): DaWaK 2017, LNCS 10440, pp. 74–87, 2017.
DOI: 10.1007/978-3-319-64283-3_6

databases running in dedicated and expensive servers. However, this approach is inappropriate for many large scale scientific applications due to its cost [14]. Besides, some scientific data is completely unstructured, or non-relational, therefore difficult to be dealt in databases. More recently, some approaches to process unstructured data over clusters of commodity hardware have appeared. These include *MapReduce* [3], *Spark* [13] and *Pegasus* [4].

In the context of scientific applications, the standard has been to express the scientific processes as workflows. Scientific workflows define a computational processing composed of *activities*. Each activity consumes input data and generates output. Activities are linked together forming a directed acyclic graph (DAG), taking in account the data dependency between them [6,8,9].

Typically, scientific workflows have been designed to run in clusters using a shared-disk model, inherited from HPC systems. For processing big datasets, however, moving data through the network jeopardize the computation.

By analysing current solutions for the parallel execution of scientific workflows, we observe that the great majority of them is concerned with the lack of data locality. In addition, currently there is not a complete coupling between the data management framework and the scientific workflow engine in most of the existing systems. Solutions based on MapReduce [3] or Hadoop [2] (like Pig Latin [10] or Hive [11]) do not analyse the entire workflow. They do not consider the chaining of activities when scheduling tasks to nodes. *Swift/K* is a C-like programming language for defining workflows that can be executed in clusters, among other architectures. The language coordinates the execution of activities defined as programs that consume and produce files [12]. However, the *Swift/K* engine does not ensure the data locality during the workflow execution.

Spark is an Apache open-source framework for processing big data in clusters. It allows storing data in memory and querying it repeatedly, therefore providing performance increase over *Hadoop* or other *MapReduce* implementations [13].

To use *Apache Spark*, users should develop their applications using the provided API, e.g., in Scala, Python or R. However, usually scientific workflows activities are defined in existing software, *e.g.*, *MAFFT*, *Align2D*, *Montage* or some scripts developed by the scientists [5,7]. It is not trivial to execute a workflow designed in *Swift/K*, or any other workflow whose activities read and write from files in *Spark*.

In this paper, we propose *TARDIS* (Task Analyser Regarding Data in Spark), a parallel scientific workflow engine that allows to run workflows using *Spark*, without needing to rewrite their activities in *Spark* API. *TARDIS* executes workflows with activities which are scientific programs, allowing them to behave in *Spark* as a code written in one of its compatible languages reading and writing data from *Spark* RDDs. In addition, it deals with the optimal partitioning of the activities inputs in the *Spark* RDDs avoiding unnecessary data transfers during workflow execution. *TARDIS* also includes new algorithms for scheduling the input data in the *Spark* cluster.

We have compared the performance of *TARDIS* with *Swift/K* for executing a *Montage* workflow to generate an image of the sky from mosaics of tiles obtained from space catalogues. The results show that *TARDIS* performs much better than *Swift/K* for the execution of workflows.

The rest of the paper is organized as follows. In the next section, we describe the problem we are facing. In Sect. 3, we present some background about *Spark*, and in Sect. 4, we present the *TARDIS* engine. In Sect. 5, we report our experimental results.

2 Problem Definition

A workflow $W = (A, F, I, O)$ is composed of a set of *activities* $A = \{a_1, ..., a_n\}$, a set of *files* $F = \{f_1, ..., f_m\}$, a set of *input dependencies* $I \subset (F \times A)$ and a set of *output dependencies* $O \subseteq (A \times F)$. Activities are defined as command-line invocations of external program that reads and writes into files. Additionally, we consider $N = \{n_1, ..., n_x\}$ a set of computer nodes and a function $\mu(F) \Rightarrow N$ that allocates a file $f_i \in F$ in a node $n_j \in N$. Furthermore, $I = \{I_1, I_2, ..., I_v\}$ is a set of sets of input dependencies such that if $(f_i, a_j) \in I_k$ and $(f_p, a_l) \in I_k$ then $j = l$, representing the set of files that are input to a single program, enabling the parallel execution of the corresponding scientific program.

Given a workflow W, let $a, a' \in A$ be activities. We define:

- **Input of an activity**: $input(a) = \{f \in F \mid \exists (f, a) \in I\}$.
- **Output of an activity**: $output(a) = \{f \in F \mid \exists (a, f) \in O\}$.
- **Activity dependency**: We say that a depends on a', $dep(a', a)$ if $output(a') \cap input(a) \neq \emptyset$.
- **Input of a workflow**: $input(W) = \{f \in F \mid \nexists\, a \in A : f \in output(a)\}$.

The output of a workflow $output(W) \subseteq F$, can be anything defined by the user.

The problem, which we address in this paper, is how to run a workflow W in a *shared-nothing* architecture pursuing data locality and using in-memory storage for intermediate data. This can be done by solving the following four sub-problems:

- **How to distribute input files in a *shared-nothing* cluster?** We want to optimize data locality by bringing the jobs close to the data and therefore minimizing the execution time of the workflow. We should consider the data transfers among nodes during the execution of the workflow. The challenge is to allocate the input files of a workflow on nodes of the shared-nothing cluster so that data locality can be achieved.
- **How to keep *Spark* scheduling under file allocation decision?** In *Spark*, parallel function scheduling is driven by RDDs partitioning. Running non-*Spark* functions and retaining *Spark* scheduling strategy requires allocating functions to nodes where files have been allocated.
- **How to benefit from local pipelines strategies in activities that should be chained together?** *Spark* executes pipelines of activities locally using memory for storing the intermediate data. This is achieved by analysing

the dependencies among the workflows activities. If possible, different activities are executed together locally in a *Spark stage*. We want to enforce *Spark* to perform this behaviour with our program-based workflows.

- **How to ensure fault-tolerance to the workflow execution?** When dealing with distributed computing, the chances of a failure become considerably high. We need to ensure that if a node fails, or a program execution terminates with error, the workflow execution should not only detect the error but also try to solve it by running again the failed fragment.

In this paper, our objective is to enable the efficient execution of existing scientific workflows in a shared-nothing execution architecture, using *Spark*, where programs run on local data and produce intermediate results using local memory.

3   Background

3.1   Spark

Apache Spark is an open-source parallel data processing framework developed and maintained by *Apache* that generalizes the *MapReduce* model [1,13]. *Spark* manages data by introducing the RDD (*Resilient Distributed Datasets*) abstraction. These are read-only in-memory collections of objects distributed into the machines of a cluster.

A workflow may be executed in *Spark* via the API provided in *Scala*, *Python*, *Java* or *R* languages. Activities of a *Spark* workflow may be *transformations* or *actions*. *Transformations* consume a RDD and output another. *Actions* consume a RDD and output objects like the ones that compose the input RDD. For example, the *collect* action returns all of the objects present in a RDD and the *map* transformation applies a function to each object of the input RDD and outputs a new RDD with the transformed objects.

In *Spark*, the transformations are executed when their output RDD is needed for an action. Actually, the *Spark* engine knows what transformations should be done to generate a given RDD. When an action is called, *Spark* will execute all transformations necessary to generate the input data to that action. RDDs do not persist in memory, but if required by another operation, they can be recreated on-the-run.

Scientific workflows are typically executed in shared-disk architectures, and this limits data locality. Conversely, *Spark* is designed to explore in memory data locality. Unfortunately, *Spark* workflows constrain activities to be coded in one of is compatible languages which limits its applicabilities to a scientific workflow already existent.

4   TARDIS Engine

TARDIS (Task Analyser Regarding Data In Spark), is a parallel scientific workflow engine developed in Python that runs existing workflows over a shared-nothing *Spark* cluster. Each *Spark* node executes some activities of the workflow, and the data is spread throughout the cluster in the RDDs.

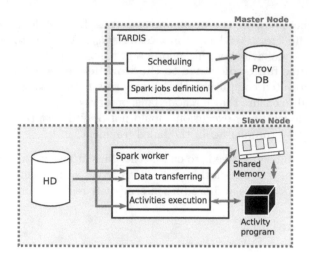

Fig. 1. TARDIS architecture.

4.1 Architecture

The *TARDIS* system is composed of two main modules: the master and slaves. The first is responsible for executing the files allocation scheduler and maintaining a *SQLite* database for provenance reasons. It connects to a *Spark* cluster and submits jobs to it. The *Spark* workers behave as *TARDIS* slave nodes. The main components of the master module are as follows:

- Job parsing: users provide a *workflow descriptor* to TARDIS using our own language. This component is responsible for parsing the job description given by the user;
- Job reformulation: User inputs his or her workflow using TARDIS defined activities types (such as *map* and *partial reduce*. During the initial workflow analysis, these activities are converted to a set of *Spark* activities. Extra ones are included in order to perform the correct execution of the workflow (i.e. to transfer data among nodes);
- Scheduling: it is responsible of deciding in which nodes the input files of activities should be consumed.

The slave module contains the following main components:

- Data placement: whenever the required files for an activity execution are not available in local shared memory, slave nodes will download them from other nodes or, if locally available, copy them from local disk;
- Workflow execution: finally, activities are executed by running external software over in-memory files.

The execution model of TARDIS achieves good performance by combining the scheduling of files to nodes and the corresponding Spark executors to achieve data locality in activities execution. Thus, once the TARDIS scheduler defines a file allocation the system must ensure that the files are in their respective nodes once the execution of activities begin. In this context, an initial job activity downloads the files to executing nodes according to the schedule.

This transfer is achieved by running a lightweight HTTP server in each node, which exposes the in-memory files to the network. The input files are loaded into its own node shared memory and then other nodes may do an HTTP request to download the required files. This may happen also for intermediate files during the execution of the workflow, especially after a *Spark* shuffle.

These downloaded files are moved to a shared-memory file system mounted at each node to enable efficient in-memory pipelined activity execution of black-box programs. The latter run under the control of the *Activity Execution* (AE) module. The execution of black-box software should privilege data locality. The AE module must identify the file allocated to its node and pass it to its associated black-box program.

At the end of execution, the master collects the output files and stores them in a local folder.

The architecture of the TARDIS engine is depicted in Fig. 1. *Prov DB* is a *SQLite* database and currently only records some data used for provenance. *HD* depicts the local storage of files in slave nodes.

4.2   TARDIS Language

TARDIS offers an adaptation of *Spark* API to enable the instrumentation of proposed execution as *TARDIS* activities. Users define their workflows using a *Python* script, calling our methods. Our system allows workflow activities to be defined as *maps*, *partial reduces* and *reduces*.

Map Activities. A *map activity* is an activity that consumes only one file of the input RDD and outputs one file. This activity will be performed in parallel over all files of the input.

It is defined using the *TARDIS* method `map_activity`. This method expects up to four parameters: the activity name or ID, the command that performs the activity, the input RDD and the pattern of files that will be affected by it. The name or ID is used only for provenance reasons. The command is given by the path to the executable that should be run over the files with all of its expected parameters. A special keyword `@!input` is replaced by the respective filename in each execution of the map.

The *pattern of files* describes in a *bash*-alike expression which files of the input RDD should be consumed by the activity. The default case is "*", which means that the command specified will be run over every file in the input RDD.

This method returns a transformed RDD with the respective output files and the files from the input that were not used.

For example, if a user wants to reverse all `txt` files in an RDD (the first line becomes the last, and do on), he or she can use the following command in Python:

```
reverseRDD = tardis.map_activity("reverse", "tac @!input >
            @!input.rev", filesRDD, "*.txt").
```

For instance, if `filesRDD` had three files: `text1.txt`, `text2.txt` and `photo.jpg`, after running the previous command, `reverseRDD` would have also three files: `text1.txt.rev`, `text2.txt.rev` and `photo.jpg`.

Partial Reduce Activities. A *partial reduce activity* consumes more than one file and outputs only one. This allows the user to specify more than one disjoint set of files that will be consumed together, so that the different sets can be executed in parallel. For instance, the user can reduce all *txt* files and all *jpg* files in parallel.

To run an activity like this, users can use the *TARDIS* method called `partial_reduce_activity`. The first 3 parameters are the same as the map method: the activity name or ID, the command that performs the activity and the input RDD. The fourth is an array of patterns of files that will be affected by it. The only difference is that in a partial reduce, the user can specify more than one pattern.

In the command string, `@!input` will be translated to all the affected filenames separated by a space, and there is a new placeholder, `@!output`, which will be translated to a filename-safe version of the current pattern.

For example, if the user wants to concatenate all `txt` files in one big file and all the `jpg` files in another, the following command can be used:

```
concatenatedRDD = tardis.partial_reduce_activity("cat", "cat
        @!input > all@!output", filesRDD, ("*.txt", "*.jpg")).
```

If we had four files in `filesRDD`: `text1.txt`, `text2.txt`, `photo1.jpg` and `photo2.jpg`, after the command `concatenatedRDD` we would have two files: `all.jpg` and `all.txt`.

Reduce Activities. A *reduce activity* is a specific case of the *partial reduce* one where only one pattern is given: `"*"`. This means that all files from the RDD will be consumed by only one instance of the command that performs the activity.

This is different from a reduce in a *MapReduce* or a *Spark* paradigm. In those cases, it can be executed in parallel, because the operation is transitive and binary (so it can be executed in a binary tree). As our operator is a black-box, we cannot assure that the operation is transitive and we cannot modify the input to receive only a pair of files.

Users can define this activity with the `reduce_activity` method. It expects three parameters: the activity name or ID, the command that performs the activity and the input RDD. The `@!input` placeholder can be used in the command to be translated to a list of all files available in the RDD. For instance, the user wants to concatenate all of the files in an RDD to a single file, the following command can be used:

```
finalRDD = tardis.reduce_activity("final_cat", "cat @!input >
                    allFiles", filesRDD).
```

In this example, the * wildcard could replace @!input with no loss of generalization. The finalRDD would have only one file, named allFiles.

4.3 Data Placement

As discussed in Sect. 4.1, TARDIS distributes the files through nodes of the shared-nothing cluster, such that the scientific black-box softwares can be scheduled to access their input files locally. In this section, we present our file allocation algorithms for mapping each input file $f \in F$ to a node $n \in N$ in the cluster.

The file allocation algorithm should take into account the size of the file, their initial allocation and the cost to transfer the file from one node to another. This cost may vary if the nodes are not in the same network. Additionally, to avoid skew, the scheduler should consider the *ideal load* of each node. This ideal is given by the ratio of the computing capability of the respective node in relation to the general computing capability of the cluster. At this moment, we consider only the quantity of cores and the respective CPU frequency as a measure of this capability.

Let c_i be the computing capability of node n_i. $\sum_j c_j$ is the sum of the capabilities of the entire cluster. Supposing that to run a workflow W, the partition p_i is allocated to node n_i, the ideal size of p_i is given by

$$ideal\_size(p_i) = \frac{c_i \times \sum_{d \in input(W)} size(d)}{\sum_j c_j}. \tag{1}$$

We propose four different file allocation algorithms: *only local, greedy allocation, locals first* and *lazy allocation*. They are presented in the next sections.

File Allocation Algorithms

Only local
This algorithm tries to minimize the data transfer during the execution of workflows by maximizing the data locality. It trivially allocates each file to be run in the same node where it is already stored. Algorithm 1 describes the *only local* allocation. The *FQDN* (fully qualified domain name) of the node is used as its identification.

Algorithm 1. Only local algorithm

1: **procedure** LOCAL(nodes,objects)
2:     **for** obj in objects **do**
3:         **for** no in nodes **do**
4:             **if** obj.node.fqdn == no.fqdn **then**
5:                 no.alloc_obj_to(obj)

Greedy allocation

The *Greedy allocation* algorithm pursues the optimal solution by reducing the skew of data according to the computing capability of each node. As discussed previously, each node is attributed an ideal fraction of the total input volume size (as computed by Eq. 1). As shown in Algorithm 2, our greedy algorithm allocates the biggest files to the nodes with highest capability.

The algorithm proceeds as follows. After sorting the nodes and the files in descending order (lines 2 and 3), we allocate the biggest files to the node with biggest capability that can still fit this file (line 6–10). If the file cannot fit any host, the host with the biggest capability, which has less tasks than it should, will receive the task (lines 11–15) even if this makes it have more tasks than what is required.

Algorithm 2. Greedy allocation algorithm

1: **procedure** GREEDY(nodes,objects)
2:     **sort** nodes **by** capability **into descending order**
3:     **sort** objects **by** size **into descending order**
4:     **for** object **in** objects **do**
5:         obj_allocated ← False
6:         **for** no **in** nodes **do**
7:             **if** obj.size < no.avail_size **then**
8:                 no.alloc_obj_to(obj)
9:                 obj_allocated ← True
10:                **break**
11:        **if not** obj_allocated **then**
12:            **for** no **in** nodes_greedy **do**
13:                **if** no.ideal_size > no.curr_size **then**
14:                    no.alloc_obj_to(obj)
15:                    **break**

Locals first

The *locals first* is a **hybrid** algorithm, in the sense that they tries to reduce both the skew and the data transfer among the nodes. The *locals first* algorithm tries to allocate the data to its local node, while this node has space (lines 2–6). After filling every node, the remained data will be allocated to the closest node with space available (lines 7–12). This is done by ordering the nodes with respect to the transfer cost to where the object currently is.

Lazy allocation

The *lazy allocation* algorithm, also hybrid, starts by allocating all data to the local node, like the *only local* algorithm (lines 2–5). Then, it redistributes the overflowing data to other nodes, as shown in Algorithm 4. This is done by ordering the files in a overflowing node by their size and removing the smallest ones until the node is close to its ideal size (lines 6–11). Then each removed file will be allocated to the closest node with available space (lines 12–16).

Algorithm 3. Locals first algorithm

1: **procedure** FIRST(nodes,objects)
2:     **for** obj in objects **do**
3:         **for** no in nodes **do**
4:             **if** obj.node.fqdn == no.fqdn **and** no.avail_size ≥ obj.size **then**
5:                 no.alloc_obj_to(obj)
6:                 objects.remove(obj)
7:     **for** obj in objects **do**
8:         **sort** nodes **by** transfer costs from *obj.node* **into ascending order**
9:         **for** no in nodes **do**
10:             **if** no.avail_size > 0 **then**
11:                 node.alloc_obj_to(obj)
12:                 **break**

Algorithm 4. Lazy allocation algorithm

1: **procedure** LAZY(nodes,objects)
2:     **for** obj in objects **do**
3:         **for** no in nodes **do**
4:             **if** obj.node.fqdn == no.fqdn **then**
5:                 no.alloc_obj_to(obj)

6:     **for** no in nodes **do**
7:         **if** no.avail_size < 0 **then**
8:             **sort** no.objs_to_run **by** size **in ascending order**
9:             **while** *no* is still overflowing **do**
10:                 obj_taken ← no.objs_to_run[0]
11:                 **remove** no.objs_to_run[0] **from** no
12:                 ordered_nodes ← **sort** nodes **by** transfer costs from *no* **into ascend-**
        ing order
13:                 **for** other_no in ordered_nodes **do**
14:                     **if** other_node.avail_size > 0 **then**
15:                         other_node.alloc_obj_to(obj_taken)
16:                         **break**

4.4 Scheduling

Given the *TARDIS* file allocation scheduling, we must conceive a strategy that would oblige *Spark* scheduling to follow the proposed solution. This is done by proceeding four main steps.

Step 1: Placeholders Distribution. The output of the previous allocation algorithms is a table, with the input files and theirs respective nodes to be allocated to. We create an RDD containing $n \times f$ unique integers, where n is the number of nodes and f is the number of input files. The $n \times f$ copies of the scheduling table reserves f slots for each of the n nodes. The f slots are place-holders for possible files at each node. Each place-holder has an individual *id*. Later, each slot may hold a pointer to file allocated by the scheduling algorithm.

This is needed because it is not possible to increase the number of elements in the RDD later without *Spark* performing a shuffling of the data.

By invoking the `parallelize` method of *Spark* with the parameter n, we create n partitions of the allocation table and send them to the cluster. We cannot assure that each node will receive a RDD partition with f elements. So we check if the number of elements allocated to each node is higher or equal to the number of files that should be allocated to that specific node. If this condition is not assured, we restart this algorithm, this time creating an RDD with $m \times n \times f$ integers. m is initialized to 2, and is increased until the above condition is satisfied. This condition is checked in the *master node* after executing a *Spark* **map** transformation over the RDD which tags the element with the current node name. After caching it in the nodes and collecting all elements to the master, it knows not only the quantity of placeholders of each node, but also which exact ids have been allocated to each node. This information will be used for fault-tolerance reasons.

Step 2: Distribution of the allocation table. At this point, the *master node* has two tables, a file allocation one, generated by the algorithms presented in the previous section and an id allocation, collected by the previous step. By joining both tables we generate (id, file) pairs, associating different ids of one node to the files it should store. This new table will be broadcast to all nodes.

Step 3: Memory allocation. Then, we run a *Spark* **map** over this RDD with a *TARDIS* method that will be executed over each element of this array. This method will copy into memory the file related to the current id. This map returns a pointer to the file allocated in memory as well as its name. This copy will be done by locally reading the disk, if the file is already available in this node, or by transferring it via the network. This is done, currently, using the `rsync` utility. To allocate and access files in the memory, we use the *shared memory* area (SHM) of the operational system (i.e. Linux).

Step 4: Unused placeholders removal. Finally, a *Spark* `filter` transformation with a *TARDIS* routine is executed over the RDD to remove the elements flagged with zero. Thus, in this step we have an RDD full of filenames with pointers to its respective positions in local memory. Besides, we also have this RDD partitioned according to our scheduling algorithm.

4.5   Collecting Output Files

After executing the workflow, the `collect_files` *TARDIS* method should be used to write the files in the output RDD to the master node hard disk.

A folder called `output<N>` will be created in the current directory with the files inside it. `<N>` is a random integer that should allow the user to execute the workflow many times and keep the outputs. *TARDIS* prints to the standard output the name of the created folder.

5   Experiments

In this section, we report the results of our experiments done for validating *TARDIS* and evaluating its performance.

We tested *TARDIS* with a workflow using *Montage* to produce an image in a cluster. This workflow consumes 3.7 GB of input data and 20 GB of intermediate data. *Montage* is a toolkit to assemble astronomical images to generate mosaics, or panoramas [5]. It can be used for generating a colorful image of a certain part of the sky. To generate a color image, *Montage* uses tiles obtained from the DSS2 (Digitalized Sky Survey) with three filters: blue, red and infra-red. These are catalogs that are generated by using the pictures taken by telescopes.

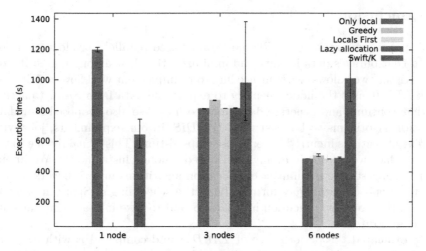

Fig. 2. Execution time in seconds for different experiments. (Color figure online)

During the required image generation, the tiles (photos obtained from telescopes) need to be normalized and unified to cover the area requested by the user. Besides, after doing this for the three filters, they need to be combined to generate a color image. Although all the tiles of the three channels are used to generate the final image, intermediate activities perform a *reduce* operation in each channel separately.

The workflow is composed of three activities. The *mProjectPP* activity projects some tiles to the coordinates of the final image. The *mAdd* concatenate the different tiles, already reprojected. This is done for three color-channels: red, blue and infrared. Finally, *mJPEG* creates a colorful picture using the three channels.

The experiments were run in a cluster running Linux (CentOS distribution). Each node has two *Intel(R) Xeon(R) CPU E5-2630 v3* running at 2.40 GHz with eight physical cores each. The RAM memory available in each node is 94 GB. The cluster nodes were used for running this workflow using six, three and one node. We compare the execution time of *TARDIS* and its different scheduling algorithms with *Swift/K*, which is the state-of-art workflow engine.

After performing 10 times each experiment, the averages, maximums and minimums of the execution time are presented in Fig. 2.

We see that *TARDIS* performs better than *Swift/K* for the execution of the workflow when the number of nodes is higher than one. With 6 nodes, the best algorithm is *TARDIS* with *locals first allocation*. *Swift/K* does not pursue data locality, thus its performance is not very good compared to *TARDIS* when the number of nodes increases. However, it performs better than *TARDIS* when running in only one node, i.e., not parallel. The reason is that *TARDIS* is optimized for running in distributed environments, particularly its data allocation and scheduling modules.

6   Conclusion

In this paper, we proposed *TARDIS*, a *Spark*-based parallel workflow execution system that pursues data locality and minimizes the skew among nodes. It executes existing workflows, without needing to re-implement workflow activities in Spark API. It uses the nodes memory to export the data from *Spark* to activities that consume and generate data from/to files. We also proposed four data allocation algorithms to be used with *TARDIS*. In our experiments, jobs with different allocations incurred in very close elapsed-time. This is mainly due to the fact that files were already evenly distributed among cluster nodes. We intend to further investigate the impact of allocation algorithms in future work. Moreover, we proposed techniques for modifying the scheduling of Spark in order to optimize the workflow execution in *TARDIS* and to take into account our data placement strategies.

We evaluated the performance of *TARDIS* and compared it with *Swift/K*. The results show that the data locality and in-memory execution of *TARDIS* is highly adequate to parallel workflow execution in distributed environments, leading to an improvement of up to 138% in execution time.

Acknowledgements. The authors would like to thank the brazilian agency CNPq for financial support. This research is partially funded by EU H2020 Program and MCTI/RNP-Brazil (HPC4e Project - grant agreement number 689772), FAPERJ (MUSIC Project E36-2013). This research made use of Montage. It is funded by the National Science Foundation under Grant Number ACI-1440620, and was previously funded by the National Aeronautics and Space Administration's Earth Science Technology Office, Computation Technologies Project, under Cooperative Agreement Number NCC5-626 between NASA and the California Institute of Technology.

References

1. Apache: Apache spark programming guide. https://spark.apache.org/docs/2.0.1/programming-guide.html
2. Apache: Hadoop. http://hadoop.apache.org/
3. Dean, J., Ghemawat, S.: Mapreduce: simplified data processing on large clusters. Commun. ACM **51**(1), 107–113 (2008). doi:10.1145/1327452.1327492

4. Deelman, E., Singh, G., Su, M.H., Blythe, J., Gil, Y., Kesselman, C., Mehta, G., Vahi, K., Berriman, G.B., Good, J., et al.: Pegasus: a framework for mapping complex scientific workflows onto distributed systems. Sci. Program. **13**(3), 219–237 (2005)
5. Jacob, J.C., Katz, D.S., Berriman, G.B., Good, J.C., Laity, A., Deelman, E., Kesselman, C., Singh, G., Su, M.H., Prince, T., et al.: Montage: a grid portal and software toolkit for science-grade astronomical image mosaicking. Int. J. Comput. Sci. Eng. **4**(2), 73–87 (2009)
6. Liroz-Gistau, M., Akbarinia, R., Pacitti, E., Porto, F., Valduriez, P.: Dynamic workload-based partitioning for large-scale databases. Database and Expert Systems Applications. doi:10.1007/978-3-642-32597-7_16
7. Ocaña, K., de Oliveira, D.: Parallel computing in genomic research advances and applications. Adv. Appl. Bioinf. Chem. **8**, 23–35 (2015). AABC
8. Oliveira, D., Boeres, C., Porto, F., Fausti, A.: Avaliaçã da localidade de dados intermediários na execuçã o paralela de workflows bigdata. In: SBBD Proceedings (2015)
9. de Oliveira, D.E.M., Boeres, C., Porto, F.: Análise de estratégias de acesso a grandes volumes de dados. In: SBBD Proceedings (2014)
10. Olston, C., Reed, B., Srivastava, U., Kumar, R., Tomkins, A.: Pig Latin: a not-so-foreign language for data processing. In: Proceedings of the 2008 ACM SIGMOD International Conference on Management of Data, SIGMOD 2008, NY, USA, pp. 1099–1110 (2008). doi:10.1145/1376616.1376726
11. Thusoo, A., Sarma, J.S., Jain, N., Shao, Z., Chakka, P., Anthony, S., Liu, H., Wyckoff, P., Murthy, R.: Hive: a warehousing solution over a map-reduce framework. Proc. VLDB Endow. **2**(2), 1626–1629 (2009). doi:10.14778/1687553.1687609
12. Wilde, M., Hategan, M., Wozniak, J.M., Clifford, B., Katz, D.S., Foster, I.: Swift: A language for distributed parallel scripting. Parallel Comput. **37**(9), 633–652 (2011)
13. Zaharia, M., Chowdhury, M., Das, T., Dave, A., Ma, J., McCauley, M., Franklin, M.J., Shenker, S., Stoica, I.: Resilient distributed datasets: a fault-tolerant abstraction for in-memory cluster computing. In: Proceedings of the 9th USENIX Conference on Networked Systems Design and Implementation (2012)
14. Zhou, J., Bruno, N., Wu, M.C., Larson, P.A., Chaiken, R., Shakib, D.: Scope: parallel databases meet mapreduce. VLDB J. **21**(5), 611–636 (2012). doi:10.1007/s00778-012-0280-z

MDA-Based Approach
for NoSQL Databases Modelling

Fatma Abdelhedi[2], Amal Ait Brahim[1]($\boxtimes$), Faten Atigui[3],
and Gilles Zurfluh[1]

[1] Toulouse Institute of Computer Science Research (IRIT), Toulouse Capitole
University, Toulouse, France
{amal.ait-brahim,gilles.zurfluh}@irit.fr
[2] CBI2 – TRIMANE, Paris, France
fatma.abdelhedi@irit.fr
[3] CEDRIC-CNAM, Paris, France
faten.atigui@cnam.fr

Abstract. It is widely accepted today that relational systems are not appropriate
to handle Big Data. This has led to a new category of databases commonly
known as NoSQL databases that were created in response to the needs for better
scalability, higher flexibility and faster data access. These systems have proven
their efficiency to store and query Big Data. Unfortunately, only few works have
presented approaches to implement conceptual models describing Big Data in
NoSQL systems. This paper proposes an automatic MDA-based approach that
provides a set of transformations, formalized with the QVT language, to translate
UML conceptual models into NoSQL models. In our approach, we build an
intermediate logical model compatible with column, document and graph oriented systems. The advantage of using a unified logical model is that this model
remains stable, even though the NoSQL system evolves over time which simplifies the transformation process and saves developers efforts and time.

Keywords: UML · NoSQL · Big data · MDA · QVT · Models transformation

1   Introduction

Big Data is one of the current and future research themes. Recently, the advisory and
research firm Gartner Group outlined the top 10 technology trends that will be strategic
for most organizations over the next five years, and unsurprisingly Big Data is mentioned in the list [13]. Relational systems that had been for decades the one solution for
all databases needs prove to be inadequate for all applications, especially those
involving Big Data [3]. Consequently, new type of DBMS, commonly known as
"NoSQL" [2], has appeared. These systems are well suited for managing large volume
of data; they keep good performance when scaling up [1]. NoSQL covers a wide
variety of different systems that can be classified into four basic types: key-value,
column-oriented, document-oriented and graph-oriented. In this paper, we focus on the
last three. The first one (key-value) is implicitly considered since all of the mentioned
systems extend the concepts of key-value [10].

© Springer International Publishing AG 2017
L. Bellatreche and S. Chakravarthy (Eds.): DaWaK 2017, LNCS 10440, pp. 88–102, 2017.
DOI: 10.1007/978-3-319-64283-3_7

To motivate and illustrate our work, we present a case study in the healthcare filed. This case study concerns international scientific programs for monitoring patients suffering from serious diseases. The main goal of this program is (1) to collect data about diseases development over time, (2) to study interactions between different diseases and (3) to evaluate the short and medium-term effects of their treatments. The medical program can last up to 3 years. Data collected from establishments involved in this kind of program have the features of Big Data (the 3 V). **Volume:** the amount of data collected from all the establishments in three years can reach several terabytes. **Variety:** data created while monitoring patients come in different types; it could be (1) structured as the patient's vital signs (respiratory rate, blood pressure, etc.), (2) semi-structured document such as the package leaflets of medicinal products, (3) unstructured such as consultation summaries, paper prescriptions and radiology reports. **Velocity:** some data are produced in continuous way by sensors; it needs a [near] real time process because it could be integrated into a time-sensitive processes (for example, some measurements, like temperature, require an emergency medical treatment if they cross a given threshold).

The lack of a model when creating a database is a key feature in NoSQL systems. In a table, attributes names and types are specified as and when the row is entered. Unlike relational systems - where the model must be defined when creating the table - the schema less appears in NoSQL systems. This property offers undeniable flexibility that facilitates the evolution of models in NoSQL systems. But this property concerns exclusively the physical level (implementation) of a database [14]. In information system, the model serves as a document of exchange between end-users and developers. It also serves as a documentation and reference for development and system evolution due to the business needs and typically deployment technologies evolution. Furthermore, the conceptual model provides a semantic knowledge element close to human logic, which guarantees efficient data management [3].

UML is widely accepted as a standard modelling language for describing complex data [3]. In the medical application, briefly presented above, the database contains structured data, data of various types and formats (explanatory texts, medical records, x-rays, etc.), and big tables (records of variables produced by sensors). Therefore, we choose the UML class diagram to design describe the medical data.

The rest of the paper is structured as follows: Sect. 2 defines our research problem and reviews previous work on models transformation; Sect. 3 introduces our MDA-based approach; two transformations processes are presented in this section, the first one creates a logical model starting from a UML class diagram, and the second one generates NoSQL physical models from this logical model; Sect. 4 details our experiments; and Sect. 5 concludes the paper and announces future work.

2 Research Problem and Related Work

Big Data applications developers have to deal with the question: how to store Big Data in NoSQL systems? To address this problem, existing solutions propose to model Big Data, and then define mapping rules towards the physical level.

In the specific context of a data warehouse, both [9, 15] have proposed to transform a multidimensional model into a NoSQL model. In [9] the authors defined a set of rules to map a star schema into two NoSQL models: column-oriented and document-oriented. The links between facts and dimensions have been converted using imbrications. Authors in [15] proposed three approaches to map a multidimensional model into a logical model adapted to column-oriented NoSQL systems.

Other studies [5, 6] have investigated the process of transforming relational databases into a NoSQL model. Li [5] have proposed an approach for transforming a relational database into HBase (column-oriented system). Vajk et al. [6] defined a mapping from a relational model to document-oriented model using MongoDB.

To the best of our knowledge, only few works have presented approaches to implement UML conceptual model into NoSQL systems. Li et al. [11] propose a MDA-based process to transform UML class diagram into column-oriented model specific to HBase. Starting from the UML class diagram and HBase metamodels, authors have proposed mapping rules between the conceptual level and the physical one. Obviously, these rules are applicable to HBase, only. Gwendal et al. [7] describe the mapping between a UML conceptual model and graph databases via an intermediate graph metamodel. In this work, the transformation rules are specific to graph databases used as a framework for managing complex data with many connections. Generally, this kind of NoSQL systems is used in social networks where data are highly connected.

Regarding the state of the art, some of the existing works [5, 6] focus on relational model that, unlike UML class diagram, lacks of semantic richness, especially through the several types of relationships that exist between classes. Other solutions, [9, 15] have the advantage to start from the conceptual level. But, the proposed models are Domain-Specific (Data Warehouses system), so they consider fact, dimension, and typically one type of links only. [7, 11] consider, each, a single type of NoSQL systems (column-oriented in [11] and graph-oriented in [7]). However, it makes more sense to choose the target system according to the user's needs. For example, if processing operations requires access to hierarchically structured data, the document-oriented system proves to be the most adapted solution.

The main purpose of our work is to assist developers in storing Big Data in NoSQL systems. For this, we propose a new MDA-based approach that transforms a conceptual model describing Big Data into several NoSQL physical models. This automatic process allows the developer to choose the system type (column, document or graph) that suits the best with business rules and technical constraints.

3 UMLtoNoSQL Approach

Our purpose is to define, to formalize and to automate the storage of Big Data by means of NoSQL systems. For this, we propose UMLtoNoSQL approach that automatically transforms a UML conceptual model into a NoSQL physical model. We introduce a logical level between conceptual (business description) and physical (technical description) levels in which a generic logical model is developed. This logical model exhibits a sufficient degree of platform-independency making possibleits mapping to one or more NoSQL platforms. This model have two main advantages: (1) it describes data according to the common features of NoSQL models, (2) it is independent of technical details of NoSQL systems, this means that the logical level remains stable, even though the NoSQL system evolves over time. In this case, it would be enough to evolve the physical model, and of course adapt the transformation rules; this simplifies the transformation process and saves time for developers.

To formalize and automate UMLtoNoSQL process, we use the Model Driven Architecture (MDA). One of the main aims of MDA is to separate the functional specification of a system from the details of its implementation in a specific platform [4]. This architecture defines a hierarchy of models from three points of view: Computation Independent Model (CIM), Platform Independent Model (PIM), and Platform Specific Model (PSM) [8]. Among these models, we use the **PIM** to describe data hiding all aspects related to the implementation platforms, and the **PSM** to represent data using a specific technical platform.

In our scenario, the UML class diagram and the generic logical model belong to the PIM level. UMLtoNoSQL process transforms the UML class diagram (conceptual PIM) into a generic logical model (logical PIM). At the PSM level, we consider three different physical models that correspond to Cassandra (column-oriented system), MongoDB (document-oriented system) and Neo4j (graph-oriented system). Figure 1 shows the different component of UMLtoNoSQL process.

UMLtoGenericModel (1) is the first transformation in UMLtoNoSQL process. It transforms the input UML class diagram into the generic logical model (2). This model is conform to the generic logical metamodel presented in Sect. 3.1. GenericModelto PhysicalModel (3) is the second transformation (Sect. 3.2) that generates the NoSQL physical models (PSMs) (4) starting from the generic logical model.

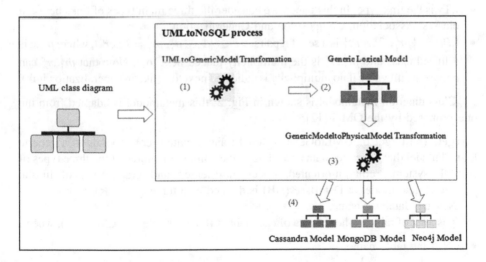

Fig. 1. Overview of UMLtoNoSQL process

3.1 UMLtoGenericModel Transformation

In this section we present the UMLtoGenericModel transformation, which is the first step in our approach as shown in Fig. 1. We first define the source (UML Class Diagram) and the target (Generic Logical Model). After that, we focus on the transformation itself.

Source. A Class Diagram (CD) is defined as a tuple (N, C, L), where:

N is the class diagram name,

C is a set of classes. Classes are composed from structural and behavioral features; in this paper, we consider the structural features only. Since the operations describe the behavior, we do not consider them. For each class $c \in C$, the schema is a tuple $(N, A, IdentO^c)$, where:

- c.N is the class name,
- c.A $= \left\{ a_1^c, \ldots, a_q^c \right\}$ is a set of q attributes. For each attribute $a^c \in A$, the schema is a pair (N,C) where "$a^c.N$" is the attribute name and "$a^c.C$" the attribute type; C can be a predefined class, i.e. a standard data type (String, Integer, Date...) or a business class (class defined by user),
- c.IdentO^c is a special attribute of c; it has a name IdentO^c.N and a type called "Oid". In this paper, an attribute which type is "Oid" represents a unique object identifier, i.e. an attribute which value distinguishes an object from all other objects of the same class,

L is a set of links. Each link l between n classes, with n >= 2, is defined as a tuple (N, Ty, Pr^l), where:

- l.N is the link name.
- l.Ty is the link type. In this paper, we consider the three main types of links between classes: Association, Composition and Generalization.
- l.$Pr^l = \{pr_1^l, \ldots, pr_n^l\}$ is a set of n pairs. $\forall i \in \{1, .., n\}$, $pr_i^l = (c, cr^c)$, where $pr_i^l.c$ is a linked class and $pr_i^l.cr^c$ is the multiplicity placed next to c. Note that $pr_i^l.cr^c$ can contain a null value if no multiplicity is indicated next to c (like in generalization link).

Class diagram metamodel is shown in Fig. 2; this metamodel is adapted from the one proposed by the OMG [12].

Target. The target of UMLtoGenericModel transformation corresponds to a generic logical model that describes data according to the common features of the three types of NoSQL systems: column-oriented, document-oriented and graph-oriented. In the generic logical model, a DataBase (DB) is defined as a tuple (N, T, R), where:

N is the database name,

T is a set of tables. The schema of each table $t \in T$ is a tuple $(N, A, IdentL^t))$, where:

- t.N is the table name,
- t.A $= \left\{ a_1^t, \ldots, a_q^t \right\}$ is a set of q attributes that will be used to define rows of t; each row can have a variable number of attributes. The schema of each attribute $a^t \in A$ is a pair (N,Ty) where "$a^t.N$" is the attribute name and "$a^t.Ty$" the attribute type.

- t.IdentL^t is a special attribute of t; it has a name IdentL^t.N and a type called "Rid". In this paper, an attribute which type is "Rid" represents a unique row identifier, i.e. an attribute which value distinguishes a row from all other rows of the same table,

R is a set of binary relationships. In the generic logical model there are only binary relationships between tables. Each relationship r ∈ R between t_1 and t_2 is defined as a tuple (N, Pr^r), where:

- r.N is the relationship name.
- $r.Pr^r = \{pr_1^r, pr_2^r\}$ is a set of two pairs. $\forall i \in \{1, 2\}$, $pr_i^r = (t, cr^t)$, where $pr_i^r.t$ is a related table and $pr_i^r.cr^t$ is the multiplicity placed next to t.

Metamodel of the proposed generic logical model is shown in Fig. 3. Note that the attribute value may be either atomic or complex (set of attributes). We represent this by using the UML XOR constraint.

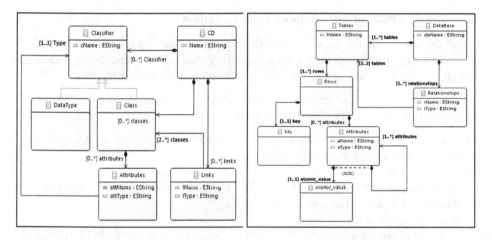

Fig. 2. Source metamodel   **Fig. 3.** Target metamodel

Transformation Rules.
R1: each class diagram CD is transformed into a database DB, where DB.N = CD.N.
R2: each class c ∈ C is transformed into a table t ∈ DB, where t.N = c.N, IdentL^t.N = IdentO^c.N.
R3: each attribute a^c ∈ c.A is transformed into an attribute a^t, where a^t.N = a^c.N, a^t.Ty = a^c.C, and added to the attribute list of its transformed container t such as a^t ∈ t.A.
R4: each binary link l ∈ L (regardless of its type: Association, Composition or Generalization) between two classes c_1 and c_2 is transformed into a relationship r ∈ R between the tables t_1 and t_2 representing c_1 and c_2, where r.N = l.N, r.Pr^r = $\{(t_1, cr^{c_1}), (t_2, cr^{c_2})\}$, cr^{c_1} and cr^{c_2} are the multiplicity placed respectively next to c_1 and c_2.

R5:   each link $l \in L$ between n classes $\{c_1, \ldots, c_n\}(n > = 3)$ is transformed into (1) a new table t^l, where $t^l.N = l.N$ and $t^l.A = \emptyset$, and (2) n relationships $\{r_1, \ldots, r_n\}$, $\forall i \in \{1, .., n\}r_i$ links t^l to another table t_i representing a related class c_i, where $r_i.N = (t^l.N)\_(t_i.N)$ and $r_i.Pr^r = \{(t^l, null), (t_i, null)\}$.

R6:   each association class c_{asso} between n classes $\{c_1, \ldots, c_n\}(n > = 2)$ is transformed like a link between multiple classes (R5) using (1) a new table t^{ac}, where $t^{ac}.N = l.N$, and (2) n relationships $\{r_1, \ldots, r_n\}$, $\forall i \in \{1, .., n\}r_i$ links t^{ac} to another table t_i representing a related class c_i, where $r_i.N = (t^{ac}.N)\_(t_i.N)$ and $r_i.Pr^r = \{(t^{ac}, null), (t_i, null)\}$. Like any other table, t^{ac} contain also a set of attributes A, where $t^{ac}.A = c_{asso}.A$.

We have formalized these transformation rules using the QVT (Query/View/Transformation), which is the OMG standard for models transformation. An excerpt from QVT rules is shown in Fig. 7.

3.2   GenericModeltoPhysicalModel Transformation

In this section we present the GenericModeltoPhysicalModel transformation, which is the second step in our approach UMLtoNoSQL (Fig. 1). This transformation creates NoSQL physical models starting from the proposed generic logical model.

Source. The source of GenericModeltoPhysicalModel transformation is the target of the previous UMLtoGenericModel transformation.

Target. To illustrate our approach, we have chosen three well known NoSQL systems: Cassandra, MongoDB and Neo4j.

Cassandra physical model

In Cassandra physical model, KeySpace (KS) is the top-level container that owns all the elements. It's defined as a tuple (N, F), where:

N is the keyspace name,

F is a set of columns-families. The schema of each columns family $f \in F$ is defined as a tuple $(N, Cl, PrimaryKey^f)$, where:

- f.N is the columns-family name,
- f.Cl $= \{cl_1, \ldots, cl_q\}$ is a set of q columns that will be used to define rows of f; each row can have a variable number of columns. The schema of each column cl $\in$ Cl is a pair (N,Ty) where "cl.N" is the column name and "cl.Ty" the column type.
- f.PrimaryKey^f is a special column of f; it has a name PrimaryKey^f.N and a type PrimaryKey^f.Ty (standard data type). PrimaryKey^f identifies each row of f.

MongoDB physical model

In MongoDB physical model, DataBase (DB^{MD}) is the top-level container that owns all the elements. It's defined as a tuple (N, Cll), where:

N is the database name,

Cll is a set of collections. The schema of each collection cll $\in$ Cll is a tuple (N, Fl, Id^{cll}), where:

- cll.N is the collection name,
- cll.Fl $= Fl^A \cup Fl^{CX}$ sis a set of atomic and complex fields that will be used to define rows, called documents, of Cll. Each document can have a variable number of fields. The schema of an atomic field $fl^a \in Fl^A$ is a tuple (N,Ty) where "fl^a.N" is the field name and "fl^a.Ty" is the field type. The schema of a complex field $fl^{cx} \in Fl^{CX}$ is also a tuple (N, Fl') where fl^{cx} N is the field name and fl^{cx}.Fl' is a set of fields where Fl'$\subset$ Fl.
- cll.Id^{cll} is a special field of cll; it has a name Id^{cll}.N and a type Id^{cll}.Ty (standard data type).Id^{cll} identifies each document of cll.

Neo4j physical model
In Neo4j physical model, Graph (GR) is the top-level container that owns all the elements. It's defined as a tuple (V, E), where:
    V is a set of vertex. The schema of each vertex v $\in$ V is a tuple (L, Pro, Id^v), where:

- v.L is the vertex label,
- v.Pro $= \{pro_1, \ldots, pro_q\}$ is a set of q properties. The schema of each property pro $\in$ Pro is a pair (N,Ty), where "pro.N" is the property name and "pro.Ty" the property type.
- v.Id^v is a special property of v; it has a name Id^v.N, a type Id^v.Ty and the constraint "Is Unique". It identifies uniquely v in the graph.

    E is a set of edges. The schema of each edge e $\in$ E is a tuple (L, v_1, v_2), where:

- e.L is the edge label,
- e.v_1 and e.v_2 are the vertexes related by e.

Transformation Rules
For some NoSQL systems, many solutions can ensure the implementation of the generic logical model. In order to choose the most suitable solution, the developer can be well guided thanks to the performance measurement shown in Sect. 4.2. These measurements concern the response time of queries that access two related tables; the relationship between these tables has being implemented according to the different solutions shown below. The developer will make his choice according to the queries features he needs to perform as well as the expected performances.
    We note that the set of solutions proposed in this section is not inclusive; more marginal solutions may be considered.

To Cassandra physical model
R1: each database DB is transformed into a keyspace KS, where KS.N = DB.N.
R2: each table t $\in$ DB is transformed into a columns-family f $\in$ KS, where f.N = t.N, PrimaryKey^f.N = IdentL^t.N.
R3: each attribute $a^t \in$ t.A is transformed into a column cl, where cl.N $= a^t$.N, cl.Ty $= a^t$.Ty, and added to the column list of its transformed container f such as cl $\in$ f.Cl.

R4:   As Cassandra does not support imbrication; the only solution we can use to express relationships between columns-families consists in using reference columns. A reference column is a monovalued or multivalued column in one columns-family whose values must have matching values in the primary key of another columns-family; we note that this constraint is not automatically managed by the system Cassandra; it remains the responsibility of the user to check it.

For each relationship r between two tables t_1 and t_2, three solutions could be considered:

Solution 1:   r is transformed into a reference column cl referencing f_2 (the columns-family representing t_2), where $cl.N = (f_2.N)$ _Ref and $cl.Ty = PrimaryKey^{f2}.Ty$, and then added to the columns list of f_1 (the columns-family representing t_1) such as $cl \in f_1.Cl$. While instantiating f_1, the value of the reference column cl will correspond to one or many values in the primary key of f_2.

Solution 2:   r is transformed into a reference column cl referencing f_1 (the columns-family representing t_1), where $cl.N = (f_1.N)$ _Ref et $cl.Ty = PrimaryKey^{f1}.Ty$, and then added to the columns list of f_2 (the columns-family representing t_2) such as $cl \in f_2.Cl$ While instantiating f_2, the value of the reference column cl will correspond to one or many values in the primary key of f_1.

Solution 3:   r is transformed into a new columns-family f composed of two reference columns referencing the columns-families f_1 and f_2 representing the related tables t_1 and t_2, where $f.N = r.N$, $f.Cl = \{cl_1, cl_2\}$, $cl_1.N = (f_1.N)$ _Ref, $cl_1.Ty = PrimaryKey^{f1}.Ty, cl_2.N = (f_2.N)$ _Ref and $cl_2.Ty = PrimaryKey^{f2}.Ty$.

A reference column can either be monovalued or multivalued. Table 1 indicates the type of the reference column according to the relationship cardinalities and the transformation solution used.

Table 1. Descriptive table of reference column types

Relationship	Solution	Reference column type
$r = (N, \{(t_1, *), (t_2, 1)\})$	Solution 1	Monovalued
	Solution 2	Multivalued
	Solution 3	Monovalued
$r = (N, \{(t_1, 1), (t_2, 1)\})$	Solution 1	Monovalued
	Solution 2	Monovalued
	Solution 3	Monovalued
$r = (N, \{(t_1, *), (t_2, *)\})$	Solution 1	Multivalued
	Solution 2	Multivalued
	Solution 3	Monovalued

To MongoDB physical model

R1: each database DB is transformed into a MongoDB database DB^{MD}, where $DB^{MD}.N = DB.N$.

R2: each table $t \in DB$ is transformed into a collection cll $\in DB^{MD}$, where cll.N = t.N et $Id^{cll}.N = IdentL^t.N$.

R3: each attribute $a^t \in t.A$ is transformed into an atomic field fl^a, where $fl^a.N = a^t.N$, $fl^a.Ty = a^t.Ty$, and added to the field list of its transformed container cll such as fl $\in$ cll.Fl^A.

R4: relationships in MongoDB could be transformed by using reference fields or embedding. A reference field is a monovalued or multivalued field in one collection whose values must have matching values in the Id of another collection; checking this constraint remains the responsibility of the user.

For each relationship r between two tables t_1 and t_2, five solutions could be considered:

Solution 1:   r is transformed into a reference field fl referencing cll_2 (the collection representing t_2), where fl.N = $(cll_2.N)$ _Ref and fl.Ty = $Id^{cll_2}.Ty$, and then added to the fields list of cll_1 (the collection representing t_1) such as fl $\in cll_1.Fl^A$.

Solution 2:   r is transformed into a reference field fl referencing cll_1 (the collection representing t_1), where fl.N = $(cll_1.N)$ _Ref and fl.Ty = $Id^{cll_1}.Ty$, and added to the field list of cll_2 (the collection representing t_2) such as fl $\in cll_2.Fl^A$.

Solution 3:   r is transformed by embedding the collection cll_2 representing t_2 in the collection cll_1 representing t_1, where $cll_2 \in cll_1.Fl^{CX}$.

Solution 4:   r is transformed by embedding the collection cll_1 representing t_1 in the collection cll_2 representing t_2, where $cll_1 \in cll_2.Fl^{CX}$.

Solution 5:   r is transformed into a new collection cll, where cll.N = r.N, cll.Fl = $\{fl_1, fl_2\}$, $fl_1.N = (cll_1.N)$ _Ref, $fl_1.Ty = Id^{cll_2}.Ty$, $fl_2.N = (cll_2.N)$ _Ref and $fl_2.Ty = Id^{cll_2}.Ty$, where cll_1 and cll_2 are the collections representing t_1 and t_2.

Each reference field used in Solution 1, 2 and 5 can either be monovalued or multivalued. Table 2 indicates the type of the reference field according to the relationship cardinalities and the transformation solution used.

To Neo4j physical model

R1: each table $t \in DB$ is transformed into a vertex $v \in V$, where v.L = t.N, $Id^v.N = IdentL^t.N$.

Table 2. Descriptive table of reference field types

Relationship	Solution	Reference field type
$r = (N, \{(t_1, *), (t_2, 1)\})$	Solution 1	Monovalued
	Solution 2	Multivalued
	Solution 5	Monovalued
$r = (N, \{(t_1, 1), (t_2, 1)\})$	Solution 1	Monovalued
	Solution 2	Monovalued
	Solution 5	Monovalued
$r = (N, \{(t_1, *), (t_2, *)\})$	Solution 1	Multivalued
	Solution 2	Multivalued
	Solution 5	Monovalued

R2: each attribute $a^t \in$ t.A is transformed into a property pro, where pro.N $= a^t$.N, pro.Ty $= a^t$.Ty, and added to the property list of its transformed container v such as pro $\in$ v.Pro.

R3: Each relationship r between two tables t_1 and t_2 is transformed into an edge e, where e.L = r.N, relating two vertex v_1 and v_2, where v_1 and v_2 are the vertex representing t_1 and t_2.

4   Experiments

In this section, we show how to transform a UML conceptual model into NoSQL physical models. As presented in Sect. 3.2, several solutions can ensure this transformation; we therefore began by implementing the UMLtoNoSQL process according to each proposed solution, and then we evaluated their performances to assist the developer in choosing the most effective one.

4.1   Implementation

Experimental environment. We carry out the experimental assessment using a model transformation environment called Eclipse Modeling Framework (EMF). It's a set of plugins which can be used to create a model and to generate other output based on this model. Among the tools provided by EMF we use: (1) _Ecore_: the metamodeling language that we used to create our metamodels. (2) _XML Metadata Interchange (XMI)_: the XML based standard that we use to create models. (3) _Query/View / Transformation (QVT)_: the OMG language for specifying model transformations.

Implementation of UMLtoGenericModel Transformation. UMLtoGenericModel transformation is expressed as a sequence of elementary steps that builds the resulting model (generic logical model) step by step from the source model (UML class diagram):

Step 1: we create Ecore metamodels corresponding to the source (Fig. 2) and the target (Fig. 3).

Step 2: we build an instance of the source metamodel. For this, we use the standard-based XML Metadata Interchange (XMI) format (Fig. 4).

Step 3: we implement the transformation rules by means of the QVT plugin provided within EMF. An excerpt from the QVT script is shown in Fig. 7; the comments in the script indicate the rules used.

Step 4: we test the transformation by running the QVT script created in step 3. This script takes as input the source model built in step 2 and returns as output the logical model. The result is provided in the form of XMI file as shown in Fig. 5.

Implementation of GenericModeltoPhysicalModel Transformation. The generic logical model that we proposed in this paper does not imply a specific system; it exhibits a sufficient degree of independence so as to enable its mapping to different NoSQL platforms. For some NoSQL systems, relationships could be transformed into several forms (monovalued or multivalued references, embedding). Lacks of place, we only present one implementation of the generic logical model that was performed on Cassandra according to Solution 1. Figure 8 shows the corresponding QVT script. This script takes as input the logical model (Fig. 5) generated by the previous transformation and return as output Cassandra physical model (Fig. 6).

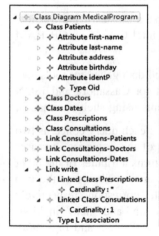

Fig. 4. Source model   **Fig. 5.** Target model   **Fig. 6.** Cassandra model

```
modeltype UML uses "http://UMLClassDiagram.com";
modeltype COLM uses "http://GenericLogicalModel.com";
transformation
TransformationUmlToColumnsOrientedModel(in Source:
UML, out Target: COLM);main()
{Source.rootObjects()[ClassDiagram] -> map toDataBase();}
-- Transforming Class Diagram to DataBase
mapping ClassDiagram::toDataBase():DataBase{name :=
self.name;table:=self.classes -> map
toClass();relationship:=self.links -> toRelationship();}
-- Transforming Class to Table
mapping UML
::Class::toClass():COLM::Table{name:=self.name;attributet:=
self.attributec -> map toAttribute();}
-- Transforming Attribute to Column
mapping UML
::Attribute::toAttribute():COLM::Attribute{name:=self.name;
typea:=self.typea -> map toType(); }
mapping UML
::Type::toType():COLM::Type{typea:=self.typea;}
mapping UML
::Link::toRelationship():COLM::RelationShip{name:=self.na
me; linkedtable:=self.linkedclass -> map toLinkedTable();
```

```
TransformationTransformationGenericModelToCassan
draModel(in Source: LogicalPIM, out Target:
CassandraPSM);
main() {
Source.rootObjects()[DataBase] -> map toKeySpace();}
-- Transforming DataBase to KeySpace
mapping DataBase::toKeySpace():KeySpace{
name := self.name;
columnsfamily:=self.table ->
map toColumnsFamily();}
-- Transforming Table to Columns-Family
mapping LogicalPIM
::Table::toColumnsFamily():CassandraPSM::
ColumnsFamily{name:=self.name;column:=self.attribute
t -> map toColumn();referencecolumn:=self.islinkedto ->
map toReferenceColumn();}
mapping LogicalPIM
::Type::toType():CassandraPSM::Type{
if(self.typea =
"Rid"){type:='Int';}endif;type:=self.typea;}
-- Transforming (1,*) RelationShip to a Monovalued Ref
mapping LogicalPIM
::LinkedToTable::toReferenceColumn():CassandraPSM
::ReferenceColumn{if(self.cardinalityoftable = "*" and
self.cardinalityoflinkedtable
```

Fig. 7. UMLtoGenericModel   **Fig. 8.** GenericModeltoCassandraModel

4.2 Evaluation

The graph-oriented system Neo4j does not offer many solutions to implement relationships; therefore, the developer does not need to choose between several solutions. For Cassandra and MongoDB, where many choices are available, we have evaluated the transformation solutions proposed in Sect. 3.2. This evaluation aims at studying the impact that the choice of the used solution may have on the queries execution time.

Experimental environment. The experiments are done on a cluster made up of 3 machines. Each machine has the following specifications: Intel Core i5, 8 GB RAM and 2 TB disk.

Data set. In order to perform our experiments, we have used data generator tools. We have generated a dataset of about 1 TB with CSV format for Cassandra and JSON format for MongoDB. These files are loaded into the systems using shell commands.

Queries set. For our experiments, we have written 6 queries; each query concerns two tables and the relationship between them. The complexity of these queries increases gradually. The simplest one applies a filter to a table and returns attributes of the other table; the most complex one applies several filters and returns attributes of the two related tables. We note that the concepts "table" and "attribute" correspond respectively to "columns-family" and "column" in Cassandra or "collection" and "field" in MongoDB. An excerpt from our experiment results is depicted in Table 3 and Figs. 9(a), (b) and (c). For each query, we indicate the response time obtained according to (1) the relationship cardinalities and (2) the solution used.

Table 3. Queries response time

NoSQL system	Relationship	Solution	Query	Time (s)
Cassandra	$r = (N, \{(t_1, *), (t_2, 1)\})$	Solution 1	Q1	140
			Q4	980
			...	
		Solution 2	Q1	830
			...	
		Solution 3	...	
			Q4	420
	$r = (N, \{(t_1, 1), (t_2, 1)\})$	Solution 1	...	
			Q5	310
		Solution 2	...	
			Q6	290
		Solution 3	...	
			Q4	735
	$r = (N, \{(t_1, *), (t_2, *)\})$	Solution 1	...	
			Q2	720
			Q5	1400
		Solution 2	...	
			Q3	1050
		Solution 3	...	
			Q5	510
			Q6	530
MongoDB	$r = (N, \{(t_1, *), (t_2, 1)\})$	Solution 1	...	
			Q4	4300
		Solution 2	...	
			Q6	6200
		Solution 3	Q5	1700
			Q6	1500
		Solution 4	Q2	870

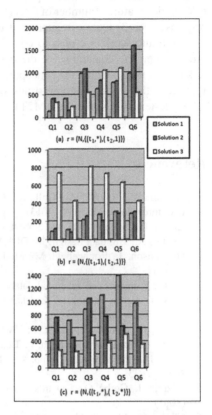

(a) $r = (N, \{(t_1, *), (t_2, 1)\})$

(b) $r = (N, \{(t_1, 1), (t_2, 1)\})$

(c) $r = (N, \{(t_1, *), (t_2, *)\})$

Fig. 9. Cassandra experimental results

5 Conclusion and Future Work

In this paper we have proposed an automatic approach that guides and facilitates the Big Database implementation task within NoSQL systems. This approach is based on MDA especially known as a framework for models automatic transformations. Our approach provides a set of transformations that generate a NoSQL physical model starting from a UML conceptual model. In our approach, we build an intermediate logical model compatible with column, document and graph oriented systems; this model uses tables and binary relationships that link them. The independence between the three physical models is ensured. The advantage of using a unified logical model is that this model remains stable, even though the NoSQL system evolves over time. In this case, it would be enough to evolve the physical models, and of course adapt the transformation rules. Furthermore, we have proposed different solutions to transform the binary relationships of the logical model under Cassandra and MongoDB. Depending on the systems functionalities, the binary relationships could be converted into different forms. We have measured the queries response time using each of the proposed solution. The developer can choose the most suited solution according to: (1) Queries features (number of filters, number of attributes to return, etc.), (2) The time response and (3) Query frequency of use.

As future work, we plan to complete our transformation process in order to take into account the constraints of the conceptual level and to preserve the semantics of links when transforming the conceptual model to the logical one. Furthermore, we want to define the transformations rules of physical models into NoSQL scripts using model-to-text transformation (M2T).

References

1. Angadi, A., Angadi, A., Gull, K.: Growth of new databases and analysis of NOSQL datastores. In: IJARCSSE (2013)
2. Cattell, R.: Scalable SQL and NoSQL data stores. ACM SIGMOD Rec. **39**(4), 12–27 (2011)
3. Abello, A.: Big data design. In: DOLAP (2015)
4. Hutchinson, J., Rouncefield, M.: Model-driven engineering practices in industry. In: ICSE (2011)
5. Li, C.: Transforming relational database into HBase: A case study. In: ICSESS (2010)
6. Vajk, T., Feher, P., Fekete, K., Charaf, H.: Denormalizing data into schema-free databases. In: CogInfoCom (2013)
7. Daniel, G., Sunyé, G., Cabot, J.: UMLtoGraphDB: Mapping conceptual schemas to graph databases. In: Comyn-Wattiau, I., Tanaka, K., Song, I.-Y., Yamamoto, S., Saeki, M. (eds.) ER 2016. LNCS, vol. 9974, pp. 430–444. Springer, Cham (2016). doi:10.1007/978-3-319-46397-1_33
8. Bézivin, J., Gerbé, O.: Towards a precise definition of the OMG/MDA framework. In: ASE (2001)
9. Chevalier, M., El Malki, M., Kopliku, A., Teste, O., Tournier, R.: Implementing multidimensional data warehouses into NoSQL. In: ICEIS (2015)
10. Abadi, D., Madden, S., Hachem, N.: Column-stores vs. row-stores: how different are they really?. In: COMAD (2008)
11. Li, Y., Gu, P., Zhang, C.: Transforming UML class diagrams into HBase based on meta-model. In: ISEEE (2014)
12. http://www.omg.org/spec/UML/2.5/
13. http://www.gartner.com/smarterwithgartner/gartners-top-10-technology-trends-2017/
14. Herrero, V., Abelló, A., Romero, O.: NOSQL design for analytical workloads: variability matters. In: Comyn-Wattiau, I., Tanaka, K., Song, I.-Y., Yamamoto, S., Saeki, M. (eds.) ER 2016. LNCS, vol. 9974, pp. 50–64. Springer, Cham (2016). doi:10.1007/978-3-319-46397-1_4
15. Dehdouh, K., Bentayeb, F., Boussaid, O., Kabachi, N.: Using the column oriented model for implementing big data warehouses. In: PDPTA (2015)

Advanced Programming Paradigms

MiSeRe-Hadoop: A Large-Scale Robust Sequential Classification Rules Mining Framework

Elias Egho[1], Dominique Gay[2(✉)], Romain Trinquart[1], Marc Boullé[1],
Nicolas Voisine[1], and Fabrice Clérot[1]

[1] Orange Labs, 2, avenue Pierre Marzin, 22307 Lannion Cédex, France
{elias.egho,romain.trinquart,marc.boulle,
nicolas.voisine,fabrice.clerot}@orange.com
[2] Université de La Réunion, 2, rue Joseph Wetzell, 97490 Sainte Clotilde, France
dominique.gay@univ-reunion.fr

Abstract. Sequence classification has become a fundamental problem in data mining and machine learning. Feature based classification is one of the techniques that has been used widely for sequence classification. Mining sequential classification rules plays an important role in feature based classification. Despite the abundant literature in this area, mining sequential classification rules is still a challenge; few of the available methods are sufficiently scalable to handle large-scale datasets. MapReduce is an ideal framework to support distributed computing on large data sets on clusters of computers. In this paper, we propose a distributed version of *MiSeRe* algorithm on MapReduce, called *MiSeRe-Hadoop*. MiSeRe-Hadoop holds the same valuable properties as *MiSeRe*, i.e., it is: *(i)* robust and user parameter-free anytime algorithm and *(ii)* it employs an instance-based randomized strategy to promote diversity mining. We have applied our method on two real-world large datasets: a marketing dataset and a text dataset. Our results confirm that our method is scalable for large scale sequential data analysis.

1 Introduction

Sequential data is widely present in several domain such as biology [7,19], medical [10] and web usage logs [20]. Consequently, sequence classification [24] has become a fundamental problem in data mining and machine learning. The sequence classification task is defined as learning a sequence classifier to map a new sequence to a class label [24]. The literature in sequence classification is split into three main paradigms: (1) feature based classification; (2) distance based classification and (3) model based classification.

The feature based classification approach aims at extracting sequential classification rules of the form $\pi : s \rightarrow c_i$ where s is the body of the rule and c_i is the value of a class attribute. Then, these rules are used as an input of a classification method to build a classifier. Due to its potential of interpretability, many approaches have been developed for mining sequential classification rules, such

© Springer International Publishing AG 2017
L. Bellatreche and S. Chakravarthy (Eds.): DaWaK 2017, LNCS 10440, pp. 105–119, 2017.
DOI: 10.1007/978-3-319-64283-3_8

a BayesFM [16], CBS [21], DeFFeD [13], SCII [27], and so on. All these methods need parameters to prune the enumeration space. Unfortunately, setting these parameters is not an easy task – each application data could require a specific setting. Egho et al. [8,9] introduce a user parameter-free approach, *MiSeRe*, for mining robust sequential classification rules. This algorithm does not require any parameter tuning and employs an instance-based randomized strategy that promotes diversity mining. *MiSeRe* works well in practice on typical datasets, but it can not provide scalability, in terms of the data size and the performance, for big data.

In modern life sciences, the sequential data can be very large; e.g., considering a document collection with millions of documents or a web site with millions of user web logs [1]. Mining massive sequential data on single computer suffers from the problems of limited memory and computing power. To solve this problem, parallel programming is an essential solution [23]. Parallel programming can be divided into two categories: shared memory system; in which processes share a single memory address space, and distributed memory system; where processes only have access to a local private memory address space [2].

A number of efficient and scalable parallel algorithms have been developed for mining sequential patterns [14]. Zaki et al. [26] present how a serial sequential approach SPADE [25] can be parallelized by using a shared memory system. Parallelizing an algorithm by using shared system architecture is easy to implement, but it does not provide enough scalability due to high synchronization among processors and memory overheads [2]. On the other hand, Guralnik et al. [12] propose a distributed memory parallel algorithm for a tree-projection based sequential pattern mining algorithm. Cong et al. [5] present Par-CSP a distributed memory parallel algorithm for mining closed sequential pattern. Par-CSP is implemented by using the Message Passing Interface (MPI) in which low level language is used for programming [11].

A recent framework for distributed memory system, Hadoop-Yahoo MapReduce, has been proposed by Google [6]. MapReduce is a scalable and fault-tolerant data processing model that allows programmers to develop efficient parallel algorithms at a higher level of abstraction [6]. Many parallel algorithms has been developed by using MapReduce framework for mining big sequential data such as: BIDE-MR [22], PLUTE [17], SPAMC [4], MG-FSM [3] and so on. However, all these algorithms need parameters to prune the enumeration space (frequency threshold, maximum length and a gap constraint) and they focus solely on mining sequential pattern.

Although a significant amount of research results have been reported on parallel implementations of sequential pattern mining, to the best of our knowledge, there is no parameter-free parallel algorithm that targets the problem of mining sequential classification rules. In this paper, we propose *MiSeRe-Hadoop*, the first scalable parameter-free algorithm for mining sequential classification rules. *MiSeRe-Hadoop* is the parallel implementation of a serial algorithm *MiSeRe* [8,9] on MapReduce framework. *MiSeRe-Hadoop* has the same features as *MiSeRe* which are: it is user parameter-free and it promotes diversity mining.

To validate our contributions, we perform an experimental evaluation on two real large datasets. The first one is marketing dataset from the French Telecom company Orange containing sequential information about the behavior of 473, 898 customers. Our second dataset includes natural language text from New York Times corpus (NYT) which consists of over 53 million sentences.

The remainder of this paper is organized as follows. Section 2 briefly reviews the preliminaries needed in our development as well as a running example. Section 3 describes the serial algorithm *MiSeRe*, while the design and implementation issues for *MiSeRe-Hadoop* are presented in Sect. 4. Section 5 presents experimental results before concluding.

2   Preliminaries

Let $\mathcal{I} = \{e_1, e_2, \cdots, e_m\}$ be a finite set of m distinct items. A **sequence** s over $\mathcal{I}$ is an ordered list $s = \langle s_1, \cdots, s_{\ell_s} \rangle$, where $s_i \in \mathcal{I}$; $(1 \leq i \leq \ell_s, \ell_s \in \mathbb{N})$. An atomic sequence is a sequence with length 1. A sequence $s' = \langle s'_1 \cdots s'_{\ell_{s'}} \rangle$ is a **subsequence** of $s = \langle s_1 \ldots s_{\ell_s} \rangle$, denoted by $s' \preceq s$, if there exist indices $1 \leq i_1 < i_2 < \cdots < i_{\ell_{s'}} \leq \ell_s$ such that $s'_z = s_{i_z}$ for all $z = 1 \ldots \ell_{s'}$ and $\ell_{s'} \leq \ell_s$. s is said to be a **supersequence** of s'. $\mathbb{T}(\mathcal{I})$ will denote the (infinite) set of all possible sequences over $\mathcal{I}$. Let $\mathcal{C} = \{c_1, \cdots, c_j\}$ be a finite set of j distinct classes. A **labeled sequential data set** $\mathcal{D}$ over $\mathcal{I}$ is

Table 1. $\mathcal{D}$: a tiny labeled sequential data set as an example.

sid	Sequence	Class
1	$\langle baadg \rangle$	c_1
2	$\langle agbe \rangle$	c_1
3	$\langle badgb \rangle$	c_2
4	$\langle eefgbg \rangle$	c_2

a finite set of triples (sid, s, c) with sid is a sequence identifier, s is a sequence $(s \in \mathbb{T}(\mathcal{I}))$ and c is a class value $(c \in \mathcal{C})$. The set $\mathcal{D}_{c_i} \subseteq \mathcal{D}$ contains all sequences that have the same class label c_i (i.e., $\mathcal{D} = \cup_{i=1}^{j} \mathcal{D}_{c_i}$). The following notations will be used in the rest of the paper: m is the number of items in $\mathcal{I}$, j is the number of classes in $\mathcal{C}$, n is the number of triples (sid, s, c) in $\mathcal{D}$, n_c is the number of triples (sid, s, c) in $\mathcal{D}_c$, ℓ_s is the number of items in the sequence s, k_s is the number of distinct items in the sequence s, $(k_s \leq \ell_s)$ and ℓ_{max} is the number of items in the longest sequence of $\mathcal{D}$.

Definition 1. *(Support of a sequence) Let $\mathcal{D}$ be a labeled sequential data set and let s be a sequence. The **support** of s in $\mathcal{D}$, denoted $f(s)$, is defined as:*

$$f(s) = |\{(sid', s', c') \in \mathcal{D} | s \preceq s'\}|$$

The value of $n - f(s)$ can be written as $\overline{f}(s)$. The support of s in $\mathcal{D}_c$ is noted $f_c(s)$ and $\overline{f}_c(s)$ stands for $n_c - f_c(s)$.

Given a positive integer σ as a minimal support threshold and a labeled sequential data set $\mathcal{D}$, a sequence s is frequent in $\mathcal{D}$ if its support $f(s)$ in $\mathcal{D}$ exceeds the minimal support threshold σ. A frequent sequence is called a *"sequential pattern"*.

Definition 2 (Sequential classification rule). *Let $\mathcal{D}$ be a labeled sequential data set with j classes. A sequential classification rule π is an expression of the form:*

$$\pi : s \rightarrow f_{c_1}(s), f_{c_2}(s), \cdots, f_{c_j}(s)$$

where s is a sequence, called body of the rule, and $f_{c_i}(s)$ is the support of s in each $\mathcal{D}_{c_i}$, $i = 1 \cdots j$.

This definition of classification rule is slightly different from the usual definition where the consequent is a class value (i.e., $s \rightarrow c_i$). It refers to the notion of distribution rule [15] and allows us to access the whole frequency information within the contingency table of a rule π – which is needed for the development of our framework.

Example 1. We use the sequence database $\mathcal{D}$ in Table 1 as an example. It contains four data sequences (i.e., $n = 4$) over the set of items $\mathcal{I} = \{a, b, d, e, f, g\}$ (i.e., $m = 6$). $\mathcal{C} = \{c_1, c_2\}$ is the set with two classes (i.e., j=2). The longest sequence of $\mathcal{D}$ is $s = \langle eefgbg \rangle$ (i.e., $\ell_s = \ell_{max}$), $\ell_{max} = 6$ while $k_s = 4$. Sequence $\langle aad \rangle$ is a subsequence of $\langle baadg \rangle$. The sequence $\langle a \rangle$ is an atomic sequence. Given the sequence $s = \langle ab \rangle$, we have $f(s) = 2$, $\overline{f}(s) = 2$, $f_{c_1}(s) = 1$, $\overline{f}_{c_1}(s) = 1$, $f_{c_2}(s) = 1$ and $\overline{f}_{c_2}(s) = 1$. $\pi : \langle ab \rangle \rightarrow f_{c_1}(\langle ab \rangle) = 1$, $f_{c_2}(\langle ab \rangle) = 1$ is a sequential classification rule.

Given a labeled sequential data set $\mathcal{D}$ and a sequential classification rule $\pi : s \rightarrow f_{c_1}(s), f_{c_2}(s), \cdots, f_{c_j}(s)$, a Bayesian criterion *level* is defined by Egho et al. [8,9] for evaluating the interestingness of sequential classification rule. This criterion is based on the a posteriori probability of a rule given the data and does not require any wise threshold setting. The *level* criterion is defined as follows:

$$level(\pi) = 1 - \frac{cost(\pi)}{cost(\pi_\emptyset)}$$

where $cost(\pi)$ is defined as the negative logarithm of the a posteriori probability of a rule given the data:

$$cost(\pi) = -log(P(\pi \mid \mathcal{D})) \propto -log(P(\pi) \times P(D \mid \pi))$$

Considering a hierarchical prior distribution on the rule models, Egho et al. [8,9] obtained an exact analytical expression of the *cost* of a rule:

$$cost(\pi) = log(m + 1) + log(\ell_{max} + 1) + log(\frac{m^{k_s}}{k_s!}) + log(k_s{}^{\ell_s})$$

$$+ log \binom{f(s) + j - 1}{j - 1} + log \binom{\overline{f}(s) + j - 1}{j - 1}$$

$$+ log(f(s)!) - \sum_{i=1}^{j} log(f_{c_i}(s)!) + log(\overline{f}(s)!) - \sum_{i=1}^{j} log(\overline{f}_{c_i}(s)!)$$

$cost(\pi_\emptyset)$ is the cost of the default rule with empty sequence body. The cost of the default rule $\pi_\emptyset$ is formally:

$$cost(\pi_\emptyset) = log(m+1) + log(\ell_{max}+1) + log\binom{n+j-1}{j-1} + log(n!) - \sum_{i=1}^{j} log(n_{c_i}!)$$

The *level* naturally highlights the border between the interesting rules and the irrelevant ones. Indeed, rules π such that $level(\pi) \leq 0$, are less probable than the default rule $\pi_\emptyset$. Then using them to explain the data by characterizing classes of sequence objects is more costly than using $\pi_\emptyset$; such rules are considered spurious. *"rule such that $0 < level(\pi) \leq 1$ is an **interesting rule**"* (see [8,9] for more details about the interpretation of these two formulas).

3   MiSeRe Algorithm

In this section we describe MiSeRe [8,9], an algorithm for mining sequential classification rules, which forms the basis for our distributed algorithm. MiSeRe is an anytime algorithm: the more time the user grants to the task, the more it learns. MiSeRe employs an instance-based randomized strategy that promotes diversity mining. MiSeRe is based on the following two-step process:

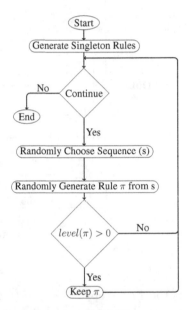

Fig. 1. Flow chart for the main procedure of the *MiSeRe* algorithm

Step 1: MiSeRe firstly performs a single data scan to gather basic statistics about the data: the number of items in $\mathcal{I}$, the number of classes in $\mathcal{C}$, the number of sequences in $\mathcal{D}$, the number of sequences in $\mathcal{D}_{c_i}$ and the length of the longest sequence of $\mathcal{D}$. In this step, MiSeRe also generates all sequential classification rules whose body is an atomic sequence, such rules with positive *level* values are chosen. These rules are selected for two reasons: first, these rules are easy to mine. Second, the short sequences are more probable a priori and preferable as the cost of the rule $c(\pi)$ is smaller for lower ℓ_s and k_s values, meeting the consensus: **"Simpler and shorter rules are more probable and preferable"**.

Step 2: In this step, MiSeRe randomly chooses one sequence **s** from the labeled sequential database $\mathcal{D}$. Then, it randomly generates a subsequence from the chosen sequence **s**. This generation is done by randomly removing z items from **s** where z is between 1 and $\ell_s - 2$. Then, the rule π is built based on the generated subsequence s'. Finally, the rule π is added to the rule set if its level value is positive and it is not already in $\mathcal{R}$. MiSeRe repeats the Step 2 until the algorithm is stopped by the user manually at some point in time. Figure 1 shows a flow chart (simplified for exposition, see [8,9] for full details) for the main procedure of the *MiSeRe* algorithm.

4   MiSeRe Hadoop Algorithm

In this section, we present *MiSeRe-Haddoop*, a distributed version of *MiSeRe*. *MiSeRe-Hadoop* has the same features as *MiSeRe* which are: (i) it is user parameter-free and (ii) anytime algorithm and (iii) it employs an instance-based randomized strategy. *MiSeRe-Hadoop* is divided into two steps as *MiSeRe* where each step is completely parallelized (Fig. 2).

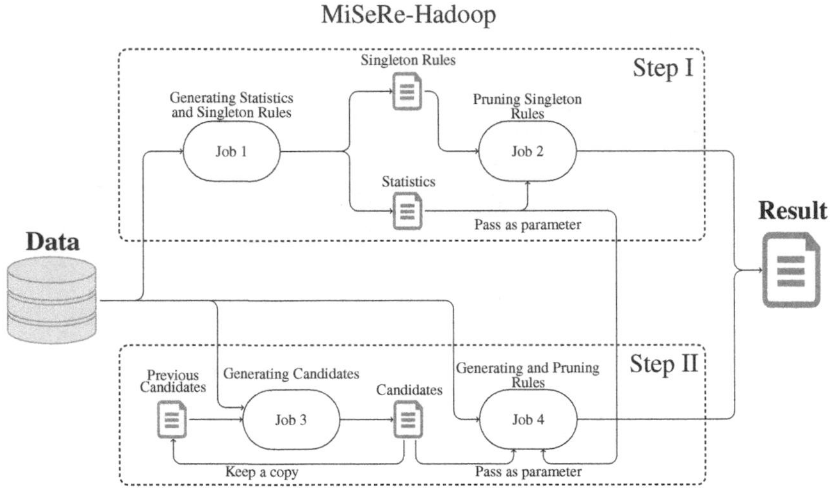

Fig. 2. MiSeRe-Hadoop

4.1   Step I:

In the first step, *MiSeRe-Hadoop* gathers basic statistics about the data and mines all sequential classification rules having only an atomic sequence in the body. This can be done efficiently in two MapReduce jobs.

The job *"Generating Statistics and Singleton Rules"* generates the singleton rules and the statistics about the data consisting of: the number of sequences in $\mathcal{D}_{c_i}$ (i.e., n_{c_i}), the number of sequences in $\mathcal{D}$ (i.e., n) and the length of the longest sequence of $\mathcal{D}$ (i.e., ℓ_{max}). During this job, the data is distributed to available mappers. Each mapper takes a pair $(s, c) \in \mathcal{D}$ as input and tokenizes it into distinct items. Then, the mapper emits the class label for each item. In order to compute the statistic values about data n_{c_i}, n and ℓ_{max}, the mappers output the pair (key,value) as follow: the mapper emits also 1 for the class label c (i.e., $(class.c, 1)$), 1 for the term n (i.e., $(notation.n, 1)$) and the number of items in the sequence for the term ℓ_{max} (i.e., $(notation.\ell_{max}, \ell_s)$).

Example 2. Given the pair $(\langle baadg \rangle, c_1)$ from our toy data set (Table 1). The mapper splits the sequence $\langle baadg \rangle$ into distinct items b, a, d and g. Then, the mapper forms the key value pairs: (b, c_1), (a, c_1), (d, c_1), (g, c_1), $(class.c_1, 1)$, $(notation.n, 1)$ and $(notation.\ell_{max}, 5)$.

The reducer forms an aggregation function and outputs its results into two files: *Singleton Rules* and *Statistics* depending on the key as follows:

- In case the key is a class label $class.c_i$ or the term $notation.n$, the reducer sums up the values associated with this key. Then, it outputs them to the *Statistics* file.
- In case the key is the term $notation.\ell_{max}$, the reducer selects the maximum values associated with the key $notation.\ell_{max}$. Then, it outputs them to the *Statistics* file.
- In case the key is an item e; $e \in \mathcal{I}$, the reducer computes the frequency of each class label appearing in its value. Then, it outputs the sequence and its support in each $\mathcal{D}_{c_i}$ to the *Singleton Rules* file.

Example 3. If the reducer receives the following (key,value) pairs from the mapper: (f, c_2), (g, c_1), (g, c_2), (g, c_2), $(notation.\ell_{max}, 5)$, $(notation.\ell_{max}, 4)$, $(notation.\ell_{max},$ $5)$, $(notation.\ell_{max}, 6)$, $(notation.n, 1)$, $(notation.n, 1)$, $(notation.n, 1)$, $(notation.n, 1)$, $(notation.c_2, 1)$ and $(notation.c_2, 1)$. Then, it results $(\langle f \rangle, f_{c_2}(\langle f \rangle) = 1)$, $(\langle g \rangle, f_{c_1}(\langle g \rangle) = 1 \; f_{c_2}(\langle g \rangle) = 2)$, $(class.c_2, 2)$, $(notation.n, 4)$ and $(notation.\ell_{max}, 6)$.

Given the output of the job **"Generating Statistics and Singleton Rules"**, we compute the rest of the statistics about the data consisting of: the number of classes in $\mathcal{C}$ (i.e., j) and the number of items in $\mathcal{I}$ (i.e., m).

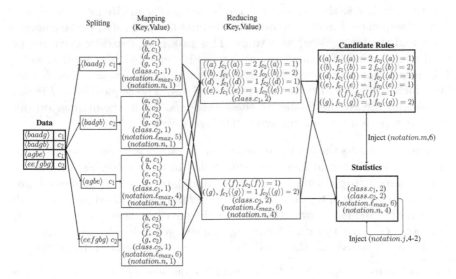

Fig. 3. Generating statistics and singleton rules job

The term j is computed by counting the number of lines in *Statistics* file then subtracting 2 from the result which represents the two lines of the terms *notation.ℓ_{max}* and *notation.n*. Then, we inject the value of j into the *Statistics* file. The term m is computed by counting the number of lines in *Singleton Rules* file and we inject it into the *Statistics* file. Figure 3 details the job *"Generating Statistics and Singleton Rules"*.

The job *"Pruning Singleton Rules"* prunes the candidate rules which are generated after applying the job *"Generating Statistics and Singleton Rules"*. This job consists only of mappers. The rules are distributed to available mappers. The basic statistics of our data (i.e., *Statistics* file) are passed as a parameter to each mapper by using the distributed cache. Then, the mapper computes the level of each rule and emits the rules having a positive level, i.e., *"interesting singleton rules"*, as a key and its level as a value.

4.2   Step II:

In the second step, *MiSeRe-Hadoop* mines all sequential classification rules having a sequence with more than one item in the body. *MiSeRe-Hadoop* iteratively repeats this step until the algorithm is stopped by the user manually at some point in time. The *Step II* can be efficiently achieved in two MapReduce jobs.

The job *"Generating Candidates"* employs an instance-based randomized strategy to generate candidate sequences from the data. This strategy can generate exactly the same candidate sequence several times. To avoid this redundancy problem, we define two kinds of mappers as follows:

- The set of first mappers take data as an input and generates new candidate sequences as an output. When the data is distributed to these mappers, each mapper randomly chooses one sequence **s** from each subset of data. From this chosen sequence **s**, the mapper randomly removes some items to generate a new subsequence. Finally, the mapper outputs the generated subsequence as a key and the term *"New"* as a value. This task is repeated by each mapper until a fixed number of candidate sequences is generated from each subset.
- The second set of mappers take the sequences from *Previous Candidates*[1] file which is generated from this step iteratively. The sequences in *Previous Candidates* file are distributed to these mappers. Then, the mapper outputs the sequence as a key and the term *"Old"* as a value.

The reducer filters the sequences based on its value i.e., it only outputs those sequences which do not include the term *"Old"* in their value. Figure 4 details the job *"Generating Candidates"*. The candidate sequences generated from this job are then copied to the *Previous Candidates* file to avoid generating the same candidate sequences from the data in the next iteration.

The job *"Generating and Pruning Rules"* generates the sequential rules from the data based on the candidate sequences generated from the job

---

[1] This file keeps a copy of all the candidate sequences generated from the job *"Generating Candidates"* in each iteration.

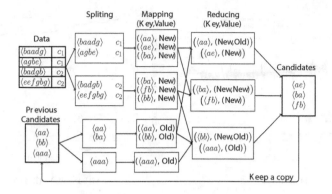

Fig. 4. Generating candidates job

"Generating Candidates". The data is distributed to available mappers while the candidates are passed as a parameter to each mapper by using the distributed cache. Each mapper takes a pair $(s, c) \in \mathcal{D}$ as input. Then, each candidate sequence is checked against the input sequence, if it is a sub-sequence of the input sequence, a key value pair is emitted where the key is the candidate sequence and the value is the class label. The reducer then generates the sequential classification rules by aggregating the frequency of the sequence in each class. Finally, the reducer computes the level of the rule and emits only the rules having a positive level, i.e., *"interesting rules"*, as a key and its level as a value.

5   Experiments

In this section, we empirically evaluate our approach on two real big datasets. *MiSeRe-Hadoop* is implemented in JAVA. The experiments are performed on a cluster of 6 computers, each with 64GB memory. One machine actes as the Hadoop master node, while the other five machines acte as worker nodes. The experiments are designed to discuss the following two points: the scalability of *MiSeRe-Hadoop* and the diversity of the mined rules with *MiSeRe-Hadoop*.

Datasets. We use two real-world datasets for our experiments. The first dataset is a large marketing database from the French Telecom company Orange containing sequential information about the behavior of **473 898 customers**. We use this dataset for predicting their propensity to churn. Each sequence represents a time-ordered set of actions (or events) categorized into two parts: (1) history of interaction between the customer and the LiveBox[2], e.g., changing the channel, rebooting the router, etc., (2) state of the box such as sending the temperature of box etc. These customers are classified into **2** classes. The first class includes **159 229** customers who terminated their contract with the company.

---

[2] Orange Livebox is an ADSL wireless router available to customers of Orange's Broadband services in several countries.

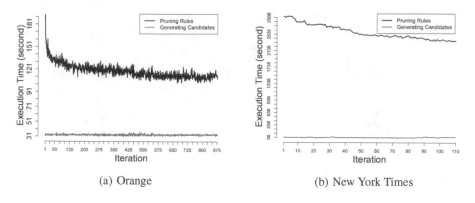

(a) Orange   (b) New York Times

Fig. 5. The scalability results for each iteration in Step 2.

The second class consists of **314 669** customers who still have the contract. This data set contains **433** distinct actions, the longest sequence is a customer having **23 759** actions while the median length of sequences is **192** actions. Our second dataset is the New York Times corpus (NYT) [18] which consists of **53 267 584** sentences from **1.8 million** articles published between 1987 and 2007. We treat each sentence as an input sequence with each word (token) as an item. All the sentences were stemmed and lemmatized using WordNet lemmatizer. These sentences are classified into **34** classes such as News, Sports, Health, etc. This data set contains **965 782** distinct words, the longest sequence is a sentence having **2 381** words while the median length of sequences is **10** words. The difference between Orange and NYT dataset is that the former does not contain a large number of sequences however each sequence contains too many number of actions creating longer sequences. While the latter consists of huge number of sequences with lesser number of words in the sequences.

Scalability. In this experiment, we explore the scalability of each job in *MiSeRe-Hadoop*. We run *MiSeRe-Hadoop* for **48 h** over *Orange* dataset. *MiSeRe-Hadoop* firstly runs the job *"generating statistics and singleton rules"* which takes **43 s** and generates **433 candidate singleton rules**. Then, *MiSeRe-Hadoop* filters these rules over the second job *"pruning singleton rules"* which takes **40 s** and returns **268 singleton interesting rules**. Then, *MiSeRe-Hadoop* passes to the second step where it iteratively repeats the two jobs *"generating candidates"* and *"generating and pruning rules"* until it is stopped after **48 h**. In each iteration, *MiSeRe-Hadoop* generates at most **2000** distinct candidate sequences then prunes them. *MiSeRe-Hadoop* repeats this step **877 times** during **48 h**. Figure 5 (a) shows the execution times of the two jobs *"generating candidates"* and *"generating and pruning rules"* over *Orange* dataset. The average execution time of the job *"generating candidates"* is **33.4 s** while for the job *"generating and pruning rules"* it is **118.51 s**.

We also run *MiSeRe-Hadoop* for **192 h** over *NYT* dataset. The job *"generating statistics and singleton rules"* takes **183 s** and generates **965 782 candidate**

singleton rules. Then, the job *"pruning singleton rules"* filters these rules which takes **47 s** and returns **7 092 singleton interesting rules**. Then, *MiSeRe-Hadoop* iteratively repeats **110 times** the two jobs *"generating candidates"* and *"generating and pruning rules"* during **192 h**. In each iteration, *MiSeRe-Hadoop* generates at most **2000** distinct candidate sequences then prunes them. Figure 5(b) shows the execution times of the two jobs *"generating candidates"* and *"generating and pruning rules"* over *NYT* dataset. The average execution time of the job *"generating candidates"* is **43.67 s** while for the job *"generating and pruning rules"* it is **2 528.4 s**.

From Fig. 5, it can be noticed that the job *"generating and pruning rules"* is the only job which takes the most time in the pipeline of *MiSeRe-Hadoop*. The performance of this job is based on two criteria. The first one is the number of candidate sequences which are generated form the job *"generating candidates"*. For this reason, the execution time of the job *"generating and pruning rules"* for NYT data set is more stable and fixed around **2 528.4 s** as the number of candidates are more stable and fixed around 2000 new candidates sequences (see Fig. 6). On the other hand, for *Orange* dataset, the number of candidate sequences is lesser at each iteration, for this reason the execution time of the job *"generating and pruning rules"* is decreasing at each iteration. The second criteria which effects on the performance of the job *"generating and pruning rules"* is the size of the dataset. For this reason, the execution time of this job over *NYT* dataset takes 20 times more than the execution time of the same job over *Orange* dataset because the size of *NYT* data set is larger than the *Orange* data set.

Diversity. In this section, we study the diversity of the mined rules by *MiSeRe-Hadoop*. The maximum number of candidate sequences to be generated by the job *"generating candidates"* at each iteration over the two datasets was set to 2000. The main goal of these experiments are as follows: (1) how many candidate sequences and interesting rules are generated by *MiSeRe-Hadoop* at each itera-

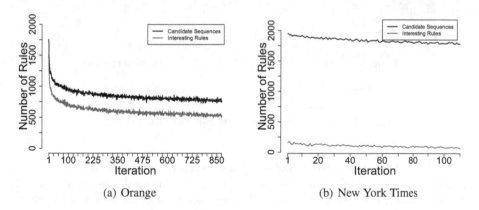

(a) Orange   (b) New York Times

Fig. 6. Number of candidate sequences and rules generated over each iteration in Step 2.

tion, (2) the number of candidate sequences and interesting rules generated by
MiSeRe-Hadoop having n-items and (3) what is the relation between the number
of items in the body of the rules and its value for the level criterion.

Figure 6 shows the number of candidate sequences generated and the interesting rules found over each iteration. From Fig. 6(a), it can be observed that
for *Orange* dataset, at each iteration around **70%** of the candidate sequences
are interesting rules which implies that these rules were easier to generate. On
the contrary, for *NYT* dataset, we have **text data** where *MiSeRe-Hadoop* has
difficulty in finding interesting rules at each iteration as the percentage of finding
rules from candidate sequence is very low, which is around **5%** (see Fig. 6(b)).
The total number of interesting rules generated from *Orange* dataset during **877
iterations** is **534 460** rules, while the total number of rules generated from *NYT*
dataset during **110 iterations** is 11 473 rules.

In Fig. 7, we study the relation between the length and the value of level of
interesting rules. We plot the length of the interesting rule against its level. For
Orange dataset, we can extract interesting rules with up to **72 items** while for
NYT, *MiSeRe-Hadoop* extracts interesting rules with at most **4 items (words)**.
From Fig. 7, it can be concluded that the level value of the shorter rules is larger
than the longer ones, meeting the consensus: ***"Simpler and shorter rules are
more probable and preferable"***.

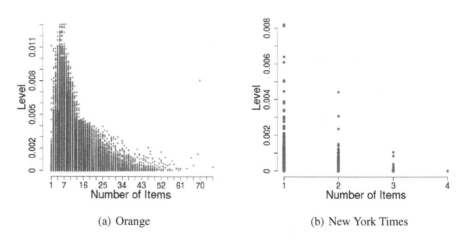

(a) Orange   (b) New York Times

Fig. 7. Length of rules against its value of level criterion.

We also study the relation between the length of candidate sequences and
mined rules. Figure 8(a) shows the number in logscale of candidate sequences and
rules mined over the step 2 of *MiSeRe-Hadoop* from *Orange* dataset. For this plot,
we limit the visualization up to **30 items** as **99.75%** of mined rules have a body
with maximum **30 items**. In *MiSeRe-Hadoop*, the job *"generating candidates"*

generates more candidate sequences with lesser number of items as compared to the longer ones because short rules are more probable and preferable. For this reason, the candidate sequences having items less than 8 represent 85% of all the generated candidate sequences. For *NYT* data set, *MiSeRe-Hadoop* generates most of the candidate sequences with 2-items because it is a text data set and has 965 782 distinct words. Thus, it can generate upto $(965782)^2$ distinct candidate sequences with 2-items. From Fig. 8(b), It can be observed that generating sequential classification rules form text data is not easy task. For example, *MiSeRe-Hadoop* generates **11 673 candidate sequences** with three words and finally finds just **157** interesting rules over these candidates.

Figures 6, 7 and 8 highlight that the randomized strategy employed in *MiSeRe-Hadoop* allows us to mine interesting rules with diversity which is highly dependent on the data.

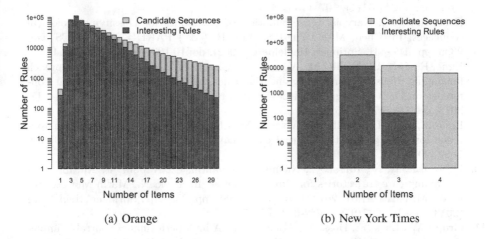

(a) Orange   (b) New York Times

Fig. 8. Length of candidate sequences and rules against the number of rules.

6   Conclusion and Future Work

This paper focuses on the important problem of mining sequential rule patterns from large scale sequential data. We propose *MiSeRe-Hadoop*, a scalable algorithm for mining sequential rules in MapReduce. *MiSeRe-Hadoop* is a parameter-free algorithm that efficiently mines interesting rules. The empirical experiments show that: (1) our method is scalable for large scale sequential data analysis and (2) the randomized strategy employed in *MiSeRe-Hadoop* allows us to mine interesting rules with diversity. As future work, we plan to extend our approach for multi label sequential data set. On the other hand, we are also planning on using the mined sequential classification rules by *MiSeRe-Hadoop* as an input of the classification method to build a classifier.

References

1. Anastasiu, D.C., Iverson, J., Smith, S., Karypis, G.: Big data frequent pattern mining. In: Aggarwal, C.C., Han, J. (eds.) Frequent Pattern Mining, pp. 225–259. Springer, Cham (2014). doi:10.1007/978-3-319-07821-2_10
2. Andrews, G.R.: Foundations of Multithreaded, Parallel, and Distributed Programming. University of Arizona, Wesley (2000)
3. Beedkar, K., Berberich, K., Gemulla, R., Miliaraki, I.: Closing the gap: sequence mining at scale. ACM Trans. Database Syst. **40**(2), 8:1–8:44 (2015)
4. Chen, C.C., Tseng, C.Y., Chen, M.S.: Highly scalable sequential pattern mining based on mapreduce model on the cloud. In: 2013 IEEE International Congress on Big Data, pp. 310–317 (2013)
5. Cong, S., Han, J., Padua, D.: Parallel mining of closed sequential patterns. In: Proceedings of the Eleventh ACM SIGKDD International Conference on Knowledge Discovery in Data Mining, pp. 562–567. ACM (2005)
6. Dean, J., Ghemawat, S.: Mapreduce: simplified data processing on large clusters. Commun. ACM **51**(1), 107–113 (2008)
7. Deshpande, M., Karypis, G.: Evaluation of techniques for classifying biological sequences. In: Chen, M.-S., Yu, P.S., Liu, B. (eds.) PAKDD 2002. LNCS, vol. 2336, pp. 417–431. Springer, Heidelberg (2002). doi:10.1007/3-540-47887-6_41
8. Egho, E., Gay, D., Boullé, M., Voisine, N., Clérot, F.: A parameter-free approach for mining robust sequential classification rules. In: 2015 IEEE International Conference on Data Mining, ICDM 2015, Atlantic City, NJ, USA, November 14–17, 2015, pp. 745–750 (2015)
9. Egho, E., Gay, D., Boullé, M., Voisine, N., Clérot, F.: A user parameter-free approach for mining robust sequential classification rules. Knowl. Inform. Syst. **52**, 1–29 (2016)
10. Egho, E., Jay, N., Raïssi, C., Nuemi, G., Quantin, C., Napoli, A.: An approach for mining care trajectories for chronic diseases. In: Peek, N., Marín Morales, R., Peleg, M. (eds.) AIME 2013. LNCS, vol. 7885, pp. 258–267. Springer, Heidelberg (2013). doi:10.1007/978-3-642-38326-7_37
11. Gropp, W., Lusk, E., Doss, N., Skjellum, A.: A high-performance, portable implementation of the MPI message passing interface standard. Parallel Comput. **22**(6), 789–828 (1996)
12. Guralnik, V., Karypis, G.: Parallel tree-projection-based sequence mining algorithms. Parallel Comput. **30**(4), 443–472 (2004)
13. Holat, P., Plantevit, M., Raïssi, C., Tomeh, N., Charnois, T., Crémilleux, B.: Sequence classification based on delta-free sequential patterns. In: ICDM 2014, pp. 170–179 (2014)
14. Itkar, S., Kulkarni, U.: Distributed sequential pattern mining: a survey and future scope. Int. J. Comput. Appl. **94**(18), 28–35 (2014)
15. Jorge, A.M., Azevedo, P.J., Pereira, F.: Distribution rules with numeric attributes of interest. In: Fürnkranz, J., Scheffer, T., Spiliopoulou, M. (eds.) PKDD 2006. LNCS, vol. 4213, pp. 247–258. Springer, Heidelberg (2006). doi:10.1007/11871637_26
16. Lesh, N., Zaki, M.J., Ogihara, M.: Mining features for sequence classification. In: ACM SIGKDD 1999, pp. 342–346 (1999)
17. Qiao, S., Li, T., Peng, J., Qiu, J.: Parallel sequential pattern mining of massive trajectory data. Int. J. Comput. Intell. Syst. **3**(3), 343–356 (2010)

18. Sandhaus, E.: The New York Times Annotated Corpus. Linguistic Data Consortium, Philadelphia (2008)
19. She, R., Chen, F., Wang, K., Ester, M., Gardy, J.L., Brinkman, F.S.L.: Frequent-subsequence-based prediction of outer membrane proteins. In: ACM SIGKDD 2003, pp. 436–445 (2003)
20. Tan, P., Kumar, V.: Discovery of web robot sessions based on their navigational patterns. Data Min. Knowl. Discov. **6**(1), 9–35 (2002)
21. Tseng, V.S., Lee, C.: CBS: a new classification method by using sequential patterns. In: SDM 2005, pp. 596–600 (2005)
22. Wang, J., Han, J.: BIDE: efficient mining of frequent closed sequences. In: ICDE 2004, pp. 79–90 (2004)
23. Wu, X., Zhu, X., Wu, G.Q., Ding, W.: Data mining with big data. IEEE Trans. Knowl. Data Eng. **26**(1), 97–107 (2014)
24. Xing, Z., Pei, J., Keogh, E.J.: A brief survey on sequence classification. SIGKDD Explor. **12**(1), 40–48 (2010)
25. Zaki, M.: Sequence mining in categorical domains: incorporating constraints, pp. 422–429 (2000)
26. Zaki, M.J.: Parallel sequence mining on shared-memory machines. J. Parallel Distrib. Comput. **61**(3), 401–426 (2001)
27. Zhou, C., Cule, B., Goethals, B.: Itemset based sequence classification. In: ECML/PKDD 2013, pp. 353–368 (2013)

An Efficient Map-Reduce Framework to Mine Periodic Frequent Patterns

Alampally Anirudh[1]($\boxtimes$), R. Uday Kiran[2], P. Krishna Reddy[1], M. Toyoda[2], and Masaru Kitsuregawa[2,3]

[1] Kohli Center on Intelligent Systems, IIIT - Hyderabad, Hyderabad, India
alampally.anirudh@research.iiit.ac.in, pkreddy@iiit.ac.in
[2] Institute of Industrial Science, The University of Tokyo, Tokyo, Japan
{uday_rage,toyoda,kitsure}@tkl.iis.u-tokyo.ac.jp
[3] National Institute of Informatics, Tokyo, Japan

Abstract. Periodic Frequent patterns (PFPs) are an important class of regularities that exist in a transactional database. In the literature, pattern growth-based approaches to mine PFPs have be proposed by considering a single machine. In this paper, we propose a Map-Reduce framework to mine PFPs by considering multiple machines. We have proposed a parallel algorithm by including the step of distributing transactional identifiers among the machines. Further, the notion of *partition summary* has been proposed to reduce the amount of data shuffled among the machines. Experiments on Apache Spark's distributed environment show that the proposed approach speeds up with the increase in number of machines and the notion of *partition summary* significantly reduces the amount of data shuffled among the machines.

Keywords: Data mining · Periodic frequent pattern mining · Map-Reduce

1 Introduction

Periodic frequent pattern (PFP) mining extracts regularities from transactional databases (TDB). A PFP is an itemset which is both frequent and periodic. For example, PFP mining extracts the knowledge about how regularly a set of items are being purchased by the customers from the super-market TDB. An example of PFP is $\{Bread, Butter\}$ $[support = 10\%, periodicity = 1\,\text{h}]$. The preceding pattern demonstrates that both 'Bread' and 'Butter' are purchased in 10% of the transactions, and the maximum time interval between any two consecutive purchases containing both of these items is no more than an hour. The predictive behavior of PFPs could be used to improve the performance of several data mining based applications in the areas of customer relation management, inventory management, recommendation systems and so on.

In the literature, Tanbeer et al. [1] proposed a pattern-growth-based algorithm by considering a single machine. Several improvements to the approach

© Springer International Publishing AG 2017
L. Bellatreche and S. Chakravarthy (Eds.): DaWaK 2017, LNCS 10440, pp. 120–129, 2017.
DOI: 10.1007/978-3-319-64283-3_9

proposed in [1] have been investigated [2–4] by considering a single machine. A Map-Reduced framework to exploit the power of thousands of machines is proposed in [5]. Encouraged by power of Map-Reduce paradigm, researchers are making efforts to propose parallel algorithms under Map-Reduce framework. A Map-Reduce based parallel FP growth approach to extract frequent patterns has been proposed in [6]. In this paper, we proposed a parallel PFP mining approach under Map-Reduce framework.

The PFP mining approach requires the processing of transaction identifiers (tids) for computing the periodicity. We have proposed a parallel approach which contains two Map-Reduce phases similar to [6]. In each phase the step to manage tids is integrated. Further, we have developed an improved approach by proposing the notion of *partition summary* in which instead of processing tid-list, the corresponding summary information is processed. Experimental results on three real-world datasets show that the proposed approach speeds up with the increase in number of machines and the notion of *partition summary* reduces the amount of data shuffled.

The rest of the paper is organized as follows. Section 2 discusses background. Proposed approach is discussed in Sect. 3. Performance evaluation is reported in Sect. 4. Finally, Sect. 5 concludes the paper.

2   Background

2.1   Mining Periodic-Frequent Patterns on a Single Machine

Model of PFPs [1]: Let $I = \{i_1, i_2, \cdots, i_n\}$, $1 \leq n$, be a set of items. A set $X = \{i_j, \cdots, i_k\} \subseteq I$ is called a pattern. A transaction $t = (tid, Y)$ is a tuple, where tid represents a TID and Y is a pattern. A transactional database (TDB) over I is a set of transactions, i.e., $TDB = \{t_1, t_2, \cdots, t_m\}$, $m = |TDB|$. If $X \subseteq Y$, it is said that t contains X and such tid is denoted as tid_j^X. Let $TID_X = \{tid_j^X, \cdots, tid_k^X\}$, be the set of all tids where X occurs in TDB. The **support** of a pattern X is the number of transactions containing X in TDB, denoted as $Sup(X)$. Therefore, $Sup(X) = |TID^X|$. Let tid_i^X and tid_j^X be two consecutive tids where X appeared in TDB. The **period** of a pattern X is the number of transactions between tid_i^X and tid_j^X. Let $P^X = \{p_1^X, p_2^X, \cdots, p_r^X\}$, $r = Sup(X) + 1$, be the complete set of periods of X in TDB. The **periodicity** of a pattern X is the maximum difference between any two adjacent occurrences of X, denoted as $Per(X) = \max(p_1^X, p_2^X, \cdots, p_r^X)$. A pattern X is a PFP if $Sup(X) \geq minSup$ and $Per(X) \leq maxPer$, where $minSup$ and $maxPer$ represent the user-specified thresholds on minimum *support* and maximum *periodicity* respectively.

The existing periodic pattern growth (PF-growth) algorithm [1] accepts a TDB, $minSup$ and $maxPer$ as inputs and outputs a complete set of PFPs. The structure of periodic-frequent pattern tree (PF-tree) consists of a PF-list and a prefix tree. Here, the prefix tree of PF-tree explicitly maintains the tids for each occurrence of the pattern only at the tail-node of every branch unlike FP-tree which maintains support in every node. The algorithm consists of two database scans.

(1) Finding one sized PFPs: In the first database scan, the PF-growth scans the entire database and discovers 1-patterns by computing support and periodicity values for each item. The final PF-list is generated after pruning the items that have failed to satisfy the *minSup* and *maxPer* constraints and sorting the items in the increasing order of support.

(2) Construction of PF-tree: In the second database scan, the PF-tree is constructed by inserting the transactions according to PF-list order with tail-node carrying the tid. After completing two database scans PF-tree is constructed.

Mining of PFPs: The mining of PFPs start by constructing prefix tree (PT_i) for the last item i in the PF-list as an initial suffix item. For each item j in PT_i, all of its nodes' tid-list is aggregated to derive the tid-list of the pattern ij. If ij is a PFP, then j is considered to be periodic-frequent in PT_i. The conditional tree is constructed by choosing every periodic-frequent item j in PT_i, and is mined recursively to discover the patterns. After finding all PFPs for a suffix item i, it is pruned from the original PF-tree and the corresponding nodes' tid-lists are pushed to their parent nodes. The above steps are repeated until the PF-list becomes NULL.

2.2   Mining PFPs with Period Summary

The concept of *period summary* [4] was introduced to mine PFPs on a single machine, where only the summary information is stored instead of the tids in the tail-node. It has been shown that the notion of *period summary* results in lower memory consumption.

2.3   Map-Reduce Framework

Map-Reduce [5] framework has been proposed to enable the processing of large datasets on a large cluster of commodity machines. Users specify the problem as a sequence of Map-Reduce steps. In each Map-Reduce step, the Map function processes key-value pairs and the reduce function merges all the values associated values with the same key.

2.4   Parallel FP-growth

Li et al. [6] proposed parallel FP-growth to extract frequent patterns using two Map-Reduce phases. The first phase constructs the F-list, which contains 1-sized frequent patterns. For each transaction, the mapper outputs key-value pairs as $\langle item, 1 \rangle$ and reducer sums up all the ones for each item to count the corresponding support. The second phase constructs independent local FP-trees on different machines. For each transaction, all the sub-patterns are generated and mapper outputs the key-value pairs as $\langle partition\text{-}id, sub\text{-}pattern \rangle$ and reducer aggregates the transactions and construct local FP-trees. Frequent patterns are extracted at each worker by mining the local FP-trees.

Table 1. A running example of a transactional database

tid	Items	tid	Items	tid	Items	tid	Items	tid	Items
1	bcdf	4	abde	7	acdef	10	cd	13	de
2	abdef	5	e	8	abc	11	bcde	14	ae
3	b	6	bc	9	bf	12	abc	15	e

3 Proposed Approaches

3.1 Parallel Periodic Frequent Pattern Growth (PPF-growth)

Both FP-growth (Sect. 2.4) and PF-growth (Sect. 2.1) mining are recursive pattern-growth approaches. The difference comes in the tree structure where additional tid information is processed which is required for computing the periodicity in PF-growth. The proposed approach consists of initial step, two Map-Reduce phases and the mining step.

(1) Initialization: Initially, the TDB is segmented into multiple partitions. As an example, consider Table 1 as a TDB which is divided into two partitions with $minSup = 5$ and $maxPer = 4$. Here, 0th and 1st partitions contain transactions with tids between 1–8 and 9–15 respectively.

(2) Map-Reduce phase 1 (first database scan): The phase is depicted in Fig. 1(a). In this step, parallel periodic frequent pattern list (PPF-list) is constructed by calculating the support and periodicity values for each item. The Map-Reduce steps are as follows.

Map: For each item in a transaction, it outputs key-value pairs where key is the *item* and value is the *tid* of the current transaction ($\langle item, tid \rangle$).

Reduce: It groups all the tids of each item in a tid-list (Algorithm 1). The tid-list is sorted and then its support and periodicity are computed (Fig. 2(a)–(c)).
  The final PPF-list is obtained by filtering the items which do not satisfy the *minSup* and *maxPer* thresholds. After the master machine obtains the PPF-list, the items are sorted in decreasing order of their supports and are assigned a rank (to simplify the distribution of transactions using a hash function). The most frequent item is assigned a rank of 0, the second most frequent item is assigned a rank of 1 and so on.

(3) Map-Reduce phase 2 (second database scan): The phase is depicted in Fig. 1(b). In this step, parallel periodic frequent pattern trees (PPF-trees) are constructed on each machine. The Map-Reduce steps are as follows.

Map: For each transaction, the items which are not present in PPF-list are filtered, translated into their ranks and are sorted in ascending order. Then all the sub-patterns (n sub-patterns will be generated) are extracted and are assigned to a partition based on a simple hash function $rank[item]\%numOfPartitions$. Here, $numOfPartitions$ is the number of partitions available and *item* is the

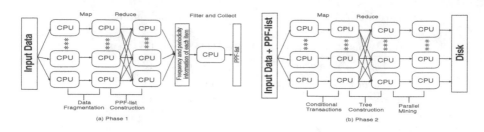

Fig. 1. Two phases of parallel periodic frequent pattern mining

---

Algorithm 1. PPF-listConstruction (TDB)

Procedure: Map($key = null, value = TDB_i$)
    for each transaction $t_{cur} \in TDB_i$ **do**
        for each item $itint_{cur}$ **do**
            Output (it, tid) // tid is the current transaction id
Procedure: Reduce($key = it, value = TID\text{-}list$)
    Sort $TID\text{-}list$ and initialize $id_l = 0$, $sup = 0$ and $per = 0$
    for each $tid \in TID\text{-}list$ **do**
        Set $sup+ = 1$, $per = max(per, tid - id_l)$ and $id_l = tid$

---

last item in a sub-pattern. The hash function gives a partition-id for which the pattern is responsible for further computation. Each sub-pattern is outputted as a key-value pair, with key as the *partition-id* and value as a tuple of *sub-pattern* and current *tid* ($\langle partition\text{-}id, (sub\text{-}pattern, tid) \rangle$).

Reduce: Independent local PPF-trees are constructed by inserting all the sub-patterns into the tree in the same order as the PPF-list with tid stored only in the tail-node of the branch. The process of tree construction (Algorithm 2) is the same as the construction of PF-tree [1]. Trees constructed on two different machines are shown in Fig. 2(d), (e).

(4) Mining of PPF-trees: Note that for any suffix item, the complete conditional tree information is available in the corresponding machine. So, during conditional pattern building, communication is not required between the machines and PFPs can be mined in parallel. Parallel mining of PFPs is similar to the

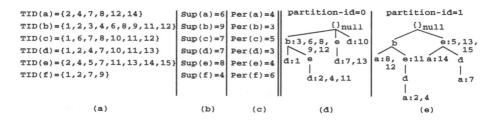

Fig. 2. Construction of PPF-list and prefix tree (a)–(c) Finding support and periodicity for each item (d) tree at partition-id 0 (e) tree at partition-id 1

mining process of PF-growth but worker machine processes only for those suffix items for which it is responsible for computation. This is checked by using the hash function $rank[it]\%numOfPartitions$. Here, it is the chosen suffix item. Mining for a suffix item (in increasing order of support) is done only if the output of hash function is equal to the partition-id of that machine. In the existing approach, the mining process of 'd' occurs only after the mining process of 'a' is completed. Whereas in the proposed approach, both the processes happen in parallel due to which the time taken to extract the PFPs is reduced.

Algorithm 2. PPF-treeConstructionMining (TDB, PPF-list)

Procedure: Map($key = null, value = TDB_i$) // TDB_i is the segment of TDB
    for each transaction $t_{cur} \in TDB_i$ **do**
        filter and sort the elements in t_{cur} which are not in PPF-list
        for $j = (|t_{cur}| - 1)$ to 0 **do**
            partition-id = getPartition($t_{cur}[j]$)
            if H does not contain partition-id **then**
                Output(partition-id; ($t_{cur}[0:j]$, tid)) // tid is the current transaction id
Procedure: Reduce($key = partition\text{-}id, value = transactions$)
    Initialize PPF-tree, T
    for t_{cur} in transactions **do**
        for it in t_{cur} **do**
            if T does not have child it **then**
                Create a new child node it and link it with the parent
            Traverse to the child it
        Add the current transaction id to the tid-list at the tail node of the transaction t_{cur}
Procedure: Map($key = partition\text{-}id, value = PPF\text{-}tree$) // Parallel Mining
    for each suffix item i in PPF-list **do**
        if current partition-id is responsible for item i **then**
            Generate PT_i and CT_i and mine recursively in CT_i for patterns with suffix i

---

3.2   PPF-growth Using Partition Summary

A straight-forward approach for the first phase was discussed where large tid-lists are shuffled across machines. In [4], the notion of *period summary* is used to reduce the memory consumption by storing the interval information instead of tids in the tail-node of the PF-tree. Similarly, instead of tid-list for each item, only interval information can be processed in a partition. The summarized interval information concerning to an item for the given partition is called *partition summary*. We proposed an improved Map-Reduce based PF mining approach based on partition summary, which is defined as follows.

Definition 1. A partition summary (PS) captures the interval information of the item occurrences, the periodicity and support of respective item within that interval. That is, $PS = \langle tid_i, tid_j, per, sup \rangle$, where tid_i and tid_j, represents the first and last tids of that interval respectively, per is the periodicity and sup is the support of a pattern within the interval whose tids are within tid_i and tid_j.

The approach with PS is as follows: With PS, the phase 1 of PPF-growth proposed in the preceding section is modified. The map function is the same but instead of a reduce function, the *combine* function is applied which consists of three steps: initialization, intra-partition and inter-partition merging steps. During the initialization step, $PS = \langle 0, 0, 0, 0 \rangle$ is initialized for each item. During the intra-partition merging step, the interval information in PS is updated for each item by iterating over its tids as explained in Algorithm 3. All the PS for an item are merged into one PS during inter-partition merging step as explained in Algorithm 3. Figure 3(a)–(c) shows the construction of PPF-list after scanning first transaction, second transaction and the entire shard of data assigned to partition 1. Similarly, Fig. 3(d)–(f) shows the construction of PPF-list in partition 2. Figure 3(g) represents the final PPF-list constructed by merging the PPF-lists built on partition 1 and 2. The elements striked off are the ones which did not satisfy $minSup$ and $maxPer$ constraints.

Algorithm 3. PPF-list with Partition Summaries

Procedure: Combine($key = it, value = TID\text{-}list$)
  Initialize $PS = \langle 0, 0, 0, 0 \rangle$ // (first tid, last tid, periodicity, support)
  intra-partition: generatingSummaries (PS, $TID\text{-}list$)
    $PS[0] = TID\text{-}list[0]$
    for each tid $\in$ $TID\text{-}list$ **do**
      $PS[2] = max(PS[2], tid - PS[1])$, $PS[1] = tid$ and $PS[3] += 1$
  inter-partition: mergingSummaries ($PS_1, PS_2, \cdots, PS_3$)
    Group all PS_i into $PS\text{-}List$
    Sort $PS\text{-}List$ based on first tid
    $Final\text{-}PS = <PS\text{-}List[0][0], 0, 0, 0 >$
    for each $PS \in PS\text{-}List$ **do**
      $Final\text{-}PS[2] = max(Final\text{-}PS[2], PS[2], PS[0] - Final\text{-}PS[1])$
      $Final\text{-}PS[1] = PS[1]$ and $Final\text{-}PS[3] += PS[3]$
    return $Final\text{-}PS$

I	PS
b	<1,1,0,1>
c	<1,1,0,1>
d	<1,1,0,1>
f	<1,1,0,1>

(a)

I	PS
b	<1,2,1,2>
c	<1,1,0,1>
d	<1,2,1,2>
f	<1,2,1,2>
a	<2,2,0,1>
e	<2,2,0,1>

(b)

I	PS
b	<1,8,2,6>
c	<1,8,5,4>
d	<1,7,4,3>
f	<1,7,5,4>
a	<2,8,4,3>
e	<2,7,4,1>

(c)

I	PS
b	<9,9,0,1>
f	<9,9,0,1>

(d)

I	PS
b	<9,9,0,1>
f	<9,9,0,1>
c	<10,10,0,1>
d	<10,10,0,1>

(e)

I	PS
b	<9,12,2,3>
f	<9,9,0,1>
c	<10,12,1,3>
d	<10,13,3,2>
e	<11,15,4,2>
a	<12,14,2,2>

(f)

I	PS
b	<1,12,3,9>
d	<1,13,3,7>
a	<2,14,4,6>
c	<1,13,5,7>
f	<1,9,6,4>

(g)

Fig. 3. Construction of PPF-list using partition summaries

4   Performance Evaluation

We have conducted the experiments to evaluate the speed up performance of the proposed approach. The algorithm is written in Python using Apache Spark architecture and the experiments are conducted on Amazon Elastic Map-Reduce (EMR) cluster with each machine of 8 GB memory. The runtime in the experiments specifies the total execution time of a Spark job. We employed 3 real-world datasets for conducting experiments, Retail store [7] (88,162 transactions with 16,470 items), Twitter dataset [2] (43,200 transactions with 44,201 items) and Online store [8] (541,909 transactions with 2,603 items).

The experiments are conducted by increasing the number of machines from 1 to 16 against the total time consumed. Figure 4(a) shows variation of total time consumed with increase in number of machines for the Retail store dataset with $minSup = 0.05\%$ and $maxPer = 10\%$. As the number of machines is increased, the total time consumed decreases rapidly due to parallel computation. This shows that it is possible to improve performance with parallel computation. Here, the time taken for 1 machine is 282 s which is reduced to 72 s for 8 machines. The speedup can be computed as $(282/72)/8 = 48.95\%$. However, the algorithm reaches a saturation point for all the datsets and is obtained when 16 machines are used.

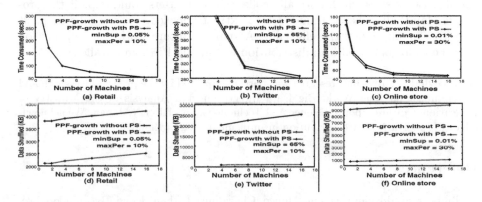

Fig. 4. Time Consumed and corresponding amount of data shuffled vs number of machines for different datasets at different $minSup$ and $maxPer$ values

Figure 4(b) shows the results for the twitter dataset by fixing $minSup = 65\%$ and $maxPer = 10\%$. It can be noticed it is not possible to extract patterns with one or two machines as it is a dense dataset and the main memory is not sufficient to carry out recursive computation of patterns from PPF-tree. The Map-Reduce model enables the extraction of patterns by splitting the task among multiple machines and processing in parallel. Figure 4(c) also shows a similar performance for Online store dataset.

Results with Partition Summary: Figure 4(d)–(f) shows how the amount of data shuffled among the machines varies with and without using the notion of PS. It can be observed that there is a significant reduction of data shuffled with PS for all datasets. Figure 4(a)–(c) shows the corresponding improvement in total time consumed with PS because of the reduced amount of data shuffled. Figure 4(a) shows no improvement in time (overlapping lines) as it is a sparse dataset (Retail) and optimization using PS did not have much effect. Overall, the reason for minor improvement with PS is that in the experimental environment of Amazon EMR the data transfer speeds among the machines is very fast. So, the influence of reduction in data shuffled is not significant. However, we believe that in the slow LAN, and WAN environments, the effect of reduction in data shuffled will lead to significant improvement in total time consumed.

5   Conclusion

In this paper, we presented a parallel periodic frequent pattern extraction approach with Map-Reduce. Further, the notion of *partition summary* was introduced to reduce the amount of data shuffled among the machines. Experiments on massive datasets show that the proposed algorithm speeds up with the increase in number of machines. The modern e-Commerce applications and social network sites which normally collect huge datasets could exploit the proposed Map-Reduce based framework to extract knowledge of periodic-frequent patterns for improving the efficiency. As part of future work, we will develop algorithms for balanced load distribution for mining PFPs in parallel.

Acknowledgment. This research was partly supported by the program Research and Development on Real World Big Data Integration and Analysis of the Ministry of Education, Culture, Sports, Science and Technology, and RIKEN, Japan. We acknowledge K. Amulya for her contribution in implementation of the idea.

References

1. Tanbeer, Syed Khairuzzaman, Ahmed, Chowdhury Farhan, Jeong, Byeong-Soo, Lee, Young-Koo: Discovering periodic-frequent patterns in transactional databases. In: Theeramunkong, Thanaruk, Kijsirikul, Boonserm, Cercone, Nick, Ho, Tu-Bao (eds.) PAKDD 2009. LNCS (LNAI), vol. 5476, pp. 242–253. Springer, Heidelberg (2009). doi:10.1007/978-3-642-01307-2_24
2. Kiran, R.U., Shang, H., Toyoda, M., Kitsuregawa, M.: Discovering recurring patterns in time series. In: EDBT, pp. 97–108 (2015)
3. Amphawan, K., et al.: Mining periodic-frequent itemsets with approximate periodicity using interval transaction-ids list tree. In: WKDD, pp. 245–248 (2010)
4. Anirudh, A., Kiran, R.U., Reddy, P.K., Kitsuregawa, M.: Memory efficient mining of periodic-frequent patterns in transactional databases. In: IEEE, SSCI, pp. 1–8
5. Dean, J., Ghemawat, S.: Mapreduce: simplified data processing on large clusters. Commun. ACM **51**(1), 107–113 (2008)

6. Li, H., et al.: Pfp: parallel fp-growth for query recommendation. In: Proceedings of the 2008 ACM Conference on Recommender Systems, RecSys 2008, pp. 107–114. ACM (2008)
7. Brijs, T., et al.: Using association rules for product assortment decisions: a case study. In: Knowledge Discovery and Data Mining, pp. 254–260 (1999)
8. Chen, D., et al.: Data mining for the online retail industry: a case study of rfm model-based customer segmentation using data mining. JDMCSM **19**(3), 197–208 (2012)

MapReduce-Based Complex Big Data Analytics over Uncertain and Imprecise Social Networks

Peter Braun[1], Alfredo Cuzzocrea[2,3], Fan Jiang[1], Carson Kai-Sang Leung[1(✉)], and Adam G.M. Pazdor[1]

[1] University of Manitoba, Winnipeg, MB, Canada
kleung@cs.umanitoba.ca
[2] University of Trieste, Trieste, TS, Italy
alfredo.cuzzocrea@dia.units.it
[3] ICAR-CNR, Rende, CS, Italy

Abstract. With advances in technology, high volumes of valuable but complex data can be easily collected and generated from various sources in the current era of big data. A prime source of these complex big data is the social network, in which users are often linked by some interdependencies such as friendships and follower-followee relationships. These interdependencies can be uncertain and imprecise. Moreover, as the social network keeps growing, there are situations in which individual users or businesses want to find those popular (i.e., frequently followed) groups of users so that they can follow the same groups. In this paper, we present a complex big data analytic solution that uses the MapReduce model to mine uncertain and imprecise social networks for discovering groups of potentially popular users. Evaluation results show the efficiency and practicality of our solution in conducting complex big data analytics over uncertain and imprecise social networks.

1 Introduction and Related Work

With advancements in technology and the popularity of social networking sites (e.g., Facebook, Twitter [30,31]), high volumes of valuable but complex data [3,9,28] can be easily collected or generated from social networks [2,36]. In general, social networks are made of social entities (e.g., individuals, corporations, collective social units, or organizations) that are linked by some specific types of interdependencies (e.g., friendships, common interest, follower-followee relationships). A social entity is connected to another entity as friend, classmate, co-worker, team member, and/or business partner. For instance, Facebook users can create a personal profile, add other Facebook users as friends, exchange messages, and join common-interest user groups. The number of (mutual) friends may vary from one Facebook user to another. It is not uncommon for a user A to have hundreds or thousands of friends. Note that, although many of the Facebook users are linked to some other Facebook users via their mutual friendship (i.e., if a user A is a friend of another user B, then B is also a friend of A), there are situations in which such a relationship is not mutual. To handle these

© Springer International Publishing AG 2017
L. Bellatreche and S. Chakravarthy (Eds.): DaWaK 2017, LNCS 10440, pp. 130–145, 2017.
DOI: 10.1007/978-3-319-64283-3_10

situations, Facebook added the functionality of "follow", which allows a user to subscribe or follow public postings of some other Facebook users without the need of adding them as friends. So, for any user C, if many of C's friends followed some individual users or groups of users, then C might also be interested in following the same individual users or groups of users. Furthermore, the "like" button allows users to express their appreciation of content such as status updates, comments, photos, and advertisements.

As another instance, Twitter users can read the tweets of other users by "following" them. Relationships between social entities are mostly defined by following (or subscribing) each other. Each user (social entity) can have multiple followers, and follows multiple users at the same time. The follow/subscribe relationship between follower and followee is not the same as the friendship relationship (in which each pair of users usually know each other before they setup the friendship relationship). In contrast, in the follow/subscribe relationship, a user D can follow another user E while E may not know D in person. In this paper, $D \rightarrow E$ denotes the follow/subscribe (i.e., "following") relationship that D is following E.

Big data analytics of these complex big social networks computationally facilitates social studies and human-social dynamics in these networks, as well as designs and uses information and communication technologies for dealing with social context. In this paper, we focus on a particular class of social networks called *imprecise and uncertain social networks*. Here, the main challenges due to the fact that edges are weighted by an *existential probability* [1,17,24,33].

With the growing number of users of these social networking sites, big data analytics has become very useful in order to extract useful and actionable knowledge from enormous data repositories (e.g., [4]), with a plethora of applications. In this methodological context, several data mining algorithms and techniques (e.g., [7,13,19,22,23]) have been proposed over the past two decades. Many of them (e.g., [18,25,29]) have been applied to mine social networks (e.g., discovery of special events (e.g., [11]), detection of communities (e.g., [27,34]), sub-graph mining (e.g., [35]), as well as discovery of popular friends (e.g., [14,25]), influential friends (e.g., [26]) and strong friends (e.g., [32])).

To discover "following" relationships among the aforementioned social networking sites, we developed a serial algorithm called FoP-miner [12] in DaWaK 2014 to mine interesting "following" patterns from social networks. To speed up the mining process, we extended the FoP-miner algorithm to get a parallel and concurrent algorithm called ParFoP-miner [21] for mining interesting "following" patterns in parallel. Moreover, in order to deal with massive numbers of these "following" relationships that are embedded in big data, we also extended the FoP-miner algorithm to get a MapReduce-based algorithm called BigFoP [20] in DaWaK 2015 for big data analytics of social networks and discovery of "following" patterns. These three algorithms work well when mining *precise* social network, in which the social analysts have precise information regarding the interdependencies among the social entities.

However, there are real-life situations in which this information is limited or unavailable. For instance, due to privacy preserving purposes, this information is not fully revealed. As such, social analysts may need to rely on their experience, expertise, or other sources to determine the likelihood of the existence of some of these interdependencies among the social entities within a complex social network with uncertainty and imprecision. In response, our *key contribution* of the current paper is our proposal of a new big data analytics and mining solution, which uses the *MapReduce* model [10] to discover *interesting popular patterns* consisting of social entities (or their social networking pages) that are frequently followed by social entities in a complex social network with *uncertain and imprecise* social data. Discovery of these patterns helps individual users find popular groups of social entities so that they can follow the same groups. Moreover, many businesses have used social network media to either (*i*) reach the right audience and turn them into new customers or (*ii*) build a closer relationship with existing customers. Hence, discovering those who follow collections of popular social networking pages about a business (i.e., discovering those who care more about the products or services provided by a business) helps the business identify its targeted or preferred customers.

The remainder of this paper is organized as follows. The next section provides some background on data science. Then, Sect. 3 describes our new data analytics solution, which uses the MapReduce model to discover interesting popular patterns from complex, big, uncertain and imprecise social networks in Sect. 4 observes and discusses evaluation results. Finally, Sect. 5 gives conclusions and future work.

2   Background: Data Science

Data science aims to develop systematic or quantitative processes to analyze and mine big data for continuous or iterative exploration, investigation, and understanding of past business performance so as to gain new insight and drive science or business planning. By applying *big data analytics and mining* (which incorporates various techniques from a broad range of fields such as cloud computing, data analytics, data mining, machine learning, mathematics, and statistics), data scientists can extract implicit, previously unknown, and potentially useful information from big data (e.g., big social network data).

In the past few years, researchers have used the high-level programming model MapReduce to process high volumes of big data by using parallel and distributed computing on large clusters or grids of nodes (i.e., commodity machines) or clouds, which consist of a master node and multiple worker nodes. As implied by its name, MapReduce involves two key functions: (*i*) the map function and (*ii*) the reduce function. Specifically, the input data are read, divided into several partitions (sub-problems), and assigned to different processors. Each processor executes the *map function* on each partition (sub-problem). The map function takes a pair of $\langle key_1, value_1 \rangle$ and returns a list of $\langle key_2, value_2 \rangle$ pairs as an intermediate result, where (*i*) key_1 and key_2 are keys in the same or different domains and (*ii*) $value_1$ and $value_2$ are the corresponding values in some domains. Afterward, these pairs are shuffled and sorted. Each processor then executes the *reduce function* on (*i*) a single key key_2 from this intermediate result $\langle key_2,$ list of $value_2 \rangle$ together with (*ii*) the list of all values that appear with this key in the intermediate result. The reduce function "reduces"—by combining, aggregating, summarizing, filtering, or transforming—the list of values associated with a given key key_2 (for all k keys) and returns a single (aggregated or summarized) value $value_3$, where (*i*) key_2 is a key in some domains and (*ii*) $value_2$ and $value_3$ are the corresponding values in some domains. An advantage of using the MapReduce model is that users only need to focus on (and specify) these "map" and "reduce" functions—without worrying about implementation details for (*i*) partitioning the input data, (*ii*) scheduling and executing the program across multiple machines, (*iii*) handling machine failures, or (*iv*) managing inter-machine communication. The construction of an inverted index as well as the word counting of a document for data processing [10] are a few examples of MapReduce applications.

3   Mining Complex Big Data in Uncertain and Imprecise Social Networks

Now, let us present our new big data analytics and mining solution—called **BigUISN**—for mining **big u**ncertain and **i**mprecise **s**ocial **n**etworks for interesting popular patterns using the MapReduce model.

3.1   Interdependencies Between Followers and Followees in Complex Big Social Networks

Social entities (i.e., users) in social networking sites like Twitter are linked by *"following" relationships* such as $A \rightarrow B$ indicating that a user A (i.e., *follower*) follows another user B (i.e., *followee*). Then, given a social network in which each social entity is *following* some other social entities, such a social network can be represented as a graph $G = (V, E)$ where (*i*) V is a set of vertices (i.e., social entities) and (*ii*) E is a set of weighted directional edges connecting some of these vertices (i.e., "following" relationships). See Example 1.

Example 1. For illustrative purpose, let us consider a small portion of a complex big social network as shown in Fig. 1. It can be represented by $G = (V, E)$, where (*i*) $V = \{Alain, Benoit, Charlot, Denis, Emile, Frederic\}$ and (*ii*) $E = \{\langle Alain, B \rangle$:0.9, $\langle Alain, E \rangle$:0.9, $\langle Benoit, A \rangle$:1, $\langle Benoit, C \rangle$:1, $\langle Benoit, E \rangle$:1, $\langle Charlot, A \rangle$:0.7, $\langle Charlot, E \rangle$:0.7, $\langle Denis, B \rangle$:1, $\langle Denis, C \rangle$:1, $\langle Denis, E \rangle$:1, $\langle Emile, A \rangle$:0.8, $\langle Emile, B \rangle$:0.8, $\langle Emile, C \rangle$:0.8, $\langle Emile, D \rangle$:0.8, $\langle Frederic, A \rangle$:1, $\langle Frederic, B \rangle$:1, $\langle Frederic, C \rangle$:1, $\langle Frederic, E \rangle$:1$\}$.   □

In contrast to the mutual friendship relationships, the "following" relationships are different in that the latter are *directional*. For instance in Example 1, *Benoit* is following *Charlot*, but *Charlot* is not following *Benoit*. This property increases the complexity of the problem because of the following reasons.

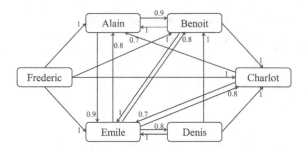

Fig. 1. A sample social network with uncertain and imprecise information about $|V| =$ 6 users.

The group of users followed by *Benoit* (e.g., *Benoit* → {*Alain, Charlot, Emile*}) may not be same group of users as those who are following *Benoit* (e.g., {*Alain, Denis, Emile, Frederic*} → *Benoit*). Hence, we need to store directional edges (e.g., ⟨*Alain, Benoit*⟩, ⟨*Benoit, Alain*⟩) instead of undirected edges (e.g., {*Alain, Benoit*} indicating that *Alain* and *Benoit* are mutual friends). Given $|V|$ social entities, there are potentially $|V|(|V| - 1)$ directional edges for "following" relationships (cf. potentially $\frac{|V|(|V|-1)}{2}$ undirected edges for mutual friendship relationships). In addition to an increase in storage space, the computation time also increases because we need to check both directions to get relationships between pairs of users (e.g., cannot determine whether or not *Charlot* → *Benoit* if we only know *Benoit* → *Charlot*).

Moreover, as we focus on imprecise and uncertain social networks in this paper, each edge (u, v) in the network is associated with an *existential probability* to indicate the likelihood for social entity u to "follow" another social entity v. For instance, due to the privacy setting of some social entities, analysts may suspect—but cannot guarantee—that a user *Alain* is likely to follow user *Benoit*. Such suspicion can be expressed in terms of existential probability. In Example 1, an edge ⟨Alain, Benoit⟩ is associated with an existential probability value in the range of $(0, 1]$—namely, 0.9, which expressed that *Alain* has a 90% chance of following *Benoit*. In other words, *Alain* has a 10% chance of *not* following *Benoit*.

Besides the privacy setting, another source of uncertainty or imprecision is the ambiguity in name. For instance, a user may know a friend by nick name or common name (say, *Johnny*), which may not be his official name or the name used in the social networking sites (say, *Jean*). In this situation, the user may be uncertain about which social entities to follow? Should that user follow *Johnny, John, Jean, Juan, Hans*, or *Ivano*? Once again, such uncertainty can be expressed in terms of existential probability with value in the range of $(0, 1]$.

3.2 Discovery of Popular Followees

With the explosive growth of the number of users in social networking sites (e.g., Twitter), the number of "following" relationships between followers and followees in complex social network are also growing. One of the important research problems with regard to this high volume of data is to discover interesting "following" patterns.

A popular pattern is a pattern representing the linkages when a significant number of users (i.e., followers) following the same combination/group of users (i.e., followees). For example, users who follow the twitter feed or tweets of UBC also follow the tweets of its President. If there are large numbers of users who follow the tweets of both UBC and its university President together, we can define this combination ({UBC, President Santa Ono}) of followees as an interesting popular pattern (i.e., a frequently followed group).

To discover interesting popular patterns (i.e., collections of social network pages that are frequently followed by users), we propose a multi-step data science solution called **BigUISN** for mining complex big data from uncertain and imprecise social networks by using sets of map and reduce functions.

3.3   The First Set of MapReduce Functions in BigUISN

In high-level abstract terms, **BigUISN** first applies a map function to each edge as follows:

$$map_1: \langle edgeID, followingRel \rangle \rightarrow \langle follower\ F, indFollowee\ f, Pr(F,f) \rangle \qquad (1)$$

where: (i) $edgeID$ is the identifier of the edge; (ii) $followingRel$ is the "following" relationship captured by the edge with ID $edgeID$; (iii) $follower$ is the follower, denoted hereinafter as F; (iv) $indFollowee$ is the individual followee, denoted hereinafter as f; and (v) $Pr(F,f)$ is the existential probability of F following f. Looking in more implementation details, the master node reads edges modeled like that, and divides big social network data in partitions. Specifically, the map_1 function is detailed by Algorithm 1.

Algorithm 1. map_1

1: **Input:** social network $G = (V,E)$
2: **Output:** list of objects $\langle F, f, Pr(F,f) \rangle$
3: **for each** edge $e = \langle F, f \rangle \in E$ **do**
4:      **emit** $\langle F, f, Pr(F,f) \rangle$;

Map function map_1 is applied to each edge $e = \langle F, f \rangle \in E$—with an existential probability $0 < Pr(F,f) \leq 1$—in the social network $G = (V,E)$, and provides as result in a list of $\langle F, f, Pr(F,f) \rangle$ capturing all potential "following" relationships (between followers and followees) in the social network. See Example 2.

Example 2. After applying the map_1 function to the social network data in Example 1, our **BigUISN** returns the following list: $\mathcal{L} = \{\langle Alain, B, 0.9 \rangle,$ $\langle Alain, E, 0.9 \rangle, \langle Benoit, A, 1 \rangle, \langle Benoit, C, 1 \rangle, \langle Benoit, E, 1 \rangle, \langle Charlot, A, 0.7 \rangle,$

$\langle Charlot, E, 0.7 \rangle$, $\langle Denis, B, 1 \rangle$, $\langle Denis, C, 1 \rangle$, $\langle Denis, E, 1 \rangle$, $\langle Emile, A, 0.8 \rangle$, $\langle Emile, B, 0.8 \rangle$, $\langle Emile, C, 0.8 \rangle$, $\langle Emile, D, 0.8 \rangle$, $\langle Frederic, A, 1 \rangle$, $\langle Frederic, B, 1 \rangle$, $\langle Frederic, C, 1 \rangle$, $\langle Frederic, E, 1 \rangle \}$. □

Hereafter, **BigUISN** applies a reduce function to group and compute the expected number of followers for each followee, as well as to list these followers for each followee. This function is named as $reduce_1$, and its general form for $reduce_{k \geq 1}$ reads as follows:

$$reduce_{k \geq 1}: \langle P_k, \text{list of } \langle F, P_k, Pr(F, P_k) \rangle \rangle$$
$$\rightarrow \text{list of } \langle P_k, \text{expCnt}[P_k], \text{list}[P_k] \rangle \qquad (2)$$

For $k = 1$, followee group P_k is an individual social entity f returned by map_1. More specifically, $\langle F, f, Pr(F, f) \rangle$-tuples from the map_1 function are shuffled and sorted (where $P_{k=1} = f$). Each processor then executes the reduce function on the shuffled and sorted pairs to compute the expected number of followers and list them for each followee. Here, the expected number of followers is computed as a product of the related existential probabilities. To speed up this big social network data mining process, **BigUISN** also allows users to specify the *interestingness* of groups of social entities by a frequency threshold τ. Here, the users can indicate the minimum number of followers for a group of followees so that the group can be considered interesting. By incorporating this user preference, **BigUISN** returns (i) a list of followers only for those popular followees (i.e., followees who are frequently followed by at least the minimum number of followers) and (ii) the expected number of each followee. The $reduce_{k \geq 1}$ function is detailed by Algorithm 2.

Algorithm 2. $reduce_{k \geq 1}$

1: **Input:** followee group P_k; list of $\langle F, P_k, Pr(F, P_k) \rangle$; min freq. threshold τ
2: **Output:** list of $\langle$interesting group P_k of k followees, expCnt$[P_k]$, list$[P_k] \rangle$
3: **for each** followee group $P_k \in \langle \_, P_k, \_ \rangle$ emitted by map_k **do**
4:     set expCnt$[P_k] \leftarrow 0$; list$[P_k] \leftarrow \oslash$;
5:     **for each** follower $F \in \langle F, P_k, Pr(F, P_k) \rangle$ emitted by map_k **do**
6:         expCnt$[P_k] \leftarrow$ expCnt$[P_k] + Pr(F, P_k)$; list$[P_k] \leftarrow$ list$[P_k] \cup F$;
7:     **if** (expCnt$[P_k] \geq \tau$) **then**
8:         **emit** $\langle P_k$, expCnt$[P_k]$, list$[P_k] \rangle$;

Reduce function $reduce_1$ results in (i) a list of followers and (ii) its expected number of each *interesting/popular* individual followee f. See Example 3.

Example 3. Let us continue with Example 2. **BigUISN** applies the $reduce_1$ function with user-specified minimum frequency threshold $\tau = 2$ followers and returns the following list: $\mathcal{L} = \{ \langle A, 3.5, \{Benoit, Charlot, Emile, Frederic\} \rangle$, $\langle B, 3.7, \{Alain, Denis, Emile, Frederic\} \rangle$, $\langle C, 3.8, \{Benoit, Denis, Emile, Frederic\} \rangle$, $\langle E, 4.6, \{Alain, Benoit, Charlot, Denis, Frederic\} \rangle \}$. Note that our

BigUISN does not return the lists for followees D or F because their corresponding counters were low (D and F were expected to be followed by only 0.8 and 0 followers, respectively).

To recap, after applying the first set of map_1 and $reduce_1$ functions, our **BigUISN** has so far discovered four interesting popular patterns—in the form of *individual* frequently followed social entities—namely: $\{\{A\}, \{B\}, \{C\} \text{ and } \{E\}\}$, who are expected to be followed by 3.5, 3.7, 3.8 and 4.6 followers, respectively. In other words, each of these four individual followees is followed by at least $\tau = 2$ followers.   □

3.4   The Second Set of MapReduce Functions in BigUISN

After applying the first set of MapReduce functions, **BigUISN** then applies a next set of map and reduce functions to mine interesting popular patterns in the form of *pairs* of frequently followed social entities based on the results from the first set of map_1 and $reduce_1$ functions. For instance, knowing that D and E are unpopular individual followees, it is guaranteed that any pairs containing followee D or E is also unpopular. By making use of this knowledge, the search space for mining interesting popular patterns can then be pruned effectively. In other words, the general form for $map_{k \geq 2}$ reads as follows:

$$map_{k \geq 2}: \text{list of} \langle P_{k-1}, \text{expCnt}[P_{k-1}], \text{list}[P_{k-1}] \rangle \rightarrow \text{list of } \langle F, P_k, Pr(F, P_k) \rangle \quad (3)$$

where $P_k = P_{k-1} \cup \{f\}$. For $k = 2$, followee group P_{k-1} is an individual followee p and followee group P_k is a followee pair $\{p, f\}$. More specifically, the map_2 function returns objects of kind $\langle F, \{p, f\}, Pr(F, \{p, f\}) \rangle$ for every follower F in the follower list of each popular/interesting individual followee p. The map_2 function is detailed by Algorithm 3.

Algorithm 3. $map_{k \geq 2}$

1: **Input:** list of $\langle P_{k-1}, \text{expCnt}[P_{k-1}], \text{list}[P_{k-1}] \rangle$
2: **Output:** list of $\langle F, P_{k-1} \cup \{f\}, Pr(F, P_{k-1} \cup \{f\}) \rangle$
3: **for each** interesting followee $P_{k-1} \in \langle P_{k-1}, \_, \text{list}[P_{k-1}] \rangle$ emitted by $reduce_{k-1}$ **do**
4:     **for each** follower $F \in \text{list}[P_{k-1}]$ **do**
5:         **for each** $\langle F, f \rangle \in E$ **do**
6:             **if** (isRelevant(f, P_{k-1})) **then**
7:                 **emit** $\langle F, P_{k-1} \cup \{f\}, Pr(F, P_{k-1} \cup \{f\}) \rangle$;

Note that $isRelevant(f, P_{k-1})$ is a Boolean function checking the relevance (e.g., consistence to the mining order) of followee f with respect to P_{k-1}. Map function map_2 results in lists of $\langle F, P_k, Pr(F, P_k) \rangle$ where $P_k = P_{k-1} \cup \{f\}$. See Example 4.

Example 4. Continue with Example 3. Recall that the first set of map_1 and $reduce_1$ functions returns four popular followees A, B, C and E. So, for popular followee A (followed by four followers: $\{Benoit, Charlot, Emile, Frederic\}$), the map_2 function emits all *relevant* followees of these four followers, which are defined as follows: $\mathcal{L} = \{\langle Benoit, \{A, C\}, 1\rangle, \langle Benoit, \{A, E\}, 1\rangle, \langle Charlot, \{A, E\}, 0.49\rangle, \langle Emile, \{A, B\}, 0.64\rangle, \langle Emile, \{A, C\}, 0.64\rangle, \langle Frederic, \{A, B\}, 1\rangle, \langle Frederic, \{A, C\}, 1\rangle, \langle Frederic, \{A, E\}, 1\rangle\}$.

Note that: (*i*) followees of *Alain* are not emitted (because it is not meaningful for *Alain* to follow himself); (*ii*) followees of *Denis* are not emitted (because *Denis* does not follow A); (*iii*) four relationships in the form $\langle \_, A, \_\rangle$ (e.g., $\langle Benoit, A, 1\rangle$) are irrelevant with respect to $p = A$ (because we already knew these four followers are following *single individual followee* A when we started this map_2 function and we aimed to find followers who follow *pairs of followees*); (*iv*) $\langle Emile, \{A, D\}, 0.64\rangle$ is also irrelevant (because followee D is unpopular).

Similarly, for popular followee B (followed by four followers: $\{Alain, Denis, Emile, Frederic\}$), the map_2 function emits all *relevant* followee of these four followers: $\{\langle Alain, \{B, E\}, 0.81\rangle, \langle Denis, \{B, C\}, 1\rangle, \langle Denis, \{B, E\}, 1\rangle, \langle Emile, \{B, C\}, 0.64\rangle, \langle Frederic, \{B, C\}, 1\rangle, \langle Frederic, \{B, E\}, 1\rangle\}$. Note that: (*i*) followees of *Benoit* are not emitted (because it is not meaningful for *Benoit* to follow himself); (*ii*) followees of *Charlot* are not emitted (because *Charlot* does not follow B); (*iii*) four relationships in the form $\langle \_, B, \_\rangle$ (e.g., $\langle Denis, B, 1\rangle$) are irrelevant with respect to $p = B$ (because we already knew these four followers are following *single individual followee* B when we started this map_2 function and we aimed to find followers who follow *pairs of followees*); (*iv*) $\langle Emile, \{B, D\}, 0.64\rangle$ is also irrelevant (because followee D is unpopular); (*v*) relationships in the form $\langle \_, \{A, B\}, \_\rangle$ (e.g., $\langle Emile, \{A, B\}, 0.64\rangle$, $\langle Frederic, \{A, B\}, 1\rangle$) are irrelevant with respect to $p = B$ (because these relationships are already processed by the map_2 function).

Then, for popular followee C (followed by four followers: $\{Benoit, Denis, Emile, Frederic\}$), the map_2 function emits all *relevant* followee of these four followers: $\{\langle Benoit, \{C, E\}, 1\rangle, \langle Denis, \{C, E\}, 1\rangle, \langle Frederic, \{C, E\}, 1\rangle\}$.

Finally, for popular followee E (followed by five followers: $\{Alain, Benoit, Charlot, Denis, Frederic\}$), the map_2 function does not emit any followee because there is no *relevant* followee for these five followers.   □

In similarity to the $reduce_1$ function, $reduce_2$—which is also a specialization of Eq. (2)—shuffles and sorts objects of kind $\langle F, \{p, f\}, Pr(F, \{p, f\})\rangle$ to find and compute followers for each followee pair $P_2 = \{p, f\}$, as detailed by Algorithm 2. Reduce function $reduce_2$ results in (*i*) a list P_k of followers and (*ii*) its expected number of each *interesting/popular* followee pair $P_2 = \{p, f\}$. See Example 5.

Example 5. Let us continue with Example 4. Our **BigUISN** applies the $reduce_2$ function with user-specified minimum frequency threshold $\tau = 2$ followers and returns the following list: $\mathcal{L} = \{\langle\{A, C\}, 2.64, \{Benoit, Emile, Frederic\}\rangle, \langle\{A, E\}, 2.49, \{Benoit, Charlot, Frederic\}\rangle, \langle\{B, C\}, 2.64, \{Denis, Emile, Frederic\}\rangle, \langle\{B, E\}, 2.81, \{Alain, Denis, Frederic\}\rangle, \langle\{C, E\}, 3, \{Benoit, Denis, Frederic\}\rangle\}$.

Hence, after applying this second set of map_2 and $reduce_2$ functions, our **BigUISN** algorithm discovered five interesting "following" patterns—in the form of pairs of frequently followed social entities—namely: $\{\{A, C\}, \{A, E\},$

$\{B, C\}$, $\{B, E\}$, $\{C, E\}\}$, who are expected to be followed by 2.64, 2.49, 2.64, 2.81 and 3 followers, respectively. Each of these five followee pairs is thus followed by at least $\tau = 2$ followers.   □

3.5   Beyond the Second Set of MapReduce Functions in BigUISN

So far, **BigUISN** has found interesting popular patterns in the form of (*i*) *individual* frequently followed social entities as well as (*ii*) *pairs* of frequently followed social entities. **BigUISN** then applies *similar* sets of map and reduce functions to find triplets, quadruplets, quintuplets and higher (i.e., k-tuplets for $k \geq 3$) of frequently followed social entities. See Example 6.

Example 6. Let us continue with Example 5. For popular followee group $\{A, B\}$ (followed by two followers: $\{Emile, Frederic\}$), the map_3 function emits the following three *relevant* followees: $\{\langle Emile, \{A, B, C\}, 0.512\rangle$, $\langle Frederic, \{A, B, C\}, 1\rangle$, $\langle Frederic, \{A, B, E\}, 1\rangle\}$.

For popular followee group $\{A, C\}$ (followed by three followers: $\{Benoit, Emile, Frederic\}$), the map_3 function emits the following two *relevant* followees: $\{\langle Benoit, \{A, C, E\}, 1\rangle, \langle Frederic, \{A, C, E\}, 1\rangle\}$.

In a similar fashion, the map_3 function emits the following two *relevant* followees: $\{\langle Denis, \{B, C, E\}, 1\rangle, \langle Frederic, \{B, C, E\}, 1\rangle\}$ for popular followee group $\{B, C\}$ (followed by three followers: $\{Denis, Emile, Frederic\}$).

Hereafter, by applying the $reduce_3$ function, **BigUISN** discovers the following two interesting "following" patterns $\varphi_1 = \{A, C, E\}$ and $\varphi_2 = \{B, C, E\}$, with their associated lists and number of followees, which are defined as follows: $\mathcal{L} = \{\langle\{A, C, E\}, 2, \{Benoit, Frederic\}\rangle, \langle\{B, C, E\}, 2, \{Denis, Frederic\}\rangle\}$.

Based on the results returned by the $reduce_3$ function, **BigUISN** applies map_4 but returns nothing because there is no relevant quadruplet of frequently followed social entities. This completes the mining process for interesting "following" patterns from our illustrative example social network. Note that key concepts and steps illustrated in this example are applicable to any uncertain and imprecise social network.   □

4   Evaluation, Observations, and Discussion

Our **BigUISN** takes advantages of the MapReduce model when discovering popular patterns over uncertain and imprecise social networks. The input complex social data are divided into several partitions (sub-problems) and assigned to different processors. Each processor executes the map_k and $reduce_k$ functions (for $k \geq 1$). On the surface, one might worry that lots of communications or exchanges of information are required among processors. Fortunately, due to the divide-and-conquer nature of our big social network data analytics solution of discovering popular patterns, once the original big social network is partitioned and assigned to each processor (e.g., one processor is assigned the followers of A, another is assigned the followers of B, a third one is assigned the followers of C), each processor handles the assigned data without any reliance on the results from other processors. As observed from the above examples, the processor assigned for the followers of a popular followee can apply the subsequent sets of map and reduce functions on data emitted by that processor. For example, a processor applies map_1 and $reduce_1$ to find popular followee A. That processor can then apply map_2 on the data emitted by $reduce_1$ from that processor to find popular followee group $\{A, B\}$ (i.e., group containing A). Similarly, the processor applies map_3 on the data emitted by $reduce_2$ from the same processor to find subsequent popular followee group $\{A, B, C\}$. Without the need of extra communications and exchanges of data among processors, our **BigUISN** discovers all interesting popular patterns efficiently.

Moreover, if a partition of the complex big social network is too big to be handled by a single processor, our **BigUISN** furthers sub-divide that partition so that the resulting sub-partitions can be handled by each of the multiple processors. Furthermore, due to the divide-and-conquer nature of our big social network data analytics solution of discovering popular patterns, the amount of data input for the map_k and $reduce_k$ functions monotonically decreases as the size of the popular group of k followees increases. Our **BigUISN** discovers all interesting popular patterns in a space effective manner.

In terms of functionality, existing algorithms like FoP-miner [12], ParFoP-miner [21] and the BigFoP algorithm [20] mine popular patterns from precise social networks. In contrast, our current **BigUISN** algorithm is capable of handling both *precise* social networks as well as *uncertain and imprecise* social networks. Note that the former can be considered as a special case of the latter when $Pr(F, f) = 1$ for all existing edges $(F, f) \in E$ in the complex big social network represented as a social graph $G = (V, E)$. In other words, the former three algorithms can handle social graphs with every edge (F, f) having a weight (i.e., existential probability) of 1, whereas the latter (i.e., **BigUISN**) is more flexible in the sense that it can handle social graphs with any edge (F, f) of different weight (i.e., $0 < Pr(F, f) \leq 1$).

In terms of accuracy, when $Pr(F, f) = 1$ for all existing edges $(F, f) \in E$ in the complex big social network represented as a social graph $G = (V, E)$, all four algorithms—namely, FoP-miner [12], ParFoP-miner [21], the BigFoP algorithm [20], and our current **BigUISN** algorithm—gives the same results. In other words, all four algorithms return the same sets of popular (i.e., frequently followed) groups of users.

As another quality measure, we also compared the runtime performance of our **BigUISN** with related works (e.g., FoP-miner [12], ParFoP-miner [21], and the BigFoP algorithm [20]). Note that FoP-miner is a serial algorithm that discovers "following" patterns from precise social networks, ParFoP-miner is a parallel algorithm that discovers "following" patterns from precise social networks, and BigFoP is a MapReduce-based algorithm that discovers "following" patterns from precise social networks. In contrast, our **BigUISN** uses the MapReduce model for the discovery of "following" patterns from uncertain and imprecise

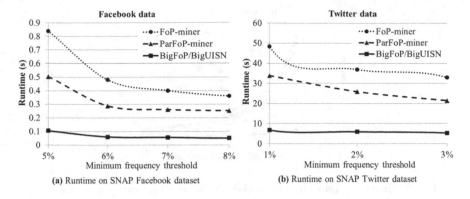

Fig. 2. Experimental results of **BigUISN** on social network data sets.

social networks. We used real-life social network data sets: *The Stanford Network Analysis Project* (SNAP) ego-Facebook data set and ego-Twitter data set[1]. The SNAP Facebook data set contains $4,039$ social entities and $88,234$ connections ("following" relationships) between these social entities. The SNAP Twitter data set contains $81,306$ social entities and $1,768,149$ connections between these social entities. All experiments were run using either (i) a single machine with an *Intel Core i7 4-core processor* (1.73 GHz) and 8 GB of main memory running a *64-bit Windows 7 operating system*, or (ii) the *Amazon Elastic Compute Cloud* (EC2) cluster—specifically, *11 High-Memory Extra Large (m2.xlarge) computing nodes*[2]. We implemented both the existing algorithms and our proposed **BigUISN** in the Java programming language. The stock version of *Apache Hadoop 2.6.5* was used. The results shown in Fig. 2, in which the x-axis shows the user-specified minimum frequency threshold (in percentage of the number of social entities) expressing the interestingness of the mined patterns, are based on the average of 10 runs. Runtime includes CPU and I/Os in the mining process of interesting "following" patterns. In particular, Fig. 2(a) shows that **BigUISN** provided a speedup of about 8 times when compared with FoP-miner, as well as a speedup of about 5 times when compared with ParFoP-miner, in mining the SNAP Facebook data set. Higher speedup is expected when using more processors. It is interesting to note that both BigFoP and **BigUISN** took almost the same runtime because the former can be considered as a special case of the latter where $Pr(F, f) = 1$ for every edge (F, f) in the complex big social networks. Figure 2(b) shows a similar result for the SNAP Twitter data set. Furthermore, our **BigUISN** is shown to be scalable with respect to the number of social entities in the big social network. In addition, we experimented with various existential probability values associated with edges. The results show that the runtime was stable regardless of the probability values.

---

[1] http://snap.stanford.edu/data/.
[2] http://aws.amazon.com/ec2/.

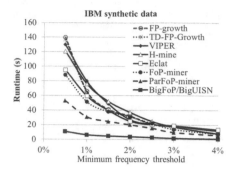

Fig. 3. Experimental results of **BigUISN** on an IBM synthetic data set.

We also experimented with other data sets. For instance, we compared **BigU-ISN** with existing frequent pattern mining algorithms. The results shown in Fig. 3 reveal the benefits of using **BigUISN**.

As ongoing work, we are conducting more experiments, including an in-depth study on the quality of discovered popular patterns.

As we can see from overall our experimental campaign, **BigUISN** not only ensures effectiveness and quality of results, but also performance. Indeed, as highlighted by many recent studies (e.g., [16]), performance is extremely important when processing social networks, especially those exposing big data repositories as underlying data layer.

5   Conclusions and Future Work

In this paper, we proposed a big data analytics and mining algorithm—called *BigUISN*—for discovering interesting popular patterns from big uncertain and imprecise social networks. **BigUISN** helps social network users to discover groups of frequently followed followees from complex big social networks with uncertain and imprecise social data by using the MapReduce model. By applying **BigUISN**, social network users (e.g., newcomers) could find popular groups of followees and follow them. Similarly, a business could find popular groups of followed products and services and incorporate customers' feedback on these products and services. Experimental results show the effectiveness of **BigUISN** as a MapReduce-based solution in this complex big data analytics task of conducting big social data analytics over uncertain and imprecise social networks for interesting popular patterns.

Future work is oriented towards enriching our framework with some innovative features, as to deal with emerging big data trends. Among these, we recall: (*i*) embedding data compression paradigms (e.g., [8]) as to improve efficiency; (*ii*) embedding data partition/fragmentation paradigms (e.g., [5,6]) as to improve distribution; (*iii*) exploring techniques to reduce the number of mapreduce functions used during the mining process; and (*iv*) exploiting alternative big data science frameworks such as the Spark framework (e.g., [15]).

Acknowledgement. This project is partially supported by Natural Sciences and Engineering Research Council of Canada (NSERC) and University of Manitoba.

References

1. Balsa, E., Troncoso, C., Diaz, C.: A metric to evaluate interaction obfuscation in online social networks. Int. J. Uncertain. Fuzziness Knowl.-Based Syst. **20**(6), 877–892 (2012)
2. Bohlouli, M., Dalter, J., Dornhöfer, M., Zenkert, J., Fathi, M.: Knowledge discovery from social media using big data-provided sentiment analysis (SoMABiT). J. Inf. Sci. **41**(6), 779–798 (2015)
3. Chen, C.L.P., Zhang, C.: Data-intensive applications, challenges, techniques and technologies: a survey on big data. Inf. Sci. **275**, 314–347 (2014)
4. Cuzzocrea, A., Bellatreche, L., Song, I.-Y.: Data warehousing and OLAP over big data: current challenges and future research directions. In: ACM DOLAP 2013, pp. 67–70 (2013)
5. Cuzzocrea, A., Darmont, J., Mahboubi, H.: Fragmenting very large XML data warehouses via k-means clustering algorithm. Int. J. Bus. Intell. Data Min. **4**(3/4), 301–328 (2009)
6. Cuzzocrea, A., Furfaro, F., Saccà, D.: Hand-OLAP: a system for delivering OLAP services on handheld devices. In: ISADS 2003, pp. 80–87 (2003)
7. Cuzzocrea, A., Leung, C.K.-S., MacKinnon, R.K.: Mining constrained frequent itemsets from distributed uncertain data. Future Gener. Comput. Syst. **37**, 117–126 (2014)
8. Cuzzocrea, A., Saccà, D., Serafino, P.: A hierarchy-driven compression technique for advanced OLAP visualization of multidimensional data cubes. In: Tjoa, A.M., Trujillo, J. (eds.) DaWaK 2006. LNCS, vol. 4081, pp. 106–119. Springer, Heidelberg (2006). doi:10.1007/11823728_11
9. Cuzzocrea, A., Saccà, D., Ullman, J.D.: Big data: a research agenda. In: IDEAS 2013, pp. 198–203 (2013)
10. Dean, J., Ghemawat, S.: MapReduce: simplified data processing on large clusters. Commun. ACM **51**(1), 107–113 (2008)
11. Dhahri, N., Trabelsi, C., Ben Yahia, S.: RssE-Miner: a new approach for efficient events mining from social media RSS feeds. In: Cuzzocrea, A., Dayal, U. (eds.) DaWaK 2012. LNCS, vol. 7448, pp. 253–264. Springer, Heidelberg (2012). doi:10.1007/978-3-642-32584-7_21
12. Jiang, F., Leung, C.K.-S.: Mining interesting "following" patterns from social networks. In: Bellatreche, L., Mohania, M.K. (eds.) DaWaK 2014. LNCS, vol. 8646, pp. 308–319. Springer, Cham (2014). doi:10.1007/978-3-319-10160-6_28
13. Jiang, F., Leung, C.K.-S.: Stream mining of frequent patterns from delayed batches of uncertain data. In: Bellatreche, L., Mohania, M.K. (eds.) DaWaK 2013. LNCS, vol. 8057, pp. 209–221. Springer, Heidelberg (2013). doi:10.1007/978-3-642-40131-2_18
14. Jiang, F., Leung, C.K.-S., Liu, D., Peddle, A.M.: Discovery of really popular friends from social networks. In: IEEE BDCloud 2014, pp. 342–349 (2014)
15. Jiang, F., Leung, C.K.-S., Sarumi, O.A., Zhang, C.Y.: Mining sequential patterns from uncertain big DNA data in the Spark framework. In: IEEE BIBM 2016, pp. 874–881 (2016)

16. Jin, S., Lin, W., Yin, H., Yang, S., Li, A., Deng, B.: Community structure mining in big data social media networks with MapReduce. Cluster Comput. **18**(3), 999–1010 (2015)
17. Liu, H., Chen, L., Zhu, H., Lu, T., Liang, F.: Uncertainty community detection in social networks. J. Softw. **9**(4), 1045–1049 (2014)
18. Kang, Y., Yu, B., Wang, W., Meng, D.: Spectral clustering for large-scale social networks via a pre-coarsening sampling based NystrÖm method. In: Cao, T., Lim, E.-P., Zhou, Z.-H., Ho, T.-B., Cheung, D., Motoda, H. (eds.) PAKDD 2015, Part II. LNCS (LNAI), vol. 9078, pp. 106–118. Springer, Cham (2015). doi:10.1007/978-3-319-18032-8_9
19. Leung, C.K.-S., Cuzzocrea, A., Jiang, F.: Discovering frequent patterns from uncertain data streams with time-fading and landmark models. In: Hameurlain, A., Küng, J., Wagner, R., Cuzzocrea, A., Dayal, U. (eds.) TLDKS VIII. LNCS, vol. 7790, pp. 174–196. Springer, Heidelberg (2013). doi:10.1007/978-3-642-37574-3_8
20. Leung, C.K.-S., Jiang, F.: Big data analytics of social networks for the discovery of "following" patterns. In: Madria, S., Hara, T. (eds.) DaWaK 2015. LNCS, vol. 9263, pp. 123–135. Springer, Cham (2015). doi:10.1007/978-3-319-22729-0_10
21. Leung, C.K.-S., Jiang, F., Pazdor, A.G.M., Peddle, A.M.: Parallel social network mining for interesting 'following' patterns. Concurr. Comput. Practice Exp. **28**(15), 3994–4012 (2016)
22. Leung, C.K.-S., MacKinnon, R.K.: BLIMP: a compact tree structure for uncertain frequent pattern mining. In: Bellatreche, L., Mohania, M.K. (eds.) DaWaK 2014. LNCS, vol. 8646, pp. 115–123. Springer, Cham (2014). doi:10.1007/978-3-319-10160-6_11
23. Leung, C.K.-S., MacKinnon, R.K., Tanbeer, S.K.: Fast algorithms for frequent itemset mining from uncertain data. In: IEEE ICDM 2014, pp. 893–898 (2014)
24. Leung, C.K.-S., Mateo, M.A.F., Brajczuk, D.A.: A tree-based approach for frequent pattern mining from uncertain data. In: Washio, T., Suzuki, E., Ting, K.M., Inokuchi, A. (eds.) PAKDD 2008. LNCS (LNAI), vol. 5012, pp. 653–661. Springer, Heidelberg (2008). doi:10.1007/978-3-540-68125-0_61
25. Leung, C.K.-S., Tanbeer, S.K.: Mining popular patterns from transactional databases. In: Cuzzocrea, A., Dayal, U. (eds.) DaWaK 2012. LNCS, vol. 7448, pp. 291–302. Springer, Heidelberg (2012). doi:10.1007/978-3-642-32584-7_24
26. Leung, C.K.-S., Tanbeer, S.K., Cameron, J.J.: Interactive discovery of influential friends from social networks. Soc. Netw. Anal. Min. **4**(1), art. 154 (2014)
27. Ma, L., Huang, H., He, Q., Chiew, K., Wu, J., Che, Y.: GMAC: a seed-insensitive approach to local community detection. In: Bellatreche, L., Mohania, M.K. (eds.) DaWaK 2013. LNCS, vol. 8057, pp. 297–308. Springer, Heidelberg (2013). doi:10.1007/978-3-642-40131-2_26
28. Madden, S.: From databases to big data. IEEE Internet Comput. **16**(3), 4–6 (2012)
29. Mumu, T.S., Ezeife, C.I.: Discovering community preference influence network by social network opinion posts mining. In: Bellatreche, L., Mohania, M.K. (eds.) DaWaK 2014. LNCS, vol. 8646, pp. 136–145. Springer, Cham (2014). doi:10.1007/978-3-319-10160-6_13
30. Rader, E., Gray, R.: Understanding user beliefs about algorithmic curation in the Facebook news feed. In: ACM CHI 2015, pp. 173–182 (2015)
31. Rajadesingan, A., Zafarani, R., Liu, H.: Sarcasm detection on Twitter: a behavioral modeling approach. In: ACM WSDM 2015, pp. 97–106 (2015)
32. Tanbeer, S.K., Leung, C.K.-S., Cameron, J.J.: Interactive mining of strong friends from social networks and its applications in e-commerce. J. Organ. Comput. Electron. Commerce **24**(2–3), 157–173 (2014)

33. Wang, Y., Vasilakos, A.V., Ma, J., Xiong, N.: On studying the impact of uncertainty on behavior diffusion in social networks. IEEE Trans. Syst. Man Cybern. Syst. **45**(2), 185–197 (2015)
34. Wei, E.H.-C., Koh, Y.S., Dobbie, G.: Finding maximal overlapping communities. In: Bellatreche, L., Mohania, M.K. (eds.) DaWaK 2013. LNCS, vol. 8057, pp. 309–316. Springer, Heidelberg (2013). doi:10.1007/978-3-642-40131-2_27
35. Yu, W., Coenen, F., Zito, M., Salhi, S.: Minimal vertex unique labelled subgraph mining. In: Bellatreche, L., Mohania, M.K. (eds.) DaWaK 2013. LNCS, vol. 8057, pp. 317–326. Springer, Heidelberg (2013). doi:10.1007/978-3-642-40131-2_28
36. Yuan, N.J.: Mining social and urban big data. In: ACM WWW 2015, p. 1103 (2015)

Non-functional Requirements Satisfaction

A Case for Abstract Cost Models
for Distributed Execution of Analytics Operators

Rundong Li[1]([✉]), Ningfang Mi[2], Mirek Riedewald[1], Yizhou Sun[3], and Yi Yao[2]

[1] CCIS, Northeastern University, Boston, USA
{rundong,mirek}@ccs.neu.edu
[2] ECE, Northeastern University, Boston, USA
{ningfang,yyao}@ece.neu.edu
[3] Department of Computer Science, UCLA, Los Angeles, USA
yzsun@cs.ucla.edu

Abstract. We consider data analytics workloads on distributed architectures, in particular clusters of commodity machines. To find a job partitioning that minimizes running time, a cost model, which we more accurately refer to as makespan model, is needed. In attempting to find the simplest possible, but sufficiently accurate, such model, we explore piecewise linear functions of input, output, and computational complexity. They are abstract in the sense that they capture fundamental algorithm properties, but do not require explicit modeling of system and implementation details such as the number of disk accesses. We show how the simplified functional structure can be exploited by directly integrating the model into the makespan optimization process, reducing complexity by orders of magnitude. Experimental results provide evidence of good prediction quality and successful makespan optimization across a variety of cluster architectures.

1 Introduction

With the ubiquitous availability of clusters of commodity machines and the ease of configuring them in the Cloud, there is growing interest in executing data analytics workloads in distributed environments such as Hadoop MapReduce and Spark. For effective use of resources, a job needs to be *partitioned* into tasks running in parallel on different workers. We will use the term *worker* to refer to a single processing unit, i.e., a single physical or virtual core. Hence a k-core machine would support up to k concurrent workers.

Given an analytics operator, our goal is to find a partitioning and degree of parallelism that minimizes total running time of the computation, also referred to as **makespan** of the corresponding set of tasks. Furthermore, we want to quantify the tradeoff between makespan and degree of parallelism. This is useful for identifying cases where a "good" makespan can be achieved with significantly fewer resources. For example, knowing that 36 concurrent workers achieve a makespan of 29.0 min, but 18 achieve 29.2 minutes, the user might decide to accept the small delay for the benefit of having 18 workers available for another application.

© Springer International Publishing AG 2017
L. Bellatreche and S. Chakravarthy (Eds.): DaWaK 2017, LNCS 10440, pp. 149–163, 2017.
DOI: 10.1007/978-3-319-64283-3_11

To analytically derive optimal parameter settings, the makespan model should have simple functional structure. Arguably the simplest approach with any hope for being practically useful is to estimate running time T of a task as a linear combination of input size (I), output size (O), and (asymptotic) number of computation steps (C):

$$T = c_0 + c_1 I + c_2 O + c_3 C. \tag{1}$$

This model is *abstract* in the sense that it reflects algorithm properties, not implementation or system aspects. The latter are captured by the parameters— representing fixcosts (c_0), data transfer rates (c_1, c_2), and processing speed (c_3)— learned through linear regression from a training set of workloads executed on the same system in advance. By learning from training runs, the parameters represent averages over a large number of low-level processing steps. Hence they automatically account for underlying processing complexities [5]. To apply Eq. 1, I, O, and C need to be expressed as functions of the *partitioning* parameters, e.g., number of tasks. (Note that the resulting function might not be linear in those parameters!) This requires human expertise, but is strictly easier than for traditional DBMS cost models. Consider the map phase of the MapReduce sort implementation, for which in Sect. 3 we derive task duration as $c_{m_0} + c_{m_1}(N/m) + c_{m_2}(N/m)\log(N/m)$. All that was needed to obtain this formula were (1) input and output size per task (N/m) and (2) complexity of sorting.

DBMS cost models are also linear in the sense that they are based on the sum of the number of operations, weighted by per-operation cost. However, they are significantly more complex than our approach, because they express cost at a lower level of abstraction. Beyond input size, output size, and asymptotic computation cost, the DBMS approach estimates the actual number of system-level operations such as random and sequential I/O. Those depend on implementation details of the underlying system. For example, a map task for sorting might perform sorting and partitioning completely in memory, or write multiple temp files that are merged on disk in one or more passes. Moreover, since DBMS cost models are concerned with resource usage, not running time or makespan, they do not take resource bottlenecks into account.

Machine learning models [2] could be trained directly for makespan prediction, but behave as "blackboxes", i.e., makespan optimization cannot exploit model structure. Intuitively, with all existing cost models, *finding the job partitioning that minimizes makespan requires trial-and-error style exploration of parameter combinations*. The search space can be very large, as it includes parameters controlling (i) number of tasks, (ii) degree of parallelism during execution, and (iii) problem-specific partitioning parameters. For matrix multiplication, there are 10 important such parameters (Sect. 4), requiring exploration of a 10-dimensional space of combinations. Our approach reduces complexity to three dimensions, because for the other seven we can derive optimal settings analytically. Assuming 4 values explored in each of those 7 dimensions, our approach reduces optimization cost by a factor of $4^7 \approx 16,000$.

But can a simple abstract makespan models capture the complexities of a distributed system, in particular resource **bottlenecks** during execution? Our

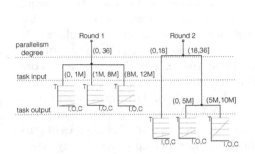

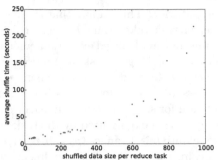

Fig. 1. Schematic illustration of piecewise linear models for a 2-round computation. The model for round 1 is partitioned on task input size only. The model for round 2 is partitioned on both parallelism degree and task output size.

Fig. 2. Shuffle time vs. Data size (MB) for round 2 of the matrix product algorithm

experiments show that for a *piecewise* linear model (Fig. 1), it only took a small number of linear pieces to be sufficiently accurate. The reason for this lies in the way resources are consumed. Consider a network link that can transmit data at a certain rate. While below capacity, doubling the amount of data transmitted approximately doubles transfer time. Once link capacity is exceeded, data is held longer in buffers, effectively decreasing transfer rate. Figure 2 shows a typical observation for a MapReduce program, where the time for shuffling data across the network increases more rapidly after about 600 MB.

The model pieces also provide insights about bottlenecks. For example, for the reduce phase of sorting (Sect. 3), model training for a cluster of quad-core machines determined that three pieces were needed when all four cores were used. Input coefficient c_1 had value 5.5, 9.9, and 12 for "small", "medium", and "large" input size, respectively. For executions using only two cores per machine, the model created only two such pieces with c_1 equal to 4.4 for "small", and 4.9 for "large" inputs. This reflects the I/O-dominated nature of sorting. With four cores competing for access to the data, larger input size stresses I/O and memory bus more than when only two cores are used. By discovering this behavior *automatically* from training data, our model can predict the effect of problem partitioning and parallelism degree.

For an initial proof-of-concept, this paper focuses on relatively "regular" problems—sorting and matrix product. This will be extended in future work through correction factors for skew.

2   Piecewise Linear Model Structure and Training

Let w denote the number of available workers, $p \leq w$ the degree of parallelism of the computation, and n the number of tasks executed during a round of

computation. This implies that there will be $\lceil n/p \rceil$ *waves* of tasks in that round.
If each task takes time T, then makespan of the round will be $\lceil n/p \rceil T$. This
represents an idealized execution, which we show in experiments to be sufficiently
accurate, as long as task interactions and bottlenecks are taken into account.

Interaction effects occur when tasks executed in parallel on a multicore
processor compete for resources, e.g., memory bus and local disk(s). This competition for scarce resources in effect causes lower rate of data transfer and local
computation experienced by the tasks. It can be represented by partitioning the
model into $k \geq 1$ ranges $(p_0 = 0, p_1], (p_1, p_2], \ldots, (p_{k-1}, p_k = w]$ of degrees of parallelism, each with a different linear model. *Bottlenecks* appear not only when
multiple tasks compete for resources. The local computation of a task might
also get delayed by I/O wait time caused by its own I/O operations, requiring a
different model for different ranges of input and output size.

The result of partitioning the design space is a family of piecewise linear models, each with its own (c_0, c_1, c_2, c_3) combination. We say that this model *covers*
the corresponding partition defined by a range of parallelism degrees, input, and
output size. The partitioning can be determined in a fully data-driven manner
from the training data, e.g., by minimizing the residual sum of squares [20] or by
using a model tree [15]. For parallelism degree, we propose a simplified approach
where the partitioning is defined by multiples of the number of worker *machines*:
For a cluster consisting of k-core machines, the interval endpoints defining a piece
based on parallelism degree are a subset of $\{p_i = i \cdot w/k \,|\, i = 1, 2, \ldots, k\}$. This
creates ranges that correspond to a degree of parallelism of 1 to k per physical
machine. Figure 1 illustrates the overall structure of the proposed models. For
each round of the computation, there is a separate piecewise linear model. A
piece is a linear model as defined in Eq. 1, which covers a partition identified by
a range of degrees of parallelism, input sizes and output sizes. For illustration
purposes, the models in the figure are shown in 1-dimensional space.

Following common practice in machine learning, models are trained based on
a set of representative instances of the given problem. As a training instance is
executed, task running times for each round are measured. To train the models
for a round, we use the *average* task running time, input size (I), and output size
(O) for this round. Given these values, computation cost (C) is derived based
on the formula expressing computation cost in terms of input and output size.

3   Makespan Model for Sorting

Sorting plays a central role in data analysis, therefore we first demonstrate how
to apply abstract piecewise linear makespan models to the classic sort algorithm
in Hadoop MapReduce.

3.1   Round-Time Estimation for Map and Reduce Phase

To apply Eq. 1, I, O, and C need to be expressed in terms of parameters controlling the problem partitioning in each round of computation. In the first round,

the map phase, each map task reads records and emits them. The record sets are partitioned and sorted by key, then transferred to the reduce tasks. Each reduce task merges the pre-sorted runs it receives from different map tasks, then emits the records. Our goal is to set number of map tasks, m, number of reduce tasks, r, and parallelism degrees p_m and p_r for map and reduce phase, respectively, to minimize makespan.

With N denoting input size, each map task receives $I = N/m$ input and writes it all out. Since $I = O$, the two separate terms $c_1 I$ and $c_2 O$ collapse to a single term $c_{m_1}(N/m)$, i.e., there is a single coefficient capturing the aggregate of data reading and writing time. (As a by-product, fewer model coefficients allow for smaller training data.) Computational complexity is $(N/m)\log(N/m)$ for sorting. (Note how this abstracts away system details such as the number of disk page accesses.) Hence map task time is modeled as $T_{map} = c_{m_0} + c_{m_1}(N/m) + c_{m_2}(N/m)\log(N/m)$. Given a degree of parallelism p_m, the map phase requires $\lceil m/p_m \rceil$ waves, resulting in round time

$$RT_{map} = T_{map} \cdot \lceil m/p_m \rceil = (c_{m_0} + c_{m_1}(N/m) + c_{m_2}(N/m)\log(N/m))\lceil m/p_m \rceil.$$

Since a reduce task pulls and merges pre-sorted files, then simply reads and emits all its records in order, it follows that all costs are linear in the reduce task's input size, i.e., $I = O = C = N/r$. (Again, system details such as the number of passes for merging of files are abstracted away.) Hence the corresponding terms in Eq. 1 collapse. Analogous to the map phase, there will be $\lceil r/p_r \rceil$ waves, resulting in round time

$$RT_{reduce} = (c_{r_0} + c_{r_1}(N/r)) \cdot \lceil r/p_r \rceil.$$

3.2 Exploiting Model Structure for Optimization

Consider finding optimal number of reduce tasks, r, and parallelism degree p_r:

$$\operatorname*{argmin}_{r,p_r} \quad RT_{reduce} = (c_{r_0} + c_{r_1}(N/r)) \cdot \lceil r/p_r \rceil. \tag{2}$$

Lemma 1. *Model $(c_{r_0} + c_{r_1}(N/r)) \cdot \lceil r/p \rceil$ covering parallelism-degree range $(p_l, p_h]$ and task input range $(s_l, s_h]$ is minimized by setting $p = p_h$ and $r = \min\{\lceil r_l/p_h \rceil \cdot p_h; r_h\}$, where $r_l = \lceil N/s_h \rceil$ and $r_h = \lfloor N/(s_l + 1) \rfloor$.*

Proof. For task input size N/r, range $(s_l, s_h]$ of input sizes implies that the model is valid for reduce task number r in range $r_l \leq r \leq r_h$ with $r_l = \lceil N/s_h \rceil$ and $r_h = \lfloor N/(s_l + 1) \rfloor$. Consider any (r, p) in the valid range, i.e., $r_l \leq r \leq r_h$ and $p_l < p \leq p_h$. For any r, $(c_{r_0} + c_{r_1}(N/r)) \cdot \lceil r/p \rceil$ is minimized by selecting the greatest possible value for p, i.e., $p = p_h$. Hence we need to find the value of r that minimizes $(c_{r_0} + c_{r_1}(N/r)) \cdot \lceil r/p_h \rceil$.

Case 1: the range of possible values for r contains a multiple of p_h. We show that the smallest such multiple minimizes time. Formally, the case condition states that there exists an integer $k \geq 1$ such that $r_l \leq kp_h \leq r_h$. For any such

---

Algorithm 1. Find p_r and r that minimize RT_{reduce} of sort

---

Input: N; M = set of models $(c_{r_0} + c_{r_1}(N/r)) \cdot \lceil r/p_r \rceil$, each covering some range
   $(p_l, p_h]$ of parallelism degrees and some range $(s_l, s_h]$ of reduce-task input sizes
1: **for all** model $m \in M$ **do** // m covers $(p_l, p_h]$ and $(s_l, s_h]$
2:    $t \leftarrow$ time returned by model m when setting $p_r = p_h$ and $r = \min\{\lceil \lceil \frac{N}{s_h} \rceil / p_h \rceil \cdot$
    $p_h; \lfloor \frac{N}{s_l+1} \rfloor \}$
3:    Keep track of smallest t
4: Return minimal time t and its (p_r, r) combination

---

k, consider all $r \in [r_l, r_h]$ that satisfy $(k-1)p_h < r \leq kp_h$. The latter implies
$\lceil r/p_h \rceil = k$ and therefore $(c_{r_0} + c_{r_1}(N/r)) \cdot \lceil r/p_h \rceil = k(c_{r_0} + c_{r_1}(N/r))$. This
formula is minimized by selecting the greatest possible r in $(k-1)p_h \leq r \leq kp_h$,
i.e., $r = kp_h$. Then $(c_{r_0} + c_{r_1}(N/r)) \cdot \lceil r/p_h \rceil = k(c_{r_0} + c_{r_1}\frac{N}{kp_h}) = kc_{r_0} + c_{r_1}N/p_h$.
This formula is minimized by setting k to the smallest possible value that satisfies
the case condition $r_l \leq kp_h \leq r_h$, i.e., $k = \lceil r_l/p_h \rceil p_h$.

Case 2: the range of possible values for r *does not* contain a multiple of p_h. Then
there exists an integer $k' \geq 1$ such that $(k'-1)p_h < r_l \leq r_h < k'p_h$. This implies
$\lceil r/p_h \rceil = k'$ for *all* values of r in $(r_l, r_h]$, and hence $(c_{r_0} + c_{r_1}(N/r)) \cdot \lceil r/p_h \rceil =$
$k'(c_{r_0} + c_{r_1}(N/r))$. This formula is minimized by selecting the greatest possible
r, i.e., $r = r_h$.

To put the solutions for both cases together, notice that for case 1
$\lceil r_l/p_h \rceil p_h \leq r_h$ and for case 2 $r_h \leq \lceil r_l/p_h \rceil p_h$. Hence, in general, $(c_{r_0} + c_{r_1}(N/r)) \cdot$
$\lceil r/p_h \rceil$ is minimized by $r = \min\{\lceil r_l/p_h \rceil \cdot p_h; r_h\}$, completing the proof.

Lemma 1 forms the foundation for Algorithm 1. Instead of exhaustively
exploring (r, p_r) combinations, optimization cost is linear in the number of model
pieces. Using more linear pieces improves model accuracy, but increases optimization cost—a directly tunable tradeoff.

To understand how the optimization process takes task interactions and bottlenecks into account, consider first the special case where a single makespan
model M covers all parallelism degree values $p_r \in (0, w]$, and all reduce-task
input sizes $s \in (0, s_h]$, where $s_h > N$. The for-loop in Algorithm 1 would be
executed once, returning $p_r = w$ and $r = \min\{w; N\} = w$. (Note that $N/s_h < 1$
and we assume $N \geq w$, i.e., the number of workers does not exceed the number
of input records.) Stated differently, the algorithm determines that the problem
should be partitioned into w tasks—one per worker—and all tasks should be
executed in a single wave in parallel.

Now consider a cluster of $w/2$ dual-core machines and assume that when
using both cores on a worker, the memory bus on the worker slows down data
transfer rate from memory to core, causing the cores to wait for data. During
model construction, our approach would automatically determine from the training data that two different linear models are needed: one covering parallelism
degree $p_r \in (0, w/2]$, and the other $p_r \in (w/2, w]$. The for-loop in Algorithm 1
now compares predicted makespan for two configurations (p_r, r): $(w/2, w/2)$ for
the model covering $p_r \in (0, w/2]$ and (w, w) for the model covering $p_r \in (w/2, w]$.

Stated differently, if the memory-bus bottleneck leads to a severe slowdown, the optimal solution may be to use only half of the cores—one per machine—and execute the reduce phase in a single wave of $w/2$ concurrently executed tasks. This perfectly captures the intuition that if the memory bus is the bottleneck (and not the CPU), then it may be better to only use one of the two cores per machine.

4   Dense Matrix Product

The second test case for our approach, dense matrix multiplication, represents a more challenging workload with high data transfer costs, but also significant CPU load in some rounds due to the large number of multiplications and additions. Furthermore, matrix partitioning increases total cost due to data replication. Dense matrix multiplication was identified as an important computation problem in a recent UC Berkeley survey on the parallel computing landscape [3]. Also note that the closed-form solution to the linear regression problem $y = X\beta + \epsilon$, given by the ordinary least squares estimator $\hat{\beta} = (X^T X)^{-1} X^T y$, involves the product of matrices that are often dense.

4.1   Makespan Model for Block-Wise Matrix Multiplication

Dense matrix-matrix multiplication can be parallelized by partitioning each matrix into blocks. We discuss the makespan model for the MapReduce implementation. (The approach for Spark is analogous.) As illustrated in Fig. 3, input matrix U with dimensions $N_0 \times N_1$ is partitioned into $B_0 \cdot B_1$ blocks, each of size N_0/B_0 by N_1/B_1; V (with dimensions $N_1 \times N_2$) is partitioned into $B_1 \cdot B_2$ blocks, each of size N_1/B_1 by N_2/B_2. Each block from U will be multiplied with the B_2 corresponding blocks from V, for a total of $B_0 \cdot B_1 \cdot B_2$ block-pair multiplication tasks. Note that each U block is duplicated B_2 times, each V block B_0 times. The data duplication (map: round 1) and local multiplication (reduce: round 2) form the **multiplication job (m-job)**. If $B_1 > 1$, then each block-pair product represents only a partial result. In that case an **aggregation job (a-job)** needs to read and re-shuffle these partial results (map: round 3) and sum them up (reduce: round 4).

For $i \in \{1, 2, 3, 4\}$, let p_i and n_i denote degree of parallelism and number of tasks, respectively, in round i. From the analysis above follows $n_2 = B_0 B_1 B_2$. Similar to the sort program, an analysis of input, output, and computation in each round results in the following round time estimators: (Note that rounds 3 and 4 are executed if and only if $B_1 > 1$.)

$$RT_1 = (c_{1_0} + c_{1_1}(N_0 N_1 + N_1 N_2)/n_1 + c_{1_2}(N_0 N_1 B_2 + N_1 N_2 B_0)/n_1) \cdot \lceil n_1/p_1 \rceil,$$

$$RT_2 = (c_{2_0} + c_{2_1}(\frac{N_0 N_1}{B_0 B_1} + \frac{N_1 N_2}{B_1 B_2}) + c_{2_2}\frac{N_0 N_2}{B_0 B_2} + c_{2_3}\frac{N_0 N_1 N_2}{B_0 B_1 B_2}) \cdot \lceil B_0 B_1 B_2/p_2 \rceil,$$

$$RT_3 = (c_{3_0} + c_{3_1} N_0 N_2 B_1/n_3) \cdot \lceil n_3/p_3 \rceil,$$

$$RT_4 = (c_{4_0} + c_{4_1} N_0 N_2 B_1/n_4 + c_{4_2} N_0 N_2/n_4) \cdot \lceil n_4/p_4 \rceil.$$

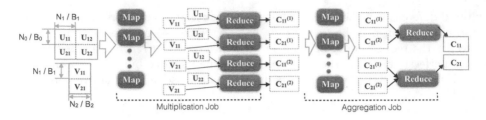

Fig. 3. Block-wise parallel matrix multiplication in 4 rounds. U is partitioned into 2×2 blocks, V into upper and lower half, i.e., $(B_0, B_1, B_2) = (2, 2, 1)$.

4.2 Optimal Partitioning

The problem partitioning that minimizes estimated makespan is defined as $\operatorname{argmin}_{B_0,B_1,B_2,p_1,p_2,p_3,p_4,n_1,n_3,n_4} RT_1 + RT_2 + RT_3 + RT_4$. With traditional cost models, this would require trial-and-error based exploration of a *10-dimensional search space*. Using our approach, we can show, like for sorting, that the optimal setting for parallelism degree is $p = p_h$ for a model covering range $(p_l, p_h]$. The optimal task number n is $\min\{\lceil \frac{n_l}{p_h} \rceil \cdot p_h; n_h\}$. Here n_l and n_h denote the lower and upper extreme of the range of possible choices for the corresponding n_i so that task input and output size are in the range covered by the model piece for round i. Hence the optimization problem simplifies to

$$\operatorname*{argmin}_{B_0,B_1,B_2} RT_1 + RT_2 + RT_3 + RT_4, \qquad (3)$$

where n_1, n_3, n_4, p_1, p_2, p_3, and p_4 are all computed directly as discussed above. This reduces optimization cost by *orders of magnitude*, from search in 10 dimensions to 3 dimensions. (Note that optimization cost is linear in the total number of linear pieces, across all rounds.)

5   Experiments

The main purpose of the experiments is to provide a *proof of concept* that abstract piecewise linear makespan models with a "small" number of pieces are accurate *enough* to rank "good" above "bad" data partitionings. Accuracy comparisons to traditional cost models, in particular DB optimizer cost formulas and blackbox models, are not included. Our abstract models trade off prediction accuracy (hence will be less accurate than a carefully designed and tuned traditional model), to gain in terms of two unique properties: (1) Make it easier to specify the model for a given data analytics operator, and (2) enable more efficient running-time optimization algorithms by exploiting the simple model structure. Note that in all experiments, the piecewise linear models had between 1 and 7 pieces per round.

Table 1. Cluster specifications

Name	#Machines	#Cores per machine	#Workers	Memory per machine	Software
9h36	10	4	36	8 GB	Hadoop 1.2
2h24	3	12 (virtual)	24	47 GB	Hadoop 2.4
20h160	21	8	160	64 GB	Hadoop 2.4
Emr10	11	1 (virtual)	10	3.75 GB	Hadoop 2.6
6s12	7	2	12	8 GB	Spark 1.6.1
Emr12s	7	2 (virtual)	12	7.5 GB	Spark 1.6.1

5.1 Basic Setup

We show representative results on six different systems with diverse properties. They include in-house clusters (9h36, 2h24, 6s12), a research cluster (20h160) provided by CloudLab [22] and two (Emr10, Emr12s) on Amazon Web Services. For details see Table 1.

For simplicity, in most experiments on Hadoop, the number of map tasks is left at the Hadoop default value, i.e., total map input size divided by Hadoop Distributed File System (HDFS) block size. Only for small data sets whose size is smaller than the product of desired parallelism degree and HDFS block size, we set the number of map tasks equal to the desired parallelism degree.

5.2 Sorting

We present measurements on clusters 9h36 and Emr10. All piecewise models for 9h36 are partitioned into ranges $(0, 18]$ and $(18, 36]$ on parallelism degree. Possible partitioning on task input and output size is determined automatically as discussed in Sect. 2. We create 15 different data sets with 100 million to 2.7 billion randomly generated records of type Long (8 bytes per record), and for each data set we use various numbers of waves (up to 10) in the reduce phase. In total, there are 54 problem instances. A subset of 41 of these is used for model training, the others for testing.

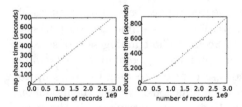 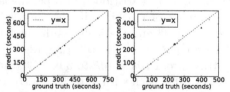

Fig. 4. Sorting: measured round time vs. Input size on 9h36 for Map (left) and Reduce (right) phase

Fig. 5. Sorting: predicted vs. Measured round time on 9h36 for Map (left) and Reduce (right) phase

Figure 4 presents measurements of the relationship between input size and round time. In particular, the y-axis reports the true value for RT_{map} and RT_{reduce}, computed as the product of average *measured* task running time and number of waves in the round. Degree of parallelism is set to the number of workers for all runs. The dotted green line shows a piecewise linear model fitted to the data.

Figure 5 compares predicted and measured round time of map and reduce phase for sorting on cluster 9h36, using either all or only half of the available cores. The red dots are for training cases, while the green triangles are for test cases. All individual times and the overall trend are captured very accurately, as the *relative errors* are mostly around 1%, and never exceed 5%.

Table 2. Degree of parallelism vs. Measured and predicted makespan on 9h36.

Number of records	Degree of parallelism = 18		Degree of parallelism = 36	
	True (sec)	Prediction	True (sec)	Prediction
1.17E + 9	790	601.96	698	564.21
1.26E + 9	835	657.36	723	629.59
1.62E + 9	1056	842.00	1050	833.66
1.80E + 9	1146	928.18	1112	926.13
2.43E + 9	1558	1254.39	1524	1288.04
2.70E + 9	1751	1408.24	1741	1465.02

Table 2 shows that our models significantly underestimate true makespan. This is caused by tasks starting and/or finishing later than others, delaying job completion. However, this bias is consistent, allowing the model to capture the trend correctly, no matter if all cores or only half of them is used per machine. For large inputs, it identifies the I/O-related bottleneck: doubling the number of cores used per machine results in virtually no improvement of makespan when data size reaches 1.6 billion records.

5.3 Matrix Multiplication

All models are partitioned into parallelism-degree ranges based on multiples of the number of machines in the cluster; partitioning on input and output size is determined automatically as discussed in Sect. 2. The training set consists of 104 problem instances, covering 12 different matrix-size combinations (square matrices from $10k \times 10k$ to $30k \times 30k$ and also extreme rectangular ones up to $200 \times 4 \cdot 10^6$), each with 3 to 20 (B_0, B_1, B_2)-combinations. We test the model on 57 independent problem instances, covering 10 different matrix sizes in the same broad range. As Fig. 6 shows, predicted and true round times are very close.

Like for sorting, our model significantly underestimates true makespan, but can still correctly separate "good" from "bad" problem partitionings. In all cases

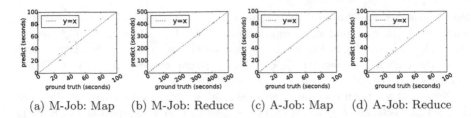

(a) M-Job: Map   (b) M-Job: Reduce   (c) A-Job: Map   (d) A-Job: Reduce

Fig. 6. Matrix product: predicted vs. measured round time on **9h36**. The test cases (red dots) are near the perfect-prediction line (blue dotted line). (Color figure online)

Table 3. Ranking quality: predicted vs. true makespan (in sec) for matrix product (Hadoop MapReduce, (a)~(c) are synthetic data, (d) and (e) are real data)

(a) $15,000 \times 15,000$ matrices on **9h36**

B_0	B_1	B_2	prediction makespan	rank	ground truth makespan	rank
6	1	6	305.40	1	400.00	1
3	3	4	330.87	2	434.00	2
4	1	4	345.89	5	440.00	3
3	3	3	350.39	7	445.67	4
3	4	3	333.55	3	448.00	5
5	1	5	356.48	9	452.00	6
3	2	3	344.69	4	453.00	7
2	6	3	348.85	6	471.00	8
4	2	4	353.48	8	479.00	9
2	9	2	385.02	11	485.00	10
2	6	2	380.85	10	497.00	11
2	4	2	403.84	13	505.00	12
2	8	2	410.55	14	525.00	13
2	7	2	446.67	15	548.00	14
4	1	8	401.43	12	556.00	15
2	2	2	614.17	16	656.00	16
1	18	1	638.41	17	713.00	17
1	36	1	941.19	18	1,290.00	18

(b) $15,000 \times 15,000$ matrices on **2h24**

B_0	B_1	B_2	prediction makespan	rank	ground truth makespan	rank
4	1	6	247.93	1	325.03	1
2	4	3	267.90	4	366.53	2
2	3	4	257.58	3	384.64	3
3	2	4	248.79	2	388.53	4
2	6	2	290.78	5	408.92	5
1	12	2	356.74	6	455.77	6
1	24	1	765.48	7	574.52	7

(c) $10,000 \times 60,000$ matrices on **20h160**

B_0	B_1	B_2	prediction makespan	rank	ground truth makespan	rank
4	10	4	124.57	1	186	1
4	8	5	128.14	2	204	2
2	20	4	132.53	3	205	3
4	5	8	134.07	4	205	3
2	8	10	137.51	5	206	5
1	20	8	141.89	6	211	6
12	1	12	171.08	7	238	7

(d) 68×2458285 matrices (1990 US Census data) on **9h36**

B_0	B_1	B_2	prediction makespan	rank	ground truth makespan	rank
1	18	1	30.93	1	95.00	1
1	36	1	37.97	2	107.25	2
3	2	3	59.05	4	127.00	3
2	9	2	51.12	3	128.5	4
3	4	3	64.20	5	132.25	5
6	1	6	85.72	6	145.00	6

(e) 481×191779 matrices (KDD Cup 1998 data) on **9h36**

B_0	B_1	B_2	prediction makespan	rank	ground truth makespan	rank
1	18	1	24.12	1	94	1
1	36	1	30.89	2	103	2
3	2	3	39.19	4	109	3
2	9	2	37.43	3	112	4
3	4	3	44.61	5	121	5
6	1	6	48.37	6	144	6

our approach would find a near-optimal configuration. Table 3 confirms this for both synthetic and real data sets (from the UCI Machine Learning Repository [13]). There our technique is applied to the step where the data matrix is multiplied with its own transpose during linear regression analysis. Table 4 confirms that this observation also holds for Spark.

Note that for both real data sets (Table 3d, e), our model correctly discovers that setting (B_0, B_1, B_2) to $(1, 18, 1)$ results in lower makespan than $(1, 36, 1)$. We confirmed that due to I/O bottlenecks, it is better to only use half of the available cores per machine, even though round 2 performs a huge number of arithmetic operations (more than $11 \cdot 10^9$ for the Census data).

Table 4. Ranking quality: predicted vs. true makespan (in sec) for matrix product (synthetic data, Spark)

(a) $800 \times 80,000$ matrices on 6s12

B_0	B_1	B_2	prediction makespan	rank	ground truth makespan	rank
2	2	3	73.81	1	88.8	1
2	3	2	74.08	3	90.67	2
1	12	1	73.88	2	96	3
1	3	4	87.84	4	101	4
1	6	2	100.10	7	101.4	5
1	4	3	96.79	6	104	6
3	1	4	133.95	9	109.5	7
1	6	1	92.80	5	113	8
2	1	3	154.30	11	134	9
1	2	3	134.22	10	141	10
1	3	2	131.48	8	154	11

(b) 6000×6000 matrices on Emr12s

B_0	B_1	B_2	prediction makespan	rank	ground truth makespan	rank
3	1	4	144.73	1	149.5	1
2	2	3	152.50	2	152	2
2	3	2	156.63	3	162	3
1	2	6	171.79	5	170.5	4
1	3	4	166.27	4	171	5
1	4	3	180.81	6	173.5	6
1	6	2	184.95	7	195	7
2	1	3	251.14	9	254	8
1	2	3	268.33	8	268.5	9
1	3	2	277.20	11	277	10
1	1	6	266.92	10	304	11
1	6	1	365.17	12	362	12

6   Related Work

Structured cost models that capture execution details are essential for query optimization in relational DBMS [16], and they can be highly accurate when tuned [23]. This motivated similar approaches for MapReduce and other distributed data analysis systems [11,14,19,21,24]. As an alternative to structured cost models, blackbox-style machine learning techniques were explored for a variety of performance prediction problems [2,5,6,8,10]. For all previous cost models, the effect of problem-partitioning parameters on makespan is relatively complex, hence makespan minimization would have to rely on trial-and-error style exploration of possible parameter settings. For dense matrix multiplication, this corresponds to a 10-dimensional space of $(B_0, B_1, B_2, p_0, p_1, p_2, p_3, n_1, n_3, n_4)$ combinations. (Note that Ernest [19] could possibly derive optimal settings for all $p_i, i = 0, \ldots, 3$, reducing complexity to 6 dimensions.) In contrast, our approach sacrifices some prediction accuracy to simplify model structure. This enables analytical derivation of optimal settings for most parameters, reducing complexity to 3 dimensions for dense matrix multiplication.

We use dense matrix multiplication to showcase model design and makespan optimization for an analytics operator with a demanding I/O and CPU profile. Previous work explored a variety of performance-related aspects for matrix multiplication on parallel architectures. This includes load balancing [9], minimizing communication cost [1,4,12,17], and optimizing for memory hierarchy [7,18].

7   Conclusions

The goal of this work was to find the "simplest possible" realistic model to predict makespan for distributed execution of data analytics operators. To this end, we proposed abstract models that are piecewise linear functions depending only on input, output, and computation complexity. Our approach has two main benefits. First, it simplifies tying problem-partitioning parameters to model variables (input, output and computation) for user-defined operators, e.g., programs written in MapReduce or Spark. Second, we showed that the linear structure can be exploited for more efficient optimization algorithms. This reduces optimization complexity from a search process in ten dimensions to only three for matrix product; for sorting the optimal solution was directly obtainable in closed form.

Our experiments indicated that a small number of pieces achieves sufficient prediction quality, enabling us to find near-optimal problem partitionings and to identify when a lower parallelism degree delivers the same (or better) makespan.

In future work, we will explore how to extend this approach to workloads that are more heterogeneous in the sense that individual tasks may vary widely in their cost. Moreover, we will consider tuning partitioning parameters along with system parameters external to user programs, by integrating our ideas into optimizers like Starfish [10].

Acknowledgments. This work was supported by a Northeastern University (NU) Tier 1 award, by the National Institutes of Health (NIH) under award number R01 NS091421, by the Air Force Office of Scientific Research (AFOSR) under grant number FA9550-14-1-0160, and by the National Science Foundation (NSF) Career Award #1741634. The content is solely the responsibility of the authors and does not necessarily represent the official views of NU, NIH, AFOSR or NSF. The authors also would like to thank the reviewers for their constructive feedback.

References

1. Agarwal, R.C., Balle, S.M., Gustavson, F.G., Joshi, M., Palkar, P.: A three-dimensional approach to parallel matrix multiplication. IBM J. Res. Dev. **39**(5), 575–582 (1995)
2. Akdere, M., Cetintemel, U., Riondato, M., Upfal, E., Zdonik, S.: Learning-based query performance modeling and prediction. In: Proceedings of the ICDE, pp. 390–401 (2012)
3. Asanovic, K., Bodik, R., Demmel, J., Keaveny, T., Keutzer, K., Kubiatowicz, J., Morgan, N., Patterson, D., Sen, K., Wawrzynek, J., Wessel, D., Yelick, K.: A view of the parallel computing landscape. Commun. ACM **52**(10), 56–67 (2009)

4. Ballard, G., Buluc, A., Demmel, J., Grigori, L., Lipshitz, B., Schwartz, O., Toledo, S.: Communication optimal parallel multiplication of sparse random matrices. In: Proceedings of the SPAA, pp. 222–231 (2013)
5. Duggan, J., Cetintemel, U., Papaemmanouil, O., Upfal, E.: Performance prediction for concurrent database workloads. In: Proceedings of the SIGMOD, pp. 337–348 (2011)
6. Duggan, J., Papaemmanouil, O., Çetintemel, U., Upfal, E.: Contender: A resource modeling approach for concurrent query performance prediction. In: Proceedings of the EDBT, pp. 109–120 (2014)
7. Elmroth, E., Gustavson, F., Jonsson, I., Kågström, B.: Recursive blocked algorithms and hybrid data structures for dense matrix library software. SIAM Rev. **46**(1), 3–45 (2004)
8. Ganapathi, A., Kuno, H.A., Dayal, U., Wiener, J.L., Fox, A., Jordan, M.I., Patterson, D.A.: Predicting multiple metrics for queries: Better decisions enabled by machine learning. In: Proceedings of the ICDE, pp. 592–603 (2009)
9. van de Geijn, R.A., Watts, J.: Summa: Scalable universal matrix multiplication algorithm. University of Texas at Austin, Technical report (1995)
10. Herodotou, H., Babu, S.: Profiling, what-if analysis, and cost-based optimization of mapreduce programs. VLDB **4**(11), 1111–1122 (2011)
11. Huang, B., Babu, S., Yang, J.: Cumulon: optimizing statistical data analysis in the cloud. In: Proceedings of the SIGMOD, pp. 1–12 (2013)
12. Irony, D., Toledo, S., Tiskin, A.: Communication lower bounds for distributed-memory matrix multiplication. J. Parallel Distrib. Comput. **64**(9), 1017–1026 (2004)
13. Lichman, M.: UCI machine learning repository (2013)
14. Morton, K., Balazinska, M., Grossman, D.: Paratimer: a progress indicator for mapreduce DAGs. In: Proceedings of the SIGMOD, pp. 507–518 (2010)
15. Quinlan, J.R., et al.: Learning with continuous classes. In: Australian Joint Conference on Artificial Intelligence, vol. 92, pp. 343–348 (1992)
16. Ramakrishnan, R., Gehrke, J.: Database Management Systems, 3rd edn. McGraw-Hill, New York (2003)
17. Solomonik, E., Demmel, J.: Communication-optimal parallel 2.5D matrix multiplication and LU factorization algorithms. In: Jeannot, E., Namyst, R., Roman, J. (eds.) Euro-Par 2011. LNCS, vol. 6853, pp. 90–109. Springer, Heidelberg (2011). doi:10.1007/978-3-642-23397-5_10
18. Valsalam, V., Skjellum, A.: A framework for high-performance matrix multiplication based on hierarchical abstractions, algorithms and optimized low-level kernels. In: Concurrency and Computation: Practice and Experience, vol. 14(10), pp. 805–839 (2002)
19. Venkataraman, S., Yang, Z., Franklin, M., Recht, B., Stoica, I.: Ernest: efficient performance prediction for large-scale advanced analytics. In: NSDI, pp. 363–378 (2016)
20. Vieth, E.: Fitting piecewise linear regression functions to biological responses. J. Appl. Physiol. **67**(1), 390–396 (1989)
21. Wang, G., Chan, C.Y.: Multi-query optimization in mapreduce framework. In: Proceedings of the VLDB, pp. 145–156 (2013)

22. White, B., Lepreau, J., Stoller, L., Ricci, R., Guruprasad, S., Newbold, M., Hibler, M., Barb, C., Joglekar, A.: An integrated experimental environment for distributed systems and networks. In: Proceedings of the OSDI, pp. 255–270 (2002)
23. Wu, W., Chi, Y., Hacígümüş, H., Naughton, J.F.: Towards predicting query execution time for concurrent and dynamic database workloads. Proc. VLDB **6**(10), 925–936 (2013)
24. Zhang, X., Chen, L., Wang, M.: Efficient multi-way theta-join processing using mapreduce. Proc. VLDB **5**(11), 1184–1195 (2012)

Pre-processing and Indexing Techniques for Constellation Queries in Big Data

Amir Khatibi[1]($\boxtimes$), Fabio Porto[1], Joao Guilherme Rittmeyer[1],
Eduardo Ogasawara[2], Patrick Valduriez[3], and Dennis Shasha[4]

[1] DEXL Lab, LNCC, Petropolis, RJ, Brazil
{ahassan,fporto,joanonr}@lncc.br
[2] C.S. Department, CEFET/RJ, Rio de Janeiro, RJ, Brazil
eogasawara@ieee.org
[3] Zenith, LIRMM, Inria, Montpellier, France
patrick.valduriez@inria.fr
[4] Courant Institute, NYU, New York, USA
shasha@courant.nyu.edu

Abstract. Geometric patterns are defined by a spatial distribution of a set of objects. They can be found in many spatial datasets as in seismic, astronomy, and transportation. A particular interesting geometric pattern is exhibited by the Einstein cross, which is an astronomical phenomenon in which a single quasar is observed as four distinct sky objects when captured by earth telescopes. Finding such crosses, as well as other geometric patterns, collectively refered to as constellation queries, is a challenging problem as the potential number of sets of elements that compose shapes is exponentially large in the size of the dataset and the query pattern. In this paper we propose algorithms to optimize the computation of constellation queries. Our techniques involve pre-processing the query to reduce its dimensionality as well as indexing the data to fasten stars neighboring computation using a PH-tree. We have implemented our techniques in Spark and evaluated our techniques by a series of experiments. The PH-tree indexing showed very good results and guarantees query answer completeness.

Keywords: Constellation queries · Geometric shapes · PH-tree indexing · Dataset pre-processing · Query pre-processing · SQL extension

1 Introduction

The availability of large datasets in science, web and mobile applications enables new interpretations about natural phenomena and human behavior. From inferring sites of touristic interest based on pictures taken in social network applications [1] to the existence of dark matter inferred from multiple occurrences of quasars [2], new knowledge emerges whenever individual observations are combined allowing for queries on patterns. This paper extends the algorithms and techniques in a type of pattern queries in spatial databases that we referred to

© Springer International Publishing AG 2017
L. Bellatreche and S. Chakravarthy (Eds.): DaWaK 2017, LNCS 10440, pp. 164–172, 2017.
DOI: 10.1007/978-3-319-64283-3_12

as constellation queries (CQ) in our previous work [3]. Constellation queries are obtained from compositions of individual elements in large datasets. CQ computation entails matching geometric pattern queries against sets of individual data observations, such that each set collectively agrees in the geometric constraints expressed by the pattern query. In particular, we are interested in efficiently finding patterns like the Einstein cross (EC). From a constellation query representing the EC, involving a set of sky objects, we should compare it's attributes with other set of sky objects in a astronomy catalog. The data scheme of an astronomy catalog such as Sloan Digital Sky Survey (SDSS)[1] is as the following relation: *SDSS (Obj_ID, RA, DEC, u, g, r, i, z, Redshift, ...)*. The attributes u, g, r, i, z refer to the magnitude of light emitted by an object measured in specific wavelength. A constellation in the SDSS scenario would be defined by a sequence of objects from the catalog whose spatial distribution forms a shape conforming to a constellation query. In this paper we focus on improving the process of executing constellation queries, as described in [3], by applying query pre-processing and indexing techniques. We propose two new steps that could be executed prior to processing a user's query: (a) query pre-processing and (b) dataset pre-processing using the PH-tree algorithm. The advantage of pre-processing is hidden in the fact that solving a constellation query in a big dataset is hard due to the numerous possible compositions from billions of observations. In general, for a big dataset D and a number k of elements in the pattern query, the number of possible candidate combinations, $\binom{|D|}{k}$ is the number of ways to choose k items from D. Dataset pre-processing aims to reduce the size of D while pre-processing the query tries to reduce the size of k in a way that without losing the quality of solutions, we process the constellation queries in a shorter time. Our main contributions in this paper are: the adoption of the PH-tree indexing algorithm; the definition of a SQL extension for the constellation queries; and a query and dataset pre-processing techniques.

The rest of the paper is organized as follows: in Sect. 2, we review the related works. Then Sect. 3 presents the problem statement. Section 4 discusses our contributions in order to improve the CQ processing. In Sect. 5, we show our experimental results. Finally, Sect. 6 concludes.

2   Related Works

The relational data model, and SQL therein, adopt set based constraints that are imposed to each individual tuple in order to appear in a query result set. There are nevertheless many practical real-world problems such as geometric pattern queries that require tuples in a set to collectively satisfy a set of constraints [4,5]. The former is concerned with topological constraints among multi-dimensional objects. In [5], the authors present package queries that enable users to express constraints over package of tuples. The approach considers local constraints, as traditional where clause in SQL, and global constraint that refer to packages of tuples. The query expressed using additional SQL clauses is rewritten into an

---

[1] http://skyserver.sdss.org/dr12/en/help/browser/browser.aspx.

expression composed of a SQL query, submitted to a relational database, and an integer linear program that solves the package constraints on top of database results. Constellation queries [3] is a class of package queries, in which a geometric shape defines the global constraints. The assessment of such constraints requires tuples in candidate packages to be labelled so that they can be referred to elements of the query according to the ordering imposed by the shape. Expressing CQ as package queries leads to self joins in the number of elements of the query that would be impractical for large datasets. Constellation queries combine quad-trees, matrix multiplication, and un-indexed join processing to discover sets of elements that match a geometric pattern within some additive factor on the pairwise distances.

3   Problem Formulation

We formulate the problem of solving spatial pattern queries, referred to as constellation queries as follows: We consider a Big Dataset D defined as a set of elements $D = \{e_1, e_2, \dots, e_n\}$, in which each e_i, $1 \le i \le n$, is an element of a domain Dom. Furthermore, $e_i = < atr_1, atr_2, \dots, atr_m >$, such that atr_j, $1 \le j \le m$, is a value describing a characteristic of e_i. Conversely, a sample query Q is defined as a set of elements $Q = \{q_1, q_2, \dots, q_k\}$, where q_j, $1 \le j \le k$, are elements of the same domain Dom as D. We further adopt the following definitions: **Definition 1:** A Boolean function **fe** ($e_i : Dom$, $Q_j : Q$, $\theta : \mathbb{R}$) verifies whether an element e_i from a domain Dom is at most at a similarity distance θ from any element q_i in Q_j. **Definition 2:** A Boolean function **fs** ($C_i : Dom$, $Q_j : Q$, $\epsilon : \mathbb{R}$) verifies whether the sets (C_i, Q_j) is at most a distance of ϵ with respect to the similarity of their composition model. Moreover, an increasing value for ϵ flexibilizes the distance evaluation. Finally, the semantics of fs evaluation considers all permutations of C_i. **Problem statement**: given a Big dataset D and a sample query Q_j, both with elements in a domain Dom, and constants $\theta = r1$ and $\epsilon = r2$, efficiently compute the set of all compositions $C = \{C_1, C_2, \dots, C_m\}$, $C_i \subset D$, with $|C_i| = |Q_j|$, such that $fe(e_u, Q_j, r1) = true$, for all $e_u \in C_i$, and $fs(C_i, Q_j, r2) = true$, for all $1 \le i \le m$.

4   CQ Processing

This section discusses our contributions in improving the performance of CQ processing. In general, query execution in large datasets is a challenge due to long execution time. However, one way to improve the user's experience of the system is pre-processing the input large dataset so that at the time of query execution, the system provides a quicker response as it has done some of the steps in advance. Likewise, another alternative to reduce the user's waiting time is by pre-processing the query itself. In the next subsections we elaborate each of these pre-processing steps.

4.1   Query Pre-processing

In constellation queries, the number of possible compositions increases exponentially with the query size k, $(N \approx |D|^k)$. One may intuitively suggest reducing the size k of a constellation query in order to save computation. As it turns out, elements in a full query Q_k may induce redundant constraints, specially those located very close to each other. In this context, $Q_{k\prime}$ can be built from subset M of elements of Q_k that only includes elements that are candidates for defining the geometric shape induced by $Q_{k\prime}$ $(k\prime \leq k)$. As an example, $Q_{k\prime}$ may only include elements that are at a certain distance apart. Once $Q_{k\prime}$ has been fixed, an anchor element q_0 is picked and pairwise distances from it to every remaining element $Q_{k\prime} \setminus q_0$ are computed.

Axiom 1. *Given constellation queries $Q_{k\prime}$ and Q_k, such that $Q_{k\prime} = Q_k - Q_v$, $Q_v \subset Q_k$, and $S_{k\prime}$ a set of all sequences $s_{k\prime}$, such that $s_{k\prime}$ matches $Q_{k\prime}$, and S_k a set of all sequences s_k, such that s_k matches Q_k, we say that $Q_{k\prime}$ is equivalent to $Q_k \leftrightarrow (1 - F - measure(S_{k\prime}, S_k))^2 < \delta$ and $shape(s_{k\prime}) \cong shape(s_k)$.*

4.2   Query Transformation

The discussion in Sect. 4.1 raises an optimization strategy for constellation queries based on query pre-processing. The intuition is that some elements would offer little contribution for constellation specification but would add to the query elapsed-time. In this context, detecting such elements, and subtracting them from the query, could reduce query elapsed-time. Query modification, however, must be subject to equivalence guarantees between the original query and its reduced version, as stated by Axiom 1. Proposition 1 suggests a verifiable condition to assess constellation query equivalence. In Sect. 5 we experimentally evaluate the Proposition 1.

Proposition 1. *Given constellation queries $Q_{k\prime}$ and Q_k, such that $Q_{k\prime} = Q_k - Q_v$, $Q_v \subset Q_k$, If for each query element $q_j \in Q_v \; \exists \; q_i \in (Q_k - Q_v)$, such that: $distance(q_j, q_i) \leq \epsilon$ then $Q_{k\prime}$ is equivalent to Q_k.*

4.3   Dataset Pre-processing

The second opportunity to reduce the complexity of constellation queries, approximately $N \approx |D|^k$, is to reduce the size of D. Additionally, we want to look for elements to build compositions that are candidates for producing shapes geometrically close to that of Q_k. The element q_0 from Q_k becomes a key to such reduction. It enables to fix an anchor for building compositions both with respect to attribute values and to shape constraints. Regarding the former, q_0 reduces the size of D to $|\sigma_{f(e_i)}(D)|$. In other words, we only test for compositions that hold a similar anchor element as q_0 in Q_k. Secondly, as we scan D, looking for anchor elements, we store each element in a PH-tree [6]. The latter

---

² F-measure = *precision/recall.*

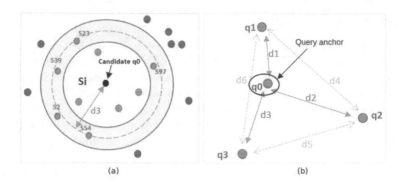

Fig. 1. (a) candidate anchor and neighboring ring elements and (b) geometric query

is used in the sequel to search for candidate elements in the neighborhood of selected e_i within a radius ρ, corresponding to the distance of the furthest query element q_i to q_0 plus ϵ as depicted in Fig. 1. The possible constellations having e as anchor are within this set.

The constellations based on an anchor e includes the neighbors within radius ρ whose distances to e match the distance d_i of some query element in Q_k. For a constellation query with k query elements, we produce $k - 1$ buckets holding neighbors of anchor e. The matching constellations are the sets of k-1 neighbors of e with one element from each bucket and having pairwise distances matching the corresponding pairwise distances of elements in Q_k. The pre-processing of D produces an intermediary relation $D\prime$, substantially smaller than D, with schema $D\prime = (e : Dom, list\ of\ neighbors < n_k, d_k >)$, where e refers to an anchor element in D, and n_k is a neighbor of e in D with distance less than the largest distance (q_i, q_0), for all q_i in Q_k. An interesting side-effect of computing $D\prime$ is that it fosters the parallelization of the constellation query processing by enabling a balanced distribution of $D\prime$ tuples over a cluster of machines to be evaluated by a Big Data processing framework, such as Spark.

5   Experiments

This section presents the experiments conducted to evaluate the pre-processing and indexing techniques in Constellation queries. We first evaluate the query pre-processing technique. Next, we evaluate the performance of the PH-tree applied to the Constellation query problem.

5.1   Query Pre-processing

In our first experiment, we randomly generate 10 queries with 7 elements each. Next, we reduce the query size, by randomly selecting one element to be deleted from each query. Figure 2(a) shows the results in elapsed-time, in seconds, for each query, sorted in ascending order by the elapsed time of the full query, k = 7.

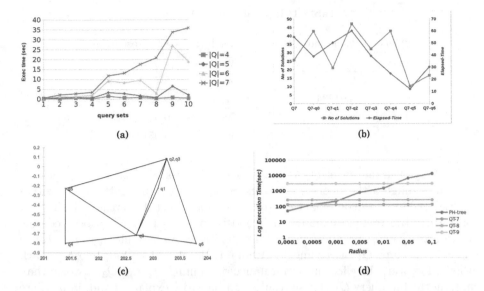

Fig. 2. (a) Query size reduction impact (b) effect on the selected query element, (c) assessing Proposition 1 (d) PH-tree versus quad-tree - log execution time

Two main observations can be drawn from this plot. Firstly, indeed, as expected, the query size impacts in its response-time. The elapsed-time of the full query dominates the ones from its reduced versions. Moreover, the decrease in computation costs follows the modification of the query once one of its elements is, randomly, subtracted. However, as it can be more clearly verified in query 8, the choice of the element to be subtracted contributes differently to the query outcome, as it can be observed by a drastic reduction on the query elapsed time.

In order to explore this last observation, we measured the effect of selecting different query elements to be subtracted from the query. Figure 2(b) plots, in the horizontal axis, the query *id* against the query elapsed-time, in the vertical axis. Curves show: (i) the number of solutions (divided by 10^4), in blue, and (ii) the query elapsed-time, in red. Query Q_7 corresponds to the full query. As it can be seen, the choice significantly impacts on both the elapsed-time and the answer set. The difference in the answer set reflects in variation of the $F-measure$. Moreover, as observed when subtracting q_2 from Q_7, the elapsed-time increases with respect to that of the full query, contrarily to the intuition. As a matter of fact, in Sect. 4.1, Proposition 1 states that modification of constellation query should be limited to query elements in close distance to one of its neighbors. Under this constraint, the results of the modified query and the full query would be equivalent. The query depicted in Fig. 2(c) was specially constructed such that $dist(q_2, q_3) \leq epsilon$. According to Proposition 1, excluding either q_2 or q_3 would produce an equivalent query answer set, compared to the full query.

Table 1 depicts the results of running the query shown in Fig. 2(c). Query $id = 1$ corresponds to the full query. The remaining queries are obtained from

Table 1. Equivalence of Queries

Query id	F-measure	Recall	No. solutions	Fs elapsed-time
1	1	1	38	0.3416
2	0.0112	1	6690	2.7499
3	0.01234	1	6117	2.6596
4	0.7102	1	69	0.0500
5	0.7446	0.9219	56	0.0512

query 1 by deleting an element corresponding to its id. The *Fs elapsed-time* column highlights the query time corresponding to running the composition function, which is the dominant constellation query cost. Unfortunately, Table 1 contradicts the premise exposed in Proposition 1. As it can be observed, the deletion of q_2 or q_3 produces many false positives. As a result, Q_2 and Q_3 are not equivalent to Q_1, according to Axiom 1, despite being very close to each other. Moreover, the effect in performance of running Q_2 and Q_3 is worse than running the full query Q_1. These results can be easily explained and, in fact, are co-related. Firstly, as the query elements are very close in space, they impose a severe restriction in the result set. Only candidate solutions that include pairs of elements, associated to $bucket_2$ and $bucket_3$, with $dist(e_2, e_3) \leq \epsilon$ are selected, expressing a very selective predicate. Conversely, modified queries caused by the deletion of either q_4 or q_5 exhibit higher precision and lower elapsed-time, with $dist(q_4, q_5) > \epsilon$. The later indicates that the original shape, as specified by the full query, would have been sensibly modified by the query modification. These experiments show that query modification must be exercised with extreme caution. An automatic decision, in line with Proposition 1, is only possible if knowledge about spatial data distribution is available such that the precision of the modified query Q_v, with respect to the full query Q_k, is above a threshold. Moreover, the user may indicate her preferences regarding:elapsed-time; shape equivalence and F-measure. Selecting a faster execution with flexibility on the remaining parameters may open opportunity to query pre-processing under Proposition 1.

5.2   PH-tree Versus Quad-Tree

In this section, we evaluate the efficiency of the PH-tree as the basic data structure used to aid retrieving the set of candidate matching neighboring stars for each given star in the dataset. In [3], a quad-tree data structure reduces the number of matching operations by adopting a representative of set of stars covered by the Quad-tree leaf nodes. In this context, node centroids are compared and, in case of successful matching, the process is carried over to stars covered by the matching nodes. Conversely, PH-tree indexes all stars, and neighboring computation must be exercised through all stars in the dataset. Thus, this experiments compares both approaches, considering the amount of memory used to build the data structure and the elapsed-time taken when retrieving the neighbors for stars in the SDSS R-12 dataset. The latter includes approximately 7 million stars and its file size is 1 GB. The Einstein cross is the query used with maximum distance between elements in order of 10^{-5}. The PH-tree implementation showed, in average, 1.5 times higher memory footprint than the quad-tree. The former consumed 4.53 GB for 3.02 GB for the latter memory space consumption.

Figure 2(d) depicts our results on searching for candidates elapsed-time (logarithmic scale). Values for the PH-tree and the Quad-tree considered the average of 10 runs, but the ones for QT-9, which only report the average of 3 runs, due to the long elapsed-time for processing the query and, at the same time, low variance among runs. Results for QT-7, QT-8 and QT-9 report on the impact on the elapsed-time for the neighbors search when the Quad-tree is built with different heights. As it can be observed, the quad-tree with heights 7 and 8 showed better results than that of the PH-tree. Height 9, however, is the inflexion point. From height 9 on, the quad-tree implementation becomes extremely costly and PH-tree is a clear winner. It is also important to note that on higher heights, the quad-tree solutions may hide candidate solutions, as they may be covered within the same node, avoiding the procedure to detect them as candidates. Thus, despite presenting better performance, high height scenarios may lead to incomplete answers.

6   Conclusion

In this paper, we proposed pre-computing and indexing techniques for processing constellation queries. The query pre-processing technique involves selecting a query element to be excluded from the pattern in order to reduce the query complexity. As our experiments have shown, this process requires further investigation. Our proposition of excluding query elements within a pre-defined threshold distance eliminates a higher selective constraint, considerably increasing the answer set and reducing its *F-measure*. Thus, we consider that further investigation is needed to determine query elements that can be deleted without compromising query answer quality. Finally, the PH-tree presented very good results in computing neighbors of stars. The technique requires larger memory space but produces very efficient neighbors computation elapsed-time with complete solution set. We intend to explore techniques that would enable the distribution of the index structure in order to cope with even larger input datasets.

References

1. Brilhante, I.R., de Macêdo, J.A.F., Nardini, F.M., Perego, R., Renso, C.: On planning sightseeing tours with tripbuilder. Inf. Process. Manage. **51**(2), 1–15 (2015)
2. Overbye, D.: Astronomers observe supernova and find they are watching reruns. New York Times, USA (2015)
3. Porto, F., Khatibi, A., Nobre, J.R., Ogasawara, E., Valduriez, P., Shasha, D.: Constellation queries over big data. eprint arXiv:1703.02638 - Bibliographic Code: 2017arXiv170302638P, March 2017

4. Papadias, N.M.D., Delis, V.: Algorithms for querying by spatial structure. In: Proceedings of the 24th VLDB Conference, pp. 546–557 (1998)
5. Brucato, A.M., Beltran, J.F., Meliou, A.: Scalable package queries in relational database systems. Proc. VLDB Endowment **9**, 576–597 (2016)
6. Zäschke, T., Zimmerli, C., Norrie, M.C.: The PH-tree: a space-efficient storage structure and multi-dimensional index. In: Proceedings of the ACM SIGMOD International Conference on Management of Data, pp. 397–408 (2014)

A Lightweight Elastic Queue Middleware for Distributed Streaming Pipeline

Weiping Qu$^{(\boxtimes)}$ and Stefan Dessloch

Heterogeneous Information Systems Group,
University of Kaiserslautern, Kaiserslautern, Germany
{qu,dessloch}@informatik.uni-kl.de

Abstract. We introduce an elastic queue middleware (EQM) in a distributed streaming processing architecture to handle drastically growing input streams at peak times and maintain resource utilization at off-peak times. EQM serves as a scalable stream buffer to solve bottlenecks of stream processing on the fly. With spikes in data rates, the stream buffer which holds the input tuples for a bottleneck operator scales out in EQM to immediately alleviate back pressure and the streaming engines can thus gradually deploy additional replicas of the bottleneck operator to cope with the increasing data rates. This differs from general elastic streaming processing where bottleneck operators scale out first and then the stream buffers are allocated. To implement a scalable buffer, EQM utilizes existing scalable data stores (e.g. HBase) to avoid re-inventing the same elasticity and scalability logic and meanwhile ensures load balancing performance. Experiment results show that stable throughput is achieved at varying data rates using EQM.

1 Introduction

With high demand for real-time business intelligence, a new generation of distributed data stream processing systems has been developed to address the *velocity* property from the 4 V's of big data. Scalability is achieved by running streaming processing jobs in a distributed cluster environment. Unbounded data streams are split into multiple contiguous small, bounded batches that are executed concurrently by parallel instances of different operators. This is called a *distributed streaming pipeline* which is a directed acyclic graph with nodes as operators and edges as streams. Several examples are given as follows. Spark Streaming [1] proposed discretized streams (*D-Streams*) which are internally cut into successive batches, each of which is a resilient distributed dataset (*RDD*, as storage abstraction) and can be executed in their underlying pull-based, batch processing engine. Flink [2] follows the dataflow processing model and uses *intermediate data streams* as the core data exchange abstraction between operators. The real data exchange is implemented as the exchange of buffers (similar to batches) of configurable sizes, which covers both streaming and batch processing applications. Additional work like Storm+Trident [3] and S-Store [4] addresses

© Springer International Publishing AG 2017
L. Bellatreche and S. Chakravarthy (Eds.): DaWaK 2017, LNCS 10440, pp. 173–182, 2017.
DOI: 10.1007/978-3-319-64283-3_13

transactional consistency properties (exactly-once processing and ordered execution properties during concurrent access to shared state by parallel operator instances) and also fault-tolerance, at batch-wise scale.

Recent work [6–9] has also contributed to achieving elastic processing capability in response to workload fluctuation in distributed streaming processing systems. On detecting spikes in data rates, bottleneck operators are scaled out to parallel operator instances across cluster nodes to keep stable throughput. After the workload spikes, over-provisioned resources are further released for better resource utilization. One use case is the Internet of Things (IoT) applications, for example, smart cars. Sensors embedded in automobiles continuously send event streams to the cloud where real-time streaming analytics is performed. Given varying input data volume, elasticity is important for the streaming pipeline to scale out at peak hours and later scale in at off-peak times. Another use case is real-time data warehousing which relies on streaming ETL [5] engines to refresh warehouse tables. With continuous updates occurring at the data source side and analytics requests (with different freshness and deadline needs) issued to the data warehouses, streaming engines need to process different amounts of input data in time windows of various sizes in an elastic manner.

With drastically increasing data rates, the streaming buffers between processing operators fill up fast, which normally results in back pressure. This back pressure impact would not be resolved until the system first detects where the bottleneck is, determines the scale out degree, lets the deployment manager apply resources, spawns threads and finally reroutes dataflow. Especially for stateful operators, additional state-shuffling cost is mandatory. To alleviate back pressure, streaming buffers can spill to disk and also scale across cluster nodes which most of the full-fledged scalable data stores have already implemented. While most of the elastic streaming processing engines first scale the operators and then set the buffers, we argue that the buffers that really *hold* the overflowed tuples should first scale and then set the operators appropriately. Instead of reinventing scalable buffer from scratch, we address the operator-after-buffer-scale logic with an *elastic queue middleware* (EQM) which utilizes existing scalable data stores to realize a scalable buffer used for elastic streaming processing. The contribution of this article is in the following:

- We briefly introduce relevant components used for elastic streaming processing and elaborate how our elastic queue middleware fits in the global architecture through a set of abstract middleware APIs.
- We show how to wrap an open-source, scalable data store HBase as an EQM and how the middleware APIs are implemented.

The paper is organized as follows. In Sect. 2, we describe the architecture of general elastic streaming processing engines and explain the role of our elastic queue middleware in this picture. Furthermore, we provide implementation details based on HBase around four middleware primitives, i.e. *put, nextBatch, split* and *merge*. Experimental results are given in Sect. 3.

2   Elastic Queue Middleware

In this section, we describe how to embed our elastic queue middleware (EQM) in a general streaming processing engine through a set of elastic queue primitives. Furthermore, we provide detailed introduction to the implementations of EQM interface primitives based on HBase [12].

2.1   The Role of EQM in Elastic Streaming Processing Engines

Elasticity components are added to existing streaming engines for runtime adaptation to changes in the workload. These components are shown in the upper part of Fig. 1. In order to observe the workload fluctuation, a *monitor* collects runtime statistics on operator execution in a streaming pipeline and sends them periodically to a *scale handler*. The statistics usually cover throughput (processed tuples per second), congestion (blocking time of tuples in the streams between operators) and resource consumption (CPU cycles, memory). To alleviate bottlenecks in time, the scale handler relies on control algorithms [6,7] to detect a bottleneck operator in an execution flow given spikes/decline in data rates and to make accurate decisions on the degree of parallelism for scale out/in. The result of the control algorithm is sent to a *deployment manager* which balances parallel computation at runtime. On receiving a scale out request, the deployment manager immediately applies for resources in specific nodes. With granted resources, it spawns new threads as replicas of the bottleneck operator and allocates memory for stream buffers between new replicas and their upstream operators. If the bottleneck is a stateless operator, after scale out/in, its upstream operator simply re-distributes (e.g. round-robin) output tuples evenly across new operator instances. A *state management* component is introduced to manage the internal processing state of stateful operators and to scale out/in the state along with operators, meanwhile guaranteeing the state consistency. The processing state is generally maintained in key-value data structures through explicit state management APIs. For scale out/in, one can pre-partition the state on the key at deployment time and only shuffle the state partitions across a varying number of operator replicas at runtime [9]. For better load balancing on skewed workload, another solution is to re-partition the local state using different partition functions at runtime before shuffling [7,8]. In both cases, the upstream operators have to know the new locations of the shuffled partitions (i.e. new partition schema) in order to reroute the dataflow. This information may be acquired from a dedicated *partitioner*. Moreover, to guarantee fault tolerance, the processing state is periodically checkpointed and restored in new nodes by a *failure handler* in case of failure.

Since the state migration (re-shuffling) happens on the fly, one has to prevent re-shuffled state partitions from being processed by inconsistent stream tuples before successful dataflow rerouting. One simple solution is to block the upstream operators and prevent them from sending output tuples while the state is being (probably repartitioned and) re-shuffled across cluster nodes. This results in back pressure, especially when the input data rates increase drastically. Another

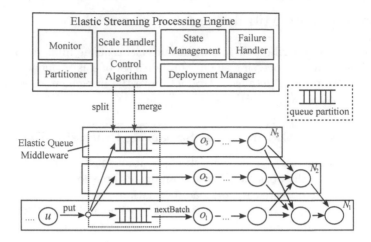

Fig. 1. Embedding elastic queue middleware in streaming processing engine

general solution is to buffer the output tuples sent from upstream operators temporarily until state migration finishes. Assuming that the workload spikes are long-time, not transient, buffers may be overloaded and frequently spill to disk. If upstream backup [8] is used, a scale out of upstream operators can be caused in a cascade fashion. Alternatively, the buffers can be split along with the processing state and placed in the same nodes where new partitioned operators (replicas of original bottleneck operator) are located and which should not be overloaded. We call this locality-affinity buffering. In this work, instead of *actively* detecting and replicating bottleneck operators, we let the operators scale along with an elastic buffer (more concisely, a *queue*[1]) in a *passive* manner. We see an example at the bottom part of Fig. 1. An operator u was initially only attached to a downstream operator o_1 through an elastic queue with one queue partition having a pre-defined maximum size as threshold. On a spike in data rates, o_1 becomes the bottleneck and the queue gradually scales out to three partitions in nodes N_{1-3} to balance the load. The deployment manager spawns new threads o_2 and o_3 in the nodes N_2 and N_3, respectively, to cope with the workload spike. To eliminate the overhead of thread construction at runtime, the threads can be created, suspended at deployment and be woke up at runtime when the queue partitions are *spread* on the same node. Moreover, to increase parallelism, the operators that can be partitioned on the same key are grouped and replicated together across cluster nodes until a new partition key is used in downstream operators, which requires a shuffle phase.

In the *era* of big data, a wide spectrum of scalable data stores [11] has been developed and extensively used in industry. Instead of writing a new, internal, elastic queue from scratch, we propose *elastic queue middleware* (EQM) as the

---

[1] Instead of buffer, we want to achieve "write once; read once" queue feature with *enqueue* and *dequeue* operation support in this context.

solution which wraps scalable data stores and exposes only a basic set of methods to the existing components of an elastic streaming engine. These EQM primitives are *put, nextBatch, split* and *merge*. In the following, we explain how we wrapped a scalable data store HBase [11] as EQM based on a brief HBase introduction and further describe the implementations of these four methods.

2.2   Implementing EQM Based on HBase

The implementation of HBase follows the architecture of Google's Bigtable [13] and is internally designed as a distributed, log-structured merge tree. Data manipulations on a table in HBase are re-directed to remote partitions where local read and write operations are carried out. We briefly illustrate the read and write paths in HBase partitions in Fig. 2. A table partition contains an in-memory data structure called *memstore*, which can be filled by key-value tuples using the *add* method. When its size reaches a threshold, a *flush* action is triggered to first sort the in-memory tuples on the key and then to write the content of memstore to disk as a *store file*. To read out the content, the *scan* method is called, which internally opens scanners on the memstore and store files and performs a merge phase based on the key.

HBase deploys Multi-Version Concurrency Control (MVCC) mechanism to isolate read and write operations. A monotonically increasing sequence-id list is maintained internally to ensure that concurrent adds are committed in sequence based on sequence ids assigned. With READ COMMITTED as the isolation level, a variable called *read point* is used to indicate the current greatest sequence id, before which all the previous add operations with smaller sequence ids

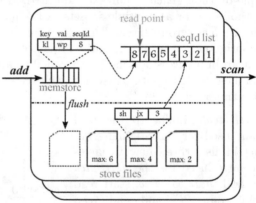

Fig. 2. HBase table partitions

have successfully committed. When multiple adds commit, the read point is advanced as much as possible. The current read point is assigned to all the scanners of a new scan request, which is used to check whether each accessed key-value tuple should be returned by comparing its sequence id and the read point. If the sequence id of a key-value tuple is smaller than the read point in a scanner, this tuple is returned. In Fig. 2, a key-value tuple (key: "*kl*"; val: "*wp*"; seqId: 8) which is inserted by an in-progress add operation with an assigned sequence id (8) is invisible to the current scan operation since its sequence id (8) is greater than the current read point (7) in a memstore scanner while another on-disk key-value tuple (key: "*sh*"; val: "*jx*"; seqId: 3) is returned by a store file scanner which shares the same read point. Moreover, to increase read performance, small store files can be *compacted*, which opens scanners on all the sorted store files and merges them to a larger one. Furthermore, when HBase detects that one partition is overloaded by current workload, it *splits* the key range of the partition down the middle and migrates two daughter partitions to separate remote nodes for load balancing. On the absence of significant load, HBase can also move two remote partitions with adjacent key ranges to the same node and perform a local *merge* by simply putting store files together in one directory and opening a new partition with a larger key range. Note that, after split or merge, on-going add or scan requests can be rerouted to the correct nodes where new partitions reside.

We wrap HBase as an EQM prototype and provide four methods *put, next-Batch, split* and *merge* to support streaming engine integration (see Fig. 1).

Put. Since most of the streaming processing engines attach batch ids to stream tuples to distinguish different batches at runtime, we assume that the incoming key-value (kv) tuples contain additional batch id (bid) information and the original add method is extended to $add(kv|bid)$. We modified HBase to replace the original sequence-id list with a new *batchId-seqId map* which maps a (streaming-level) batch id to multiple (HBase-specific) sequence ids. Depending on the upstream operators, all the tuples belonging to one batch can be added tuple-by-tuple (each requires an add method invocation), all together (just one method call) or as multiple subsets. All sequence ids assigned to the adds from one batch are grouped together as the values of that batch-id key which represent a contiguous subset of the entire sequence-id list. The example in Fig. 3 shows that an in-progress add method is called to append a kv tuple to the memstore and meanwhile to put its assigned sequence id (8) in the sequence list associated with the batch-id key b_4.

NextBatch. Similar to the read point used for the original sequence-id list, three new variables are introduced for the *nextBatch* method: *next batch id*, start read point ($startPt$) and end read point ($endPt$). Moreover, a new *DequeueScanner* is implemented to scan the memstore and store files in a queue partition. When a dequeueScanner is instantiated, it tries to fetch the next available batch id in the batchId-sequenceId map, as the map is

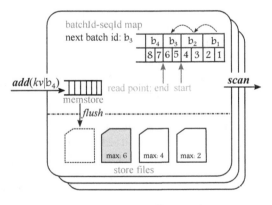

Fig. 3. EQM-specific partitions in HBase

sorted on the increasing batch id. If any add operation containing sequence id associated with the fetched batch id has not committed yet, the dequeueScanner blocks the current calling thread, which also suspends the downstream operator thread, until all adds of this batch have committed. The dequeueScanner uses the fetched batch id to find its sequence-id list in the map and assign the minimum and maximum sequence ids in the list to the startPt and endPt,

respectively. Memstore and store-file scanners are modified so that they only
return the tuples that have their sequence ids falling in the range of [startPt,
endPt]. It is important to note that a store file contains a max sequence id
which indicates the greatest sequence id among all its key-value tuples. Given a
[startPt, endPt] range and a list of store files sorted on their max sequence ids,
only required store files are scanned, thus significantly reducing disk I/Os. In
Fig. 3, invoking the nextBatch creates a memstore scanner and multiple store-
file scanners with [5, 6] as the read point range after looking up the map with
b_3 as the key. Only the store file with max sequence id (6) is hit for scanning. A
subsequent nextBatch call would stall until the $add(kv—b_4)$ has its sequence id
as (8) commits.

Split. HBase supports automatic splitting and manual splitting. One can set the
maximal allowed size of a store file at deployment time and let HBase automatically split those queue partitions that have one of its store files exceeding this
threshold at runtime. Automatic splitting has the main advantage that there is
no need to implement extra elasticity control algorithms and monitoring component in streaming engines. However, the drawback is that the decision of scaling
out/in is limited to the HBase-internal metrics (e.g. max file size), thus losing the
flexibility of tuning scaling using different metrics, e.g. processing throughput,
resource consumption, etc. In addition, if the max file size is set too high, EQM
would not react to workload variations immediately and not scale out in time. If
the max file size is set too low, EQM would be load-sensitive and be busy with
splitting, thus incurring lots of unnecessary network I/Os. All these drawbacks
can be compensated by letting the elasticity control algorithms invoke HBase-
internal *split* methods to manually split partitions on given split keys, whereas
new overhead is incurred that the streaming engines need to keep track of load
distributions across partitions. The choice of two splitting scheme is a trade off
between software engineering costs and performance.

Merge. We see that the nextBatch method does not immediately remove the
scanned tuples from the storage, although the performance is not impacted as
the dequeueScanner selectively scans store files with given read point range. If
the workload shows a decline, we can merge queue partitions to less number of
partitions by calling the merge method in HBase. Before that, a local compaction
is processed in each partition to reduce the amount of store files transferred
through the network. As mentioned in Subsect. 2.2, the working principle of the
compaction process is to first read out tuples from the store files and write them
to a single one. Hence, another new *QueueCompactor* is implemented. It fetches
the minimum sequence id from the sequence-id list of the current batch id to
filter out unnecessary store files with their maximum sequence id smaller than
the fetched one, since they have been scanned and are not needed any more.
Original version of memstore and store-file scanners are then used to finish the
rest of the compaction and generate a much smaller store file. In the example,
the store files with max sequence id (4) and (2) are skipped during compaction.
However, HBase does not support automatic merging. The right time of scale in
thus has to be determined by the control algorithm of elastic streaming engines.

3   Experiments

The experiments were based on a streaming ETL pipeline introduced in our previous work [14]. The streaming ETL pipeline is a directed acyclic graph with nodes as operators and edges as streams of change data (deltas). Delta streams are split into contiguous delta batches, each of which updates data warehouse tables to a consistent state at specific points in time. With spikes in data rates, slow bottleneck operators restrict overall throughput. Therefore, we implemented an elastic queue middleware (EQM) using HBase and embedded it into the streaming ETL pipeline to alleviate the bottlenecks. In this section, we present initial experimental results based on a small EQM prototype, which proves that EQM ensures throughput when facing workload variations.

The experiments were set as follows. HBase ran on a 6-node cluster where the master process was running in a master node (2 Quad-Core Intel Xeon Processor E5335, 4×2.00 GHz, 8 GB RAM, 1 TB SATA-II disk) and the rest five nodes (2 Quad-Core Intel Xeon Processor X3440, 4×2.53 GHz, 4 GB RAM, 1 TB SATA-II disk, Gigabit Ethernet) were used to run the slave processes. Our streaming ETL pipeline ran in one slave node and hosted a data maintenance flow which refreshes the *store sales* (SF 10) table in TPC-DS benchmark [15]. The EQM was set explicitly before the bottleneck operator (a lookup operator on an *item* table) and utilized a scalable HBase table which replaced the original in-memory stream buffer. It is important to note that the downstream operator of the bottleneck was a join operator in the data flow which required shuffling at runtime after the bottleneck operator scaled out. We simulated the workload fluctuation by continuously increasing the data rates by adding more clients and gradually slowing down the inserting speed. In both cases, we compared the throughputs of running the streaming pipeline in single node with those of running the same pipeline with embedded EQM.

Figure 4 shows the throughput comparison in the case of scale out. We see that single-node pipeline suffered from back pressure caused by the bottleneck operator (max. 320 tuples/second) while EQM helped the streaming pipeline scale out with the automatic splitting feature in HBase, thus increasing the throughput. Each time after one splitting finished, the throughput climbed

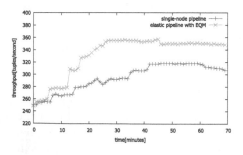

Fig. 4. Scale out

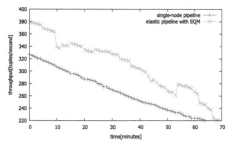

Fig. 5. Scale in

up significantly and slowly converged until the next splitting occurred. In the end, the throughput converged at around 340 tuples/s due to limited resources. Figure 5 presents the results of scale in. At the beginning, EQM still performed well due to high parallelism. When the workload decline occurred, we manually invoked the merge methods at specific points in time to reduce the degree of parallelism, which also reduces the shuffling impact caused by the downstream join operator as explained before. By using the QueueCompactor, only required store files were transferred through network, which avoids unnecessary network I/Os. The results show that scale-in achieved better throughput than running in a single node.

4 Related Work

There is much work addressing elastic scaling in streaming processing engines with particular goals. Schneider and Gedik et al. [6] described how to make stateless operators adaptive to changing workload by dynamically balancing load across working threads through a global tuple queue. A control algorithm was designed to identify the peak rate (maximum processing throughput) and to decide on correct number of parallel working threads at runtime. They improved this algorithm in [7] by additionally considering the tuple block time in the buffers between operators. Besides, they addressed the state migration problem for partitioned stateful operators through a state management API. When a scale out is triggered, local state is re-partitioned to multiple packages according to new partition schema and shuffled across new partitioned operators through a backend database. During the state migration, the upstream operators are blocked to send output tuples, which results in back pressure. Fernandez et al. [8] addressed the state migration problem by backing up the processing state of a bottleneck operator periodically in its upstream operator and repartitioned for scale out. Buffers are used in upstream operators to temporarily ensure consistent state during scale out, which might trigger a scale out of the upstream operator if the data rates still rise. In contrast, Wu et al. [9] pre-partition the processing state before runtime and distribute state partition replicas across cluster nodes. Transparent workload migration is achieved by spawning new threads on remote nodes where backup partition replicas are located, releasing original ones and rerouting outputs from upstream operators. However, skewed workload may result in bottlenecks when state is only pre-partitioned at deployment time, which is addressed by EQM where large partitions are automatically split under load. Besides, EQM scales bottleneck operators along with the queue partitions without blocking upstreaming operators.

Other work also exists which covers queue or buffer data structures in dataflow systems. Karakasidis et al. [10] proposed a framework for the implementation of active data warehousing. ETL activities are considered as queues while an ETL flow is transformed to a queue network, where queue theory can be performed to predict the mean delay and the queue length of ETL queues, given the source production rates and the processing power of the staging area.

They did not address the scalability of queues, although the idea of pipelining blocks of tuples in ETL queues was already there. Carbone et al. introduced the principles of Apache Flink in [2] where intermediate data streams are categorized into pipelined intermediate streams and blocking streams. For pipelined intermediate streams, intermediate buffer pools are used to compensate for short-term throughput fluctuations. In our work, we mainly focused on wrapping existing scalable data stores as stream buffers for long-term workload variations.

5   Conclusion

In this paper, we studied how current elastic streaming processing engines react to changes in the workload and showed the functionality of relevant components in a general architecture. To realize elasticity in distributed streaming pipeline, we proposed our elastic queue middleware (EQM) implementation as a scalable locality-affinity buffer by wrapping HBase and exposing four elasticity methods to elastic streaming engine integration, thus significantly reducing the software engineering costs. Experimental results from a small prototype proved that EQM can help streaming processing flow keep stable throughput.

References

1. Zaharia, M., Das, T., et al.: Discretized streams: fault-tolerant streaming computation at scale. In: SOSP (2013)
2. Carbone, P., Katsifodimos, A., et al.: Apache flink: stream and batch processing in a single engine. In: Data Engineering (2015)
3. Trident Tutorial. http://storm.apache.org/documentation/Trident-tutorial.html
4. Meehan, J., Tatbul, N., et al.: S-store: streaming meets transaction processing. In: VLDB (2015)
5. Meehan, J., Aslantas, C., et al.: Data ingestion for the connected world. In: CIDR (2017)
6. Schneider, S., Andrade, H., et al.: Elastic scaling of data parallel operators in stream processing. In: IPDPS (2009)
7. Gedik, B., Schneider, S., et al.: Elastic scaling for data stream processing. IEEE TPDS **25**(6), 1447–1463 (2014)
8. Fernandez, R.C., Migliavacca, M., et al.: Integrating scale out and fault tolerance in stream processing using operator state management. In: SIGMOD (2013)
9. Wu, Y., Tan, K.L.: ChronoStream: Elastic stateful stream computation in the cloud. In: ICDE (2015)
10. Karakasidis, A., Vassiliadis, P., et al.: ETL queues for active data warehousing. In: IQIS (2005)
11. Cattell, R.: Scalable SQL and NoSQL data stores. SIGMOD **39**(4), 12–27 (2011)
12. https://hbase.apache.org/
13. Chang, F., Dean, J., et al.: Bigtable: a distributed storage system for structured data. ACM TOCS 26(2) (2008)
14. Qu, W., Basavaraj, V., Shankar, S., Dessloch, S.: Real-time snapshot maintenance with incremental ETL pipelines in data warehouses. In: Madria, S., Hara, T. (eds.) DaWaK 2015. LNCS, vol. 9263, pp. 217–228. Springer, Cham (2015). doi:10.1007/978-3-319-22729-0_17
15. http://www.tpc.org/tpcds/default.asp

Modeling Data Flow Execution
in a Parallel Environment

Georgia Kougka[1]([✉]), Anastasios Gounaris[1], and Ulf Leser[2]

[1] Department of Informatics, Aristotle University
of Thessaloniki, Thessaloniki, Greece
{georkoug,gounaria}@csd.auth.gr
[2] Institute for Computer Science, Humboldt-Universität
zu Berlin, Berlin, Germany
leser@informatik.hu-berlin.de

Abstract. Although the modern data flows are executed in parallel and distributed environments, e.g. on a multi-core machine or on the cloud, current cost models, e.g., those considered by state-of-the-art data flow optimization techniques, do not accurately reflect the *response time* of real data flow execution in these execution environments. This is mainly due to the fact that the impact of parallelism, and more specifically, the impact of concurrent task execution on the running time is not adequately modeled. In this work, we propose a cost modeling solution that aims to accurately reflect the *response time* of a data flow that is executed in parallel. We focus on the single multi-core machine environment provided by modern business intelligence tools, such as Pentaho Kettle, but our approach can be extended to massively parallel and distributed settings. The distinctive features of our proposal is that we model both time overlaps and the impact of concurrency on task running times in a combined manner; the latter is appropriately quantified and its significance is exemplified.

1 Introduction

Nowadays, data flows constitute an integral part of data analysis. The modern data flows are complex and executed in parallel systems, such as multi-core machines or clusters employing a wide range of diverse platforms like Pentaho Kettle[1], Spark[2] and Stratosphere[3] to name a few. These platforms operate in a manner that involves significant time overlapping and interplay between the constituent tasks in a flow. However, there are no cost models that provide analytic formulas for estimating the *response time (wallclock time)* of a flow in such platforms. Cost models, apart from being useful in their own right, are encapsulated in cost-based optimizers; currently, cost-based optimization solutions for task ordering in data flows employ simple cost models that may

---

[1] http://community.pentaho.com/projects/data-integration.
[2] http://spark.apache.org/.
[3] http://stratosphere.eu/.

© Springer International Publishing AG 2017
L. Bellatreche and S. Chakravarthy (Eds.): DaWaK 2017, LNCS 10440, pp. 183–196, 2017.
DOI: 10.1007/978-3-319-64283-3_14

not capture the flow execution running time accurately, as shown in this work. This results in an execution cost computation that may deviate from the real execution time, and the corresponding optimizations may not be reflected on response time.

Typically, cost models rely on the existence of appropriate metadata regarding each task, which are combined using simple algebraic formulas with the sum and max operations. Most often, task metadata consider the cost of each task, which is commensurate with the task running time if executed in a stand-alone manner. The main challenges in devising a cost model for running time that is appropriate for modern data flow execution stem from the following factors: (i) many tasks are executed in parallel employing all three main forms of parallelism, namely, partitioned, pipelined and independent, and the resulting time overlaps, which entail that certain task executions do not contribute to the overall running time, need to be reflected in the cost model; and (ii) computation resources are shared among multiple tasks, and the concurrent execution of tasks using the same resource pool impacts on their execution costs.

In this work, we focus on devising a cost model that can be used to estimate the response time, when the dataflows are executed in parallel and distributed execution environments. To this end, we extend existing cost modeling techniques that tend to consider time overlapping (e.g., [1,2,4,11,16,18]) but not the interplay between task costs. In order to achieve this, we propose a solution in which the cost of each task is weighted according to the number of concurrent tasks taking into account constraints of execution machines, such as capacity in terms of number of cores. More specifically, we initially focus on a single multicore machine environment, such as Pentaho Data Integration (PDI, aka Kettle), and the contribution is as follows:

1. We explain and provide experimental evidence on why the existing cost models provide estimates that widely deviate from the real execution time of modern workflows.
2. We propose a model that not only considers overlapping task executions but also quantifies the correlation between task costs due to concurrent allocation to the same processing unit. The model is execution engine software- and data flow type-independent.
3. We show how our model applies to example flows in PDI, where inaccuracies of up to 50% are observed if the impact of concurrency is not considered.

In the remainder of this section we provide background on flow parallelization, the assumptions regarding the execution environment that we consider and a discussion about the inadequacy of cost models employed in data flow optimization. We continue the discussion of related work in Sect. 2. In Sect. 3 we introduce the notation. Our modeling proposal is presented in detail in Sect. 4 and we conclude in Sect. 5.

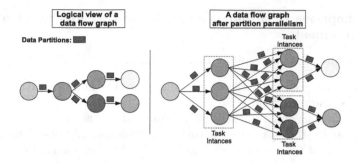

Fig. 1. A data flow graph before and after partitioned parallelism; circles with the same color correspond to partitioned instances of the same flow task. (Color figure online)

1.1 Parallelizing Data Flows

The parallel execution of a data flow exploits three types of parallelism, namely *inter-operator*, *intra-operator* and *independent* parallelism. Here, we use the terms *task*, *activity* and *operator* interchangeably. These types of parallelization are well-known in query optimization [6], and used to decrease the *response time* of a data flow execution.

The *intra-operation parallelism* considers the parallelization of a single task of a data flow. This type of parallelization is defined by the instantiation of a logical task as a set of multiple physical instances, each operating on a different portion of data, i.e. each task can be executed by several processors and data is partitioned. An example of partitioned parallelism is depicted in Fig. 1. There is a set of different methods of partitioning, such as round-robin and hash-partitioning. In this work, we assume that the degree of *intra-operation* parallelism is fixed; e.g., in Fig. 1, the degree for the green task is set to 3.

The *independent* parallelism is achieved when the tasks of the same data flow may be executed in parallel because there are no dependency constraints or communication between them. An example is the two branches at the right part of the flow in Fig. 1(left).

The *pipeline* parallelism takes place when multiple tasks are executed in parallel with a producer-consumer link and each producer sends part of its output, which is a collection of records, as soon as this output is produced without waiting the processing of its input to complete and, therefore, the whole output to be produced.

In this work, we present a cost model for data flow execution plans that accurately estimates the *response time* considering the *pipeline parallelism* and *independent* types of parallelism, which are relevant to a single machine Kettle execution. However, it is straightforward to extend our work to cover partitioned parallelism as well, as briefly discussed in Sect. 4.

1.2 Assumptions Regarding a Single Multi-core Machine Execution Environment

Our main assumptions are summarized as follows:

- Data flows utilize all the available machine cores. The number of cores depends on the execution machine.
- The execution machine is exclusively dedicated to the data flow execution. I.e., we assume that an execution machine executes only one data flow and the execution of the next flow can be started only after the completion of the previous flow. So, the available machine executes tasks and stores data for a single data flow at a time.
- Multiple tasks of a data flow are executed simultaneously. Each task spawns a separate thread, running on a core decided by the underlying operating system scheduler. Obviously, if two task threads share the same core, they are executed concurrently but not simultaneously.
- The execution engine exploits pipeline and independent parallelism to the largest possible extent; i.e., the default engine configuration regarding task execution operates in a mode, according to which flow tasks are aggressively scheduled as soon as their input data is available.

The assumptions above hold also for massive parallel settings. The main difference is that, in massive parallelism settings, partitioned parallelism typically applies.

1.3 Motivation for Devising a New Cost Model

A main application of cost models is in cost-based optimization. One of forms of data flow optimization that has been largely explored in the data management literature is task re-ordering. Taking this type of optimization as case study, we can observe from the survey in [9] that the corresponding techniques target one of the following optimization objectives:

1. *Sum Cost Metric of the Full plan (SCM-F)*: minimize the sum of the task and communication costs of a data flow [7,8,10,13,15,16,20].
2. *Sum Cost Metric of the Critical Path (SCM-CP)*: minimize the sum of the task and communication costs along the flow's critical path [1,2].
3. *Bottleneck*: minimize the highest task cost [1,2,16,18].
4. *Throughput*: maximize the throughput (number of records processed per time unit) [5].

The first three metrics and the associated cost models can capture the *response time* under specific assumptions only. The *response time* represents the wall-clock time from the beginning until the end of the flow execution. *SCM-F* defines the *response time* when the tasks of a data flow are executed sequentially; for example when all tasks are blocking. Another case is when tasks are pipelined but are executed on the same CPU core (processor). In that case, the *SCM-F* may serve as a good approximation of the *response time*. *SCM-CP* reflects the *response time* when the data flow branches are executed independently and the tasks of each branch are executed sequentially. Finally, *bottleneck* represents the *response time* when all the tasks of the flow are executed in a pipelined manner and each task is executed on a different processor assuming enough cores are available.

So, why do we need another cost model? PDI, Flink, Spark and similar environments aggressively employ pipeline parallelism potentially on multiple processors. Consequently, the *SCM-F* and *SCM-CP* cost metrics do not correspond to the *response time* of the flow execution. *Bottleneck* cost metric is not appropriate either. This is because there are pipelined tasks that are executed on the same processor, but also there are tasks that are blocking, e.g., sort. So, for estimating *response time*, we need to employ a cost metric that explicitly considers parallelism and the corresponding overlaps in task execution. A cost model that computes this cost metric is therefore needed.

Furthermore, a more accurate cost model for describing the response time is significant in its own right even when not used to drive optimizations. It allows us to better understand the flow execution and provides better insights into the details involved. Moreover, as will be shown in the subsequent section, merely considering time overlaps does not suffice, because the task costs are correlated during concurrent task execution.

2 Other Related Work

The main limitation of existing cost models is that, even if they consider overlapped execution, they assume that the cost of each task remains fixed independently of whether other tasks are executing concurrently sharing CPU, memory and other resources. Examples that fall in this category are the work in [11], which targets a cloud environment for scientific workflow execution, and in [4]. The cost model in the latter considers that the flow is represented by a graph with multiple branches (or paths), where the tasks in each path are executed sequentially and multiple branches are executed in parallel. In contrast, we cover more generic cases.

Additionally, several proposals based on the traditional cost models have been presented in order to capture the execution of MapReduce jobs. For example, a performance model that estimates the job completion time is presented for ARIA Framework in [19]; this solution accounts for the fact that the map and reduce phases are executed sequentially employing partitioned parallelism but do not take into account the effect of allocation of multiple map/reduce tasks on the same core. The same rationale is also adapted by cost models introduced in proposals, such as [17, 21]. Nevertheless, an interesting feature of these models is that they model the real-world phenomenon of imbalanced task partition running times. In the MapReduce setting, the authors in [14] propose the Produce-Transporter-Consumer model to define the parallel execution of MapReduce tasks. The key idea is to provide a cost model that describes the tradeoffs of four factors (namely, map and reduce waves, output compression of map tasks and copy speed during data shuffling) considering any overlaps. As previously, the impact of concurrency is neglected. Other works for MapReduce, such as [3], suffer from the same limitations.

3   Preliminaries

A data flow is represented as a *Directed Acyclic Graph (DAG)*, where each vertex corresponds to a task of the flow and the edges between vertices represent the communication among tasks (intermediate data shipping among tasks). In data flows, the exchange of data between tasks is explicitly represented through edges. We assume that the data flows can have multiple *sources* and multiple *sinks*. A *source* of a data flow corresponds to a task with no incoming edges, while a *sink* corresponds to a task with no outgoing edges. The main notation and assumptions are as follows:

Let $G = (V, E)$ be a *Directed Acyclic Graph (DAG)*, where $V = v_1, v_2, ..., v_n$ denotes the vertices of the graph (data flow tasks) and E represents the edges (flow of data among the tasks); n is the total number of vertices. Each vertex corresponds to a data flow task and is responsible for one or both of the following: (i) reading or storing data, and (ii) manipulating data. The tasks of a data flow may be complex data analysis tasks, but may also execute traditional relational operations, such as union and join. Each edge equals to an ordered pair (v_j, v_k), which means that task v_j sends data to task v_k.

Each data flow is characterized by the following metadata:

- *Cost* (c_i), which applies to each task. The c_i corresponds to the cost for processing all the input records that the v_i task receives taking into consideration the required CPU cycles and disk I/Os. In distributed systems, the cost of network traffic needs to be considered as well, and may be the most important factor. An essentially similar consideration is c_i to denote the cost per single input record; in that case the *selectivity* information of all tasks is needed in order to derive the task cost for its entire input.
- *Communication Cost* $(cc_{i \to j})$, which may apply to edges. The communication cost of data shipping between the v_i and v_j depends on either the forward local pipelined data transfer between tasks or the data shuffling between parallel instances of the same data flow. It does not include any communication-related cost included in c_i; it includes only the cost that is dependent on both v_i and v_j rather than only on v_i.
- *Parallelism Type of Task* (pt_i), which describes the type of parallelism of a task i, when the task is executed. More specifically, the parallelism type characterizes if a data flow task is executed in a pipelined, denoted as p or no pipelined manner (*blocking* task), denoted as np. A *blocking* task requires all the tuples of the input data in order to start producing results; i.e., the parallelism type of a task reflects the way a task process the input data and produces its output.

4   Our Cost Model

First, we describe the main formula template of our model, then we explain how it applies to a single-machine setting and finally, we generalize to distributed settings.

4.1   A Generalized Cost Model for Response Time

We define the following cost model for estimating the response time:

$$Response\ Time\ (RT)\ =\ \sum z_i w^c c_i + \sum z_{ij} w^{cc} cc_{i \to j} \qquad (1)$$

where variable $z_i = \{0, 1\}$ is binary and defined as 1 only for tasks that determine the RT. The c_i factor denotes the cost of the ith task, where $i = \{1, \ldots, n\}$. The w^c and w^{cc} weights cover a set of different factors that are responsible for the increase/decrease of RT during the task execution and communication between two tasks (data shipping), respectively. The z variables capture the time overlapping of different tasks, whereas w^c and w^{cc} quantify the impact of the execution of one task on all the other tasks that are concurrently executed, i.e., they capture the correlation between the execution of multiple concurrent tasks.

In general, the weights aim to abstract the impact of multi-threading in a single metric. Multi-threading may lead to performance overhead due to several factors, such as the *context switching* between threads, as the flow tasks are executed concurrently and need to switch from one thread to another multiple times. An additional factor for response time increase is due to the *locks* that temporarily restrict tasks sharing memory to write to the same memory location. Finally, the most significant factor in the terms of affecting the response time is the *contention* that captures the interference of the multiple interactions of each data flow task with memory and disk. Specifically, when there are multiple requests to memory, this may result in exceeding memory bandwidth and consequently, to RT increase. Finally, allocating and scheduling threads incurs some overhead, which, however, is negligible in most cases.

Nevertheless, multi-threading execution leads to execution cost improvement because of the parallel task execution. So, we may observe RT minimization, when all or more of the available cores are exploited by the data flow tasks and one copy of data is used by multiple threads at the same time. Also, the delays occurred by transferring data from memory and disk are overlapped by the task execution, when the number of tasks is higher than the available execution units.

The cost model in Eq. (1) generalizes the traditional ones. For example, based on the proposed formula, if we consider w^c and w^{cc} set to 1 and that all the tasks have $z_i = 1$, then the cost model actually corresponds to *SCM-F* and defines the RT under the specific assumptions discussed previously. If only the tasks that belong to the critical path have $z_i = 1$, then the cost model corresponds to *SCM-CP*. Similarly, if we want to consider the bottleneck cost metric, we set $z_i = 1$ for the most expensive task and $z_i = 0$ for all the other tasks.

4.2   Models Without Considering the Communication Cost

Firstly, we examine simple flows and we gradually extend our observations to larger and more complicated ones. In all cases, given that we target single-machine environments, it is reasonable to consider that the communication cost $cc_{i \to j}$ is set to 0.

A Linear Flow with a Single Pipelined Segment of n Tasks

A pipeline segment is defined by a sequence of n tasks in a chain, where the first task is a child of either a source or a blocking task, and the last task is either a sink or a blocking task; additionally, the tasks in between are all of p type. Also, pipeline segments do not overlap with regards to the vertices they cover. The key point of our approach is to account for the fact that there is non-negligible interference between tasks, captured by the variable α. Let us suppose that our machine has m cores. In the case where $n \leq m$, each task thread can execute on a separate core exclusively. The cost model that estimates the *response time (RT)* of a data flow execution is defined as follows, which aims to capture the fact that the running times of tasks overlap. So, we set (i) $z_i = 0$ for all tasks, apart from the task with the maximum cost, for which z is set to 1, since it determines the RT; and (ii) $w^c = \alpha$:

$$Response\ Time\ (RT) = \alpha \max\{c_1, ..., c_n\} \qquad (2)$$

Let us consider now the case where $n > m$ and the task threads need to share the available cores in order to be executed. In this case, each core may execute more than one task and the RT is determined by all the flow tasks, so $z_i = 1$ for all the flow tasks. An exception is when there is a single task with cost higher than the sum of all the other costs (similarly to the modeling in [19]):

$$Response\ Time\ (RT) = \alpha \max\{\max\{c_1, .., c_n\}, \frac{\sum\{c_1, ..., c_n\}}{m}\} \qquad (3)$$

In Eq. (3), w^c equals either to α, as in Eq. (2), or to α/m.

Experiments in PDI

In the following, we present a set of experiments that we conducted in order to understand the role of α in RT estimation. We consider synthetic flows in PDI with $n = 1, \ldots, 26$ tasks and an additional source task. The input ranges from 2.4M to 21.8M records. Two machines are used, with (i) a 4-core/4-thread i5 processor; and (ii) a 4-core/8-thread i7 processor, respectively. Finally, the task types are two, either homogenous or heterogeneous. In the former case, all tasks have the same cost (denoted as *equal*). In the latter case (denoted as *mixed*), half of the tasks have the same cost as in the *equal* case, and the other tasks have another cost, which is lower by an order of magnitude. All the tasks apply filters to the input data, but these filters are not selective in the sense that they produce the same data that they receive. The data input is according to the TPC-DI Benchmark [12] and we consider records

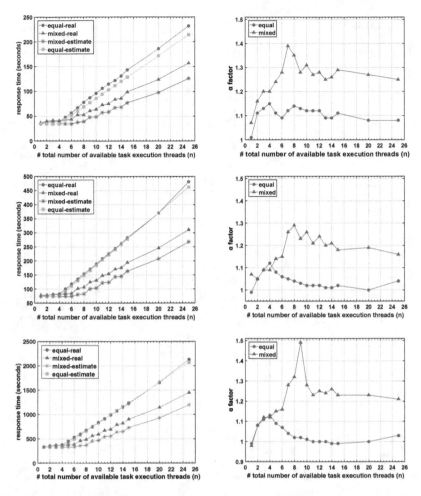

Fig. 2. Response Time (RT) and the α factor of linear flows with same and different task costs for $n \in [1, 25]$ executed by the 4-core/4-thread i5 machine for 2.4 (top), 4.8 (middle) and 21.8 M (bottom) input records.

taken from the implementation in http://www.essi.upc.edu/dtim/blog/post/ tpc-di-etls-using-pdi-aka-kettle. Each experiment run was repeated 5 times and the median times are reported; in all experiments the standard deviation was negligible.

The left column of Figs. 2 and 3 shows how the response time of the two different types of data flows evolves as the number of tasks, and consequently the number of execution threads, increases. It also shows what the cost model estimates would be if no weights were considered. The main observation is twofold. First, the response time, as expected from Eqs. (2) and (3), stays approximately stable when $n \leq m$, and then, grows linearly when $n > m$. This behavior does not change with the increase in the data size. Second, estimates with no weights

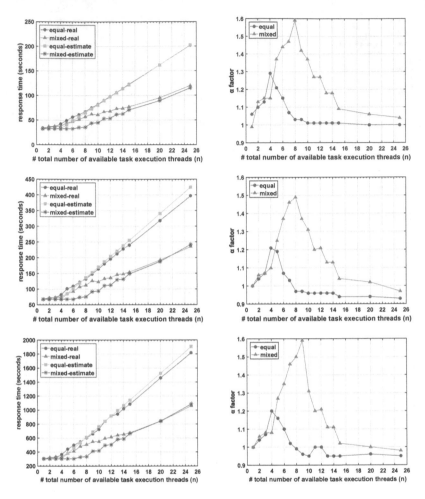

Fig. 3. Response Time (RT) and the α factor of linear flows with same and different task costs for $n \in [1, 25]$ executed by the 4-core/8-thread i7 machine for 2.4 (top), 4.8 (middle) and 21.8 M (bottom) input records.

can underestimate the running time by up to 50%, whereas there are also cases when they overestimate the running times by a smaller factor (approx. 5%). More importantly, the main inaccuracies are observed in the mixed-cost case, which is more common in practice.

The α factor is shown in the right column of Figs. 2 and 3. Values both lower and higher than 1 are observed. Although α captures the combination of overhead and improvement causes described in the previous section, the importance of each cause varies. In values greater than 1, resource contention is dominating; whereas, in values lower than 1, the fact that waits for resources are hidden outweighs any overheads. The main observations are as follows: (i) the α factor

Table 1. Comparison of running times between flows with the same number of tasks but (a) with 2 independent and (b) a single segment (in seconds).

	2.4 million records				4.8 million records				21.8 million records			
n	4cores-4threads		4cores-8threads		4cores-4threads		4cores-8threads		4cores-4threads		4cores-8threads	
	2 paths	1 path	2 path	1 path	2 paths	1 path	2 paths	1 path	2 paths	1 path	2 path	1 path
2	38.5	38.1	35.1	35.6	78	78	69	71	346	356	315	317
4	42	39.4	42.1	41.9	84	83	82	82	369	374	370	377
6	59.3	56.3	56	56.1	118	118	109	109	530	533	497	499
8	78	78	67	67	155	154	134	132	694	677	606	594
10	96	96	83	82	192	189	166	164	851	837	735	727
12	115	115	99	98	229	226	197	196	1040	991	879	916
14	134	131	115	114	265	262	229	228	1210	1151	1006	1019
16	153	152	131	129	304	305	260	256	1382	1311	1146	1152
18	164	170	145	145	327	323	287	289	1486	1493	1296	1299
20	187	186	160	162	380	370	323	318	1720	1662	1454	1437

varies significantly for the same dataset when the number of tasks is modified; (ii) α can be of significant magnitude corresponding to more than 50% increase in the task costs; (iii) for flows that consist of up to 4 tasks with equal cost, the α factor continuously grows (i.e., contention is dominating) and then, when the number of tasks further increases, the behavior differs between cases; and (iv) for data flows with different task costs and $n > m$, the α factor increases sharply for flows with up to 7–9 tasks depending on the input data size.

A Linear Flow with Multiple Independent Pipelined Segments

In Table 1, we show the running times of flows with the same number of tasks when all tasks belong to a single pipelined segment and when there are two segments belonging to two different paths originating from the same source. We can observe that the running times are similar. From this observation, we can draw the conclusion that the magnitude of the weights (i.e., the w^c and the corresponding α factors) depend on the number of concurrent tasks and need not be segment-specific; that is, it is safely to assume that all concurrent tasks share the same factors.

Estimating the Response Time of a Flow: The Complete Case

In the previous sections, we showed how we can estimate the response time of a single pipelined segment in data flows. Now, we leverage our proposal to more generic data flows with multiple pipeline segments, in order to estimate the response time of flows that consist of multiple pipeline segments. To this end, we employ a simple list scheduling simulator. The steps of this methodology are described, as follows:

1. Receive as input the flow DAG, the cost (c_i) of all the tasks of a dataflow, the number of available cores, and the α factors.
2. Isolate all the single-pipeline segments of the flow with the help of the parallelism type task metadata.
3. Split the input in blocks of a fixed size B.

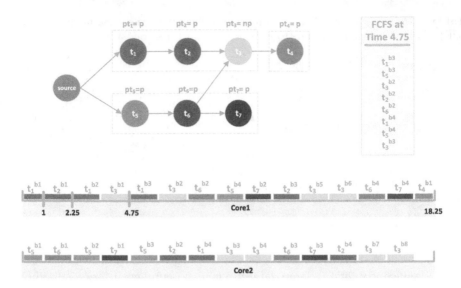

Fig. 4. Example of running a generic flow on 2 cores; the dotted borders denote pipeline segments.

4. Create a copy for the first block for each task directly connected to a source and insert it in a FCFS (First Come First Serve) queue.
5. Schedule blocks arbitrarily to cores until there are no blocks in the queue under the condition that a task can process at most one block at time at any core:
   (a) when a block finishes its execution, re-insert it in the queue annotated by the subsequent tasks in the DAG;
   (b) if the task is a child of a source, insert its next block in the queue;
   (c) if a blocking task has received its entire input, start scheduling the corresponding blocks for the segment initiating from this task.
6. The response time is defined by the longest sequence of block allocations to a core.

In Fig. 4, we present an example with a flow running on 2 cores, where all task costs per block are 1, each task receives as input 4 blocks and emits 4 other processed blocks except t_3, which outputs a single block. The α factor is 1, when there are up to two concurrent tasks, and 1.25 otherwise. Concurrent tasks are those for which there is at least 1 block either in the queue or being executed. In this example, the running time is when the work on the first is completed.

4.3   Considering Communication Costs

We need to consider communication only in settings where multiple machines are employed. Broadly, we can distinguish among the following three cases:

1. On each sender, there is a single thread for computation and transmission. In this case, both z_i and $z_{i,j}$ in Eq. (1) are 1 to denote that computation and transmission occur sequentially.
2. On each sender, there is a separate thread for data transmission, regardless of the number of outgoing edges. In this case, depending on which type of cost dominates, only one of z_i and $z_{i,j}$ is set to 1, since computation and transmission overlap in time.
3. On each sender or receiver, there is a separate thread for each edge. If all edges share the same network, then we can follow the same approach as in the case of multiple pipelined tasks sharing a single core.

The first two cases assume a push based data communication model, whereas the third one applies to both push and pull models.

4.4   Considering Partitioned Parallelism

Paritioned flows running on multiple machines can be covered by our model as well. More specifically, we can model and estimate the DAG flow instance on each machine independently using the same approach, and then take the maximum running time as the final one. The factors may differ between partitioned tasks. Finally, if a DAG instance does not start its execution immediately, we need to add the time to receive its first input (which kicks-off its execution) to its estimated running time.

5   Conclusions and Future Work

In this work, we show that up to date the existing cost models do not estimate accurately the *response time* of real data flow execution, which heavily depends on parallelism. We propose a model that considers the time overlaps during the task execution, while it is capable of quantifying the impact of concurrent task execution. The latter is an aspect largely overlooked to date and may lead to significant inaccuracies if neglected, e.g., we provided simple examples of deviations up to 50%.

Our work can be extended in several ways. Applying the proposed model relies on the existence of accurate weight information; deriving efficient ways to approximate the weights before flow execution is an open issue. More thorough validating experiments (e.g., using the complete TPC-DI benchmark) are required. Also, instead of using a list scheduler simulation, one could develop analytic cost models that directly estimate the response time. Another direction for future work is to make a deep dive into the low-level resource utilization and wait measurements to establish the detailed cause of contention. Finally, there is a lack of flow cost-based optimization techniques that directly target the minimization of response time as modeled in this work.

References

1. Agrawal, K., Benoit, A., Dufossé, F., Robert, Y.: Mapping filtering streaming applications with communication costs. In: SPAA, pp. 19–28 (2009)
2. Agrawal, K., Benoit, A., Dufossé, F., Robert, Y.: Mapping filtering streaming applications. Algorithmica 62(1–2), 258–308 (2012)
3. Boehm, M., Tatikonda, S., Reinwald, B., Sen, P., Tian, Y., Burdick, D.R., Vaithyanathan, S.: Hybrid parallelization strategies for large-scale machine learning in systemML. Proc. VLDB Endow. 7(7), 553–564 (2014)
4. Chirkin, A.M., Belloum, A.S.Z., Kovalchuk, S.V., Makkes, M.X.: Execution time estimation for workflow scheduling. In: WORKS 2014, pp. 1–10 (2014)
5. Deshpande, A., Hellerstein, L.: Parallel pipelined filter ordering with precedence constraints. ACM Trans. Algorithms 8(4), 41:1–41:38 (2012)
6. DeWitt, D.J., Gray, J.: Parallel database systems: the future of high performance database systems. Commun. ACM 35(6), 85–98 (1992)
7. Hueske, F., Peters, M., Sax, M., Rheinländer, A., Bergmann, R., Krettek, A., Tzoumas, K.: Opening the black boxes in data flow optimization. PVLDB 5(11), 1256–1267 (2012)
8. Kougka, G., Gounaris, A.: Optimization of data-intensive flows: is it needed? Is it solved? In: Proceedings of the DOLAP, pp. 95–98 (2014)
9. Kougka, G., Gounaris, A., Simitsis, A.: The many faces of data-centric workflow optimization: a survey. CoRR, abs/1701.07723 (2017)
10. Kumar, N., Kumar, P.S.: An efficient heuristic for logical optimization of ETL workflows. In: Castellanos, M., Dayal, U., Markl, V. (eds.) BIRTE 2010. LNBIP, vol. 84, pp. 68–83. Springer, Heidelberg (2011). doi:10.1007/978-3-642-22970-1_6
11. Pietri, I., Juve, G., Deelman, E., Sakellariou, R.: A performance model to estimate execution time of scientific workflows on the cloud. In: WORKS 2014, pp. 11–19. IEEE Press (2014)
12. Poess, M., Rabl, T., Caufield, B.: TPC-DI: the first industry benchmark for data integration. PVLDB 7(13), 1367–1378 (2014)
13. Rheinlnder, A., Heise, A., Hueske, F., Leser, U., Naumann, F.: SOFA: an extensible logical optimizer for UDF-heavy data flows. Inf. Syst. 52, 96–125 (2015)
14. Shi, J., Zou, J., Jiaheng, L., Cao, Z., Li, S., Wang, C.: MRTuner: a toolkit to enable holistic optimization for mapreduce jobs. Proc. VLDB Endow. 7(13), 1319–1330 (2014)
15. Simitsis, A., Vassiliadis, P., Sellis, T.K.: State-space optimization of ETL workflows. IEEE Trans. Knowl. Data Eng. 17(10), 1404–1419 (2005)
16. Simitsis, A., Wilkinson, K., Dayal, U., Castellanos, M.: Optimizing ETL workflows for fault-tolerance. In: ICDE, pp. 385–396 (2010)
17. Singhal, R., Verma, A.: Predicting job completion time in heterogeneous MapReduce environments. In: IEEE IPDPSW, pp. 17–27 (2016)
18. Srivastava, U., Munagala, K., Widom, J., Motwani, R.: Query optimization over web services. In: Proceedings of the PVLDB, pp. 355–366 (2006)
19. Verma, A., Cherkasova, L., Campbell, R.H.: ARIA: Automatic resource inference and allocation for MapReduce environments. In: ICAC 2011, pp. 235–244. ACM (2011)
20. Yerneni, R., Li, C., Ullman, J., Garcia-Molina, H.: Optimizing large join queries in mediation systems. In: Beeri, C., Buneman, P. (eds.) ICDT 1999. LNCS, vol. 1540, pp. 348–364. Springer, Heidelberg (1999). doi:10.1007/3-540-49257-7_22
21. Zhang, Z., Cherkasova, L., Loo, B.T.: Performance modeling of MapReduce jobs in heterogeneous cloud environments. In: CLOUD 2013, pp. 839–846 (2013)

Machine Learning

Accelerating K-Means by Grouping Points Automatically

Qiao Yu and Bi-Ru Dai$^{(\boxtimes)}$

Department of Computer Science and Information Engineering,
National Taiwan University of Science and Technology,
No. 43, Section 4, Keelung Road, Da'an District, Taipei 106, Taiwan, ROC
M10415803@mail.ntust.edu.tw,
brdai@csie.ntust.edu.tw

Abstract. K-means is a well-known clustering algorithm in data mining and machine learning. It is widely applicable in various domains such as computer vision, market segmentation, social network analysis, etc. However, k-means wastes a large amount of time on the unnecessary distance calculations. Thus accelerating k-means has become a worthy and important topic. Accelerated k-means algorithms can achieve the same result as k-means, but only faster. In this paper, we present a novel accelerated exact k-means algorithm named Fission-Fusion k-means that is significantly faster than the state-of-the-art accelerated k-means algorithms. The additional memory consumption of our algorithm is also much less than other accelerated k-means algorithms. Fission-Fusion k-means accelerates k-means by grouping number of points automatically during the iterations. It can balance these expenses well between distance calculations and the filtering time cost. We conduct extensive experiments on the real world datasets. In the experiments, real world datasets verify that Fission-Fusion k-means can considerably outperform the state-of-the-art accelerated k-means algorithms especially when the datasets are low-dimensional and the number of clusters is quite large. In addition, for more separated and naturally-clustered datasets, our algorithm is relatively faster than other accelerated k-means algorithms.

Keywords: K-means · Clustering · Accelerating k-means

1 Introduction

The standard k-means algorithm [1] is well known and widely used on large datasets because of its simplicity and applicability. It is one of the most popular clustering algorithms in data mining and machine learning. The applications of standard K-means include computer vision, market segmentation, social network analysis, etc. The standard k-means algorithm includes two steps. The assignment step assigns each point to its closest cluster and the update step updates each cluster center after points are assigned to their closest clusters. However, the time complexity of Lloyd's k-means [1]

© Springer International Publishing AG 2017
L. Bellatreche and S. Chakravarthy (Eds.): DaWaK 2017, LNCS 10440, pp. 199–213, 2017.
DOI: 10.1007/978-3-319-64283-3_15

is quite large. The bottleneck of time complexity is to identify the closest center for each input point by computing the distance from each point to each center. Thus, in order to accelerate the exact k-means, there are several researches concentrating on avoiding the unnecessary distance calculations of k-means while keeping the same clustering results as standard k-means under the same initialization.

Previous works on accelerating k-means can be classified into two groups. Instead of accelerating the exact k-means, the algorithms in the first group focus on speeding up k-means by keeping an approximate solution of k-means, such as the algorithms in [2–4]. Actually, these algorithms can accelerate k-means, but they cannot guarantee that the final clustering results are the same as k-means. The second group of algorithms contributes to achieving the exact same results as k-means via some techniques. For example, in [5, 6], they accelerate the nearest neighbor search by using efficient data structures. In [7–14], these algorithms maintain the bounds of distance to avoid the unnecessary distance calculations by using the triangle inequality or calculating the distances between centers. Those algorithms we mentioned all make significant contributions to the progress of accelerating k-means.

In this paper, we extend the way of accelerating the exact k-means based on the triangle inequality. According to aforementioned algorithms, the purpose of using triangle inequality is to maintain the lower bounds and upper bounds for each point so that distance calculations can be eliminated to decrease the computational time. However, most of these techniques are based on using triangle inequality on the distances to cluster centers for each point. In fact, the datasets we use k-means to analyze have much more natural groups. Therefore, we can maintain bounds only for each group of data points instead of keeping bounds for each data point. If the bounds we kept are effective for each group of data points, enormous computational time will be reduced. In the experiments, real world datasets are used to verify that our algorithm is significantly faster than the existing state-of-the-art accelerated algorithms, especially when the number of clusters is quite large. Thus, for many applications requiring a large number of clusters, such as clustering online encyclopedia Wikipedia into thousands of topics, our algorithm is extremely more efficient than other algorithms.

Our contributions in this paper can be summarized below.

First, we propose a new method that can divide the data points into small groups automatically during the iterations of k-means. By this way, we can maintain lower bounds and upper bounds for groups instead of maintaining bounds for each point. The bounds we kept for each group can avoid unnecessary distance calculations for the points in the groups. Our method is significantly faster than the current state-of-the-art accelerated exact k-means algorithms in most of cases.

Second, our method can extremely reduce the additional space cost than the prior works. The additional space cost in prior algorithms is based on the number of points, while it is based on the number of groups produced in our method.

The rest of this paper is organized as follows. In Sect. 2, we introduce related work about accelerating the exact k-means and the improvements they achieve. In Sect. 3, we discuss our proposed algorithm in detail and show how it is faster than the state-of-the-art accelerated k-means. In Sect. 4, results and discussions of experiments on different datasets will be presented. We conclude the entire paper in Sect. 5.

2 Related Work

In this section, we discuss recent accelerated k-means algorithms using the triangle inequality. Previous accelerated k-means algorithms are introduced at first and we describe other important enhanced accelerated k-means algorithms at second.

The algorithm of [7] takes advantage of lower bounds and upper bounds to reduce the redundant distance calculations. It fully utilizes triangle inequality to maintain bounds for each point. Both of the additional space cost and time complexity of this algorithm are $O(nk)$, where n is the number of points and k is the number of clusters. The algorithm of [8] only keeps one lower bound on the distance between each point and its second closest center instead of keeping k lower bounds per point. Actually, it is a simplified version of [7], but it is more efficient than [7] when the dimension of data is low. The algorithm from [9] extends the approach of [8] to remain variable ordering closest centers and Annulus algorithm from [10] prunes the search space for each point by annular region. The algorithm of [12] proposes a few methods to accelerate previous algorithms, such as producing tighter lower bounds, finding neighbor centroids and accelerating k-means in the first iteration.

Other efficient ways to speed up k-means are based on Yinyang k-means from [11]. Yinyang k-means accelerates standard k-means by grouping number of cluster centers, which significantly balances well between the time of filtering and the time of distance calculations. Fast Yinyang k-means of [13] creates an extension of Yinyang k-means algorithm to approximate Euclidean distances by using block vectors, which can achieve good improvements when the dimension of data is high. In addition, the algorithm of [14] first simplifies Yinyang k-means to make it faster, and makes bounds tighter during the update of bounds. It also proposes an Exponion algorithm performing better when the dimension of data is low.

3 Proposed Method

In this section, we introduce our proposed algorithm named Fission-Fusion k-means which is a novel way to accelerate the standard k-means by grouping number of points automatically during the iterations. The framework of our algorithm is presented in Sect. 3.1. Section 3.2 describes the approach of filtering for clusters of points. In Sect. 3.3 we devise a method to group points automatically. In Sect. 3.4 our approach of filtering for groups of points is introduced and we further improve the method of grouping points in Sect. 3.5. The whole algorithm is provided in Sect. 3.6.

3.1 The Framework of Our Algorithm

K-means algorithm is to find similar points by computing the Euclidean distance from each point to each center. However, it wastes large amounts of time on the distance calculations. Therefore, we create two filters to avoid unnecessary distance calculations. We first filter a number of centers that cannot be the closest center for each cluster of points. Then, groups produced by Fission-Fusion step will provide more effective bounds for filtering. For the rest of centers, we filter them for each group of points in each cluster. Therefore, distance calculations can be avoided between points and those filtered centers. By these two filters, our algorithm can achieve better speedup than state-of-the-art accelerated k-means algorithms. The framework is illustrated in Fig. 1 and each step will be introduced in the following subsections.

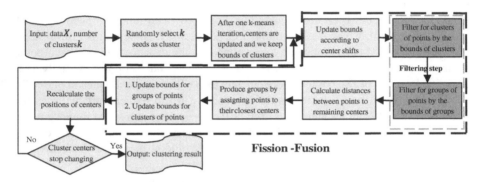

Fig. 1. The framework of Fission-Fusion k-means.

3.2 Filtering for Clusters of Points

The challenge of accelerating k-means is how we balance the time between distance calculations and filtering. Our method can avoid many distance calculations and simultaneously use less filtering time. Fission-Fusion k-means is inspired by the concept of Yinyang k-means [11] which groups cluster centers, but groups number of points instead. The points change their clusters due to the assignment step. We keep one upper bound and $k - 1$ lower bounds for each cluster. The definitions of the upper bound, lower bounds, and how they are updated will be introduced in Definition 1.

Definition 1. The bounds of clusters: The upper bound and lower bounds of each cluster are decided by all points in each cluster. k is the number of clusters. For point x belonging to cluster C_i, $i \in \{1 \dots k\}$, $d(x, c_i)$ denotes the distance between point x and its cluster center c_i. After one k-means iteration, The upper bound for points in cluster C_i is kept as $ub(C_i) = max_{x \in C_i} d(x, c_i)$, and $k - 1$ lower bounds for points in cluster C_i

to each other center c_j, where $j \in \{1 \ldots k\}$ and $i \neq j$, are denoted as $lb(C_i, c_j) = min_{x \in C_i} d(x, c_j)$. Let c_i' and C_i' represent the corresponding entities in the next iteration. $\delta(c_i)$ represents $d(c_i, c_i')$ which is the shift of cluster center c_i after it is updated. The upper and lower bounds are updated across iterations by $ub'(C_i) = ub(C_i) + \delta(c_i)$ and $lb'\left(C_i, c_j'\right) = lb(C_i, c_j) - \delta(c_j)$.

According to Definition 1, how we use these bounds to filter cluster centers that cannot be the closest center for each cluster of points will be presented in Lemma 1.

Lemma 1. *All points in cluster C_i will not be assigned to cluster C_j after the update of centers if*

$$ub(C_i) + \delta(c_i) \leq lb(C_i, c_j) - \delta(c_j). \tag{1}$$

Proof. c_i is the cluster center of point x, $\forall x \in C_i$. c_i' is the new cluster center after the update of center c_i. Similarly, Let c_j be a cluster center, where $c_j \neq c_i$, and c_j' is the new cluster center of c_j, where $c_j' \neq c_i'$.

Applying the triangle inequality, we know that

$$d(x, c_i') \leq d(x, c_i) + d(c_i, c_i') \leq d(x, c_i) + \delta(c_i)$$
$$\leq max_{x \in C_i} d(x, c_i) + \delta(c_i) \leq ub(C_i) + \delta(c_i)$$
$$d\left(x, c_j'\right) \geq d(x, c_j) - d\left(c_j, c_j'\right) \geq d(x, c_j) - \delta(c_j)$$
$$\geq min_{x \in C_i} d(x, c_j) - \delta(c_j) \geq lb(C_i, c_j) - \delta(c_j)$$

Therefore, if $ub(C_i) + \delta(c_i) \leq lb(C_i, c_j) - \delta(c_j)$, $d\left(x, c_j'\right)$ will be larger than $d(x, c_i')$ and all points in cluster C_i will not be assigned to cluster C_j. □

By Lemma 1, we can prune cluster center c_j that cannot be the closest center for the points in cluster C_i. For the remaining centers, they will be filtered for groups of points in Sect. 3.4. However, the advantage of this filtering is that it just takes $O(k^2)$ time to filter many cluster centers for the points. Therefore, the proposed filtering method is much faster than the state-of-the-art algorithm Yinyang k-means $O(nt)$, where t is the number of groups of centers, and even requires less space cost $O(k^2)$ than that of Yinyang k-means $O(nt)$.

Example 1. Figure 2 demonstrates the mechanism of filtering for clusters of points. Among all points in cluster C_1, suppose that $d(a, c_1)$ is the maximum distance to c_1 and $ub'(C_1) = d(a, c_1) + d(c_1, c_1')$. In addition, $d(b, c_2)$, $d(b, c_3)$ and $d(c, c_4)$ are the minimum distances to c_2, c_3 and c_4 for all points in cluster C_1, so $lb'(C_1, c_2') = d(b, c_2) - d(c_2, c_2')$, $lb'(C_1, c_3') = d(b, c_3) - d(c_3, c_3')$ and $lb'(C_1, c_4') = d(c, c_4) -$

$d(c_4, c_4')$, respectively. According to Lemma 1, $lb'(C_1, c_3') > ub'(C_1)$ and $lb'(C_1, c_4') > ub'(C_1)$, we can prune centers c_3' and c_4' that cannot be the closest centers for all points in cluster C_1. Center c_2' will not be filtered because $lb'(C_1, c_2') < ub'(C_1)$, but it will be filtered for groups of points in Example 2.

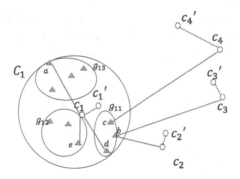

Fig. 2. An example of filtering for clusters of points. Cluster C_1 contains groups g_{11}, g_{12} and g_{13}. a, b, c, d and e represent points in cluster C_1. This space includes four centers c_1, c_2, c_3 and c_4. c_1', c_2', c_3' and c_4' are the corresponding entities in the next iteration. Dotted line is the shift of a center. Red line is the distance between a point to a center. Cluster C_1 and groups g_{11}, g_{12} and g_{13} are surrounded by circles respectively.

3.3 Fission Step: Grouping Points Automatically

According to Sect. 3.2, filtering for clusters of points can take less filtering time and simultaneously avoid many distance calculations. However, in real world applications, we can observe that when cluster centers are updated, some points will change their cluster assignments together. Based on this phenomenon, our method can be further enhanced by grouping points to get tighter bounds for each group of points. We utilize the assignment of points during the iterations to produce natural groups of points automatically. Each cluster can contain many groups. Note that Yinyang k-means groups cluster centers at the beginning of clustering and the number of groups is fixed at one tenth of the number of centers. However, our way of grouping is totally different from that of Yinyang k-means. We first give the definition of groups in Definition 2 and explain how they work in the following paragraph and Example 2.

Definition 2. The groups of points: Each new group of points is generated by splitting out from another group. M_i is the number of groups in cluster C_i, $i \in \{1 \dots k\}$. g_{im} is the mth group of cluster C_i, where $m \in \{1 \dots M_i\}$, $g_{im} \subseteq C_i$ and g_{i1} is equal to C_i initially. For each C_i and each C_j, $i \neq j$, a new group $g_{jm'}$, where m' is an integer increased with new groups being produced, will be generated by points which are split out from group g_{im} and join C_j after the update of cluster centers. Finally, g_{im} will be updated across iterations by removing points which belong to g_{im} but are assigned to other clusters.

In the assignment step of k-means, each point x finds out its closest center and will be assigned to the corresponding cluster. Thus we use this characteristic to group points. For points from the same cluster C_i, if they are assigned to the same cluster C_j in this iteration, they will be as a new group for C_j. Initially, after the first iteration, k clusters are generated and they are also regarded as the first group for each cluster at the beginning. In the following iteration, some points in g_{im} are changed to their new closest cluster C_j due to the update of centers. We can produce new group $g_{jm'}$ naturally from the points which are split out from their previous group g_{im} to their new closest cluster C_j for $j \in \{1...k\}$ and $i \neq j$. It is the *fission step* of our algorithm because the whole process is similar to the nuclear fission. Therefore, during the iterations, points are divided into groups according to the assignment of points.

Example 2. Figure 3 shows the grouping process. We choose three initial centers c_1, c_2, c_3 and each point is assigned to its closest center. As shown in Fig. 3(a), three groups g_{11}, g_{21} and g_{31} are produced. In Fig. 3(b), after the shift of centers, three groups g_{11}, g_{21} and g_{31} are updated according to Definition 2. In addition, g_{12} and g_{13} contains the points which are original from cluster C_2 and C_3 respectively in the 1st iteration and are assigned to C_1 in the 2nd iteration. Figure 3(c) shows that in the 3rd iteration, g_{32} which contains the points from g_{13} in the 2nd iteration is a new group of cluster C_3. Figure 3(d) shows that two groups g_{22} and g_{23} which are split out from g_{12} and g_{13} respectively in the 3rd iteration are produced in C_2. Thus we can produce more natural groups of points automatically by this way.

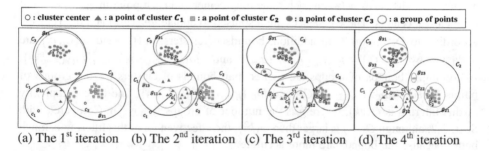

(a) The 1st iteration   (b) The 2nd iteration   (c) The 3rd iteration   (d) The 4th iteration

Fig. 3. The grouping process of Fission-Fusion k-means. Cluster C_1 contains groups g_{11}, g_{12} and g_{13}. Cluster C_2 contains groups g_{21}, g_{22} and g_{23}. Cluster C_3 contains groups g_{31} and g_{32}. These groups are produced naturally from the assignment of points.

3.4   Filtering for Groups of Points

Under the way of grouping automatically, we keep bounds for each generated group.

Definition 3. The bounds of groups: The upper bound is kept as $ub_{C_i}(g_{im}) = max_{x \in g_{im}} d(x, c_i)$, which is the upper bound for the points in the group g_{im} to their center c_i, and $lb_{C_i}(g_{im}, c_j) = min_{x \in g_{im}} d(x, c_j)$, denotes the lower bound to center c_j. These upper and lower bounds are updated across iterations by $ub'_{C_i}(g_{im}) = ub_{C_i}(g_{im}) + \delta(c_i)$ and $lb'_{C_i}(g_{im}, c'_j) = lb_{C_i}(g_{im}, c_j) - \delta(c_j)$.

Lemma 2 introduces how we use Definition 3 to filter cluster centers that cannot be the closest centers for each group of points.

Lemma 2. *All points in group g_{im} will not be assigned to cluster C_j after the update of centers if*

$$ub_{C_i}(g_{im}) + \delta(c_i) \leq lb_{C_i}(g_{im}, c_j) - \delta(c_j). \tag{2}$$

Proof. $\forall x \in g_{im}$, applying the triangle inequality as in Lemma 1,

$$
\begin{aligned}
d(x, c'_i) &\leq d(x, c_i) + d(c_i, c'_i) \leq d(x, c_i) + \delta(c_i) \\
&\leq max_{x \in g_{im}} d(x, c_i) + \delta(c_i) \leq ub_{C_i}(g_{im}) + \delta(c_i) \\
d(x, c'_j) &\geq d(x, c_j) - d(c_j, c'_j) \geq d(x, c_j) - \delta(c_j) \\
&\geq min_{x \in g_{im}} d(x, c_j) - \delta(c_j) \geq lb_{C_i}(g_{im}, c_j) - \delta(c_j)
\end{aligned}
$$

Therefore, if $ub_{C_i}(g_{im}) + \delta(c_i) \leq lb_{C_i}(g_{im}, c_j) - \delta(c_j), d(x, c'_j)$ will be larger than $d(x, c'_i)$. In other words, all points in group g_{im} will not be assigned to cluster C_j. $\square$

By this way, we can prune cluster center c_j that cannot be the closest center for the points in group g_{im}. After filtering for groups of points, for the remaining centers, we calculate the distances between each remaining center and each point in g_{im} to find out their closest centers. These distances are also used to update $ub_{C_i}(g_{im})$ and $lb_{C_i}(g_{im}, c_j)$ according to Definition 3. In addition, bounds of clusters, $ub(C_i)$ and $lb(C_i, c_j)$ are updated by $ub(C_i) = max_{g_{im} \subseteq C_i} ub_{C_i}(g_{im})$ and $lb(C_i, c_j) = min_{g_{im} \subseteq C_i} lb_{C_i}(g_{im}, c_j)$. Accordingly, bounds of groups are tighter than bounds of clusters for the groups of points because $ub_{C_i}(g_{im}) \leq ub(C_i)$ and $lb_{C_i}(g_{im}, c_j) \geq lb(C_i, c_j)$. The time complexity of this filtering is $O(gk)$, where g is the number of groups, $k \leq g \leq n$. Compared with Sect. 3.2, although $O(gk)$ is larger than $O(k^2)$, the effectiveness of this filtering is much better because the bounds are more accurate.

Example 3. As shown in Fig. 2, $d(d, c_1)$ is the maximum distance to c_1 among all points in group g_{11} and $ub'_{C_1}(g_{11}) = d(d, c_1) + d(c_1, c'_1)$. From Example 1, for the remaining center c'_2, $d(b, c_2)$.is the minimum distance to c_2 for all points in g_{11}, so $lb'_{C_1}(g_{11}, c'_2) = d(b, c_2) - d(c_2, c'_2)$. Because $lb'_{C_1}(g_{11}, c'_2) > ub'_{C_1}(g_{11})$ by Lemma 2, c'_2 will be filtered since it cannot be the closest center for all points in g_{11}.

Algorithm 1. Fission-Fusion k-means

INPUT: data X, $x \in X$, number of clusters k
OUTPUT: cluster centers $c_1 \ldots c_k$
1: **procedure** Fission-Fusion k-means (X, k)
2: **repeat**
3:   Update $ub(C_i)$, $lb(C_i, c_j)$, $ub_{C_i}(g_{im})$, $lb_{C_i}(g_{im}, c_j)$ according to **center shifts**
4: **for** $i = 1 \ldots k$ **do**
5:    **do Filtering 1: Filtering for clusters of points**
6:     $Set1 = \{j \in \{1 \ldots k\}\}$ // $Set1$ contains the remaining centers after **Filtering 1**
7:    **if** $Set1$ is not empty **then**
8:     $Setpoints = \{\{\}_1 \ldots \{\}_k\}$ // $Setpoints$ will be used to contain groups of points after
9:     **for** $g_{im} \in G_i$ **do**   // G_i denotes all groups in cluster C_i
10:      **do Filtering 2: Filtering for groups of points**
11:       $Set2 = \{j \in \{1 \ldots k\}\}$ // $Set2$ contains the remaining centers after **Filtering 2**
12:      **if** $Set2$ is not empty **then**
13:       **for each** x in group g_{im} **do**
14:        $ub(x) = d(x, c_i)$
15:        $b(x) = i$ // $b(x)$ denotes the assigned cluster center of point x
16:        $b(x)_{old} = b(x)$
17:        **for** $j \in Set2$ **do**
18:         **if** $ub(x) > d(x, c_j)$ **then**
19:          $ub(x) = d(x, c_j)$
20:          $b(x) = j$
21:        **if** $b(x) \neq b(x)_{old}$ **then**
22:         x is out of g_{im} and is included in the $\{\}_{b(x)}$ in $Setpoints$ $\{\{\}_1 \ldots \{\}_k\}$
23:          Update $\{ub_{b(x)}(\{\}_{b(x)}), lb_{b(x)}(\{\}_{b(x)}, c_j)\}$; $x \in C_{b(x)}$
24:        **else:**
25:          Update $\{ub_{C_i}(g_{im}), lb_{C_i}(g_{im}, c_j)\}$; $x \in C_i$
26:      **if** $\{\}_k$ is not empty in $Setpoints$:
27:       $\{\}_k$ is added in group g_{ki} of cluster C_k
28:       Keep $\{ub_{C_k}(\{\}_k), lb_{C_k}(\{\}_k, c_j)\}$
29: **for** $i = 1, \ldots, k$ **do**
30: **for** $g_{im} \in G_i$ **do**
31:   Use $\{ub_{C_i}(\{\}_k), lb_{C_i}(\{\}_k, c_j)\}$ to update $\{ub_{C_i}(g_{im}), lb_{C_i}(g_{im}, c_j)\}$
32:   Use $\{ub_{C_i}(g_{im}), lb_{C_i}(g_{im}, c_j)\}$ to update $\{ub(C_i), lb(C_i, c_j)\}$
33: //**Update cluster centers and center shifts**
34: **for** $i = 1 .. k$ **do**
35: $c_i' = c_i$; $c_i = $ mean$(\{x \in X \,|\, b(x) = i\})$; $\delta(c_i) = d(c_i', c_i)$
36: **end procedure** until cluster centers stop changing

3.5 Fusion Step: Limiting the Increasing Number of Groups

Although the bounds of groups a pretty effective, the filtering time probably gets closed to $O(nk)$ in the worst case, where each point becomes a group by itself. Therefore, we devise the *fusion step* to further improve the efficiency of our framework.

During the iterations, as we mentioned in Sect. 3.3, groups will be split into many groups to their new clusters due to the assignment step. In order to get better performance, if groups are split out from the same clusters and are changed to the same new clusters, they will be merged into new groups for the new clusters. For example, as shown in Fig. 3(d), in the *fusion step*, g_{22} and g_{23} will be merged into same group for C_2 because g_{22} and g_{23} are split out from g_{12} and g_{13} respectively in the 3rd iteration, where g_{12} and g_{13} both belong to C_1. Therefore, by this way, the maximum number of groups for each cluster will be limited to k during the iterations. The filtering time complexity will be reduced to $O(g'k)$, where g' is the number of groups after the *fusion step* and $k \le g' \le k^2 \le g$. Thus our algorithm can be more efficient by controlling the number of groups.

3.6 Algorithm

We combine two filters together to create our Fission-Fusion k-means. In filtering for clusters of points, some centers are pruned for some clusters of points. Then, in filtering for groups of points, we filter the rest of centers for each group of points in each cluster. Thus these two filters can avoid many distance calculations and use less filtering time. Algorithm 1 presents the pseudo code of Fission-Fusion k-means.

4   Experiment and Analysis

In this section, we design our experiment in Sect. 4.1 and we show the cost comparison and relative speedup with other four algorithms in Sect. 4.2. The analysis about separability is given in Sect. 4.3 and avoided distance calculations of our algorithm will be shown in Sect. 4.4.

4.1   Experiment Design

We compare our proposed method with four algorithms: standard k-means [1], Elkan [7], Yinyang k-means [11], Syinyang k-means [14], where [7, 11, 14] are the state-of-the-art accelerated k-means algorithms. All the implements ran on Intel(R) Core(TM) i7-4770 CPU 3.40 GHz machine with 8G RAM. All algorithms are executed on the same datasets with the same randomly initial center seeds, so the clustering results of all algorithms are the same. We use ten real world datasets from these three websites (1) the UCI machine learning repository [15]; (2) clustering datasets – Joensuu homepage [16]; (3) the libsvm homepage [17]. In the experiment, our algorithm can run faster than state-of-the-art accelerated k-means in low-dimensional datasets regardless of the number of clusters. For the middle-dimensional and high-dimensional datasets, our algorithm can still achieve better a performance when the number of clusters is large.

4.2   Cost Comparison and Relative Speedup

The time and memory cost comparisons at the filtering step are summarized in Table 1. As shown in Table 1, we can know that Fission-Fusion k-means just takes additional memory cost $O(gk)$ and time cost $O(k^2) \sim O(gk)$ in the filtering step. Both of cost are far less than other three algorithms because g and k are certainly less than n. Table 2 shows the relative speedup of Fission-Fusion k-means over competitors in different datasets. According to our experiments, when k is under 100, we do not limit the number of groups of our algorithm because it can achieve better performances. In most of cases, Fission-Fusion k-means can perform better than other algorithms, especially in low-dimensional datasets. In addition, when k is large, Fission-Fusion k-means can still achieve the best speedup in all datasets, because the groups are small and the bounds are effective for filtering centers.

Table 1. Cost comparison (n is the mber of points; k is the number of clusters; g is the number of groups of points; t is the number of groups of clusters for Yinyang k-means. Generally, $n \gg g \geq k \geq t$)

Algorithm	Additional memory cost	Filtering time cost
Fission-Fusion k-means	$O(gk)$	$O(k^2) \sim O(gk)$
Syinyang k-means	$O(nt)$	$O(n) \sim O(nt)$
Yinyang k-means	$O(nt)$	$O(n) \sim O(nt)$
Elkan	$O(nk)$	$O(nk)$

4.3 Separability

For more separated and naturally-clustered data, our method can perform much better than state-of-the-art accelerated k-means because the bounds we maintained are very effective for points in these situations.

Therefore, in this test, we evaluate all algorithms with four Synthetic 2-d datasets (s1, s2, s3, s4) from [16]. Each dataset contains 5000 points and 15 Gaussian clusters, but with different degree of cluster overlapping. Dataset s1 is more separated than others, and dataset s4 is less separated. We show the time speedup comparison with these datasets in Fig. 4(a). As shown in Fig. 4(a), for the less separated datasets s3 and s4, our algorithm is less efficient because it is more difficult to avoid unnecessary distance calculations. However, for the more separated and naturally-clustered datasets s1 and s2, Fission-Fusion k-means achieves the best performance than other three algorithms. The reason is that the bounds we maintained are very effective and lots of centers are successfully filtered. Elkan can have better performance than Syinyang and Yinyang in these four datasets since the number of clusters is small and the bounds of Elkan are effective in this case.

4.4 Avoided Distance Calculations

Our algorithm avoids many unnecessary distance calculations during the iterations. We compare the avoided distance calculations of our algorithm with other three algorithms on three datasets (1) 3Droad network (low-dimensional); (2) Gassensor (middle-dimensional); (3) Mnist784 (high-dimensional) with 500 clusters. As shown in Fig. 4(b), because Elkan keeps bounds for each point, it can avoid the largest number of distance calculations. However, our algorithm still avoids more distance calculations

Table 2. Relative speedup to standard k-means

Dataset	n	d	k	Mean iteration	Relative speed up to standard k-means			
					Fission-Fusion	Syinyang	Yinyang	Elkan
Userlocations (Joensuu)	6,014	2	10	25	**2.83**	1.19	1.11	2.39
			50	58	**4.74**	2.21	2.07	3.35
			100	51	**5.80**	2.98	2.79	3.51
			500	25	**10.05**	5.25	4.86	3.32
			800	17	**9.57**	4.21	4.47	3.00
Userlocations (Finland)	13,467	2	10	16	**2.50**	1.27	1.05	1.95
			50	69	**5.04**	2.34	2.14	3.37
			100	79	**10.46**	4.04	3.79	3.29
			500	26	**12.59**	3.32	3.09	1.88
			800	33	**7.45**	4.95	5.40	3.98
Skin_nonskin	245,057	3	50	40	**5.32**	5.28	4.93	4.28
			100	67	**9.47**	8.54	7.89	5.70
			500	78	**25.81**	17.56	16.43	5.20
			800	64	**26.61**	17.93	16.92	6.01
			1000	45	**23.58**	14.50	14.48	4.62
3Droad network	434,874	4	50	34	**15.41**	4.72	4.45	5.40
			100	19	**13.77**	6.51	6.19	5.27
			500	16	**12.67**	10.31	9.96	4.95
			800	13	**12.74**	10.57	9.86	4.55
			1000	19	**16.23**	11.98	11.95	4.66
Poker	25,010	10	50	119	5.40	8.13	**8.56**	7.86
			100	119	7.75	10.37	**11.04**	8.72
			500	45	**10.21**	9.93	9.95	7.99
			800	36	**10.50**	9.95	10.13	7.07
			1000	29	**9.47**	8.72	9.01	6.48
Kegg network	65,000	28	50	56	3.78	**6.27**	6.21	2.69
			100	69	6.61	**7.62**	7.49	4.89
			500	40	**17.47**	9.32	9.13	5.96
			800	27	**14.82**	8.48	8.53	5.82
			1000	25	**15.17**	7.93	8.05	6.06
Kddcup98	95,412	56	50	211	3.86	**6.49**	6.48	5.64
			100	220	5.63	**11.11**	11.02	9.64
			500	166	**20.42**	14.30	13.92	10.08
			800	126	**26.98**	17.30	15.31	9.13
			1000	159	**32.10**	23.17	25.02	8.68
Gassensor	14,000	128	50	73	7.04	6.42	6.48	**9.35**
			100	40	10.25	6.99	7.05	**20.01**
			500	29	**19.98**	11.33	11.43	12.43
			800	35	**26.78**	14.07	13.63	11.46
			1000	32	**22.70**	10.91	11.26	11.90
Usps	7,291	256	10	35	1.37	3.07	3.10	**5.19**
			50	50	3.54	5.71	5.79	**12.83**
			100	65	5.49	9.53	9.60	**15.67**
			500	20	**7.62**	5.79	5.85	7.33
			800	16	**8.31**	5.96	5.87	5.56
Mnist784	60,000	784	50	89	3.20	5.90	5.67	**12.17**
			100	48	3.79	6.10	6.16	**9.70**
			500	24	6.44	5.54	5.59	**6.76**
			800	12	**5.38**	4.57	4.47	5.03
			1000	14	**5.40**	4.58	4.67	4.88

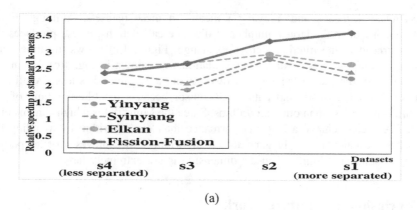

(a)

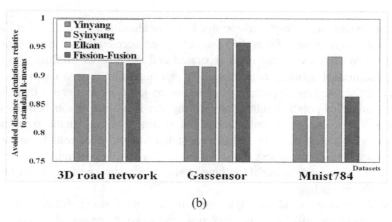

(b)

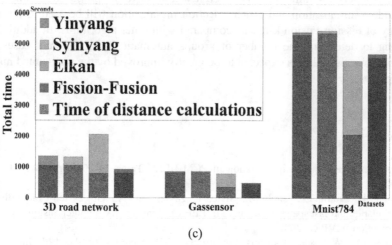

(c)

Fig. 4. Performance comparisons. (a) Relative speedup to standard k-means with four different degree of separated datasets. (b) Comparison of avoided distance calculations (c) Time comparison including the time of distance calculations.

than Syinyang k-means and Yinyang k-means in these datasets. Although Elkan algorithm can reduce the largest number of distance calculations, it also produces the largest overhead in the filtering step as exchange. Figure 4(c) shows the time comparison including the time of distance calculations. As shown in Fig. 4(c), Elkan can take less time of distance calculations. However, our algorithm takes less overall time in dataset 3Droad network and dataset Gassensor because the filtering time of our algorithm is much less than other algorithms. Since the Mnist784 is a high-dimensional dataset, Elkan can achieve a better performance than our algorithm by avoiding the largest distance calculations. In general, our algorithm still achieves a better performance than other algorithms when the dimension is not extremely large.

5 Conclusion and Future Work

In this paper, we propose a novel accelerated k-means algorithm named Fission-Fusion k-means that is significantly faster than state-of-the-art accelerated k-means algorithm especially when the datasets are more separated and the number of clusters is quite large. The additional memory consumption of our algorithm is also much less than other accelerated k-means algorithms. Fission-Fusion k-means accelerates k-means by grouping number of points automatically during the iterations. It can balance these expenses well between distance calculations and filtering time cost. In the experiments, Fission-Fusion k-means can run faster than state-of-the-art accelerated k-means in low-dimension datasets regardless of the number of clusters. For the middle-dimension and high-dimension datasets, our algorithm can still achieve best performance when the number of clusters is large.

In the future, we intend to implement our algorithm using different selection methods to determine initial centers, since it would make the proposed method faster. Both random initialization and center selection initialization will be used to judge the efficiency of Fission-Fusion k-means compared with other algorithms. In addition, we are trying to determine the number of groups automatically in different cases. The speedup of our algorithm is expected to be greatly improved by the well-suited number of groups.

References

1. Lloyd, S.: Least squares quantization in PCM. IEEE Trans. Inf. Theory **28**(2), 129–137 (1982)
2. Philbin, J., Chum, O., Isard, M., Sivic, J., Zisserman, A.: Object retrieval with large vocabularies and fast spatial matching. In: IEEE Conference on Computer Vision and Pattern Recognition (CVPR) (2007)
3. Sculley, D.: Web-scale k-means clustering. In: Proceedings of the 19th International Conference on World Wide Web, pp. 1177–1178 (2010)
4. Wang, J., Wang, J., Ke, Q., Zeng, G., and Li, S.: Fast approximate k-means via cluster closures. In: IEEE Conference on Computer Vision and Pattern Recognition (CVPR), pp. 3037–3044 (2012)

5. Pelleg, D., Moore, A.: Accelerating exact k-means algorithms with geometric reasoning. In: Proceedings of the 5th ACM SIGKDD International Conference on Knowledge Discovery and Data Mining (KDD), pp. 277–281 (1999)

6. Kanungo, T., Mount, D.M., Netanyahu, N.S., Piatko, C.D., Silverman, R., Wu, A.Y.: An efficient k-means clustering algorithm: Analysis and implementation. IEEE Trans. Pattern Anal. Mach. Intell. **24**, 881–892 (2002)

7. Elkan, C.: Using the triangle inequality to accelerate k- means. In: Proceedings of the 20th International Conference on Machine Learning (ICML), pp. 147–153 (2003)

8. Hamerly, G.: Making k-means even faster. In: SIAM International Conference on Data Mining (SDM), pp. 130–140 (2010)

9. Drake, J., Hamerly, G.: Accelerated k-means with adaptive distance bounds. In: 5th NIPS Workshop on Optimization for Machine Learning, pp. 579–587 (2012)

10. Drake, J.: Faster k-means clustering (2013). Accessed online 19 August 2015

11. Ding, Y., Zhao, Y., Shen, X., Musuvathi, M., Mytkowicz, T.: Yinyang k-means: A drop-in replacement of the classic k-means with consistent speedup. In: Proceedings of the 32nd International Conference on Machine Learning (ICML), pp. 579–587 (2015)

12. Ryšavý, P., Hamerly, G.: Geometric methods to accelerate k-means algorithms. In: SIAM International Conference on Data Mining (SDM), pp. 324–332 (2016)

13. Bottesch, T., Bühler, T., Kächele, M.: Speeding up k-means by approximating euclidean distances via block vectors. In: Proceedings of the 33rd International Conference on Machine Learning, New York (2016)

14. Newling, J., Fleuret, F.: Fast K-means with accurate bounds. In: Proceedings of the 33rd International Conference on Machine Learning, New York (2016)

15. Bache, K., Lichman, M.: UCI machine learning repository (2013). url: http://archive.ics.uci.edu/ml/

16. Joensuu: clustering datasets – Joensuu homepage url: https://cs.joensuu.fi/sipu/datasets/

17. Rong-En, F.: LIBSVM homepage url:https://www.csie.ntu.edu.tw/~cjlin/libsvmtools/datasets/

A Machine Learning Trainable Model to Assess the Accuracy of Probabilistic Record Linkage

Robespierre Pita[1]($\boxtimes$), Everton Mendonça[1], Sandra Reis[2], Marcos Barreto[1,3], and Spiros Denaxas[3]

[1] Computer Science Department,
Federal University of Bahia (UFBA), Salvador, Brazil
pierre.pita@gmail.com, evertonmj@gmail.com
[2] Centre for Data and Knowledge Integration for Health (CIDACS),
Oswaldo Cruz Foundation (FIOCRUZ), Rio de Janeiro, Brazil
ssreis02@gmail.com
[3] Farr Institute of Health Informatics Research,
University College London, London, UK
{m.barreto,s.denaxas}@ucl.ac.uk

Abstract. Record linkage (RL) is the process of identifying and linking data that relates to the same physical entity across multiple heterogeneous data sources. Deterministic linkage methods rely on the presence of common uniquely identifying attributes across all sources while probabilistic approaches use non-unique attributes and calculates similarity indexes for pair wise comparisons. A key component of record linkage is accuracy assessment — the process of manually verifying and validating matched pairs to further refine linkage parameters and increase its overall effectiveness. This process however is time-consuming and impractical when applied to large administrative data sources where millions of records must be linked. Additionally, it is potentially biased as the gold standard used is often the reviewer's intuition. In this paper, we present an approach for assessing and refining the accuracy of probabilistic linkage based on different supervised machine learning methods (decision trees, naïve Bayes, logistic regression, random forest, linear support vector machines and gradient boosted trees). We used data sets extracted from huge Brazilian socioeconomic and public health care data sources. These models were evaluated using receiver operating characteristic plots, sensitivity, specificity and positive predictive values collected from a 10-fold cross-validation method. Results show that logistic regression outperforms other classifiers and enables the creation of a generalized, very accurate model to validate linkage results.

1 Introduction

Record linkage (RL) is a methodology to aggregate data from disparate data sources believed to pertain to the same entity [21]. It can be implemented using deterministic and probabilistic approaches, depending on the existence (first case) or the absence (second case) of a common set of identifier attributes in

© Springer International Publishing AG 2017
L. Bellatreche and S. Chakravarthy (Eds.): DaWaK 2017, LNCS 10440, pp. 214–227, 2017.
DOI: 10.1007/978-3-319-64283-3_16

all data sources. In both cases, these attributes are compared through some similarity function that decides if they match or not.

Literature has a wide range of sequence- and set-based similarity check functions providing very accurate results. On the other hand, there are no gold standards widely assumed to assess the accuracy of probabilistic linkage, as the resulting data marts are specific to each domain and influenced by different factors, such as data quality and choice of attributes. So, manual review is frequently used in these cases, being dependent of common sense or the reviewer experience and, as such, prone to misunderstanding and subjectivity [9].

Our proposal is to use a set of supervised machine learning techniques to build a trainable model to assess the accuracy of probabilistic linkage. We aim at to eliminate manual review as it is limited by the amount of data to be revised, as well we believe it is less reliable than a computer-based solution. In order to choose the most appropriate techniques, we made experiments with decision trees, naïve Bayes, logistic regression, random forest, linear support vector machine (SVM), and gradient boosted trees.

Training data came from an ongoing Brazil-UK project in which we built a huge population-based cohort comprised by 114 million individuals receiving cash transfer support from the government. This database is probabilistically linked with several databases from the Public Health System to generate "data marts" (domain-specific data) for various epidemiological studies. Accuracy is assessed through established statistical metrics (sensitivity, specificity and positive predictive value) calculated during the manual review phase. So, these data marts together with their accuracy results were used to train our models. Our results show that SVM presents better sensitivity but logistic regression outperforms the remaining methods presenting better overall results.

The main contribution of our proposal is a workflow to preprocess data marts obtained from probabilistic linkages and use them as training data sets for different machine learning classifiers. Scenarios comprising fuzzy, approximate and probabilistic decisions on matching can benefit from this workflow to reduce or even eliminate manual review specially in big data applications.

This paper is structured as follows: Sect. 2 presents some related work focusing on accuracy assessment and different approaches to improve it. Section 3 presents some basic concepts related to accuracy assessment and details on our data linkage scenario. Section 4 describes the machine learning techniques used in this work. Section 5 presents the proposed trainable model targeted to eliminate manual review during the probabilistic linkage of huge data sets. Our experimental results are discussed in Sect. 6 and some concluding remarks and future work are given in Sect. 7.

2 Related Work

Record linkage is a research field with significant contributions present in the literature, covering from data acquisition and quality analysis to accuracy assessment and disclosure methods (including a vast discussion on privacy). In this section, we emphasize some works presenting different approaches to validate the accuracy of probabilistic data linkage as well as the use of machine learning techniques on linkage applications.

The authors in [28] have proposed a generalizable method to validate probabilistic record linkage consisting of three phases (sample selection, data collection and data analysis) performed by different teams on a double-bind manner. They used more than 30.000 records from a newborn registry database linked against 408 records produced by pediatricians based on external data sources. The results obtained showed a high accuracy rate with less than 1% of errors.

Some approaches have applied machine learning to improve pairwise classification [12,32,33], presenting accuracy, precision and recall measures above 90% using synthetic and real-world data. The work described in [26] explores the use of machine learning techniques in linking epidemiological cancer registries. The authors have used neural networks, support vector machines, decision trees and ensemble of trees to classify records. Ensemble techniques outperformed the other approaches by achieving 95% of classification rate.

Learned models were also used to scale up record linkage by using blocking schemes [6,20]. In [30], neural networks were applied to record linkage and the results compared to a naïve Bayes classifier, measuring the accuracy and concluding they outperform Bayesian classifiers in this task.

The need of using data mining techniques for ease or eliminate manual review was pointed by [10]. An unsupervised learning approach has been adopted to analyze record linkage results [17]. The author established a gold standard by running a deterministic merge over the involved databases before the record linkage procedure. Transformed attributes (first name, last name, gender, date of birth and a common primary key between the bases) were submitted to several iterations of the Expectation Maximization algorithm in order to improve the agreement of true positive pairs. The estimated review showed results very similar to manual observed verification.

We have been involved with probabilistic data linkage and subsequent accuracy assessment for more than four years. We have discussed the implementation of our first probabilistic linkage tool in [23], followed by a deeper discussion on different ways to implement probabilistic linkage routines and their accuracy assessment in controlled (databases with known relationships) and uncontrolled scenarios [22]. These works used socioeconomic and public health data from Brazilian governmental databases.

The dataset used to train our models in this work is derived from the results reported on these previous works. Our proposal comprises a workflow which can be used to assess accuracy of either record linkage or deduplication procedures in a way to reduce or eliminate the manual effort of this validation process, as well the subjectivity often associated to this verification phase.

3   Assessing the Accuracy of Record Linkage

Since Fellegi and Sunter [13] provided a formal foundation for record linkage, several ways to estimate match probabilities raised [31]. One way to do matching estimation is using similarity indices capable of dealing with different kinds of data (e.g. nominal, categorical, numerical). These indices provide a measure, which can be probability-based [11] or cost-based [16], between attributes from two or more data sets.

Attributes are assumed to be a "true match" if their measure pertains to a given interval or a "true unmatch" if their measure pertains to another interval. These intervals are delimited by upper and lower cut-off points: a similarity index above the upper cut-off point means a true positive (matched) pair, while an index below the lower cut-off point means a true negative (unmatched) pair. All pairs of records classified in between these cut-off points (the "gray area") are subject to manual review for reclassification.

Sensitivity, specificity and positive predictive values (PPV) are summary measures commonly used to evaluate record linkage results [27]. These measures take into consideration the number of pairs classified as true positive (TP), true negatives (TN), false positives (FP), and false negatives (FN). Thus, the accuracy function is usually defined as $(true\ pairs)/(all\ pairs)$.

The PPV measure, calculated by the equation $TP/(TP + FP)$, brings the proportion of true positive matches against all positive predictions, representing the ability of a given method to raise positive predictions [3]. Sensitivity represents the proportion of pairs correctly identified as true positives, as depicted by the equation $TP/(TP + FN)$. In contrast, specificity represents the proportion of pairs correctly identified as true negatives [1], defined by $TN/(TN + FP)$.

Validation of accuracy in deterministic scenarios is relatively easy due to existence of common key attributes and well-known relationships between the data sources being linked. This favors the definition of gold standards or other forms of validation even if some uncertainty is present. Probabilistic data linkage faces two major challenges regarding accuracy ascertainment: the first is to establish a gold standard, which may use external data to validate linked pairs, whereas the second refers to defining cut-off points in order to enhance the ability of finding true positive and true negative pairs.

Given a cut-off point, all linked pairs are separated as matched or unmatched. The expected behavior of probabilistic linkage results is to contain a significant number of matched pairs with higher similarity indices, as well a set of unmatched pairs undoubtedly classified as such. The gray area (dubious records) appears in situations where two or more cut-off points are used, leading to the need of manual review or other form of reclassification over these dubious records.

Probabilistic record linkage, specially in big data scenarios, lacks of gold standards as they are hard to set up considering the idiosyncrasies of each application and its data. Common scenarios do not provide additional data to reviewers do their verification work, which makes this process based on common sense, intuition and personal expertise [9]. Manual (or clerical) review is also limited by the amount of data to be revised.

In our experimental scenario, we assess the accuracy of our data linkage tool through the use of data marts generated by linking individuals from a huge socioeconomic database to their health outcomes. These data marts are used in several epidemiological studies assessing the impact of public policies, so their accuracy is really a huge concern.

4   Machine Learning Algorithms

Usually, machine learning algorithms can be divided in three categories: *supervised learning*, where a training data set is used to train the classification algorithm; *unsupervised learning (or clustering)*, where the algorithm does not have a prior knowledge (labeled data) about the data and relies on similar characteristics to perform classification; and *semi-supervised learning*, where some parts of data are labeled and some are not, being a mixture of the two previous methods. Our trainable model was developed using some supervised classification methods, which are described in this section.

4.1   Decision Trees

Decision trees are used to classify instances by splitting their attributes from the root to some leaf node. They use some *if-then* learned rules to provide disjunctions of conjunctions on the attribute values [19].

Let C be a number of classes and f_i be a frequency of some class in a given node. The Gini impurity, given by

$$Gini = \sum_{i=1}^{C} f_i(1 - f_i), \tag{1}$$

refers to the probability of some sample be correctly classified. The entropy, given by

$$Entropy = \sum_{i=1}^{C} -f_i \log_2(f_i), \tag{2}$$

measures the impurity within a set of examples. The most popular implementations of decision trees use either Gini or entropy impurity measures to calculate the data information gain, mostly getting similar results [25].

The information gain determines the effectiveness of some attribute to classify the training data [19]. Splitting data using this measure may reduce impurity of samples. The information gain calculation considers some attribute A in a sample S, where Imp can be either the Gini or entropy impurity measure of S, $Values(A)$ represents all possible values of A, and S_v is the subset of S in which the attribute A has the value v [19]. So, the information gain can be obtained by

$$IG(S, A) = Imp(S) - \sum_{v \in Values(A)} \frac{|S_v|}{|S|} Imp(S_v). \tag{3}$$

4.2   Gradient Boosted Trees

Gradient boosted trees (GBT) refers to iteratively train different random subsets of training data in order to build an ensemble of decision trees and minimize some loss function [14]. Lets N be the number of instances in some subsample, y_i the label of an instance i, x_i keeps the features of an instance and $F(x_i)$ brings a predicted label, for instance, i by the model. So, the equation

$$logloss = 2 - \sum_{i=1}^{N} log(1 + exp(-2y_i F(x_i))) \tag{4}$$

illustrates the log loss function used by GBT on classification problems

4.3   Random Forests

Random forests combine a number of tree-structured classifiers to vote for the most popular class of an instance [7]. The training of each classifier takes an independent, identically distributed random subset of the training data to decide about the vote. This randomness often reduce over-fitting and produce competitive results on classification in comparison to other methods [7].

4.4   Naïve Bayes

The naïve Bayes assumes that a target value is the product of the probabilities of the individual attributes because their values are conditionally independent [19]. It is calculated as shown in Eq. 5.

$$v_{NB} = \arg \max_{v_j \in V} P(v_j) \prod_i P(a_i|v_j). \tag{5}$$

4.5   Linear Support Vector Machine

Given a training data set with n points $(\overrightarrow{x_1}, y_1), ..., (\overrightarrow{x_n}, y_n)$, where y_1 may assume 1 or -1 values to indicate which class the point $x_1^{\prime}$ belongs to, and $\overrightarrow{x_1}$ is a p-dimensional vector $\in \mathbb{R}$, the linear support vector machine (LSVM) aims to find a hyperplane that divides these points with different values of y [8].

4.6   Logistic Regression

The logistic regression classifier aims to model the probability of the occurrence of an event E depending on the values of independent variables x [24]. The Eq. 6

$$p(x) = Pr\{E|x\} = 1/[1 + \exp\{-\alpha - \beta'x\}] \tag{6}$$

can be used to classify a new data point x with a vector of independent variables w, being (α, β) estimated from the training data. Let z be the odds ratio of positive or negative outcome class given x and w. If $z > 0.5$, the outcome class is positive; otherwise is negative.

$$f(z) = \frac{1}{1 + e^{-z}} \tag{7}$$

4.7   Comparative Analysis

All these methods have different advantages and disadvantages when applied to different scenarios. By using decision trees, the user do not need to worry with data normalization as it does not highly affect tree construction. Also, decision trees are easier to visualise, explain and manipulate, and do not require a large data set for training.

Gradient boosted trees usually have a good performance, but require a bigger time to learn because the trees are built sequentially. Usually, they are more prone to overfitting, so it is important to be careful in the pre-processing stage.

Random forests have a good performance and are more robust than single decision trees, giving more accurate results. Also, they suffer less from overfitting and can handle thousands of input variables without variable deletion. For categorical data with more than one level, random forests could be biased to the attributes with a bigger number of levels.

Naïve Bayes classifiers are fast and highly scalable. The classifier provides good results and is simple to implement, well fit with real and discrete data and is not sensitive to irrelevant features. As main disadvantage, this classifier assumes independence of features on training data, being unable to learn interactions between features.

Linear support vector machine (SVM) has a regularization parameter that helps the developer to reduce the impact of overfitting and get good results. SVMs use kernels, so it is possible to build expert knowledge by adjusting these kernels. SVM is defined by a convex optimization problem and there are different efficient methods to deal with this, for example, the Sequential Minimal Optimization (SMO).

Logistic regression is a simple method and is very fast. Usually, ot requires a large data set than other methods to achieve stability and works better with a single decision boundary. Also, logistic regression is less prone to overfitting.

5   Proposed Trainable Model

The input data of our trainable model must contain features that can simulate what a statistician often use to evaluate linkage results. Our methodology consists of construct a data set to show (i) how different are the nominals, and (ii) the equality of either categorical and numerical attributes used by the linkage algorithm. A categorization based on medians is made in order to assure some data balance.

Figure 1 shows the proposed pipeline to build a trainable model to accuracy assessment of probabilistic record linkage. This pipeline submits a data mart produced by the linkage tool to data cleansing, generation of a training data set to build models, evaluation and use. There is a possibility to rearrange some pre-processing, transformation and model selection settings by re-executing these steps.

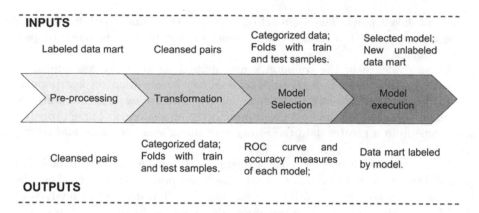

Fig. 1. Workflow for the proposed trainable model.

5.1 Pre-processing

The pre-processing step consists of (i) providing a descriptive analysis of data to select eligible common attributes within pairs and discard their missing values; (ii) select attributes to be used to build the model, usually the same attributes used by the linkage algorithm; and (iii) data cleansing and harmonization to guarantee that those selected attributes will have the same format.

The eligible common attributes to be used are: *name, mother's name, date of birth, gender* and *municipality of residence*. Attributes are chosen by their capacity of identifying an individual and their potential use by statisticians to manually verification about pairwise matching. Nominal attributes usually have a more discriminative power to determine how different two records are, followed by date of birth, gender and municipality code. Converging all different formats of date of birth, gender and municipality code into an unique one is an important task due to the diversity and heterogeneity of Brazilian information systems. The approach applied to nominal attributes is to deal with special characters, double spaces, capitalization, accentuation and typos (imputation errors).

5.2 Transformation

Statisticians verify the differences between attribute values in each pair during the accuracy assessment step despite the use of the similarity values provided by the linkage algorithms. In order to simulate this verification, a data set must reflect either equality, dissimilarity or cost between linked records. Both categorical and numerical attribute types output a binary value that represents their equality. A different approach is taken with nominal values which the degree of the dissimilarity may be useful.

A Levenshtein distance metric [16] is used to calculate how much deletions and insertions need to be done to equalize two strings. In the transformation step, this metric calculates the distance between the first, the last and the whole names in linked pairs. The approach of splitting the name attribute in given name and surname is to observe the influence of each part of the name on pair verification.

As an evidence of the common sense applied on accuracy assessment, a reviewer usually tolerates less errors on common names than on less popular names. To map this reviewer's empirical behavior, we use two new attributes to associate with each first name a given probability of occurence. These probabilities come from a greater data repository containing socioeconomic and census data.

A categorization using medians of distances (from the attribute name) and probabilities is made to promote data balance and prevent bias. Therefore, the transformation step is responsible for making a shallow descriptive analysis of data before categorization. The transformation step results on 12 features comprising: the similarity index, the distance among full, given name and surname (the same approach for mother's name), the probability of first names, equality of date of birth (day, month and year) and gender.

5.3   Model Selection

The model selection phase refers to find the best classifier to our data set. One of the best methods to evaluate and select classifiers is cross-validation [15]. The general idea of this method is to split the data set into n-folds and make n iterations setting a different fold as the test model. The remaining folds are set as training data to be used by different models and their several parameters. Accuracy measures are collected to evaluate the model at each iteration.

In addition to general accuracy, the capacity of correctly classify true positive pairs is the most important part to this work. Thus, accuracy, PPV and sensitivity become the main measures to be collected from each iteration of a cross-validation process. Furthermore, the balance between specificity and sensitivity and their interpretation by ROC curve plots [2] may be useful to model selection.

5.4   Model Execution

The model execution phase allows the reuse of some evaluated method with a new input data mart. This step outputs the classification as true or false based on the selected learned model. Also, the results from this step could increase the training data after some verification effort.

A high performance processing approach can be required due to the size of the databases involved. To meet this requirement, we use the distributed implementation of classification algorithms available in the Spark MLlib [18] tool.

6   Experimental Results

To train and evaluate supervised learning models, we used a data sample containing 13.300 pairs resulting from the linkage of a Brazilian longitudinal

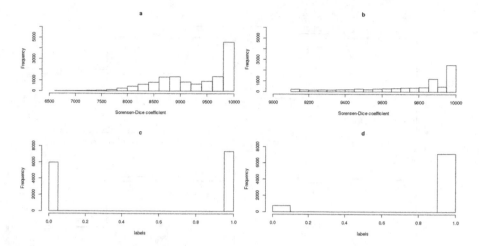

Fig. 2. Figures *a* and *b* refer to the similarity index distribution before and after the establishment of the cut-off point, respectively. As well, Figures *c* and *d* illustrate the difference with labels distribution after cut-off.

socioeconomic database with more than 100 million records (CadastroÚnico or CADU) with records from hospitalization, disease notification and mortality databases. For each pair, there is a similarity index calculated by the linkage algorithm and a label to determine if the pair is a true or false match. This label also indicates this pair already passed by a statistician evaluation [5] and can be used to train the models.

After discarding pairs with missing values and defining a cut-off point (Sorensen-Dice similarity index) [11] of 9100 (0.91%), the data sample was reduced to 7.880 pairs. This value was chosen based on several previous works and analyzes we done during these four years working on probabilistic linkage and taking into account the characteristics of our Brazilian databases.

Figures 2*b* and *d* show the data balancing of similarity index and labels. Experiments with different cut-off points obtained lower accuracy results than those showed in Figs. 3 and 4.

Several executions of machine learning algorithms with different settings are necessary to select the best model. Accuracy estimation and ROC curves may be used to choose the best model with available training data [4,15]. Figure 3 shows the accuracy, PPV, sensibility and specificity results of tested models. These measures are described in Sect. 3 and their interpretation may serve to assess the performance of these models.

Boxplots are used to allow the study of results variation for each fold in cross-validation. These plots can summarize and make comparisons between groups of data by using medians, quartiles and extremes data points [29]. A good model must get uppermost boxplots with closest quartiles, which means either a low variation of results on each fold or satisfactory model generalization.

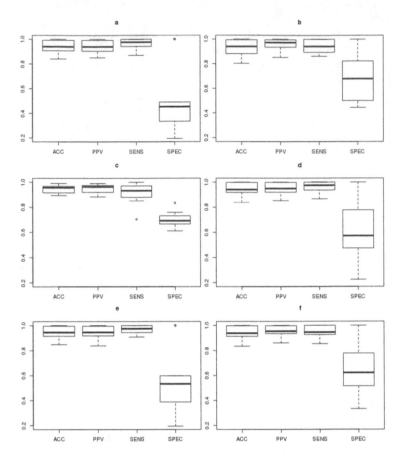

Fig. 3. Boxplots of 10-fold accuracy, PPV, sensibility and specificity measures in different machine learning algorithms: a = decision trees, b = naïve Bayes, c = logistic regression, d = random forest, e = linear support vector machine, f = gradient boosted trees.

Figure 3 shows the best results of each model. The use of entropy to split data and set the maximum depth of trees as 3 achieves the best results, showed in Fig. 3a. Results of naïve Bayes classifier are showed in Fig. 3b. Figure 3c presents logistic regression results with 1.000 iterations. Random forest achieved best results by setting 1.000 trees for voting, Gini impurity to split data and the maximum depth of tree as 5, as shown in Fig. 3d. LSVM results with 50 iterations to well fit the hyperplane are illustrated in Fig. 3e. Figure 3f brings the results for gradient boosted trees with a depth of at most 3 and 100 iterations to minimize the log loss function.

Figure 3c shows that logistic regression outperforms the other models by comparing accuracy, PPV and specificity medians. Despite the better sensibility performance of LSVM, the best specificity result is achieved by logistic regression.

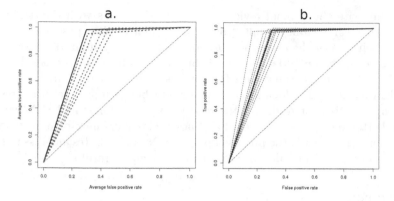

Fig. 4. ROC curves depicting the average true and false positive rates of 10-fold cross-validation. Different curve color represents an algorithm: dark green for decision trees, blue for naive Bayes, black for logistic regression, gray for random forests, orange for linear support vector machine and purple gradient boosted trees (best viewed in color). (Color figure online)

ROC curves allow the accuracy study by drawing the relation between true and false positive rates. Figure 4a shows the average true and false positive rates of each fold in cross-validation. The unbroken black line in Fig. 4a brings the average ROC curve to the 10-fold cross-validation. This curve shows the logistic regression superiority in comparison to other curves in terms of sensibility and sensitivity. The performance variation of all logistic regression curves on folds are showed in Fig. 3b.

7 Conclusions and Future Work

Accuracy assessment of record linkage refers to a time-consuming process that becomes impractical when huge databases are involved. This manual review may be reduced or even eliminated by using trainable models since this validation process can be assumed as a binary classification problem [9]. The proposed pipeline has initial steps capable of establishing a dataset with features used to build and evaluate models. The final steps allow building models by using different machine learning classifiers and their settings in order to evaluate and use them to validate new data marts.

The logistic regression outperformed others classifiers using the available dataset under a 10-fold cross-validation approach. Other models may achieve better results due to new preprocessing, transformation and categorization approaches. Different results may also occur depending on the increase or decrease of data size.

The proposed workflow is suitable to be used either in record linkage or deduplication scenarios where fuzzy, approximate and probabilistic decisions about pairs of record matching should be made. However, a trainable model could not always eliminate the manual review, mainly in situations with tiny train data sets or with lower accuracy results from cross-validation. It is possible to adopt a feedback behavior of the proposed workflow, where newly submitted data marts can increase the training data set since this new result becomes labeled.

The use of deep learning classification algorithms such as artificial neural networks with several hidden layers may achieve better model accuracy results. Increasing iterations of gradient boosted trees, random forest and SVM can also provide better results. New classical and novel classifiers may be used to verify their performance within the proposed pipeline. New attributes and dissimilarity metrics may be proposed in order to get more accurate results.

References

1. Altman, D.G., Bland, J.M.: Diagnostic tests 1: Sensitivity and specificity. BMJ Br. Med. J. **308**(6943), 1552 (1994)
2. Altman, D.G., Bland, J.M.: Diagnostic tests 3: receiver operating characteristic plots. BMJ Br. Med. J. **309**(6948), 188 (1994)
3. Altman, D.G., Bland, J.M.: Statistics notes: diagnostic tests 2: predictive values. BMJ **309**(6947), 102 (1994)
4. Antonie, M.L., Zaiane, O.R., Holte, R.C.: Learning to use a learned model: a two-stage approach to classification. In: Sixth International Conference on Data Mining, ICDM 2006, pp. 33–42. IEEE (2006)
5. Barreto, M.E., Alves, A., Sena, S., Fiaccone, R.L., Amorim, L., Ichihara, M., Barreto, M.: Assessing the accuracy of probabilistic record linkage of huge brazilian healthcare databases, vol. 1, p. 12. Oxford (2016)
6. Bilenko, M., Kamath, B., Mooney, R.J.: Adaptive blocking: Learning to scale up record linkage. In: Sixth International Conference on Data Mining, ICDM 2006, pp. 87–96. IEEE (2006)
7. Breiman, L.: Random forests. Mach. Learn. **45**(1), 5–32 (2001)
8. Burges, C.J.: A tutorial on support vector machines for pattern recognition. Data Min. Knowl. Discov. **2**(2), 121–167 (1998)
9. Christen, P., Goiser, K.: Quality and complexity measures for data linkage and deduplication. In: Guillet, F.J., Hamilton, H.J. (eds.) Quality Measures in Data Mining, pp. 127–151. Springer, Heidelberg (2007)
10. Christen, P., et al.: Parallel techniques for high-performance record linkage (data matching). Data Mining Group, Australian National University, Epidemiology and Surveillance Branch, pp. 1-27 (2002). Project web page: http://datamining.anu. edu.au/linkage.html
11. Dice, L.R.: Measures of the amount of ecologic association between species. Ecology **26**(3), 297–302 (1945)
12. Elfeky, M.G., Verykios, V.S., Elmagarmid, A.K.: Tailor: a record linkage toolbox. In: 18th International Conference on Data Engineering, 2002, Proceedings, pp. 17–28. IEEE (2002)
13. Fellegi, I.P., Sunter, A.B.: A theory for record linkage. J. Am. Stat. Assoc. **64**(328), 1183–1210 (1969)
14. Friedman, J.H.: Stochastic gradient boosting. Comput. Stat. Data Anal. **38**(4), 367–378 (2002)
15. Kohavi, R., et al.: A study of cross-validation and bootstrap for accuracy estimation and model selection. In: IJCAI, vol. 14, pp. 1137–1145, Stanford, CA (1995)

16. Levenshtein, V.I.: Binary codes capable of correcting deletions, insertions, and reversals. Sov. Phys. Dokl. **10**, 707–710 (1966)
17. McDonald, C.J.: Analysis of a probabilistic record linkage technique without human review (2003)
18. Meng, X., Bradley, J., Yavuz, B., Sparks, E., Venkataraman, S., Liu, D., Freeman, J., Tsai, D., Amde, M., Owen, S., et al.: Mllib: machine learning in apache spark. J. Mach. Learn. Res. **17**(34), 1–7 (2016)
19. Michalski, R.S., Carbonell, J.G., Mitchell, T.M.: Machine Learning: An Artificial Intelligence Approach. Springer Science & Business Media, Heidelberg (2013)
20. Michelson, M., Knoblock, C.A.: Learning blocking schemes for record linkage. In: AAAI, pp. 440–445 (2006)
21. Newcombe, H.B., Kennedy, J.M., Axford, S., James, A.P.: Automatic linkage of vital records. Science **130**(3381), 954–959 (1959)
22. Pinto, C., Pita, R., Melo, P., Sena, S., Barreto, M.: Correlação probabilística de bancos de dados governamentais, pp. 77–88 (2015)
23. Pita, R., Pinto, C., Melo, P., Silva, M., Barreto, M., Rasella, D.: A spark-based workflow for probabilistic record linkage of healthcare data. In: EDBT/ICDT Workshops, pp. 17–26 (2015)
24. Press, S.J., Wilson, S.: Choosing between logistic regression and discriminant analysis. J. Am. Stat. Assoc. **73**(364), 699–705 (1978)
25. Raileanu, L.E., Stoffel, K.: Theoretical comparison between the gini index and information gain criteria. Ann. Math. Artif. Intell. **41**(1), 77–93 (2004)
26. Siegert, Y., Jiang, X., Krieg, V., Bartholomus, S.: Classification-based record linkage with pseudonymized data for epidemiological cancer registries. IEEE Trans. Multimed. **18**(10), 1929–1941 (2016)
27. da Silveira, D.P., Artmann, E.: Accuracy of probabilistic record linkage applied to health databases: systematic review. Rev. Saúde Pública **43**(5), 875–882 (2009)
28. Tromp, M., Ravelli, A., Meray, N., Reitsma, J., Bonsel, G., et al.: An efficient validation method of probabilistic record linkage including readmissions and twins. Methods Inf. Med. **47**(4), 356–363 (2008)
29. Williamson, D.F., Parker, R.A., Kendrick, J.S.: The box plot: a simple visual method to interpret data. Ann. Intern. Med. **110**(11), 916–921 (1989)
30. Wilson, D.R.: Beyond probabilistic record linkage: using neural networks and complex features to improve genealogical record linkage. In: The 2011 International Joint Conference on Neural Networks, pp. 9–14, July 2011
31. Winkler, W.E.: The state of record linkage and current research problems. In: Statistical Research Division, US Census Bureau. Citeseer (1999)
32. Winkler, W.E.: Methods for record linkage and bayesian networks. Technical report, Statistical Research Division, US Census Bureau, Washington, DC (2002)
33. Winkler, W.E., et al.: Machine learning, information retrieval and record linkage. In: Proceedings of Section on Survey Research Methods, American Statistical Association, pp. 20–29 (2000)

An Efficient Approach for Instance Selection

Joel Luís Carbonera[(✉)]

IBM Research, Rio de Janeiro, Brazil
joelc@br.ibm.com

Abstract. Nowadays, the volume of data that is produced challenges
our capabilities of converting it in useful knowledge. Due to this, data
mining approaches have been applied for extracting useful knowledge
from this big data. In order to deal with the increasing size of datasets,
techniques for instance selection have been applied for reducing the data
to a manageable volume and, consequently, to reduce the computational
resources that are necessary to apply data mining approaches. However,
most of the proposed approaches for instance selection have a high time
complexity and, due to this, they cannot be applied for dealing with big
data. In this paper, we propose a novel approach for instance selection
called XLDIS. This approach adopts the notion of local density for selecting the most representative instances of each class of the dataset, providing a reasonably low time complexity. The approach was evaluated on
20 well-known datasets used in a classification task, and its performance
was compared to those of 6 state-of-the-art algorithms, considering three
measures: accuracy, reduction, and effectiveness. All the obtained results
show that, in general, the XLDIS algorithm provides the best trade-off
between accuracy and reduction.

Keywords: Instance selection · Data reduction · Data mining · Machine
learning · Big data

1 Introduction

Recent years have witnessed a dramatic increase in our capability to collect a
huge volume of data from different devices, in different formats, in unprecedented
rates. These resulting large collections of data are usually called *Big Data*, and
due to their large size and complexity, they challenge even the capabilities of
our current data mining approaches [12]. In this scenario, *instance selection* has
been viewed as an alternative to overcome the challenges raised by the size of
these datasets.

Instance selection (IS) is a data-mining (or machine learning) pre-processing
task that consists in choosing a representative subset of instances among the total
available data, which allows to carry out a data mining task *with no performance
loss* (or, at least, a reduced performance loss) [13]. Thus, every IS strategy
faces a *trade-off* between the *reduction rate* of the dataset and the resulting
classification quality [10].

© Springer International Publishing AG 2017
L. Bellatreche and S. Chakravarthy (Eds.): DaWaK 2017, LNCS 10440, pp. 228–243, 2017.
DOI: 10.1007/978-3-319-64283-3_17

Besides its capability of reducing a big dataset to a manageable subset, instance selection can also be used for improving the learned models through the deletion of useless (redundant), erroneous or noisy instances, before applying learning algorithms [13,19].

Most of the proposed algorithms for instance selection, such as [1,17,18,20, 23,24], are designed for preserving the *boundaries* between different classes in the dataset. This is a reasonable strategy, because border instances provide relevant information for supporting discrimination between classes [17]. However, in general, these algorithms have a *high time complexity*, and this is not a desirable property for algorithms that should deal with *Big Data*. The high time complexity of these algorithms results from the fact that they usually should search for the border instances *within the whole dataset* and, due to this, they usually need to perform comparisons between each pair of instances in the dataset.

In this paper, we propose an algorithm for instance selection, called XLDIS (e*X*tended *L*ocal *D*ensity-based *I*nstance *S*election)[1], which analyses the instances of each class separately and focuses on keeping only the most representative instance in a given (arbitrary) neighborhood. Due to this strategy, the resulting time complexity of XLDIS is reasonably low. Besides that, it is important to notice that XLDIS assumes that the *representativeness* of a given instance x is proportional to its *local density ordering*, which, intuitively, determines how many instances are less dense than x, within the class of x.

Our approach was evaluated on 20 well-known datasets and its performance was compared with the performance of 6 important algorithms provided by the literature, according to 3 different performance measures: accuracy, reduction and effectiveness (proposed in [8]). The accuracy was evaluated considering the KNN algorithm [11]. The results show that, compared to the other algorithms, XLDIS provides the best trade-off between accuracy and reduction, while presents a reasonably low time complexity.

Section 2 presents some related works. Section 3 presents the notation that will be used throughout the paper. Section 4 presents our approach. Section 5 discusses our experimental evaluation. Finally, Sect. 6 presents our main conclusions and remarks.

2 Related Works

In this section, we discuss some important instance reduction methods. In this discussion, we consider T as the original set of instances in the training set and S, with $S \subseteq T$, as the reduced set of instances, resulting from the instance selection process.

The *Condensed Nearest Neighbor* (CNN) algorithm, introduced by [16], randomly selects one instance that belongs to each class from T and puts them in S. Then, each instance $\in T$ is classified using only the instances in S. If an

---

[1] The source code of the algorithm is available in https://www.researchgate.net/publication/317040648_XLDIS_eXtended_Local_Density-based_Instance_Selection.

instance is misclassified, it is added to S, in order to ensure that it will be classified correctly. This process is repeated until there is no instance in T that is misclassified. CNN can assign *noisy* and *outlier* instances to S, causing harmful effects in the classification accuracy. Also, CNN is dependent on instance order in the training set T. The time complexity of CNN is $O(|T|^2)$, where $|T|$ is the size of the training set.

The *Reduced Nearest Neighbor* algorithm (RNN) [14], assigns all instances in T to S first. After, it removes each instance from S, until further removal causes no other instances in T to be misclassified by the remaining instances in S. RNN is less sensitive to noise than CNN and produces subsets S that are smaller than the subsets produced by CNN. However, the main drawback of RNN is its cubic time complexity, which is higher than the time complexity of CNN.

The *Generalized Condensed Nearest Neighbor* (GCNN) was proposed by [10]. Considering $d_N(x)$ as the distance between x and its nearest neighbor, and $d_E(x)$ as the distance between x and its nearest enemy (instance of a class that is different from the class of x); x is included by GCNN in S if $d_N(x) - d_E(x) > \rho$, where ρ is an arbitrary threshold. In general, GCNN produces sets S that are smaller than the sets produced by CNN. The main negative point regarding GCNN is that determining the value of ρ can be a challenge.

The *Edited Nearest Neighbor* (ENN) algorithm [24] assigns all training instances to S first. Then, each instance in S is removed if it does not agree with the label of the majority of its k nearest neighbors. This strategy is effective for improving the classification accuracy of the learned models, because it removes noisy and outlier instances. However, since it keeps internal instances, it cannot reduce the dataset as much as other reduction algorithms. The literature provides some extensions of this method, such as [22].

In [23], the authors present 5 approaches, named the *Decremental Reduction Optimization Procedure* (DROP). These algorithms assume that each instance x has k nearest neighbors ($k \in \mathbb{N}$), and those instances which have x as one of their k nearest neighbors are called the *associates* of x. Among the proposed algorithms, DROP3 has the best trade-off between the reduction of the dataset and the accuracy of the classification. As an initial step, it applies a noise-filter algorithm such as ENN. Then, it removes an instance x if its associates in the original training set can be correctly classified without x. The main drawback of DROP3 is its high time complexity.

The *Iterative Case Filtering algorithm* (ICF) [1] is based on the notions of *Coverage set* and *Reachable set*. The coverage set of an instance x is the set of instances in T whose distance from x is less than the distance between x and its nearest enemy (instance with a different class). This notion is analogous to the notion of *local set*, which we adopt in our algorithm. The Reachable set of an instance x, on the other hand, is the set of instances in T that have x in their respective coverage sets. In this method, a given instance x is removed from S if $|Reachable(x)| > |Coverage(x)|$, i.e. when the number of other instances that can classify x correctly is greater than the number of instances that x can correctly classify.

In [17], the authors adopted the notion of *local sets* for designing three complementary methods for instance selection. In this context, the local set of a given instance x is the set of instances contained in the largest hypersphere centered on x such that it does not contain instances from any other class. The first algorithm, called *Local Set-based Smoother* (LSSm), was proposed for removing instances that are harmful for the classification accuracy, i.e. instances that misclassify more instances than those that they correctly classify. It uses two notions for guiding the process: *usefulness* and *harmfulness*. The usefulness $u(x)$ of a given instance x is the number of instances having x among the members of their local sets, and the harmfulness $h(x)$ is the number of instances having x as the nearest enemy. For each instance x in T, the algorithm includes x in S if $u(x) \geq h(x)$. Since the primary goal of LSSm is to remove harmful instances, its reduction rate is lower than most of the instance selection algorithms. The second algorithm, called *Local Set-based Centroids Selector method* (LSCo), firstly applies LSSm to remove noise and then applies LS-clustering [2] to identify clusters in T. The algorithm keeps in S only the centroids of the resulting clusters. Finally, the *Local Set Border selector* (LSBo) first uses LSSm to remove noise, and then, it computes the local set of every instance $\in T$. Next, the instances in T are sorted in the ascending order of the cardinality of their local sets. In the last step, LSBo verifies, for each instance $x \in T$ if any member of its local set is contained in S, thus ensuring the proper classification of x. If that is not the case, x is included in S to ensure its correct classification. The time complexity of the three approaches is $O(|T|^2)$. Among the three algorithms, LSBo provides the best balance between reduction and accuracy.

In [8], the authors proposed the *Local Density-based Instance Selection* (LDIS) algorithm. This algorithm selects the instances with the highest density values in their neighborhoods. The LDIS algorithm searches for representative instances only among the instances of each class (separately). For this reason, it is not necessary to perform a *global search* in the whole dataset. As a consequence, the time complexity of the LDIS algorithm is lower than the time complexity of most of the instance selection algorithms. Besides that, this surprisingly simple algorithm is able to produce representative subsets of data that results in high accuracies in classification tasks. The XLDIS algorithm, proposed in this paper, can be viewed as an extension of the LDIS algorithm.

Other approaches can be found in surveys such as [13, 15].

3   Notations

In this section, we introduce a notation adapted from [8] that will be used throughout the paper.

- $T = \{x_1, x_2, ..., x_n\}$ is the non-empty set of n instances (or data objects), representing the original dataset to be reduced in the instance selection process.
- Each $x_i \in T$ is an $m - tuple$, such that $x_i = (x_{i1}, x_{i2}, ..., x_{im})$, where x_{ij} represents the value of the j-th feature of the instance x_i, for $1 \leq j \leq m$.

- $L = \{l_1, l_2, ..., l_p\}$ is the set of p class labels that are used for classifying the instances in T, where each $l_i \in L$ represents a given class label.
- $l \colon T \to L$ is a function that maps a given instance $x_i \in T$ to its corresponding class label $l_j \in L$.
- $c \colon L \to 2^T$ is a function that maps a given class label $l_j \in L$ to a given set C, such that $C \subseteq T$, which represents the set of instances in T whose class is l_j. Notice that $T = \bigcup_{l \in L} c(l)$. In this notation, 2^T represents the *powerset* of T, that is, the set of all subsets of T, including the empty set and T itself.
- $d \colon T \times T \to \mathbb{R}$ is a *distance function* (or dissimilarity function), which maps two instances to a real number that represents the distance (or dissimilarity) between them.
- $S = \{x_1, x_2, ..., x_q\}$ is a set of q instances, such that $S \subseteq T$. It represents the reduced set of instances resulting from the instance selection process.

4   The XLDIS Algorithm

In [8], the authors propose the LDIS algorithm. Instead of selecting the border points between classes, the LDIS algorithm selects instances that have a *high concentration* of instances near to them. That is, the algorithm assumes that the *denser* instance within a given neighborhood is able to represent more information about that neighborhood than its less dense neighbors. Besides that, instead of searching for these instances in the whole dataset, the LDIS deals with the instances of each class of the dataset separately, searching for the representative instances within the set of instances of each class. The adoption of this *local search strategy* (performed within each class) results in a time complexity that is lower than the time complexity of approaches that adopts a *global search* strategy.

In this Section, we will describe the XLDIS algorithm, which extends the original LDIS. The main difference between both algorithms is that, instead of selecting the denser instance within a neighborhood, the XLDIS algorithm selects the instance that have the higher *local density ordering* (LDO) within its neighborhood.

Before presenting the XLDIS algorithm, it is important to discuss the notion of *density* of a given instance, which plays an important role in our approach.

Definition 1. *The density of a given instance $x \in P$, where $P \subseteq T$, is a measure of the spatial concentration of other instances within P around the instance x. Intuitively, it can be viewed as a measure of the average similarity between x and all the instances within P. Thus, the density is formalized by the function $Dens \colon T \times 2^T \to \mathbb{R}$, such as:*

$$Dens(x, P) = -\frac{1}{|P|} \sum_{y \in P} d(x, y) \tag{1}$$

where d is a given distance function. Notice that $Dens(x, P)$ provides the density of the instance x relatively to the set P of instances. In this way, when P

is a subset of the whole dataset, $Dens(x, P)$ represents the local density *of x, considering the set P. This notion of density was adapted from* [3–5,9].

The XLDIS algorithm also adopts the notion of *partial k-neighborhood* from [8].

Definition 2. *The* partial k-neighborhood *is formalized by the function $pkn\colon T \times \mathbb{N}_1 \to 2^T$ that maps a given instance $x \in T$ and a given $k \in \mathbb{N}_1$ ($k \geq 1$) to a given set C, such that $C \subseteq (c(l(x)) - \{x\})$, which represents the set of the k nearest neighbors of x, in $c(l(x))$ (excepting x_i). Since the resulting set C includes only the neighbors that have a given class label, it defines a partial k-neighborhood.*

Finally, the XLDIS algorithm also adopts the notion of *local density ordering* (LDO).

Definition 3. *The* local density ordering *is formalized by the function $LDO\colon T \to \mathbb{N}$ that maps a given instance $x \in T$ to a value $o \in \mathbb{N}$ that represents the ordering of the instance x when the set $c(l(x))$ is sorted according to the* local density *of the instances within the set $c(l(x))$. Thus, the LDO of a given instance x represents how dense x is, in comparison with the other instances within $c(l(x))$.*

Notice that when all instances within a set $c(l(x))$ have different local densities, the local density ordering of an instance x represents the number of instances within $c(l(x))$ whose local density is less than $Dens(x, c(l(x)))$. However, when the instances in a set $G \subseteq c(l(x))$ have the same local density, the function LDO assigns a different local density ordering for each of them. That is, supposing that $G = \{y_0, y_1, ..., y_s\}$, the LDO of the instances in G would be: $LDO(y_0) = ld_G$, $LDO(y_1) = ld_G + 1,..., LDO(y_s) = ld_G + s$; where $ld_G = |\{i|i \in c(l(x)) \wedge Dens(i, c(l(x))) < Dens(y_0, c(l(x)))\}|$. Thus, the LDO is able to distinguish instances with the same local density.

The XLDIS algorithm assumes that the *local density ordering*, $LDO(x)$, of a given instance x is proportional to its capability of representing information about its neighborhood. Considering this, for each $x \in c(l)$, the algorithm defines $toAvoid_x$ as false, indicating that this instance should be considered as a candidate for selection. Then, the algorithm sorts the $c(l)$ set, in a descendent order, according to the LDO of the instances. In the next step, for each $l \in L$, the XLDIS algorithm verifies if x should be considered as a candidate for selection (if $toAvoid_x = true$). If this is the case, it verifies if there is some instance $y \in pkn(x, k)$ (where k is arbitrarily chosen by the user), such that $LDO(y) > LDO(x)$. If this is not the case, this means that x has the highest *local density ordering* in its *partial k-neighborhood* and, due to this, x should be included in S. Besides that, the algorithm considers that every instance $y \in pkn(x, k)$, such that $LDO(y) < LDO(x)$ should be avoided in further analysis, because they have a neighbor (x) with a higher LDO that was already included in the final set. The Algorithm 1 formalizes this strategy. Notice that when $|c(l(x))| \leq k$, it is necessary to consider $k = |c(l(x))| - 1$, for calculating the *partial k-neighborhood* of x.

It is important to notice that the XLDIS algorithm has 3 main differences regarding the original LDIS. The first one is that XLDIS considers *LDO* as a measure of the importance of the instances. The second one is that XLDIS analyses the instances in a specific order, from the instance with the highest *LDO* to the instance that has the smallest *LDO*, and selects first the instances that have a higher *LDO*. Finally, the third one is that XLDIS keeps a list of instances that are not necessary to analyse for being included in the final set S. This strategy reduces the set of instances that should be analysed and, as a consequence, it reduces the processing time of XLDIS.

The most expensive steps of the algorithm involve determining the *local density* and the *partial k-neighborhood* of each instance of a given set $c(l)$ (for some class label l). Determining the *partial density* of every instance of a given set $c(l)$ is a process whose time complexity is proportional to $O(|c(l)|^2)$. The time complexity of determining the *partial k-neighborhood* is equivalent. An efficient implementation of the Algorithm 1 could calculate the *partial k-neighborhood* and the *partial density* of each instance of a given set $c(l)$ (for some class label l) just once, in a first step within the main loop, and use this information for further calculations. Considering this, the time complexity of XLDIS would be proportional to $O(\sum_{l \in L} |c(l)|^2)$. Thus, the XLDIS algorithm has a time complexity equivalent to the time complexity of the original LDIS algorithm.

Algorithm 1. XLDIS algorithm

Input: A set instances T and the number k of neighbors.
Output: A set S, such as $S \subseteq T$.
begin

  $S \leftarrow \emptyset$;
  foreach $l \in L$ **do**
    $toAvoid_x \leftarrow false$, for every $x \in c(l)$;
    Sorting $c(l)$ in a descending order, according to the *LDO* of its instances;
    foreach $x \in c(l)$ **do**
      if $toAvoid_x = false$ **then**
        $foundHigher \leftarrow false$;
        foreach $y \in pkn(x, k)$ **do**
          if $LDO(x) < LDO(y)$ **then**
            $foundHigher \leftarrow true$;
        if $\neg foundHigher$ **then**
          $S \leftarrow S \cup \{x\}$;
          foreach $y \in pkn(x, k)$ **do**
            if $LDO(x) > LDO(y)$ **then**
              $toAvoid_y \leftarrow true$;
  return S;

---

5   Experiments

For evaluating our approach, we compared the XLDIS algorithm in a classification task, with 6 important instance selection algorithms provided by the literature: DROP3, ENN, ICF, LSBo, LSSm and the original LDIS. We considered 20 well-known datasets: breast cancer, cardiotocography, cars, dermatology, diabetes, E. Coli, glass, iris, landsat, letter, lung cancer, mushroom, parkinson, genetic promoters, segment, soybean[2], splice-junction gene sequences, congressional voting records, wine and zoo. All datasets were obtained from the UCI Machine Learning Repository[3]. In Table 1, we present the details of the datasets that were used.

We use three standard measures to evaluate the performance of IS algorithms: *accuracy, reduction* and *effectiveness* (proposed in [8]). Following [8, 17],

Table 1. Details of the datasets used in the evaluation process.

Dataset	Instances	Attributes	Classes
Breast cancer	286	10	2
Cardiotocography	2126	21	10
Cars	1728	6	4
Dermatology	358	35	6
Diabetes	768	9	2
E. Coli	336	8	8
Glass	214	10	7
Iris	150	5	3
Landsat	4435	37	6
Letter	20000	17	26
Lung cancer	32	57	3
Mushroom	8124	23	2
Parkinson	195	23	2
Promoters	106	58	2
Segment	2310	20	7
Soybean	683	36	19
Splice	3190	61	3
Voting	435	17	2
Wine	178	14	3
Zoo	101	18	7

---

[2] This dataset combines the large soybean dataset and its corresponding test dataset.
[3] http://archive.ics.uci.edu/ml/.

Table 2. Comparison of the *accuracy* achieved by the training set produced by each algorithm, for each dataset.

Algorithm	LDIS	LSBo	DROP3	ICF	ENN	LSSm	XLDIS	Average
Breast cancer	0.69	0.61	0.73	0.72	**0.74**	**0.74**	0.69	0.70
Cardiotocography	0.54	0.56	0.61	0.60	0.66	**0.68**	0.53	0.60
Cars	**0.76**	0.65	0.75	**0.76**	**0.76**	**0.76**	**0.76**	0.74
Dermatology	0.71	0.73	0.78	0.78	**0.82**	0.84	0.69	0.77
Diabetes	0.69	0.71	**0.73**	0.70	**0.73**	0.72	0.70	0.71
E. Coli	0.85	0.80	0.83	0.81	**0.86**	0.85	0.85	0.84
Glass	0.61	0.56	0.62	0.62	0.65	**0.71**	0.60	0.62
Iris	0.96	0.94	**0.97**	0.94	**0.97**	0.96	0.95	0.95
landsat	0.87	0.86	0.89	0.85	**0.90**	**0.90**	0.87	0.88
Letter	0.77	0.75	0.87	0.80	**0.92**	**0.92**	0.75	0.83
Lung cancer	0.38	0.32	0.31	0.41	0.34	**0.45**	0.43	0.38
Mushroom	**1.00**	**1.00**	**1.00**	**1.00**	**1.00**	**1.00**	**1.00**	**1.00**
Parkinsons	0.80	**0.85**	0.83	0.81	**0.85**	**0.85**	0.81	0.83
Promoters	0.76	0.72	0.79	0.72	0.79	0.82	0.77	0.77
Segment	0.91	0.87	0.93	0.91	**0.95**	**0.95**	0.89	0.92
Soybean	0.78	0.63	0.85	0.84	0.90	**0.91**	0.73	0.80
Splice	0.75	0.75	0.71	0.70	0.73	**0.76**	0.74	0.73
Voting	0.91	0.89	**0.92**	0.91	**0.92**	**0.92**	0.91	0.91
Wine	0.69	**0.78**	0.70	0.73	0.74	0.76	0.71	0.73
Zoo	0.82	0.71	**0.91**	0.88	0.88	0.90	0.66	0.82
Average	0.76	0.73	0.79	0.77	0.81	**0.82**	0.75	0.78

we assume:

$$accuracy = \frac{Sucess(Test)}{|Test|} \qquad (2)$$

and

$$reduction = \frac{|T| - |S|}{|T|}, \qquad (3)$$

where $Test$ is a given set of instances that are selected for being tested in a classification task, and $Success(Test)$ is the number of instances in $Test$ correctly classified in the classification task. Besides that, we consider *effectiveness* as a measure of the degree to which an instance selection algorithm is successful in producing a small set of instances that allows a high classification accuracy of new instances. Thus, we consider *effectiveness = accuracy × reduction*, as per [8].

To evaluate the classification *accuracy* of new instances in each respective dataset, we adopted the *k-Nearest Neighbors* (KNN) algorithm [11], considering

Table 3. Comparison of the *reduction* achieved by each algorithm, for each dataset.

Algorithm	LDIS	LSBo	DROP3	ICF	ENN	LSSm	XLDIS	Average
Breast cancer	0.88	0.73	0.77	0.85	0.29	0.14	**0.90**	0.65
Cardiotocography	0.86	0.70	0.70	0.71	0.31	0.13	**0.88**	0.61
Cars	0.86	0.74	0.88	0.82	0.18	0.12	**0.89**	0.64
Dermatology	0.88	0.73	0.65	0.68	0.16	0.12	**0.89**	0.59
Diabetes	0.91	0.76	0.77	0.86	0.31	0.13	**0.92**	0.66
E. Coli	0.90	0.82	0.71	0.86	0.16	0.09	**0.92**	0.64
Glass	0.90	0.72	0.76	0.69	0.32	0.14	**0.91**	0.63
Iris	0.89	0.93	0.72	0.58	0.04	0.06	**0.91**	0.59
landsat	0.92	0.88	0.72	0.90	0.10	0.05	**0.93**	0.64
Letter	0.83	0.83	0.68	0.80	0.05	0.04	**0.88**	0.59
Lung cancer	0.85	0.52	0.70	0.74	0.59	0.17	**0.86**	0.63
Mushroom	0.86	**0.99**	0.86	0.94	0.00	0.00	0.90	0.65
Parkinsons	0.81	**0.87**	0.71	0.75	0.15	0.12	**0.87**	0.61
Promoters	0.84	0.60	0.60	0.70	0.18	0.05	**0.86**	0.55
Segment	0.82	**0.91**	0.69	0.83	0.04	0.04	0.87	0.60
Soybean	0.78	0.83	0.68	0.58	0.09	0.05	**0.86**	0.55
Splice	0.81	0.59	0.66	0.76	0.23	0.05	**0.86**	0.57
Voting	0.77	0.89	0.79	**0.92**	0.07	0.04	0.90	0.63
Wine	0.87	0.78	0.71	0.78	0.23	0.10	**0.88**	0.62
Zoo	0.65	**0.88**	0.65	0.33	0.06	0.06	0.82	0.49
Average	0.84	0.78	0.72	0.76	0.18	0.08	**0.89**	0.61

$k = 3$, as assumed in [8,17]. Besides that, following [8], the accuracy and reduction were evaluated in an *n-fold cross-validation* scheme, where $n = 10$. Thus, firstly a dataset is randomly partitioned in 10 equally sized subsamples. From these subsamples, a single subsample is selected as validation data (*Test*), and the union of the remaining 9 subsamples is considered the *initial training set* (*ITS*). Next, an instance selection algorithm is applied for reducing the *ITS*, producing the *reduced training set* (*RTS*). At this point, we can measure the *reduction* of the dataset. Finally, the *RTS* is used as the training set for the KNN algorithm, to classify the instances in *Test*. At this point, we can measure the accuracy achieved by the KNN, using *RTS* as the training set. This process is repeated 10 times, with each subsample used once as *Test*. The 10 values of accuracy and reduction are averaged to produce, respectively, the *average accuracy* (*AA*) and *average reduction* (*AR*). The *average effectiveness* is calculated by considering *AA* and *AR*. Tables 2, 3 and 4 report, respectively, the resulting *AA*, *AR*, and *AE* of each combination of dataset and instance selection algorithm; the best results for each dataset being marked in bold typeface.

Table 4. Comparison of the *effectiveness* achieved by each algorithm, for each dataset.

Algorithm	LDIS	LSBo	DROP3	ICF	ENN	LSSm	XLDIS	Average
Breast cancer	0.61	0.45	0.56	0.61	0.21	0.10	**0.62**	0.45
Cardiotocography	0.46	0.39	0.43	0.42	0.20	0.09	**0.47**	0.35
Cars	0.65	0.49	0.66	0.62	0.14	0.09	**0.68**	0.47
Dermatology	**0.63**	0.54	0.51	0.53	0.13	0.10	0.62	0.44
Diabetes	0.62	0.54	0.56	0.60	0.22	0.09	**0.64**	0.47
E. Coli	**0.77**	0.66	0.59	0.70	0.14	0.08	**0.77**	0.53
Glass	**0.55**	0.40	0.47	0.42	0.21	0.10	**0.55**	0.39
Iris	0.86	0.87	0.69	0.55	0.04	0.06	**0.87**	0.56
landsat	0.79	0.76	0.64	0.77	0.09	0.05	**0.80**	0.56
Letter	0.64	0.62	0.59	0.65	0.05	0.03	**0.66**	0.46
Lung cancer	0.32	0.17	0.22	0.30	0.20	0.08	**0.37**	0.24
Mushroom	0.86	**0.99**	0.86	0.94	0.00	0.00	0.90	0.65
Parkinsons	0.65	0.74	0.59	0.61	0.13	0.10	**0.70**	0.50
Promoters	0.64	0.43	0.47	0.51	0.14	0.04	**0.66**	0.41
Segment	0.75	0.79	0.64	0.75	0.04	0.04	**0.78**	0.54
Soybean	0.61	0.52	0.58	0.49	0.08	0.04	**0.64**	0.42
Splice	0.61	0.45	0.47	0.53	0.17	0.04	**0.63**	0.41
Voting	0.71	0.79	0.72	**0.84**	0.07	0.04	0.82	0.57
Wine	0.60	0.61	0.50	0.57	0.17	0.07	**0.62**	0.45
Zoo	0.53	0.62	0.59	0.29	0.05	0.05	**0.54**	0.38
Average	0.64	0.59	0.57	0.59	0.12	0.06	**0.67**	0.46

In all experiments, following [8], we adopted $k = 3$ for DROP3, ENN, ICF, and LDIS. Besides that, we adopted the following distance function $d \colon T \times T \to \mathbb{R}$:

$$d(x, y) = \sum_{j=1}^{m} \theta_j(x, y), \qquad (4)$$

with

$$\theta_j(x, y) = \begin{cases} \alpha(x_j, y_j), & \text{if } j \text{ is a categorical feature,} \\ |x_j - y_j|, & \text{if } j \text{ is a numerical feature,} \end{cases} \qquad (5)$$

where

$$\alpha(x_j, y_j) = \begin{cases} 1, & \text{if } x_j \neq y_j, \\ 0, & \text{if } x_{yj} = y_j. \end{cases} \qquad (6)$$

Table 2 shows that LSSm achieves the highest *accuracy* in most of the datasets. However, since that LSSm was designed for removing noisy instances, it does not provide high reduction rates. Besides that, it is important to notice that, for most of the datasets, the difference between the accuracy of XLDIS and the accuracy achieved by the other algorithms is not large. In cases where the achieved accuracy is lower than the accuracy provided by other algorithms, this can be compensated by a higher reduction. The Table 2 also shows that, regarding the accuracy, the results achieved by XLDIS are very similar to those of LDIS. On the other hand, in the *wine* dataset, XLDIS provided an accuracy that is significantly higher than that of LDIS. Table 3 shows that XLDIS achieves the highest *reduction* in most of the datasets, and achieves also the highest average reduction rate. This table also shows that, in some datasets (such as zoo, soybean, voting), the XLDIS algorithm achieved a reduction rate that is significantly higher than that of LDIS. Finally, Table 4 shows that XLDIS has the highest *effectiveness* in most of the datasets, as well as the highest average effectiveness. Thus, these results demonstrate that although XLDIS does not provide always the highest accuracies, it provides the highest reduction rates and the best trade-off between both measures (represented by the effectiveness).

We also carried out experiments for evaluating the impact of the parameter k in the performance of XLDIS. The Fig. 1 represents the average accuracy, average reduction and average effectiveness achieved by XLDIS as a function of the parameter k (as an average of the measures achieved in all datasets), with k assuming the values 1, 2, 3, 5, 10 and 20. These results show that the variation of k has a significant impact on the performance of the algorithm. The results suggest that, in general, as the value of k increases, the accuracy decreases, the reduction increases, and the effectiveness increases up to a point from which it begins to decrease. This chart also shows that, with $k = 1$, the XLDIS algorithm achieves an average accuracy that is comparable to that of DROP3 (as presented in Table 2), for example, while achieves a higher reduction rate than the reduction rate of DROP3.

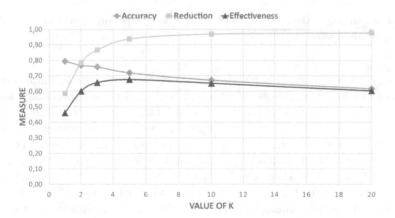

Fig. 1. Chart representing the average accuracy, average reduction and average effectiveness achieved by XLDIS as a function of the parameter k.

We also carried out a comparison of the running times of the instance selection algorithms considered in our experiments. In this comparison, we applied the 7 instance selection algorithms to reduce the 3 largest datasets considered in our experiments: letter, splice-junction gene sequences and mushroom. For conducting the experiments, we used an Intel® Core™ i5-5200U laptop with a 2.2 GHz CPU and 8 GB of RAM. The Fig. 2 shows that, considering these three datasets, the LDIS algorithm has the lowest running times compared to the other algorithms. This result is a consequence of the fact that LDIS deals with the set of instances of each class of the dataset separately, instead of performing a global search in the whole dataset. However, the XLDIS algorithm has running times that are very similar to those of LDIS, with an additional gain in effectiveness (trade-off between accuracy and reduction).

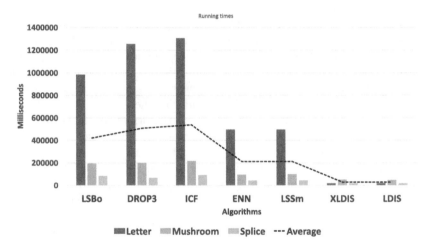

Fig. 2. Comparison of the running times of 7 instance selection algorithms, considering the 3 largest datasets.

Finally, we also carried out a specific experiment in order to emphasize the difference in the performances of LDIS and XLDIS. In this experiment, we considered 4 artificial datasets: D1, D2, D3 and D4. The four datasets have 20k 2D points, uniformly distributed in a squared area, classified in 4 classes (where each class is a quadrant of the whole area covered by the dataset). The main difference regarding the datasets is regarding the degree of redundancy. The D1 has no artificial redundancy. The D2 has 2 copies of each of 10 K basic points. The D3 has 5 copies of 4 K basic points. And D4 has 10 copies of 1 K basic points. We evaluated the performance of the algorithms in a classification task, using the same *n-fold cross validation* schema previously discussed, and adopting as the classifier the KNN with $k = 3$. We also considered $k = 3$ for both, LDIS and XLDIS. The description of the datasets and the reduction rates achieved by both algorithms are presented in Table 5. Since the accuracy achieved by

Table 5. Description of the artificial datasets adopted in the experiment and the resulting reduction rates of LDIS and XLDIS algorithms.

Dataset	Number of classes	Total number of instances	Number of basic instances	Redundancy	Reduction	
					LDIS	XLDIS
D1	4	20 K	20 K	0%	0.87	0.89
D2			10 K	50%	0.57	0.82
D3			4 K	80%	0.03	0.66
D4			1 K	90%	0	0.33

both the algorithms in all datasets was of 100%, our discussion is focused in the reduction rates.

The results in Table 5 show that the XLDIS algorithm achieves reduction rates that are significantly higher than those achieved by the LDIS algorithm, when the dataset has a high volume of redundant instances. This result agrees to our intuition, since redundant instances have the same density, and usually have a high density (redundancy reinforces the density). Due to this, the LDIS algorithm has a tendency of selecting the redundant instances. The XLDIS algorithm, on the other hand, can differentiate the promising instances, because each instance has a different *LDO*. And, once the instance with the higher *LDO* in a neighborhood is selected, the neighbors with a lower LDO are not even analysed in further steps.

6   Conclusion

In this paper, we proposed an efficient algorithm for instance selection, called XLDIS (Extended Local Density-based Instance Selection). It identifies the most representative instances of each class adopting a *local search* procedure. Due to this, its time complexity is lower than the time complexity of approaches that identifies boundary instances. Besides that, XLDIS assumes that the most representative instance in a given neighborhood has the higher *local density ordering* (LDO) within that neighborhood. The LDO of a given instance x is basically the number of instances whose local density is lower than the local density of x.

Our experiments show that XLDIS provides the best reduction rates and the best balance between accuracy and reduction, with a low time complexity, compared with other algorithms available in the literature. The experiments also showed that the XLDIS algorithm is much more aggressive in removing redundant instances than LDIS. These features make XLDIS a promising algorithm for dealing with *Big Data.*

In future works, we plan to investigate strategies for automatically estimating the best value of the parameter k for each problem. Besides that, the performance of XLDIS encourages the investigation of novel instance selection strategies that are based on other *local properties* of the dataset. Regarding this point, we plan to investigate how to use *centrality measures*, typically used in *Network Science*, for identifying the instances that should be selected.

Moreover, we also plan to investigate the possibility of applying the notion of *subspace*, usually adopted in *subspace clustering* approaches [6,7], for instance selection. The underlying hypothesis is that different classes can be viewed as subspaces whose dimensions have different degrees of importance for their instances. Thus, the instance selection algorithm could take into account this information for selecting the most important instances of each class.

Finally, we also plan to investigate how to use *Locality-Sensitive Hashing* [21] for accelerating the identification of the *partial k-neighborhood* of the instances. This can improve the performance of the XLDIS algorithm.

Acknowledgment. The author would like to thank IBM Research for the support to this work.

References

1. Brighton, H., Mellish, C.: Advances in instance selection for instance-based learning algorithms. Data Min. Knowl. Disc. **6**(2), 153–172 (2002)
2. Caises, Y., González, A., Leyva, E., Pérez, R.: Combining instance selection methods based on data characterization: an approach to increase their effectiveness. Inf. Sci. **181**(20), 4780–4798 (2011)
3. Carbonera, J.L., Abel, M.: A cognition-inspired knowledge representation approach for knowledge-based interpretation systems. In: Proceedings of 17th ICEIS, pp. 644–649 (2015)
4. Carbonera, J.L., Abel, M.: A cognitively inspired approach for knowledge representation and reasoning in knowledge-based systems. In: Proceedings of the 24th International Joint Conference on Artificial Intelligence (IJCAI 2015) (2015)
5. Carbonera, J.L., Abel, M.: Extended ontologies: a cognitively inspired approach. In: Proceedings of the 7th Ontology Research Seminar in Brazil (Ontobras) (2015)
6. Carbonera, J.L., Abel, M.: Categorical data clustering: a correlation-based approach for unsupervised attribute weighting. In: 2014 IEEE 26th International Conference on Tools with Artificial Intelligence (ICTAI), pp. 259–263. IEEE (2014)
7. Carbonera, J.L., Abel, M.: An entropy-based subspace clustering algorithm for categorical data. In: 2014 IEEE 26th International Conference on Tools with Artificial Intelligence (ICTAI), pp. 272–277. IEEE (2014)
8. Carbonera, J.L., Abel, M.: A density-based approach for instance selection. In: 2015 IEEE 27th International Conference on Tools with Artificial Intelligence (ICTAI), pp. 768–774. IEEE (2015)
9. Carbonera, J.L., Abel, M.: A novel density-based approach for instance selection. In: 2016 IEEE 28th International Conference on Tools with Artificial Intelligence (ICTAI), pp. 549–556. IEEE (2016)
10. Chou, C.H., Kuo, B.H., Chang, F.: The generalized condensed nearest neighbor rule as a data reduction method. In: 18th International Conference on Pattern Recognition, ICPR 2006, vol. 2, pp. 556–559. IEEE (2006)
11. Cover, T.M., Hart, P.E.: Nearest neighbor pattern classification. IEEE Trans. Inf. Theory **13**(1), 21–27 (1967)

12. Fan, W., Bifet, A.: Mining big data: current status, and forecast to the future. ACM sIGKDD Explor. Newsl. **14**(2), 1–5 (2013)
13. García, S., Luengo, J., Herrera, F.: Data Preprocessing in Data Mining. Springer, Cham (2015)
14. Gates, G.W.: Reduced nearest neighbor rule. IEEE Trans. Inf. Theory **18**(3), 431–433 (1972)
15. Hamidzadeh, J., Monsefi, R., Yazdi, H.S.: IRAHC: instance reduction algorithm using hyperrectangle clustering. Pattern Recogn. **48**(5), 1878–1889 (2015)
16. Hart, P.E.: The condensed nearest neighbor rule. IEEE Trans. Inf. Theory **14**, 515–516 (1968)
17. Leyva, E., González, A., Pérez, R.: Three new instance selection methods based on local sets: A comparative study with several approaches from a bi-objective perspective. Pattern Recogn. **48**(4), 1523–1537 (2015)
18. Lin, W.C., Tsai, C.F., Ke, S.W., Hung, C.W., Eberle, W.: Learning to detect representative data for large scale instance selection. J. Syst. Softw. **106**, 1–8 (2015)
19. Liu, H., Motoda, H.: On issues of instance selection. Data Min. Knowl. Disc. **6**(2), 115–130 (2002)
20. Nikolaidis, K., Goulermas, J.Y., Wu, Q.: A class boundary preserving algorithm for data condensation. Pattern Recogn. **44**(3), 704–715 (2011)
21. Slaney, M., Casey, M.: Locality-sensitive hashing for finding nearest neighbors [lecture notes]. IEEE Signal Process. Mag. **25**(2), 128–131 (2008)
22. Tomek, I.: An experiment with the edited nearest-neighbor rule. IEEE Trans. Syst. Man. Cybern. **SMC–6**(6), 448–452 (1976)
23. Wilson, D.R., Martinez, T.R.: Reduction techniques for instance-based learning algorithms. Mach. Learn. **38**(3), 257–286 (2000)
24. Wilson, D.L.: Asymptotic properties of nearest neighbor rules using edited data. IEEE Trans. Syst. Man Cybern. **SMC–2**(3), 408–421 (1972)

Search Result Personalization in Twitter Using Neural Word Embeddings

Sameendra Samarawickrama[(✉)], Shanika Karunasekera, Aaron Harwood,
and Ramamohanarao Kotagiri

Department of Computing and Information Systems,
The University of Melbourne, Melbourne, Australia
ssamarawickr@student.unimelb.edu.au,
{karus,aharwood,kotagiri}@unimelb.edu.au

Abstract. In recent years, Twitter has become one of the most popular microblogging avenues. Today it has more than 300 million monthly active users generating more than 500 million tweets everyday. Twitter users both post messages as well as search for messages. Current search results given by Twitter are chronologically ordered and often users have to manually scan through an overwhelming number of the tweets to find content of interest. This process can quickly become infeasible. Personalization techniques address this problem by learning the user interests and tailoring search results by matching them with the user's interests. Recent research on neural word embedding models, which represents each word in the vocabulary as a vector of real values, has gained much attention. These models learn word embeddings in such a way that contextually similar words have similar vectors. In this paper we propose a novel approach, PWEBA, for personalizing Twitter search, which uses neural word embeddings to model user interests. Our experimental results show that PWEBA outperforms existing approaches for all the evaluation metrics we have considered in this paper.

Keywords: Twitter · Personalization · Content-mining · Word-embeddings

1 Introduction

Use of microbloging platforms has become a daily activity of people for real-time information sharing. Twitter, one of the most popular microbloging service providers, allows its users to publish brief text abstracts, called tweets, which are limited to only 140 characters. Today, Twitter hosts about 320M monthly active users and about 6000 tweets are tweeted every second, which is about 500M tweets per day[1]. Finding relevant information becomes exceedingly difficult with such high volumes of data.

There are two types of users in Twitter: content producers and content seekers [6]. As the name suggests, content producers are the users who publish or share

---

[1] http://www.internetlivestats.com/twitter-statistics/.

© Springer International Publishing AG 2017
L. Bellatreche and S. Chakravarthy (Eds.): DaWaK 2017, LNCS 10440, pp. 244–258, 2017.
DOI: 10.1007/978-3-319-64283-3_18

content of a particular interest which could be in the form of emerging news, trending events, gossip or even personal opinions, experiences or emotions. Content seekers can further be categorized into two types based on their information seeking behaviours [5]: "asking for information" and "retrieving information". The former refers to the behaviour whereby a user broadcasts a question hoping that someone would answer their question. Hashtags such as #asktwitter, #help, #ineedanswers, #justasking are used when posting these kind of tweets to find the right audience [13]. The latter refers to conducting searches over microblog data similarly to traditional web searching, which is the focus of our work.

It is estimated that over 2 billion search queries[2] are submitted to Twitter on a daily basis using the search interface that Twitter provides which allows users to enter keywords and find matching tweets - in reverse chronological order. But given the high volume of data it contains, users can be quickly overloaded by the number of microblogging posts it returns based on existing keyword based Information Retrieval (IR) approaches. The way the users judge the relevance of results to the same query varies significantly depending on the individual's information needs. Two users might enter the same search query but actually the underlying intentions might be quite different. For example, a "tech" user may issue the query "apple" and expect to see results related to Apple computers, phones and in general technological news, while an "agricultural" user might expect to see results related to apple growing and news related to the apple fruit. Simple keyword based IR approaches fail in these cases as they do not consider the information about the user issuing the query. This calls for the need for personalization of search results using information about the individual to provide the most relevant search results for him/her. This typically involves building a user profile which contains information about the user.

There has been a tremendous amount of work done in the domain of web search results personalization ([9, 14, 20]) but research on personalization in microblogging environments is very sparse. Search in microblog environments differ from traditional web search in several ways: microblog documents are very short that give very little statistical information compared to traditional web pages, they are noisy, language is informal, rich in social interactions and also a single tweet tends to be about a single topic. In the literature, there is research in Twitter space that studies the query-tweet relevance. Twinder [15] is a search engine which analyzes various features to determine the relevance of tweets to a given query. Lau et al., [8] used topical features (e.g., named entities) as well as query expansion to retrieve tweets to a given query. TweetMotif [11] groups resulting tweets by topics which allows users to browse tweets by a topic and find tweets of interest. There has been some research which brings the personalization aspect into Twitter based on collaborative learning techniques ([17, 19]). In this paper our focus is towards building content based personalization systems using neural word embeddings.

Word embeddings (a.k.a. word vectors) represent each word in the vocabulary as a vector of real numbers. When neural networks are used to "predict" these

---

[2] http://www.statisticbrain.com/twitter-statistics/.

word embeddings, they are called neural word embeddings. Neural word embedding techniques such as word2vec [10], have significant interest in NLP research recently due to its capability of generating meaningful word embeddings which are related both syntactically and semantically.

The major contribution of this paper is a new approach to personalize search results in Twitter using word embeddings. We find contextually similar words to the user's query from user's own tweets using neural word embeddings which are then used to model the user. To the best of our knowledge, neural word embeddings have not been used elsewhere for this purpose. Our goal in this paper is to re-rank a set of generic search results such that the tweets relevant to a user are ranked higher. Experimental results with several baseline methods evaluated using two different approaches have shown that our proposed approach performs better than the baseline methods.

The rest of the paper is organized as follows. We begin by describing the related work on this topic in Sect. 2. In Sect. 3, we explain our proposed approach of user modeling and results re-ranking. In Sect. 4, we discuss the use of Twitter lists as an offline evaluation method. Our experimental results are presented in Sect. 4.1. Finally, we draw the conclusions of this paper in Sect. 5.

2   Related Work

2.1   Twitter Search

We first discuss related work in the domain of Twitter search followed by approaches towards personalizing the search results in Twitter.

Twitter search systems discussed here are based on query-tweet relevance where it is addresses the problem of finding relevant tweets to a query. Tweet-Motif [11] is a Twitter search engine which groups tweets by frequent significant terms, which allows users to search different topics and find topics of their interest. For identifying topic phrases they used a language modeling approach using unigram, bigram and trigram features. Twitter contains a massive amount of duplicate messages such as retweets, bots repeating the same tweet such as advertisements, spam, news feeds, weather reports etc. When grouping tweets by topics they identified near-duplicate posts through pairwise comparison of tweets using Jaccard similarity. Lau et al. [8] proposed a tweet retrieval system based on topical features as well as query expansion through pseudo-relevance feedback. First they index the tweet corpus using term features as well as topical features such as named entities. When the user enters a query it will be matched with the tweet index to retrieve an initial result set. From this initial result set they take all the terms in the top ten tweets to expand the initial query. Then this expanded query is used to retrieve the final result. Twinder [15] is a search engine which analyzes various features to determine the relevance of tweets to a given query. They considered 13 different features which included topic sensitive features, syntactical features and various semantic contextual features. They evaluated their results on TREC 2011 dataset using a logistic regression to classify tweets as relevant or irrelevant to a given topic. Since it was trained on a dataset containing 49 topics, the deployability of this system in practice is questionable.

2.2 Personalized Twitter Search

There are some personalized Twitter search systems found in the literature. Ushiama and Tominaga [17] proposed a personalized item ranking approach using Twitter. The user has to specify the item (e.g., name of a book) he/she likes and then through Twitter Search API, similar users are found (called supporters) that share the same item. For each supporter they retrieve his/her most recent 1000 tweets and estimate the personalized ranking score based on similarities between supporters and the user. Another recent study, [19], proposed a model which exploits the user's social connections, for delivering personalized content in response to a user's search query. They represent both the user's and his/her friends' interests using topic models. Since different friends might have different influence on a user they assigned a weight to each friend, composed of four factors.

Our research deviates from aforementioned studies where we use a content based personalization approach to re-rank search results given by a baseline search engine. We use neural word embeddings, which to the best of our knowledge, has not yet been applied into the search personalization in the Twitter space.

3 Our Approach

Our personalization technique consists of two main components: (1) user modeling, and (2) results re-ranking. User modeling (a.k.a. user profiling) gathers information about the user and creates a model of the user (user profile) which is then used to re-rank generic (unpersonalized) search results, in a way that the highly ranked results are most relevant to the user. This Section explains our approach towards user modeling and results re-ranking.

3.1 User Modeling

User modeling, which captures the user's interests to build a profile of the user, is a primary task in any personalization system. A user profile is usually built using information about the user's past activities (e.g., messages posted, search queries) and is unique to a given user. In this paper we use user's past tweets as a source for user modeling.

There are primarily two approaches for building user profiles: vector based user profiles and hierarchical user profiles. Vector based approaches represent a user in a linear form which contains word-value pairs while hierarchical user profiles have tree like structures. In this paper, we propose a novel vector based user modeling approach using word embeddings and compare it with several other approaches. We discuss them in more detail in the following sub-sections. We use the terms, "word" and "term" interchangeably to convey the same meaning of a string of finite length.

Algorithm 1. Generating user profile $P(u)$ of user u in PWEBA

Input: Word embeddings for words $w \in W_u$
Output: User profile $P(u)$ of user u
1: **for** each $w_i \in W_u$ **do**
2:     $E = []$
3:     **for** each $w_j \in W_u$ and $i \neq j$ **do**
4:         $weight = \frac{v_{w_i} \cdot v_{w_j}}{\|v_{w_i}\| \cdot \|v_{w_j}\|}$
5:         $E = E \cup (w_j, weight)$
6:     **end for**
7:     Sort tuples in E on $weight$ in descending order
8:     $P(u) = P(u) \cup (w_i, E)$
9: **end for**
10: **return** $P(u)$

Term Based Approach (TBA). This is the most basic user modeling approach. It considers all user u's past tweets, T_u, and calculates the term frequencies of all the unique terms in T_u. This weighting vector is then considered as the user profile of that particular user.

Personalized Word Embeddings Based Approach (PWEBA). This is our proposed personalization approach in this paper. Word embeddings associate each term in the vocabulary $w \in W$ with a real valued d-dimensional vector $v_w \in \mathbb{R}^d$. User profile construction is done in two phases.

1. In the first phase, we train a Neural Network language Model (NNLM) on user u's past tweets T_u. Let W_u be the set of terms observed in T_u which we call as the vocabulary of user u. Thus for each term $w \in W_u$, we get a word embedding $v_w = < x_i | x_i \in \mathbb{R}, i = 1..d >$.
2. In the second phase, we create a word-synonym table for the user using the word vectors from the first phase. For each word $w \in W_u$ we calculate the cosine similarity between the vector representation of that word and all the other word vectors and rank them in descending order based on cosine similarity. This resulting word-synonym table, which contains words and their most similar words, is considered to be the user profile which is used for results re-ranking. This procedure is defined in Algorithm 1.

Generic Word Embeddings Based Approach (GWEBA). Rather than training a NNLM on the user's past tweets, in this approach, we make use of pre-trained word vectors which are already trained on a large text corpora and freely available. Thus for each term in the user's vocabulary, $w \in W_u$, we represent w with the corresponding word vector taken from the pre-trained word vectors. This is equivalent to the first phase of PWEBA. Next, we create a word-synonym table for the user following the second phase of PWEBA, using Algorithm 1. This resulting word-synonym table is considered as the user profile of that particular user.

Algorithm 2. Re-ranking results of user u based on $P(u)$ in PWEBA

Input: User Profile $P(u)$, user's query q, generic search results set S
Output: Re-ranked search results set S'
1: Get the corresponding entry E where $(q, E) \in P(u)$
2: Construct query vector, $V(q) = < \alpha_1, .., \alpha_i, .., \alpha_{|E|} >$, where α_i is the corresponding *weight* for each w in $(w, weight) \in E$
3: $S' = []$
4: **for** each document $doc \in S$ **do**
5:     Construct document vector $V(doc)$ with tf-idf weights for each word w (where $(w, weight) \in E$) in doc
6:     $score = \frac{V(q) \cdot V(doc)}{\|V(q)\| \cdot \|V(doc)\|}$
7:     $S' = S' \cup (doc, score)$
8: **end for**
9: Sort tuples in S' on $score$ in descending order
10: **return** S'

3.2 Results Re-ranking

Result re-ranking takes a generic list of search results as input and re-ranks them by calculating a similarity score with the user profile and each individual document in the search results list, thus the documents that are most relevant to the user profile are ranked highly.

For re-ranking, we take the weighted most similar words to the user's query from the user's profile, $P(u)$, and calculate the cosine similarity between the most similar words and each document in the generic list of search results. In PWEBA and GWEBA, this re-ranking process for a user u's search query q, utilizing his/her user profile $P(u)$ is explained in Algorithm 2. In TBA, the query vector $V(q)$, is populated with term frequency values. Finally, this re-ranked list of tweets which is personalized, is presented to the user.

4   Evaluation

There are two methods being used in the literature for evaluating personalized search systems: (i) online evaluation, where a certain number of users are asked to use the system and their feedback is used for evaluation, and (ii) offline (formal) evaluation, which is performed offline using a publicly available or a specifically designed dataset. Online evaluation has its own drawbacks such as constraints on the number/types of test queries, number of participants, mood of the participants which could bias evaluation results [4] while offline evaluation can be performed at a low cost and on a larger scale. Offline evaluation is becoming more and more popular and in fact almost all of the methods surveyed in [1] involve an offline evaluation.

We evaluate the proposed personalization techniques using two different offline evaluation methods to make it more rigorous and complete which are

Table 1. Statistics of the original dataset.

List name	Count of users	Avg tweets/user	Total tweets
Entertainment (L1)	231	2296.86	530574
Fashion (L2)	334	2290.94	765173
Fitness (L3)	17	1092.18	18567
Photography (L4)	91	1693.20	154081
Technology (L5)	516	2319.26	1196738
Travel (L6)	154	2170.45	334249

explained in the following sub-sections. Under each sub-section, we first introduce the motivation behind that particular evaluation method, how the ground truth is derived, experimental details and finally the results of the experiments.

4.1 Twitter Lists Based Evaluation

Here we evaluate our results based on Twitter lists. A Twitter list is a manually curated list of users who have a particular topic of interest (e.g., "politics", "photography"). Lists can be used as very strong indicators of topical homophily in Twitter [7]. In fact, this idea of topical homophily in Twitter lists has been exploited by some of the recent research such as [3,12] as well.

In this paper we utilize Twitter lists as a way of evaluating our results. Our evaluation is based on the assumption that a list is composed of users who typically share the same interests and if a user who is a member of a particular list was to make a search, matching tweets (tweets that contain query word/s) by users in the same list are relevant. For example, in our experiments, if a user who is a member of the "photography" list initiates a search with a particular query, we consider matching tweets by users (including the original user who initiated the search) in the "photography" list are relevant to the search query of the original user who initiated the search. Based on this assumption, we generate the ground truth results for the search queries used in our experiments.

Experimental Setup. In this Section we introduce the dataset that we have collected for offline evaluation. Our approach for creating the test dataset is as follows.

First we manually select six different Twitter lists that belong to different domains. Then for each list, we obtain all users who are members of that list using the Twitter's REST API[3]. Finally for each user, we retrieve upto 3200[4] most recent tweets from the user's timeline, again using the REST API[5]. Statistics of our original dataset created using six different Twitter lists is shown in Table 1,

---

[3] https://dev.twitter.com/rest/reference/get/lists/members.
[4] limitation imposed by Twitter API.
[5] https://dev.twitter.com/rest/reference/get/statuses/user_timeline.

where list name is an alias to represent the domain of that particular list, number of users is the membership count of that particular list and average number of tweets per user is calculated by dividing the total number of tweets from all the users of the list by the membership count of that particular list.

Pre-processing and Filtering. Every tweet downloaded from the Twitter API comes as a JSON object which contains all the information about that tweet, such as the tweet ID (`id_str`), date posted (`created_at`), actual tweet text (`text`), entities mentioned in the tweet text (`entities`) among many others[6]. Pre-processing was done on a per tweet basis on the tweet text (`text`). Below we list our pre-processing steps:

1. convert all text into lower case and split them by white space,
2. remove user mentions (tokens starting with '@') and URLs,
3. strip off leading and trailing non ASCII characters from tokens returned from the previous step (except '#' as we want to preserve hashtags),
4. drop tokens with less than three characters,
5. remove stopwords using a stop word list of 544 words, and
6. discard tweets with less than 3 tokens from the dataset.

Selecting Queries and Users. In our work we focus on queries that are ambiguous in the sense that if a user was to issue such a query, tweets that are returned would come from different topics. For example for the query "show" (which is one of the queries that we have used), a user who is interested in entertainment (member of L1) would like to see entertainment shows while a user who is interested in fashion (member of L2) might expect to see tweets about fashion shows specifically. Therefore in this paper, we generate a list of query terms which are equally likely to be on several lists. Below we discuss the approach we use to generate queries as well as to find corresponding users.

For our experiments we consider combinations of four lists, thus we have considered 15 list combinations that were drawn from the 6 lists. To model user queries, for a given list combination, we first take tweets (after removing noisy and stopwords as explained above) from all members from all four lists and find common words between the lists. Next, from these common words, we find words that are occurring with a high frequency in all four lists as queries for that particular list combination.

Once a set of queries are selected as potential queries, next we find an assumed set of users who would use these queries. We assume that users who mention that query term more often in his/her tweets are better candidates than those who don't. For each query and for each list in the list combination, we get the top 10 users based on the frequency of that query term appearing in the user's tweets. For example for the list combination L1, L2, L3, L4 and for the query 'show', we will have 40 users all together where 10 users are from each of the lists L1, L2, L3 and L4 respectively. We also make sure a given user belongs only to a single list.

---

[6] dev.twitter.com/overview/api/entities-in-twitter-objects.

Indexing Tweets for Searching. In the previous sub-sections we discussed about our dataset, pre-processing, selecting queries to simulate searching as well as assumed users on behalf of the selected queries. Here we explain how we index tweets for searching.

One important thing to note here is that when a user initiates a search on Twitter with a specific query term, all the tweets that Twitter API returns contain the query term that the user initially used. In the ideal case we would have used the queries and retrieved results from the Twitter's search API. But then it becomes very challenging to evaluate our personalization approaches in that case as there is no way of formulating the ground truth of results as the tweets retrieved from the search API would not contain the tweets from the same user who originated the query. Thus, in order to properly evaluate our personalization approach as well as to be consistent with the nature of Twitter's original search results, we build a separate list of tweets which we call the *search list*. The *search list* is used for retrieval purposes for a given query and all the tweets in the *search list* contain that specific query term.

We do this as follows. For each selected user (as discussed in Subsect. 4.1) we take all of his/her tweets that have collected and add 50% of his/her tweets which contain the query term, into the *search list*. Thus each user will contribute half of his/her tweets which contain the query term into the *search list*. In this way, we make sure all the tweets in the *search list* contain the query term. For each user, the remaining 50% of tweets along with his/her other tweets are used for constructing the user profile.

We use Lucene[7] to index tweets in the *search list*. It is used as a proxy to the Twitter search engine here. Lucene provides functionalities for document indexing as well as searching and ranking. The choice of a particular information retrieval system is irrelevant here as we are interested in comparing different personalization approaches and for that purpose, Lucene works as a non-personalized search engine (baseline). We apply our personalization approaches (TBA, PWEBA, GWEBA) on the results returned by Lucene.

Search Results Personalization. Here we are interested in using the user modeling approaches that we described in Subsect. 3.1 for re-ranking the search results returned by Lucene.

In our experiments we use a python implementation of word2vec[8] which is provided in the gensim toolkit. For each user, we train the word2vec model on his/her tweets and generate his/her user profile as explained in Subsect. 3.1. We train the model for 5 iterations using the skip-gram model with negative sampling and generate 20-dimensional word embeddings. The skip-gram model has shown to perform better than the corresponding continuous bag of words model [10]. Selection of the number of training iterations and vector dimension were based on some preliminary experiments that we did on a representative dataset of tweets.

---

[7] https://lucene.apache.org/.
[8] https://radimrehurek.com/gensim/models/word2vec.html.

Next, for each query and for all the users who are interested in this query, we first retrieve a non-personalized list of search results using Lucene. We used cosine as the similarity metric used in Lucene. Given this non-personalized list of tweets, we re-rank them using the user profiles (TBA, PWEBA and GWEBA) that have been created. For GWEBA, we use `GoogleNews-vectors-negati-ve300.bin` as our pre-trained word vectors which are trained on a part of Google News dataset containing about 100 billion words[9] with vector dimensionality as 300. We ignore the words that do not have a corresponding word vector when creating user profiles. Then we use the cosine similarity for results re-ranking as explained in Subsect. 3.2.

Results and Evaluation. In this Section we discuss the results of our experiments. We compare the results of our personalization approaches (TBA, PWEBA, GWEBA) with the non-personalized approach which we name as NP. In NP we simply use the generic search results given by Lucene for a user's query without re-ranking. We evaluate our results using some of the popular metrics used in the literature: Precision@k (P@k), Average Precision (AP) and Reciprocal Rank (RR).

Table 2 shows the performance of different approaches, evaluated using P@20, P@50, AP and RR for the 15 different list combinations that we have considered in our experiments. To get a single value for a given list combination, as shown in the table, we averaged the scores of the 40 users we have considered for a particular query and then averaged this value across the 10 queries that we have considered for a given list combination. The top scoring approach based on P@20, P@50, AP or RR is highlighted in bold for each row.

Table 3 shows the overall performance comparison between the different approaches using P@20, P@50, AP and RR. We create this table by column-wise averaging the scores in Table 2. Results clearly show that our purposed approach PWEBA, outperforms all of the other approaches for all evaluation metrics. TBA follows our approach and performs better than GWEBA and NP. PWEBA performing better than GWEBA, clearly shows the impact of training NNLM on user's past tweets rather than using pre-trained word vectors. It's interesting that TBA performs better than GWEBA. This could be because words are used in different contexts in the user's tweets and in the large text corpora where the pre-trained vectors are obtained or due to the vocabulary mismatch where there are no pre-trained vectors for some twitter specific terms such as hashtags thus losing the information contained in such terms.

4.2 Hashtags Based Evaluation

In this section we explain about offline evaluation using hashtags, experimental setup and the results. Hashtags in Twitter are used by users when writing a tweet to provide a topical context for the tweet. For this reason, hashtags

---

[9] https://code.google.com/archive/p/word2vec/.

Table 2. Performance comparison between NP, GWEBA, TBA and PWEBA across the 15 list combinations using different evaluation metrics.

List Comb.	P@20				P@50				AP				RR			
	NP	GWEBA	TBA	PWEBA	NP	GWEBA	TBA	PWEBA	NP	GWEBA	TBA	PWEBA	NP	GWEBA	TBA	PWEBA
L2, L3, L1, L6	0.2366	0.4109	0.5151	**0.7396**	0.2385	0.3947	0.4803	**0.6555**	0.2534	0.3073	0.3227	**0.3918**	0.3787	0.6142	0.7522	**0.9431**
L4, L2, L1, L6	0.2375	0.3904	0.5152	**0.755**	0.2379	0.3828	0.4732	**0.6729**	0.2605	0.2999	0.3191	**0.3869**	0.3873	0.6088	0.7087	**0.949**
L4, L2, L3, L1	0.2368	0.3892	0.461	**0.6777**	0.2378	0.3784	0.4287	**0.5973**	0.2575	0.3028	0.3094	**0.3761**	0.3823	0.5844	0.6824	**0.9104**
L4, L2, L3, L6	0.2396	0.4115	0.484	**0.7241**	0.2418	0.3917	0.4411	**0.6361**	0.2596	0.3089	0.3176	**0.3907**	0.3819	0.6209	0.7092	**0.9394**
L4, L3, L1, L6	0.24	0.4055	0.478	**0.695**	0.2422	0.3855	0.4388	**0.6095**	0.2583	0.3035	0.3159	**0.3816**	0.3897	0.633	0.7126	**0.9203**
L5, L2, L1, L6	0.2346	0.4009	0.5308	**0.7935**	0.2369	0.3898	0.4866	**0.7174**	0.2566	0.3055	0.3235	**0.3997**	0.3855	0.6365	0.7358	**0.9583**
L5, L2, L3, L1	0.2348	0.4135	0.4889	**0.7371**	0.2376	0.3934	0.451	**0.6568**	0.2552	0.3125	0.3209	**0.3962**	0.3795	0.6332	0.7317	**0.9309**
L5, L2, L3, L1	0.2366	0.4325	0.5059	**0.7656**	0.2411	0.4035	0.4604	**0.6786**	0.2581	0.3152	0.3236	**0.4032**	0.3689	0.6881	0.7419	**0.9509**
L5, L3, L1, L6	0.237	0.4216	0.5044	**0.7395**	0.2416	0.3986	0.4644	**0.6606**	0.2552	0.3128	0.3272	**0.3993**	0.3822	0.6662	0.7479	**0.9379**
L5, L4, L1, L6	0.2361	0.3944	0.5082	**0.752**	0.2406	0.3796	0.4645	**0.6707**	0.2555	0.3024	0.3233	**0.3933**	0.3971	0.6172	0.7006	**0.9418**
L5, L4, L2, L1	0.235	0.393	0.5034	**0.7566**	0.2368	0.3807	0.4612	**0.6772**	0.2562	0.3041	0.3203	**0.3932**	0.3891	0.5872	0.7039	**0.9377**
L5, L4, L2, L3	0.2379	0.4136	0.499	**0.7426**	0.2414	0.3907	0.4553	**0.6534**	0.257	0.3149	0.3337	**0.4074**	0.3839	0.6302	0.729	**0.9322**
L5, L4, L2, L6	0.2366	0.4003	0.5381	**0.792**	0.24	0.3856	0.5023	**0.7076**	0.2574	0.3051	0.3416	**0.4081**	0.3872	0.624	0.72	**0.9516**
L5, L4, L3, L1	0.2359	0.4022	0.4624	**0.6817**	0.2393	0.3825	0.4272	**0.6037**	0.2532	0.3083	0.3193	**0.3843**	0.4044	0.6295	0.7035	**0.8997**
L5, L4, L3, L6	0.2311	0.4145	0.4666	**0.7254**	0.2355	0.3901	0.423	**0.6399**	0.2592	0.3115	0.3149	**0.3967**	0.4194	0.6432	0.7202	**0.924**

Table 3. Overall performance comparison between NP, GWEBA, TBA and PWEBA using P@20, P@50, AP and RR with lists based evaluation.

Approach	P@20	P@50	AP	RR
NP	0.2364	0.2393	0.2569	0.3878
GWEBA	0.4063	0.3885	0.3076	0.6278
TBA	0.4974	0.4572	0.3222	0.7200
PWEBA	**0.7385**	**0.6558**	**0.3939**	**0.9351**

are used frequently as search queries and in fact, in a recent study by Teevan et al., [16] showed that 21.28% queries in Twitter contained a hashtag.

User's own hashtags give a good indication of the topics that the user is interested in. In this evaluation method, we use such hashtags as queries to the system. We assume that, when a user does a search with a particular hashtag, his/her own tweets with that particular hashtag are relevant to the user. This assumption has also been used in the literature as a way of deriving the ground truth when evaluating personalized search systems [2,18].

Selecting Queries and Users. We have used the same dataset (as in Table 1) to select users and respective hashtags to be used as search queries. For each unique hashtag that appears in the dataset, we first select only the hashtags where there is at least a single user who has used that particular hashtag at least 20 times. Next from the remaining hashtags, we consider only the hashtags where there are at least 10 different users who have used that particular hashtag. Thus for each hashtag, we have at least 10 different users who have used it and each of those users have used that particular hashtag at least 20 times in their tweets. This makes sure that we do not consider any spam hashtags, tags are accepted among a set of users and that tags are sensical as search queries. Finally, our dataset contained 115 hashtags and 287 users who would use them as queries; meaning on average, 2.496 users share the same hashtag as the query.

Indexing Tweets for Searching. A similar tweet indexing approach as discussed under Sect. 4.1 is used here as well. Given a user and a query (hashtag) that the user is interested in, we add 90% of the tweets (as in [18]) which contain the query, into the *search list* and the remaining 10% of tweets along with the user's other tweets for creating the user profile. If a particular user is interested in more than a single query, that user will have a user profile for each of the query that he/she is interested in. Once again, we have used Lucene for indexing tweets in the *search list*.

Search Results Personalization. Here we use the different user modeling approaches (TBA, PWEBA and GWEBA) to re-rank the search results returned by Lucene. Word2vec model is trained with the same parameters on the user's

256 S. Samarawickrama et al.

Table 4. Overall performance comparison between NP, GWEBA, TBA and PWEBA using P@20, P@50, AP and RR with hashtag based evaluation.

Approach	P@20	P@50	AP	RR
NP	0.0128	0.014	0.0316	0.049
GWEBA	0.0342	0.0238	0.1263	0.1801
TBA	0.0624	0.0422	0.1393	0.2228
PWEBA	**0.1069**	**0.0603**	**0.2778**	**0.503**

tweets as explained earlier. Given a user and a query, we first retrieve a non-personalized search results list using Lucene and apply the different personalization approaches to re-rank the initial results list. Finally, different personalization approaches are evaluated based on how they re-rank the non-personalized results list.

Results and Evaluation. Figure 1 shows the variation of AP and RR scores of 10 users selected randomly for two manually selected queries: #london and #bigdata. Although there are a few users where the TBA works better (e.g., userB3 and userB7 for the query #bigdata for both AP and RR and userL4 for #london), PWEBA outperforms the other approaches for the rest of the users. Table 4 shows the average scores across the 287 users for all the queries evaluated using P@20, P@50, AP and RR. Results clearly show that our proposed approach outperforms the other approaches for all of the evaluation metrices.

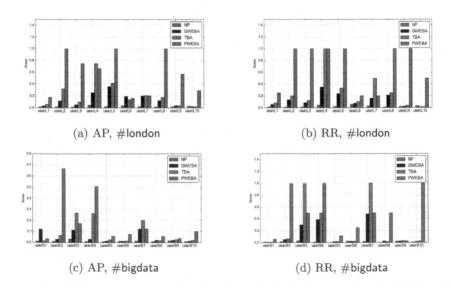

(a) AP, #london

(b) RR, #london

(c) AP, #bigdata

(d) RR, #bigdata

Fig. 1. Variation of AP and RR scores for two queries (#london, #bigdata) for randomly selected 10 users

5   Conclusions

In this paper, we presented a novel approach, PWEBA, for personalizing search results in Twitter using neural word embeddings. We evaluated our proposed technique using two different offline evaluation methods: using Twitter lists and using hashtags. They differed in the ways how search queries are chosen on behalf of users as well as what tweets are considered relevant for a given query for a given user.

Our experimental results based on three well known metrics used in the literature - P@k, Average Precision (AP) and Reciprocal Rank (RR) - and using two different offline evaluation methods - Twitter lists and hashtags - showed that our proposed approach, PWEBA, outperformed all the other approaches in P@k, AP and RR.

We plan to conduct future research in several different directions. We will focus on how to expand tweet text with the help of a search engine such as Google to make tweets more informative. Another future area would be to study how to incorporate friends in the user's social network when building the user profile.

References

1. Bouadjenek, M.R., Hacid, H., Bouzeghoub, M.: Social networks and information retrieval, how are they converging? a survey, a taxonomy and an analysis of social information retrieval approaches and platforms. Inf. Syst. **56**, 1–18 (2016)
2. Carman, M.J., Baillie, M., Crestani, F.: Tag data and personalized information retrieval. In: Proceedings of the 2008 ACM Workshop on Search in Social Media, pp. 27–34. ACM (2008)
3. Culotta, A.: Training a text classifier with a single word using twitter lists and domain adaptation. Soc. Netw. Anal. Min. **6**(1), 1–15 (2016)
4. Dou, Z., Song, R., Wen, J.-R., Yuan, X.: Evaluating the effectiveness of personalized web search. TKDE **21**(8), 1178–1190 (2009)
5. Efron, M.: Information search and retrieval in microblogs. J. Am. Soc. Inf. Sci. Technol. **62**(6), 996–1008 (2011)
6. Java, A., Song, X., Finin, T., Tseng, B.: Why we twitter: understanding microblogging usage and communities. In: WebKDD/SNA-KDD, pp. 56–65. ACM (2007)
7. Kang, J.H., Lerman, K.: Using lists to measure homophily on Twitter. In: AAAI Workshop on Intelligent Techniques for Web Personalization and Recommendation, pp. 26–32 (2012)
8. Lau, C.H., Li, Y., Tjondronegoro, D.: Microblog retrieval using topical features and query expansion. In: TREC (2011)
9. Leung, K.-T., Lee, D.L., Lee, W.-C.: Personalized web search with location preferences. In: ICDE, pp. 701–712. IEEE (2010)
10. Mikolov, T., Sutskever, I., Chen, K., Corrado, G.S., Dean, J.: Distributed representations of words and phrases and their compositionality. In: NIPS, pp. 3111–3119 (2013)
11. O'Connor, B., Krieger, M., Ahn, D.: Tweetmotif: exploratory search and topic summarization for twitter. In: ICWSM, pp. 384–385 (2010)

12. Rakesh, V., Singh, D., Vinzamuri, B., Reddy, C.K.: Personalized recommendation of twitter lists using content and network information. In: ICWSM, pp. 416–425 (2014)
13. Rzeszotarski, J.M., Spiro, E.S., Matias, J.N., Monroy-Hernández, A., Morris, M.R.: Is anyone out there? unpacking Q&A hashtags on twitter. In: CHI, pp. 2755–2758. ACM (2014)
14. Sontag, D., Collins-Thompson, K., Bennett, P.N., White, R.W., Dumais, S., Billerbeck, B.: Probabilistic models for personalizing web search. In: WSDM, pp. 433–442. ACM (2012)
15. Tao, K., Abel, F., Hauff, C., Houben, G.-J.: Twinder: a search engine for twitter streams. In: Brambilla, M., Tokuda, T., Tolksdorf, R. (eds.) ICWE 2012. LNCS, vol. 7387, pp. 153–168. Springer, Heidelberg (2012). doi:10.1007/978-3-642-31753-8_11
16. Teevan, J., Ramage, D., Morris, M.R.: # twittersearch: a comparison of microblog search and web search. In: WSDM, pp. 35–44. ACM (2011)
17. Ushiama, T., Tominaga, K.: A method for personalized ranking of items based on similarity between twitter users. In: ICUIMC, pp. 44:1–44:4. ACM (2014)
18. Vallet, D., Cantador, I., Jose, J.M.: Personalizing web search with folksonomy-based user and document profiles. In: Gurrin, C., He, Y., Kazai, G., Kruschwitz, U., Little, S., Roelleke, T., Rüger, S., Rijsbergen, K. (eds.) ECIR 2010. LNCS, vol. 5993, pp. 420–431. Springer, Heidelberg (2010). doi:10.1007/978-3-642-12275-0_37
19. Vosecky, J., Leung, K.W.-T., Ng, W.: Collaborative personalized twitter search with topic-language models. In: SIGIR, pp. 53–62. ACM (2014)
20. Wang, H., He, X., Chang, M.-W., Song, Y., White, R.W., Chu, W.: Personalized ranking model adaptation for web search. In: SIGIR, pp. 323–332. ACM (2013)

Diverse Selection of Feature Subsets
for Ensemble Regression

Arvind Kumar Shekar[1(✉)], Patricia Iglesias Sánchez[1], and Emmanuel Müller[2]

[1] Robert Bosch GmbH, Stuttgart, Germany
{arvindkumar.shekar,patricia.iglesiassanchez}@de.bosch.com
[2] Hasso Plattner Institute, Potsdam, Germany
emmanuel.mueller@hpi.de

Abstract. Regression tasks such as forecasting of sensor values play a principal role in industrial applications. For instance, modern automobiles have hundreds of process variables which are used to predict target sensor values. Due to the complexity of these systems, each subset of features often shows different type of correlations with the target. Capturing such local interactions improve the regression models. Nevertheless, several existing feature selection algorithms focus on obtaining a single projection of the features and are not able to exploit the multiple local interactions from different subsets of variables. It is still an open challenge to efficiently select multiple subsets that not only contribute for the prediction quality, but are also diverse, i.e., subsets with complementary information. Such diverse subsets enrich the regression model with novel and essential knowledge by capturing the local interactions using multiple views of a high-dimensional feature space. In this work, we propose a framework to select multiple diverse subsets. First, our approach prunes the feature space by using the properties of multiple correlation measures. The pruned feature space is used to systematically generate new diverse combinations of feature subsets without decrease in the prediction quality. We show that our approach outperforms prevailing approaches on synthetic and several real world datasets from different application domains.

1 Introduction

Regression models for predicting sensor values assist engineers to test the system response before stepping into production phase. For example, multiple information sources such as process variables, other sensor values and driving characteristics are used as predictors to predict the target values of a sensor in the automotive industry. In such a high-dimensional space, feature selection is a pivotal pre-processing step to shrink dimensionality. This procedure increases both efficiency and the prediction accuracy by removal of irrelevant variables [13,25].

The challenge arises when the predictor variables stem from multiple sources and exhibit complex relationship amidst them. Such features show different properties between each other due to the heterogeneity of the sources. For instance,

L. Bellatreche and S. Chakravarthy (Eds.): DaWaK 2017, LNCS 10440, pp. 259–273, 2017.
DOI: 10.1007/978-3-319-64283-3_19

let us consider the task of predicting the values of a temperature sensor y in an automobile. A subset of predictor variables x representing the air system of the vehicle is related to the target by a function $f : (x) \mapsto y$. However, we observe another subset of predictor variables m representing the fuel system of an automobile which is related to y by a different function $g : (m) \mapsto y$. In a real world scenario, several such interactions are hidden in the database. Conventional feature selection techniques [10,13,17,25] consider only a single projection of the feature space. Hence, the effect of these intricate local interactions cannot be captured.

For a high-dimensional feature space, several feature combinations are possible. An exhaustive search of every possible combination is inefficient. Additionally, the selected subsets have to be diverse and non-repetitive in nature. Diverse subsets contain new knowledge from which the regression algorithm can harness the local interactions. Repetitive subsets containing redundant information are undesirable as considering similar projections multiple times does not contribute to discover any new underlying patterns. This calls for an efficient strategy to generate diverse and non-redundant subsets.

Our work Diverse Subset Selection Strategy (DS3) provides a framework for multiple subsets selection and involves two key components.

(1) *A technique for feature selection based on multiple correlation properties.*
(2) *A search strategy for the selection of multiple diverse feature subsets for enhancing ensemble regression.*

Our strategy prunes the non-essential variables and generate multiple projections of the feature space. Due to heterogeneity of the interactions, each projection of the feature space has different influence on the target prediction. Hence, each of them is evaluated based on a particular property they exhibit. To address this, the first component of our approach extracts initial candidate sets based on multiple correlation measures following the filter-based paradigm. The second component aims to generate novel dissimilar subsets that also contribute for the prediction quality. Hence, each subset provides not only complementary but also essential information for the target prediction. Finally, an ensemble regression model for each subset is trained and a composite hypothesis that maps each diverse subset to the target values is identified. The final predictions are based on the unified results of each individual composite hypothesis.

In our experiments, we compare our approach to several existing feature selection algorithms and regression models on synthetic and real world data sets. The results of DS3 show better scalability and an improvement of the prediction quality.

2   Related Work

Feature selection techniques can be classified in two paradigms: *single projection* such as traditional *Wrappers* or *Filter* methods and *multiple projections* based on *Ensemble Feature Selection*. Below we discuss the shortcomings of each methodology.

A large number of methods for feature selection based on a single projection have been proposed [17,25]. Wrapper schemes use the interface of the regression algorithm itself to evaluate the relevance of each feature for the target prediction [2,10,20]. However, wrappers are computationally inefficient.

On the other hand, the filter schemes evaluate the features using a correlation function and are independent of the underlying regression algorithm [17,25]. As an unsupervised technique, *Principle Component Analysis* (PCA) is an efficient technique for dimensionality reduction [24]. Nevertheless, all aforementioned techniques focus on the selection of a single projection of the feature space. On contrary, our proposed framework aims to identify multiple and diverse projections of the feature space.

An ensemble feature selection technique for classification tasks has been proposed in the work of [16]. This approach selects a feature subset for each class using a single correlation measure. However, our approach uses different correlation measures as each of them enumerate the importance of a feature based on different properties. In contrast to the work of [5,19], we do not aggregate the results of multiple correlation measures. We generate initial candidates based on multiple correlation measures simultaneously. The initial candidates are later used for generation of subsets that have complementary information. The embedded selection technique Random Forests [3] bags the results of multiple decision trees from random subsets. On contrary, our approach select subsets by enhancing diversity and considering multiple intrinsic relationships between the variable and target.

3   Diverse Subset Selection Strategy (DS3)

3.1   Problem Overview

In this section, we formally define the problem that we aim to solve. Given a d-dimensional feature space $\mathcal{A} = \{x_1, x_2, \ldots, x_d\}$ containing N instances and a target y (of continuous values). We aim to identify a family of m feature subsets, such that:

$$\mathcal{P} = \{s_1, s_2, \ldots, s_m\}.$$

Each set $s_i \subset \mathcal{A} \mid i = 1, \ldots, m$ is selected under the fulfillment of the following constraints:

(1) Only relevant features are selected in s_i based on a function space $C = \{c_1, \ldots, c_k\}$ of k correlation measures:

$$\{c_1(x, y) \ldots c_k(x, y)\},$$

such that each correlation measure $c_i : (x, y) \mapsto \mathbb{R}$ and for an unsupervised measure (e.g. PCA) $c_i : (x) \mapsto \mathbb{R}$.

(2) To avoid subsets with similar features, we define diversity of subsets based on a difference criterion:

$$\forall(s_i, s_j) : \textbf{\textit{diff}}\,(s_i, s_j),$$

where the $\textbf{\textit{diff}}\,(s_i, s_j)$ function returns a diversity enhanced feature set. The choice of the difference function will be elaborated in the forthcoming sections. As a hybrid approach, to evaluate the quality of the subsets, we need a regression algorithm $\textbf{\textit{Reg}}$. A regression algorithm approximates a function between the feature set and the target $\textbf{\textit{Reg}} : (s) \mapsto \hat{y}$, where $\hat{y}$ is the predictions of y. The $\textbf{\textit{Err}}$ function quantifies the fit errors $e \in \mathbb{R}$ of the regression model, i.e., $\textbf{\textit{Err}} : (y, \textbf{\textit{Reg}}(s)) \mapsto e$. For a family of m subsets $\mathcal{P}$, the collection of fit errors of each subset is represented as an m-tuple, i.e., $\epsilon = \{e_1, e_2, \ldots, e_m\}$. Hence, a subset $s_i \in \mathcal{P}$ corresponds to the fit error $e_i \in \epsilon$.

For the selection of relevant variables in each s_i, we need multiple correlation functions that quantifies the importance of each feature for prediction of y. After this step, diverse projections of relevant features that have both high difference $\textbf{\textit{diff}}\,(s_i, s_j)$ in the feature sets and contribution for target predictions based on the fit errors $\textbf{\textit{Err}}(y, \textbf{\textit{Reg}}(s_i))$ are to be efficiently identified.

3.2   Solution Overview

To address the first task of selecting relevant features, we consider a function space with k correlation measures such that each measure assesses different correlation properties shared between a feature x and the target y. Hence, the correlation measures are chosen in such a way that they make diverse decisions. The features selected by each correlation measure serve as initial candidates for addressing the second challenge. Then, different combinations of the initial candidates are efficiently generated based on a difference criterion. The initial candidates are updated by the newly generated subsets that have lower prediction errors. The process of generation and updating is repeated over multiple iterations. Hence, the initial candidates are improved in each iteration. Finally, an ensemble regression model is trained using the chosen feature sets for obtaining the final prediction.

3.3   Relevance Based Generation of Initial Candidates

DS3 has two major phases in the selection of subsets. The first phase prunes the non-contributing features from the feature space and create subsets which exemplify different properties. To achieve this, each feature is evaluated using a set of correlation functions. We compute the following $d \times k$ matrix,

$$\mathbf{M} = \begin{bmatrix} c_1(x_1, y) & c_2(x_1, y) & \ldots & c_k(x_1, y) \\ c_1(x_2, y) & c_2(x_2, y) & \ldots & c_k(x_2, y) \\ \vdots & \vdots & \ddots & \vdots \\ c_1(x_d, y) & c_2(x_d, y) & \ldots & c_k(x_d, y) \end{bmatrix}.$$

The matrix, depicts the relevancy magnitude of each feature based on several correlation functions. However, one of our aim is to reduce dimensionality by neglecting irrelevant features. In order to prune them, we calculate a threshold value. This value defines the magnitude of correlation below which the features have to be neglected. We calculate thresholds for each correlation function by multiplying the user defined parameter $\alpha \in [0, 1]$ with the maximum value in each column of the matrix $\mathbf{M}$. For each correlation function, all features with a correlation magnitude greater than or equal to the threshold are selected. As a result, each correlation measure selects a subset of features. That is, for k correlation functions, we obtain $s_1, \ldots, s_k$ subsets. We represent the collection of subsets as $\mathcal{P}$. Each set in $\mathcal{P}$ has features that exemplifies an intrinsic property based on the correlation function. This is the first essential step for ensuring diversity in the second component of our strategy. Following the paradigm of hybrid approach we calculate the fit error e_i for each feature subset in $\mathcal{P}$ using a regression algorithm, i.e., $e_i = \boldsymbol{Err}(y, \boldsymbol{Reg}(s_i))$ and the fit errors are updated in $\epsilon = \{e_1, e_2, \ldots, e_k\}$. Each correlation measure in DS3 may have different range of minimum and maximum values. The parameter α is independent of the range of individual correlation measures. For example, let us consider two bi-variate correlation measures Maximal Information Coefficient (MIC) and Spearman's rank correlation (ρ) [7]. The minimum and maximum value of MIC lies between $[0, 1]$. In contrast, the values of ρ are in the interval of $[-1, 1]$. After pruning, the dimensionality of each subset s_i is implicitly dependent on α. A factor of $\alpha = 1$ leads to the omission of all features except those which have the maximum correlation in each column. On the other hand, a value of zero tends to retain all the values.

By extracting feature subsets using multiple correlation functions, we address the first requirement defined in Sect. 3.1. This is a necessary step for obtaining diverse subsets with complementary information. However, this is not a necessary condition for obtaining the best prediction quality. Therefore, we need a strategy for subset generation that increases diversity and contributes for quality in parallel.

Algorithm 1. Choose initial candidates

Input: $\mathcal{A}, y, \alpha, \boldsymbol{C}$
1: **for** $m = 1 \to |\boldsymbol{C}|$ **do**
2:     **for** $p = 1 \to |\mathcal{A}|$ **do** $\mathbf{M}_{pm} = c_m(x_p, y)$
3:     **end for**
4:     $s_m = \{\mathcal{A} \mid \mathbf{M}_{*m} \geq (\boldsymbol{max}(\mathbf{M}_{*m}) * \alpha)\}$
5:     $e_m = \boldsymbol{Err}(y, \boldsymbol{Reg}(s_m))$
6: **end for**
7: $\mathcal{P} = \{s_1, \ldots, s_m\}$
8: $\epsilon = \{e_1, e_2, \ldots, e_m\}$
9: **return** $\mathcal{P}$ and ϵ

3.4   Multiple Feature Sets Based on Difference and Quality

A preliminary selection of subsets based on different correlation functions decreases the dimensionality. Nonetheless, there are several other combinations that were not considered and they may increase the prediction quality. Therefore, we have to efficiently search for new subsets that satisfy the difference criterion (c.f. Sect. 3.1). The difference criterion aims at generation of subsets with complementary or new information to the regression model. Considering a difference criterion based on complex measures such as information gain requires a higher time complexity w.r.t database size. As these computations have to be evaluated for a large number of subsets, it is necessary to choose a criterion that performs efficiently. Hence, we choose to apply concepts from set theory. The different operations for combining sets are, Union, Intersection and Symmetric difference. Let's assume two sets of features: $s_1 = \{x_3, x_9, x_{14}\}$ and $s_2 = \{x_1, x_2, x_3\}$. By performing union operation between two sets, we obtain $s_1 \cup s_2 = \{x_3, x_9, x_{14}, x_1, x_2\}$. The union operation generates a larger set which does not capture the local interactions and ends in a full-dimensional feature set over several iterations. Performing intersection operation $s_1 \cap s_2 = \{x_3\}$ does not enhance diversity. It creates a subset based on a feature whose role has been captured by multiple feature combinations. Symmetric difference ($\triangle$) between two sets returns the objects that belong to one of the sets but not to their intersections, i.e., $s_1 \triangle s_2 = \{x_1, x_2, x_9, x_{14}\}$. The new subset generated by applying $\triangle$ operation has led to elimination of feature that exists in both sets. Thus it partially contributes for dimensionality reduction. Secondly, it enhances diversity by eliminating features whose contribution has already been captured in both sets. The core idea of our approach is on learning new information from diverse feature combinations. To achieve this, we select non-intersecting elements and avoid generation of subsets with redundant features (w.r.t. initial candidates).

Symmetric difference is associative and commutative [15], i.e., $s_1 \triangle s_2 = s_2 \triangle s_1$. Hence, a family of m sets generate $\binom{m}{2}$ number of new offspring subsets by applying symmetric difference between itself and each of the other candidates. For example, the initial candidate subsets $\mathcal{P} = \{s_1, s_2, s_3\}$ generates 3 offspring subsets, i.e., $(s_1 \triangle s_2), (s_1 \triangle s_3), (s_2 \triangle s_3)$. Using all the $\binom{m}{2}$ offsprings is inefficient and not all offsprings may contribute for the prediction quality. Thus, we follow a wrapper scheme and quantify the significance of an offspring based on its prediction quality. As shown in Algorithm 2, the fit errors of each new offspring subset $s_{new} \subset \mathcal{A}$ is estimated using a regression algorithm, i.e., $\textbf{Err} : (y, \textbf{Reg}(s_{new})) \mapsto e_{new}$. If an offspring subset outperforms the quality (fit errors) of an initial candidate subset s_i, i.e., $e_{new} < e_i \in \epsilon$, the particular offspring s_{new} replaces the worst performing set from $\mathcal{P}$. The corresponding fit error is updated as $\epsilon = \{\epsilon \setminus e_i\} \bigcup \{e_{new}\}$. The process of symmetric difference and quality check is repeated with the updated candidates (Lines 5–8). In parallel, $\mathcal{P}_{past}$ keeps track of subsets of the previous iteration. The process iterates until none of offspring subsets outperforms the quality of subsets from previous iteration ($\mathcal{P}_{past}$). In this case, the Boolean variable *improvement* switches to *false* and the algorithm ends.

Algorithm 2. Generation of quality constrained diverse subsets

---

Input: $\mathcal{P}, y,$ and ϵ

 1: Initialize $improvement = true$
 2: **while** $improvement{=}true$ **do**   ▷ Iterate for every updated family of subsets
 3:     Set $\mathcal{P}_{past} = \mathcal{P}$
 4:     **for each** tuple$(s_d, s_e) \in \mathcal{P}$ **do**
 5:         $s_{new} = \{s_d\} \triangle \{s_e\}$; $e_{new} = \boldsymbol{Err}(y, \boldsymbol{Reg}(s_{new}))$
 6:         **if** $((\exists e_i \in \epsilon) > e_{new})$ **then**   ▷ If offspring outperforms initial subsets in $\mathcal{P}$
 7:             $\mathcal{P} = \{\mathcal{P} \setminus s_i\} \bigcup \{s_{new}\}$ such that, $\boldsymbol{Err}(y, \boldsymbol{Reg}(s_i)) = max(\epsilon)$
 8:             $\epsilon = \{\epsilon \setminus e_i\} \bigcup \{e_{new}\}$ ▷ Replace the corresponding fit errors in ϵ
 9:         **end if**
10:     **end for each**
11:     **if** $(\mathcal{P}_{past} = \mathcal{P})$ **then**   ▷ If no improvement by offspring subset
12:         $improvement = false$ ▷ Reset bit to exit while loop
13:     **end if**
14: **end while**
15: **Return:** $\mathcal{P}$

3.5 Unifying Multiple Subsets by Ensemble Regression

After having described our strategy for the selection of multiple subsets, our goal is to unify the information from each subset. The unified decision of DS3 is computed by following the idea of incremental learning [12]. Nevertheless, our framework enables the use of other techniques for unifying the individual hypothesis (e.g. using Dempster-Schafer Technique [5]). Incremental learning aims to learn from newly available data of the same source. For example, in weather forecasting applications, new data from temperature sensors is used to update the existing prediction models. Thus, the source of data, i.e., temperature sensor remains the same. For each of the newly available data from the temperature sensor, an ensemble is trained. The final predictions are formed by the combined decision of the ensembles. Hence, the concept is based on building ensemble of ensembles [12]. We employ the same principle to combine the predictions of multiple subsets. For each $s_i \in \mathcal{P}$ we train an ensemble regression model. From each ensemble model, we obtain a set of predicted values of the target. The final predictions are the weighted average based on the relative errors (on training data) of each ensemble.

3.6 Time Complexity

To analyze the time complexity, we begin with the analysis of Algorithm 1. The algorithm starts with the computation of a $d \times k$ matrix. The run times for this step depends on the number and types of correlation measures used. The features are pruned based on the parameter α. In the worst-case, a user sets the parameter α to 0. This means, that Algorithm 1 does not prune the features. This leads to k subsets with dimensionality d. The fit errors of each subset has to be computed by a regression algorithm. In our experiments we use OLS which has a time complexity of $\mathcal{O}(d^2 \cdot N)$ [23]. Thus, the total complexity of Algorithm 1 is given by the worst time complexity of k correlation measures and the time complexity for estimating fit errors of k subsets, i.e., $\mathcal{O}(k \cdot d^2 \cdot N)$.

In Algorithm 2, the symmetric difference between two sets have a linear time complexity $\mathcal{O}(d)$. The symmetric difference between k subsets generate $\binom{k}{2}$ offspring subsets. The fit errors are computed for each of these newly generated offspring subset. Thus, the total complexity of Algorithm 2 is represented as $\mathcal{O}(\binom{k}{2} \cdot d^2 \cdot N)$. However, the worst-case scenario of $\alpha = 0$ will not generate any new subsets in Algorithm 2. In Sect. 4, experiments show that our algorithm ends in a fewer iterations with the best quality.

4   Experiments

In this section we compare the quality and run times of our approach with several existing techniques using synthetic and real wold datasets. We consider different selection paradigms as competitors: filter-based approaches using forward selection (SFS) [13] with the following cost functions: MIC [17] and dCor [22], FCBF with symmetric uncertainty as cost function [25], ReliefF weights [18], wrapper techniques based on Genetic Algorithms (GA) [2] with Support Vector Machines (SVM) [21] as cost function, a wrapper method with a more directed search strategy called Particle Swarm Optimization (PSO) [20] with Artificial Neural Networks (ANN) as cost function and finally PCA which is a renowned technique for dimensionality reduction [8]. After each feature selection algorithm, we apply two type of regression techniques: a single learner (sparse Gaussian processes [11]) and an ensemble learner (AdaBoost [6] of Decision trees). As a techniques based on the paradigm of multiple projections, we consider the well-established Random Forest [3].

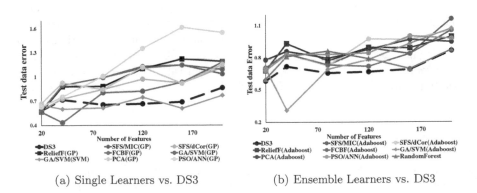

(a) Single Learners vs. DS3   (b) Ensemble Learners vs. DS3

Fig. 1. Quality (NRMSE) comparison of the competitor approaches vs. DS3 with increasing dimensionality (20, 40, 80, 120, 160, 200) and fixed database size of 1000 samples

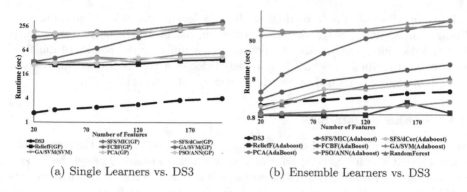

(a) Single Learners vs. DS3   (b) Ensemble Learners vs. DS3

Fig. 2. Run time (selection + training times) comparison of the competitor approaches vs. DS3 with increasing dimensionality (20, 40, 80, 120, 160, 200) and fixed database size of 1000 samples

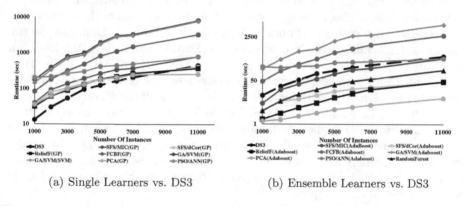

(a) Single Learners vs. DS3   (b) Ensemble Learners vs. DS3

Fig. 3. Run time (selection + training times) comparison with increasing database size (1000, 2000, 4000, 6000, 8000, 12000) and fixed dimensionality of 80

Some of the aforementioned baselines require a user defined threshold value. For a fair comparison, we performed a grid search of the parameter for baselines and DS3 and only best results are used for comparison. Each data set is split such that 60% of the samples are used for selection/training and 40% for testing. As quality measure, we use the Root Mean Square Error normalized by the variance (σ^2) of the target instances. $NRMSE$ value of zero indicates that the predicted values follow the recorded values accurately with no deviations. The run times of experiments denote the feature selection summed up with training time.

Our approach requires a set of correlation functions and a wrapper method (**Reg**) (c.f. Sect. 3.2). In this work, we have considered the following four correlation functions: MIC [17], dCor [22], PCA [8,24] and Relieff weights [18]. For faster computations, we chose Ordinary Least Squares (OLS) as the wrapper technique (**Reg**) for our hybrid approach. The OLS is used only as a wrapper within DS3 and not for estimating the final predictions. Unification the diverse subsets (c.f. Sect. 3.5) is to be done with weak learner capable of handling non-linear dependencies. Since the competitor approaches have been tested using boosted decision trees, we also employ them for unifying the results of multiple subsets as described in Sect. 3.5. A boosted decision tree is trained for each s_i and predictions are unified by weighted average of their errors [12].

4.1 Synthetic Data Sets

We analyze the efficiency of our approach w.r.t database size and the dimensionality. The analysis have been performed on synthetic datasets. The synthetic data generation program of NIPS [14] is employed to generate continuous feature sets with normal distribution and in any proportion of relevant features.

Figure 1 compares the quality between DS3 and existing competitors increasing the dimensionality, i.e., increasing the number of irrelevant variables. In these experiments, DS3 achieves better results than traditional filter-based approaches and random forests that randomly select different projections of the data. Wrappers schemes (GA with SVM) obtain similar quality results as DS3, but the run times are significantly large in comparison to DS3. In particular, DS3 has been approximately 60 times faster in comparison to GA/SVM(SVM) (c.f. Fig. 2).

Figures 2 and 3 show the run times of DS3 regarding increasing database size and dimensionality. Without significantly trading off the quality, DS3 shows to be efficient. PCA, ReliefF, SFS/dCor, SFS/MIC and RandomForest show better run times with higher prediction errors w.r.t. DS3. These approaches do not address diversity in the subset selection process. However, DS3 by applying symmetric difference aims to generate multiple diverse subsets and enhance diversity. Thus, DS3 consistently achieves better prediction accuracy (c.f. Fig. 1) in comparison to these approaches.

4.2 Real-World Data Sets

The proposed selection strategy is tested on five different real world data sets from different areas of application which include sensor data for prediction of relative humidity [26] (17 features/4137 samples), social media data [9] (77 features/13000 samples), data set on controlling the ailerons of an F16 aircraft [4] (40 features/14050 samples), stock exchange data [1] (159 features/1813 samples) and automotive sensor data from our application scenario at Bosch (179 features/14537 samples). The automotive data from Bosch is obtained from multiple sensor sources of a car and the objective is to predict a particular sensor's value. For confidentiality reasons, we do not provide explicit information about the Bosch data.

Quality. Table 1 shows the results w.r.t. the prediction errors by each model. In comparison to the use of the entire feature space, we observe that the application of feature selection algorithms improves the quality. However, existing feature selection techniques do not show significant improvements in the prediction quality for the scenario of stock data. DS3, by exploiting the hidden local

Table 1. Comparison of prediction errors (NRMSE) of competitor techniques versus DS3 on real-world datasets on test data

Feature selection	Regression	Stock data	Social media Data	Bosch data	Humidity data	Ailerons
Full dimensional	GP	0.94	1.42	0.86	1.06	1.06
	Adaboost	0.83	0.66	0.86	0.61	0.78
SFS/MIC	GP	0.86	1.09	2.07	0.51	0.71
	Adaboost	0.85	0.66	0.40	0.55	0.73
SFS/dCor	GP	0.88	1.09	1.05	0.69	0.91
	Adaboost	0.84	0.68	0.39	0.44	0.82
Relieff	GP	0.88	0.93	4.49	0.83	0.85
	Adaboost	0.86	0.65	0.42	0.62	0.68
FCBF	GP	0.89	0.85	1.24	1.33	1.16
	Adaboost	0.86	0.65	1.22	0.60	1.17
GA/SVM	GP	0.87	0.97	0.64	0.83	0.69
	SVM	1.15	2.53	0.83	0.45	0.51
	Adaboost	0.87	0.51	0.3	0.65	0.65
PCA	GP	0.85	0.93	1.16	1.19	1.12
	Adaboost	0.86	0.65	0.95	0.89	0.98
PSO/ANN	GP	0.83	0.92	2.41	0.99	1.25
	Adaboost	0.85	0.76	1.08	0.54	0.98
Random forest		0.86	0.52	0.74	0.44	0.62
DS3/OLS	Decision tree	**0.66**	**0.45**	**0.24**	**0.44**	**0.51**

interactions of the variables has the best prediction accuracy. In the other application scenarios, feature selection algorithms increase the quality of the model, but the best results are obtained by our approach.

The wrapper scheme GA/SVM was adept in identifying the non-linearities of the data in comparison to other filter-based techniques. Thereby, it obtains the best prediction accuracies amidst the competitors in the social media, Bosch and ailerons data sets. For the humidity data set, we observe that SFS/dCor and DS3 have the same prediction accuracy, showing that the information from multiple views is not significant for such lower dimensional data sets. Overall, enhancing the diversity of the subsets and using them for a combined final hypothesis contributes for higher prediction quality.

Run times. Table 2 shows that DS3 is highly efficient than the conventional iterative and evolutionary search strategies. We observe that PCA achieves the best run times. However, the prediction errors of PCA are relatively higher in comparison to DS3. Overall, DS3 prediction model achieves efficient results

Table 2. Comparison of run times of competitor techniques versus DS3 on real-world datasets

Feature selection	Run time in seconds					
	Regression	Stock data	Social media data	Bosch data	Humidity data	Ailerons
Full Set	GP	87.83	822.76	956.32	151.38	386.45
	Adaboost	4.21	6.07	26.27	1.06	2.83
SFS/MIC	GP	91.38	852.45	7392.92	363.05	361.8
	Adaboost	29.11	260.32	6992.58	235.13	7.27
SFS/dCor	GP	149.98	1111.47	293.42	238.49	665.37
	Adaboost	27.49	157.49	242.02	20.75	50.27
Relieff	GP	107.41	527	879.96	144.93	423.02
	Adaboost	10.68	136.37	347.73	6.24	43.59
FCBF	GP	459.37	762.43	576.17	149.25	434.1
	Adaboost	375.35	78.68	99.24	11.11	5.92
GA/SVM	GP	175.7	874.35	4765.87	728.31	2051.69
	SVM	62.28	380.53	4276.52	589.98	1666.75
	Adaboost	65.89	383.58	4279.02	590.78	1667.69
PCA	GP	117.32	598.88	549.01	141.55	434.55
	Adaboost	2.15	19.25	6.49	1.13	5.26
PSO/ANN	GP	1492.16	5949.19	5898.85	4025.28	3598.13
	Adaboost	1421.86	5467.89	5454.19	3882.26	3138.18
Random forest		4.7	67.9	117.66	2.21	17.40
DS3/OLS	Decision tree	**41.11**	**385.03**	**824.31**	**22.13**	**65.66**

w.r.t. both quality and run time because of breaking down the complexities of the feature space with the multiple diverse subsets.

4.3 Parameter Analysis

DS3 has one user defined parameter, i.e., the prune factor α. We experimentally analyze the influence of α on prediction errors on artificial and real world datasets. As discussed in Sect. 3, $\alpha = 1$ will choose only the features with maximum correlation as initial candidates. With such small subsets, Algorithm 2 has fewer tuples of feature subsets (for computing symmetric difference) that cannot generate new diverse offspring subsets. On the contrary, $\alpha = 0$ tends to choose the full-dimensional feature set which is inefficient. Figure 4 shows that choosing only features with best correlation does not ensure best prediction accuracies ($\alpha = 1$). Likewise, choosing all the features as initial candidates is also not contributing to improved prediction quality ($\alpha = 0$). However, performing grid search over all values of α is tedious. From experiments on artificial datasets

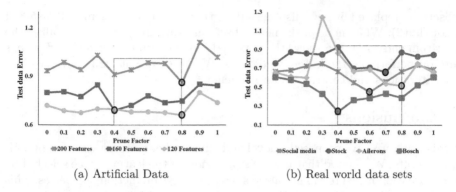

(a) Artificial Data   (b) Real world data sets

Fig. 4. Analysis of the influence of α on test data with different dimensionality. The circled points denote the minimal test data error

(c.f. the highlighted window of Fig. 4(a)), we observe that α values in the range $[0.4, 0.8]$ has the minimal test errors. In Fig. 4(b), we show that the range is also practically applicable for real world datasets. A factor less than 0.4 includes many noisy features as initial candidates. On the other hand, a factor greater than 0.8 generates small subsets with highly relevant features as initial candidates. Thus, the symmetric difference search space is also minimal. $\alpha \in [0.4, 0.8]$ gives reasonable initial candidates for which enhancing the diversity improves the prediction quality.

4.4 Iterations

DS3 prunes the feature space by adherence to multiple correlation measures and evaluate the symmetric difference search space. Figure 5 shows maximum fit error (i.e. $max(\epsilon)$) of the initial candidates for each iteration. In each iteration, the

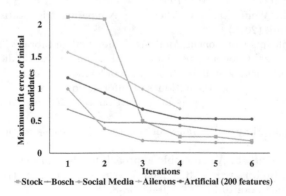

Fig. 5. Maximum fit error (using OLS) of initial candidates in each iteration of the symmetric difference search space: $\alpha = 0.9$

subsets are replaced by new offspring subset that have lower fit errors (Line 5-8 of Algorithm 2). When none of the new subsets improve the prediction quality, the algorithm ends. Figure 5 shows the maximal iteration up to which the offsprings were outperforming the initial candidates. We observe that DS3 converges in a fewer iterations and contributes for efficiency.

5   Conclusions

In this work, we have proposed a novel and efficient heuristic for selection of multiple subsets. We showed that using a single view of the feature space selected by one correlation measure, do not always have the best prediction accuracies. This corroborates the importance of diversity enhancing multiple subsets search strategy. Considering such multiple subsets enhances the prediction model by combining the underlying patterns hidden in multiple views of the high-dimensional feature space. Our experiments show that the proposed heuristic is efficient and has improved prediction quality in comparison to several state of the art techniques.

References

1. Stock closing prices for 156 companies and 3 indexes from 2000 to 2007 (2011). http://mldata.org/repository/data/viewslug/stockvalues/
2. Babatunde, O., Armstrong, L., Leng, J., Diepeveen, D.: A genetic algorithm-based feature selection. Br. J. Math. Comput. Sci. 4(21), 889–905 (2014)
3. Breiman, L.: Random forests. Mach. Learn. 45(1), 5–32 (2001)
4. Camacho, R.: Delta ailerons (1997). http://www.dcc.fc.up.pt/~ltorgo/Regression/DataSets.html
5. Fei, T., Kraus, D., Zoubir, A.M.: Contributions to automatic target recognition systems for underwater mine classification. IEEE Trans. Geosci. Remote Sens. 53(1), 505–518 (2015)
6. Freund, Y., Schapire, R., Abe, N.: A short introduction to boosting. J.-Jpn. Soc. Artif. Intell. 14(771–780), 1612 (1999)
7. Hollander, M., Wolfe, D.A., Chicken, E.: Nonparametric Statistical Methods. Wiley, New York (2013)
8. Jolliffe, I.: Principal Component Analysis. Wiley Online Library (2002)
9. Kawala, F., Douzal-Chouakria, A., Gaussier, E., Dimert, E.: Prédictions d'activité dans les réseaux sociaux en ligne. In: 4ième conférence sur les modèles et l'analyse des réseaux: Approches mathématiques et informatiques, p. 16 (2013)
10. Kohavi, R., John, G.H.: Wrappers for feature subset selection. Artif. Intell. 97(1), 273–324 (1997)
11. Lázaro Gredilla, M.: Sparse Gaussian processes for large-scale machine learning (2010)
12. Lewitt, M., Polikar, R.: An ensemble approach for data fusion with learn++. In: Windeatt, T., Roli, F. (eds.) MCS 2003. LNCS, vol. 2709, pp. 176–185. Springer, Heidelberg (2003). doi:10.1007/3-540-44938-8_18
13. Molina, L.C., Belanche, L., Nebot, À.: Feature selection algorithms: a survey and experimental evaluation. In: 2002 IEEE International Conference on Data Mining, 2002, ICDM 2003, Proceedings, pp. 306–313. IEEE (2002)

14. NIPS: Workshop on variable and feature selection (2001). http://www.clopinet.com/isabelle/Projects/NIPS2001/
15. Olson, J.E.: On the symmetric difference of two sets in a group. Eur. J. Combin. **7**(1), 43–54 (1986)
16. Oza, N.C., Tumer, K., Norwig, P.: Dimensionality reduction through classifier ensembles (1999)
17. Reshef, D.N., Reshef, Y.A., Finucane, H.K., Grossman, S.R., McVean, G., Turnbaugh, P.J., Lander, E.S., Mitzenmacher, M., Sabeti, P.C.: Detecting novel associations in large data sets. Science **334**(6062), 1518–1524 (2011)
18. Robnik-Šikonja, M., Kononenko, I.: Theoretical and empirical analysis of relieff and rrelieff. Mach. Learn. **53**(1–2), 23–69 (2003)
19. Saeys, Y., Abeel, T., Van de Peer, Y.: Robust feature selection using ensemble feature selection techniques. In: Daelemans, W., Goethals, B., Morik, K. (eds.) ECML PKDD 2008. LNCS, vol. 5212, pp. 313–325. Springer, Heidelberg (2008). doi:10.1007/978-3-540-87481-2_21
20. Sharkawy, R., Ibrahim, K., Salama, M., Bartnikas, R.: Particle swarm optimization feature selection for the classification of conducting particles in transformer oil. IEEE Trans. Dielectr. Electr. Insulation **18**(6), 1897–1907 (2011)
21. Smola, A., Vapnik, V.: Support vector regression machines. Adv. Neural Inf. Process. Syst. **9**, 155–161 (1997)
22. Székely, G.J., Rizzo, M.L., Bakirov, N.K.: Measuring and testing dependence by correlation of distances. Ann. Stat. **35**(6), 2769–2794 (2007)
23. Tan, P.N.: Introduction to Data Mining. Pearson Education, Noida (2006)
24. Yoon, H., Yang, K., Shahabi, C.: Feature subset selection and feature ranking for multivariate time series. IEEE Trans. Knowl. Data Eng. **17**(9), 1186–1198 (2005)
25. Yu, L., Liu, H.: Feature selection for high-dimensional data: a fast correlation-based filter solution. In: ICML, vol. 3, pp. 856–863 (2003)
26. Zamora-Martínez, F., Romeu, P., Botella-Rocamora, P., Pardo, J.: On-line learning of indoor temperature forecasting models towards energy efficiency. Energy Build. **83**, 162–172 (2014)

K-Means Clustering Using Homomorphic Encryption and an Updatable Distance Matrix: Secure Third Party Data Clustering with Limited Data Owner Interaction

Nawal Almutairi[✉], Frans Coenen[✉], and Keith Dures[✉]

Department of Computer Science, University of Liverpool, Liverpool, UK
{n.m.almutairi,coenen,dures}@liverpool.ac.uk

Abstract. Third party data analysis raises data privacy preservation concerns, therefore raising questions as to whether such outsourcing is viable. Cryptography allows a level of data confidentiality. Although some cryptography algorithms, such as Homomorphic Encryption (HE), allow a limited amount of data manipulation, the disadvantage is that encryption precludes any form of sophisticated analysis. For this to be achieved the encrypted data needs to coupled with additional information to facilitate third party analysis. This paper proposes a mechanism for secure k-means clustering that uses HE and the concept of an Updatable Distance Matrix (UDM). The mechanism is fully described and analysed. The reported evaluation shows that the proposed mechanism produces identical clustering results as when "standard" k-means is applied, but in a secure manner. The proposed mechanism thus allows the application of clustering algorithms to encrypted data while preserving both correctness and data privacy.

Keywords: Homomorphic encryption · k-means clustering · Privacy preserving data mining · Secure k-means clustering

1 Introduction

Data mining techniques have been exploited to improve quality of service and guide decision makers with respect to many application domains. However, the exponential growth in data availability often makes in-house data analysis impractical and expensive. This has fuelled the idea of Data Mining as a Service (DMaaS); third parity analysis using cloud computing facilities [3]. This also opens the door to collaborative data mining where a number of data owners pool their data so as to gain some (for example commercial) mutual advantage. Although DMaaS reduces the cost of managing and analysing data, and facilitates collaborative data mining, it introduces several challenges, the most significant of which is data privacy preservation. Hence the research domain of Privacy Preserving Data Mining (PPDM) [1,9,16]. The typical approach to

© Springer International Publishing AG 2017
L. Bellatreche and S. Chakravarthy (Eds.): DaWaK 2017, LNCS 10440, pp. 274–285, 2017.
. DOI: 10.1007/978-3-319-64283-3_20

PPDM is to use some form of data transformation, applied either to the entire data set (for example data perturbation) or selective sensitive data attributes (for example value anonymisation) [4]. In perturbation, individual attributes are distorted or randomised by adding noise, whereas data anonymity relies on modifying the data by replacing sensitive values with proxy values or removing the values all together. However, although perturbed data may look quite different from the original data, an adversary may take advantage of properties such as correlations and patterns in the original data to identify the real values [16]. It has also been observed that data cannot be 100% anonymised [2]. In addition the anonymisation/perturbation of data may adversely effect the quality of the analysis [1,16].

Data privacy and security can be substantially guaranteed when data is encrypted. However, standard forms of encryption do not support data mining activities, which typically require the comparison of records to determine their similarity (or otherwise). Several encryption schemes have been developed to enforce privacy and confidentiality while at the same time supporting some manipulation of encrypted data. Searchable Encryption (SE) [17] is an example of an encryption scheme that preserve data privacy while supporting different kinds of search operations over encrypted data. However, such search capabilities still do not readily support data mining activities. One potential solution is Homomorphic Encryption (HE), a family of emerging encryption mechanism that supports a limited number of simple mathematical computations. The precise nature of the mathematical capabilities supported by HE schemes is dependent on the nature of the adopted scheme. However, there is no HE scheme that provides all the mathematical operations that we might want so as to perform data mining activities; for example the similarity calculation operations required in the case of data clustering [12,15]. The solutions that have been proposed to date all entail a significant element of data owner participation so that operations not supported by the selected HE mechanism can be conducted using unencrypted data by the data owner. The degree of data owner participation is such that the only reason for using a third party is to support cooperative data mining. For example in the context of k-means cooperative clustering, as described in [7,13], local cluster centroids are shared on each iteration so that global centroids can be calculated on each k-means iteration, the majority of the work is conducted by the data owners. In the case of a single data owner there seems little point in conducting third party k-means clustering if most of the work needs to be conducted by the data owner. The challenge is therefore to limit the amount of data owner participation.

Given the above, this paper presents a k-means clustering mechanism that limits interaction with data owners by using the concept of an *Updatable Distance Matrix* (UDM). More specifically the idea is that data is stored, using third party storage, in an encrypted form, together with an associated UDM. The data owner can then request the data to be clustered, specifying the number of clusters k, as required. For the clustering to operate correctly the UDM matrix needs to be updated on each iteration of the k-means algorithm. To do this the differences between the centroids of iteration i and $i-1$ are calculated in encrypted form and returned to the data owner for decryption, after which the decrypted differences (as real numbers) are returned to the third party in the form of a *Shift Matrix* (SM) which is then used to update the UDM ready for iteration i; this is the only data owner participation that is required. Using this mechanism, as will be made clear in the paper, the amount of data owner participation is significantly reduced (by a factor of n, where n is the number of records in the dataset). Furthermore, although the idea is presented using the well known k-means clustering algorithm [12], it can, with some adjustment, be extend to encompass alternative clustering mechanisms. It should also be noted that the nature of the encryption used is Liu's scheme [10], which supports addition and subtraction of cypher texts, and division and multiplication of cypher texts by real numbers.

The rest of this paper is organised as follows, in Sect. 2 a review of related work is presented. Section 3 provides the fundamental background concerning standard k-means clustering and Liu's Homomorphic Encryption scheme. Section 4 explains the UDM concept as proposed in this paper. The next section, Sect. 5, presents the proposed Secure k-means cluster over the homomorphically encrypted data founded on the UDM concept introduced in the previous section. Method and results of experimenting the proposed algorithm have been discussed in Sect. 6. Finally, Sect. 7 concludes the paper with some ideas concerning future work to improve proposed mechanism.

2   Related Work

The main challenge of third party (k-means) data privacy preserving clustering using encryption is that, although the HE schemes used support a range of arithmetic operations that can be applied to cypher-data, there are some operations that are not supported. Mechanisms have been developed to address this challenge. These can be categorised as featuring either Multi-Parity Computation (MPC) or Partial Third-party Computation (PTC). Both feature significant data owner participation.

MPC is only applicable where the desired clustering is to be undertaken with respect to data belonging to two or more data owners. The basic idea is that the majority of the processing is conducted in-house by the data owners, only the centroids are shared on each k-means iteration [6]. Examples can be found in [7,13]. In [13] the data owners, on each k-means iterations, generate local clusters using their individual datasets. Each party then calculates the "centroid" of their local cluster, encrypts this and shares the encrypted centroid with the other owners. In [7] a similar mechanism is proposed except that only one of the users calculates the global centroids and then shares this. Using MPC data confidentiality is preserved since data is stored locally and is not shared with other parities. However, data owners are expected to do much of the work and are thus required to have the expertise and resource to carry out the desired clustering.

Using PTC most of the processing is conducted by a third party, but with recourse to the data owner (or owners) for the similarity calculations required as the k-means clustering proceeds. A disadvantage is thus that typically a great many of such calculations will be required. Using PTC data is encoded using some form of HE and sent to a third party where the homomorphic properties of the encryption are used to manipulate the data so far as possible (centroid calculation, data aggregation, and so). For example in [5] a PTC setting is used to cluster data belonging to social network users into k clusters. The data is passed to a "Semi Honest" third party where the clustering is performed, whilst similarity is determined with reference to randomly selected users. The mechanism proposed in this paper falls into the second category, but seeks to significantly minimise the interaction with the data owner(s).

There has also been work directed at incorporating the idea of order preservation into homomorphic encryption schemes to support secure k-means clustering. One example can be found in [11] where the authors describe a mechanism whereby partial order preservation can be included in Liu's encryption scheme [10]. Using this mechanism the third party uses *dynamic trap-doors* that are calculated on each iteration to convert cypher-text to order preserving text[1]. However, data owner participation is still significant because the trapdoors need to be recalculated on each iteration. Liu's scheme as used in [11], is also the encryption scheme adopted with respect to the work presented in this paper, because of its particular homomorphic properties; it is therefore discussed in further detail in Subsect. 3.2.

3   Preliminaries

Before considering the proposed secure k-means third party clustering mechanism and the proposed UDM concept, some preliminaries are presented in this section. Firstly, although the k-means algorithm is well understood [12], for completeness it is briefly presented in Subsect. 3.1. Liu's homomorphic encryption scheme [10] is then described in Sect. 3.2.

3.1   K-Means Clustering

K-means is an iterative clustering algorithm where n data items are grouped into k clusters [12]; k is specified by the user. Each cluster is represented by its *centroid*. On the first iteration the first k records are usually used as the centroids. The remaining records are then assigned to clusters according to the shortest "distance" from each record to each centroid. At the end of the first iteration the centroids are recalculated; this is typically, but not necessarily, done using the mean values of the attribute values for the records present in each cluster. The algorithm then again assigns the records to the clusters and continues in this iterative manner until the centroids become fixed.

---

[1] A "trapdoor" in this context is a function that can be simply computed on one direction but is difficult to compute in the reverse direction without additional knowledge; the concept is widely used in cryptography.

Algorithm 1. Encrypt(v, K(m))

1: **procedure** ENCRYPT($v, K(m)$)
2:     Uniformly generate m arbitrarily real random numbers $r_1, ..., r_m$
3:     Declare E as a real value array of m elements
4:     $e_1 = k_1 * t_1 * v + s_1 * r_m + k_1 * (r_1 - r_{m-1})$
5:     **for** $i = 2$ to $m - 1$ **do**
6:         $e_i = k_i * t_i * v + s_i * r_m + k_i * (r_i - r_{i-1})$
7:     **end for**
8:     $e_m = (k_m + s_m + t_m) * r_m$
9:     Exit with E
10: **end procedure**

Algorithm 2. Decrypt(E, K(m))

1: **procedure** DECRYPT($E, K(m)$)
2:     $T = \sum_{i=1}^{m-1} t_i$
3:     $S = e_m/(k_m + s_m + t_m)$
4:     $v = (\sum_{i=1}^{m-1}(e_i - S * s_i)/ki)/T$
5:     Exit with v
6: **end procedure**

3.2   Liu's Homomorphic Encryption Scheme

As noted above, the adopted encryption scheme with respect to the work presented in this paper was Liu's homomorphic scheme [10]. Using Liu's scheme each data attribute value is encrypted to m sub-cyphers $E = \{e_1, e_2, \ldots, e_m\}$ where m is the *public key* ($m \geq 3$) and $K(m)$ is a list of secret keys. $K(m) = [(k_1, s_1, t_1), \ldots, (k_m, s_m, t_m)]$ where k_i, s_i and t_i are real numbers. The secret values satisfy the following conditions: (i) $k_i \neq 0$ ($1 \leq i \leq m - 1$), (ii) $k_m + s_m + t_m \neq 0$ and (iii) there exists only one i ($1 \leq i \leq m$) such that $t_i \neq 0$ (this last so as to preserve a partial ordering of the data). Algorithm 1 shows the pseudo code for the $Encrypt(v, k(m))$ encryption function that converts a real value v to a set of sub-cyphers $E = \{e_1, \ldots, e_m\}$. Given a set of sub-cyphers $E = \{e_1, \ldots, e_m\}$ and the key $k(m)$ Algorithm 2 gives the pseudo code for the $Decrypt(E, k(m))$ decryption function to return the value v.

   Liu's scheme has both security and homomorphic properties. In terms of security, the scheme is probabilistic, therefore it produces different cypher-texts for the same plain-text each time it is applied even when using the same secret key. This feature makes the scheme semantically secure and guards against "Chosen Plain-text Attacks" (CPAs) because if $E = \{e_1, \ldots, e_m\}$ is a cypher-text which encrypts v_1 or v_2, the adversary will not know which one E encrypts. The homomorphic properties of the scheme support addition and subtraction of cypher texts, and multiplication and division of cypher texts with a real number c.

$$E + E' = \{e_1 + e'_1, \ldots, e_m + e'_m\} = v + v'$$
$$E - E' = E + (E' \times -1) = \{e_1 + (e'_1 \times -1), \ldots, e_m + (e'_m \times -1)\} = v - v'$$
$$c \times E = \{c \times e_1, \ldots, c \times e_m\} = c \times v$$
$$c \div E = \{c \div e_1, \ldots, c \div e_m\} = c \div v$$

In the context of k-means clustering, as will become clear, addition is required for summing the attribute values of records in a cluster and division to consequently arrive at a centroid for the cluster. Subtraction is required to determine the difference between the centroids arrived at on iteration i with those from iteration $i - 1$.

4   The Updatable Distance Matrix Concept

Liu's scheme, described in Sect. 3.2, transfers plain-texts values randomly to cypher-texts, hence any ordering that might feature in the plain-text data is not transferred to the cypher-texts. Therefore, cypher-texts cannot be directly compared and hence k-means clustering cannot be directly applied. The propose idea is to guide the k-means clustering using what we refer to as an *Updatable Distance Matrix* (UDM) that holds the distances between attribute values in records with the corresponding attribute values in every other record. A UDM is thus a 3D matrix with the first two dimensions corresponding to the records in the data set and the third to the set of attributes that feature in the dataset. Thus, more formerly, given a data set $D = \{r_1, r_2, \ldots r_n\}$ where each record r_x is a feature vector comprised a set of values $\{v_{x_1}, v_{x_2}, \ldots v_{x_m}\}$ (each value corresponding to an attribute in the set of attributes $A = \{a_1, a_2, \ldots, a_m\}$ featured in the data set), the associated UDM $\mathbf{U}$ will be comprises of a set of elements $u_{[x,y,z]}$ where x and y indicates the record numbers ($x < n$ and $y < n$), and z the attribute number ($z < m$). Given an element $u_{[x,y,z]} \in \mathbf{U}$ the value held will then be calculated using:

$$v_{x_z} \sim v_{y_z} \tag{1}$$

Note that where $x \equiv y$ the UDM values will equate to 0. Note also that UDMs are symmetric about there leading diagonal, and hence only the 3D leading triangle needs to be considered.

The UDM for a particular dataset is generated by the data owner and sent to the third party data miner together with the encrypted data set. The owner can then request a clustering specifying k. On start up the first k records are used as the centroids; then, as the k-means algorithm progresses, the centroids are updated. However, it is not necessary for the UDM to be sent back to the data owner as in the case of other PTC approaches, such as that described in [5]. It is only necessary for differences between the newly identified centroids C' on iteration i and the centroids C from iteration $i - 1$ to be returned to the data owner so that offsets, referred to as *shift values*, can be calculated and stored in a Shift value Matrix (SM) $\mathbf{S}$, to be returned to the third party data miner who can then update the UDM $\mathbf{U}$ to generate an updated version of the UDM $\mathbf{U}'$ ($\mathbf{U}' = \mathbf{U} + \mathbf{S}$). The process whereby the proposed UDM concept, together with the use of SMs, is utilised in the context of secure k-means clustering is discussed in further detail in the following section.

5   Secure K-Means Clustering Using the UDM Concept

This section presents the proposed *UDM* based on secure k-means clustering process. The process has two parts, a data preparation part and a clustering part. The first is conducted by the data owner and is detailed in Subsect. 5.1, while the second is conducted by the third party and is detailed in Subsect. 5.2. Note that the proposed process delegates some of the processing to the data owners, but this is limited to the generation of SMs and is significantly less than in the case of more conventional PTC approaches such as that presented in [5].

5.1   Data Owner Process

Data sets are prepared for outsourcing by generating a UDM and translating the raw data into a suitable encrypted format. The pseudo code for the process is given in Algorithm 3. The input to the algorithm is a dataset D, a set of attributes A that are featured in D and the number of required sub-cyphers m, the later required for Liu's encryption scheme adopted with respect to the work presented in this paper. Note that k-means clustering works only on numerical (or pseudo numerical) data, because similarity measurements are central to the operation of the mechanism. Therefore categorical (or labelled) data needs to be replaced with discrete integers values before any further processing can be conducted, this is done in line 2 of the algorithm. A list of secret keys $K(m)$ is then defined (line 3). The processed data set is then encrypted to form a cypher data set D' using Algorithm 1 (lines 5 to 7). Next the desired UDM, $\mathbf{U}$, is dimensioned (line 8) and populated (lines 9 to 15); recall from Sect. 4 that v_{x_z} is the value of the zth attribute in the feature vector describing the xth record in the dataset, while v_{y_z} is the value of the zth attribute in the feature vector describing the yth record in the dataset. On completion (line 16) the process returns the encrypted data set D' and the generated UDM $\mathbf{U}$ ready to be sent to the third party data miner.

5.2   Third Party Process

The secure k-means clustering is conducted by the third party data miner following a processes very similar to the traditional k-means algorithm as described in Subsect. 3.1. The pseudo code presented in Algorithm 4 describes this process. The input is the encrypted data set D', the UDM $\mathbf{U}$ (previously submitted to the third party) and the number of desired clusters k. We commence by dimensioning the set of clusters $C = \{C_1, C_2, \ldots, C_k\}$ and assigning the first k encrypted records from D' to it (lines 2 and 3). We then define a second set $Cent = \{c_1, c_2, \ldots, c_k\}$ to hold the current centroids (line 4). Next the remainder of D' is processed and assigned to clusters using the *populateClusters* sub-process (line 5) given at the end of the algorithm. Records are assigned to clusters according to the similarity between the record and centroids as calculated using

---

Algorithm 3. Data encryption and UD matrix generation

1: **procedure** OUTSOURCEDATA(D,A,m)
2:     ProcessedData = Dataset D converted to numeric data set where necessary
3:     $K(m) = LiuScheme(m)$
4:     $D' = \emptyset$
5:     **for all** $r \in ProcessedData$ **do**
6:         $D' = D' \cup Encrypt(r, K(m))$ (Algorithm 1)
7:     **end for**
8:     $\mathbf{U}$ = Empty UDM dimensioned according to $|D|$, $|D|$ and $|A|$
9:     **for** $x = 1$ to $x = |ProcessedData|$ **do**
10:         **for** $y = x$ to $y = |ProcessedData|$ **do**
11:             **for** $z = 1$ to $z = |A|$ **do**
12:                 $u_{[x,y,z]} = v_{x_z} \sim v_{y_z}$ $(u_{[x,y,z]} \in \mathbf{U})$
13:             **end for**
14:         **end for**
15:     **end for**
16:     **return** D' and $\mathbf{U}$
17: **end procedure**

---

the UDM $\mathbf{U}$ as follows (where r_x is record $x \in D$, and r_y is record $y \in D$ representing cluster centroid y ($1 \le y \le k$):

$$sim(\mathbf{U}, r_x, r_y) = \sum_{z=1}^{z=|A|} \mathbf{U}_{[x,y,z]} \qquad (2)$$

We then calculate the new centroids (line 6). Next we enter into a loop (lines 7 to 14) which repeats until $Cent$ and $Cent'$ are the same. At the start of each iteration we create an encrypted SM $\mathbf{S}'$ (line 8). However, to update $\mathbf{U}$ we need real values, the content therefore needs to be decrypted. This can only be done by the owner (line 9). Note that (not shown in the algorithm) when updating $\mathbf{U}$ we only need to update the records in the second dimension representing cluster centroids. With the new centroids we again assign all records to C using the *populateClusters* sub-process (line 12) in the same manner as before. We continue in this manner till the process terminates.

6   Evaluation

The evaluation of the proposed method is presented in this section using ten datasets from the UCI data repository [8]. Table 1 gives some statistical information concerning these data sets. The Data sets have integer, real and categorical attribute types. The number of classes in each case was used as the value for k and the class column omitted from the data set. The proposed process was implemented using the Java object oriented programming language. Cluster configuration correctness was measures by comparing the results obtained with results obtained using standard (unencrypted) k-means clustering. The metric

---

Algorithm 4. Secure k-means clustering algorithm

1: **procedure** SECURE K-MEANS(D', **U**, k)
2:      C = Set of k empty clusters
3:      Select first k records in D' and assign to C (one per cluster)
4:      $Cent$ = Set of first K records in D' (the k cluster centroids)
5:      $C = populateClusters(k + 1, \mathbf{U}, C, D', Cent)$
6:      $Cent'$ = CalculateCentroids(C)
7:      **while** $Cent \neq Cent'$ **do**
8:          $\mathbf{S'}$ = SM obtained from comparing $Cent$ and $Cent'$
9:          $\mathbf{S} = \mathbf{S'}$ decrypted by data owner
10:         $\mathbf{U} = \mathbf{U} + \mathbf{S}$
11:         C = Set of k empty clusters
12:         $C = populateClusters(1, \mathbf{U}, C, D', Cent')$
13:         $Cent = Cent'$
14:         $Cent'$ = CalculateCentroids(C)
15:     **end while**
16:     Exit with C
17: **end procedure**
18: **procedure** POPULATECLUSTERS(x,**U**,C,D',$Cent$)
19:     $id = null$
20:     **for** $x = x$ to $x = |D'|$ **do**
21:         **for** $y = 1$ to $y = |C|$ **do**
22:             $sim = sim(\mathbf{U}, r_x, c_y,)$ where $r_x \in D'$ and $c_y \in Cent$ (Eq. 2)
23:             id = cluster identifier with lowest sim value so far
24:         **end for**
25:         $C_{id} = C_{id} \cup r_x$ $(C_{id} \in C)$
26:     **end for**
27:     **return** C
28: **end procedure**

---

used was the Silhouette Coefficient (Sil. Coef.) [14] calculated using Eq. 3 where $a(x_j)$ describes the cohesion between each record x_j and the other records in the same cluster, whilst $b(x_j)$ describes the separation between each record x_j and records in other clusters. The Silhouette coefficient is a real number between -1 and 1 where the closer the coefficient is to 1 the better the clustering. We were also interested in overall runtime, the time to complete the clustering.

$$OverallSil = \frac{\sum_{i=1}^{k} \frac{\sum_{j=1}^{j=|c_i|} Sil(x_j)}{|C_i|}}{k}$$

$$Sil(x_j) = \frac{b(x_j) - a(x_j)}{max(a(x_j), b(x_j))}$$

(3)

The results are presented in Table 2, the runtime is the time to complete the clustering excluding encryption and UDM generation. From the table it can be seen firstly that the cluster configurations produced using the proposed secure k-means clustering were identical to those produced using standard k-means clustering as evidenced by the Silhouette Coefficients produced (columns 4 and

Table 1. Statistical information for the data sets used in the evaluation

Datasets	Num. records	Num. attributes	Num. clusters	Attributes description
Banknote authent.	1372	4	2	Real
Breast cancer	198	33	2	Real
Breast tissue	106	9	6	Integer and real
Lung cancer	32	56	3	Integer
Blood transfusion	748	4	2	Integer
Cardiotocography	2126	36	3	Integer and real
Chronic-kidney	400	24	2	Categorical, int. and real
Iris	150	4	3	Categorical, int. and real
Pima disease	768	8	2	Integer and real
Seeds	210	7	3	Integer and real

Table 2. Execution time for secure k-means clustering

Datasets	Secure k-means			Standard k-means		
	Runtime (Mili. Secs.)	Num. Iterations	Sil. Coef.	Runtime (Mili. Secs.)	Num. Iterations	Sil. Coef.
Banknote authent.	87	17	0.207	58	17	0.207
Breast cancer	31	9	0.020	11	9	0.020
Breast tissue	26	19	0.787	19	19	0.787
Lung cancer	17	9	0.645	3	9	0.645
Blood transfusion	27	13	0.370	19	13	0.370
Cardiotocography	333	25	0.065	216	25	0.065
Chronic-kidney	29	9	0.009	14	9	0.009
Iris	14	15	0.836	9	15	0.836
Pima disease	30	9	0.000	19	0	0.000
Seeds	10	7	0.706	9	7	0.706

7 in the table) which were identical in all cases. Thus it can be concluded that the homomorphic properties used to calculate the centroids and distances between centroids do not effect the accuracy of the clustering. In terms of run time the same number of iterations were required with respect to each dataset, although from the table it can be seen that the bigger the dataset the more iterations are required. The overall run time required for the secure k-means approach to produce the desired cluster configurations, as expected, was longer than in the case of standard k-means clustering. Inspection of the table indicates that this was not significant. The runtime required by the data owner to calculate a UDM and encrypt the data did not add a significant overhead. Even for the largest data set, Cardiotocography, the time to create the UDM was 912 ms. and to encrypt the data was 25 ms. Note that the time complexity for generating an

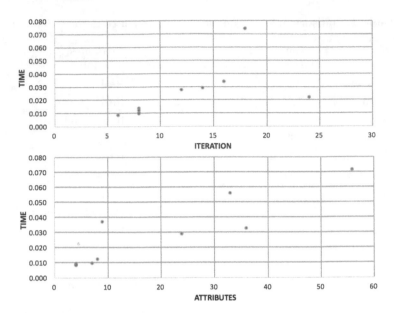

Fig. 1. Time required by data owner to decrypt shift (distance) between two centroids

initial UDM will be in the order of $O(|D| \times |C| \times k)$. The time to decrypt a SM, on each iteration, will be negligible. However, it should be noted that with respect to the experiments reported here the data owner and third party were both hosted on the same machine, thus there was no "message passing" overhead; in "real life" the data owner and third party will be hosted on separate machines, time for message passing would add to the run time although not significantly so. Figure 1 shows two plots. The first plots each data set against overall runtime for the secure k-means clustering and number of iterations, the second against overall runtime and $|A|$. The plots confirm, that although not necessarily linear, the run time increases with the number of iterations and attributes.

7   Conclusion

In this paper a secure k-means clustering mechanism has been described that uses the concept of an Updatable Distance Matrix (UDM). The advantage offered is that, unlike comparable secure k-means clustering mechanisms from the literature, data owner participation is negligible, restricted to decrypting a Shift Matrix (SM) on each iteration. The presented evaluation demonstrates that the encryption and processing does not adversely effect the quality of the clusters produced, they are the same as when k-means is applied in a standard manner. The processing time, as was to be expected, is greater but not significantly so. In an ideal situation the data owners should be able to package their data so that it is secure, send it to a third party for storage and analysis, and receive analysis results when required without the need for any further communication whilst the analysis is taking place. However, given that at present, there is no suitable encryption scheme that will support all the necessary mathematical operations for this to happen, the proposed mechanism significantly reduces the amount of data owner participation required. For future work the authors intend to investigate the utility of the UDM concept with respect to alternative clustering algorithms, and other data mining techniques (such as classification and pattern mining techniques).

References

1. Agrawal, R., Srikant, R.: Privacy-preserving data mining. SIGMOD Rec. **29**(2), 439–450 (2000)
2. Berinato, S.: There's no such thing as anonymous data. Harvard Business Review, February 2015
3. Chen, T., Chen, J., Zhou, B.: A system for parallel data mining service on cloud. In: 2012 Second International Conference on Cloud and Green Computing, pp. 329–330, November 2012
4. Chhinkaniwala, H., Garg, S.: Privacy preserving data mining techniques: Challenges and issues. In: Proceedings of International Conference on Computer Science and Information Technology, CSIT, pp. 609, July 2011
5. Erkin, Z., Veugen, T., Toft, T., Lagendijk, R.L.: Privacy-preserving user clustering in a social network. In: 2009 First IEEE International Workshop on Information Forensics and Security (WIFS), pp. 96–100. IEEE, December 2009
6. Goldreich, O.: Foundations of Cryptography: Volume 2, Basic Applications. Cambridge University Press, New York (2009)
7. Jha, S., Kruger, L., McDaniel, P.: Privacy preserving clustering. In: Vimercati, S.C., Syverson, P., Gollmann, D. (eds.) ESORICS 2005. LNCS, vol. 3679, pp. 397–417. Springer, Heidelberg (2005). doi:10.1007/11555827_23
8. Lichman, M.: UCI machine learning repository (2013)
9. Lindell, Y., Pinkas, B.: Privacy preserving data mining. J. Cryptol. **15**(3), 177–206 (2002)
10. Liu, D.: Homomorphic encryption for database querying, December 2013
11. Liu, D., Bertino, E., Yi, X.: Privacy of outsourced k-means clustering. In: Proceedings of the 9th ACM Symposium on Information, Computer and Communications Security, pp. 123–134. ACM, June 2014
12. MacQueen, J.: Some methods for classification and analysis of multivariate observations. In: Proceedings of the Fifth Berkeley Symposium on Mathematical Statistics and Probability, Oakland, CA, USA, pp. 281–297, June 1967
13. Mittal, D., Kaur, D., Aggarwal, A.: Secure data mining in cloud using homomorphic encryption. In: 2014 IEEE International Conference on Cloud Computing in Emerging Markets (CCEM), pp. 1–7. IEEE, October 2014
14. Rousseeuw, P.J.: Silhouettes: a graphical aid to the interpretation and validation of cluster analysis. J. Comput. Appl. Math. **20**, 53–65 (1987)
15. Singh, M.D., Krishna, P.R., Saxena, A.: A privacy preserving jaccard similarity function for mining encrypted data. In: TENCON 2009–2009 IEEE Region 10 Conference, pp. 1–4. IEEE, January 2009
16. Vaidya, J., Clifton, C.W., Zhu, Y.M.: Privacy Preserving Data Mining, vol. 19. Springer, Heidelberg (2006)
17. Yang, Y., Maode, M.A.: Semantic searchable encryption scheme based on lattice in quantum-era. J. Inf. Sci. Eng. **32**(2), 425–438 (2016)

Reweighting Forest for Extreme Multi-label Classification

Zhun-Zheng Lin and Bi-Ru Dai[(✉)]

Department of Computer Science and Information Engineering,
National Taiwan University of Science and Technology, No. 43, Section 4,
Keelung Road, Daan District, Taipei 106, Taiwan, ROC
M10415025@mail.ntust.edu.tw,
brdai@csie.ntust.edu.tw

Abstract. In recent years, data volume is getting larger along with the fast development of Internet technologies. Some datasets contain a huge number of labels, dimensions and data points. As a result, some of them cannot be loaded by typical classifiers, and some of them require very long and unacceptable time for execution. Extreme multi-label classification is designed for these challenges. Extreme multi-label classification differs from traditional multi-label classification in a number of ways including the need for lower execution time, training at an extreme scale with millions of data points, features and labels, etc. In order to enhance the practicality, in this paper, we focus on designing an extreme multi-label classification approach which can be performed on a single person-al computer. We devise a two-phase framework for dealing with the above issues. In the reweighting phase, the prediction precision is improved by paying more attention on hard-to-classify instances and increasing the diversity of the model. In the pretesting phase, trees with lower quality will be removed from the prediction model for reducing the model size and increasing the prediction precision. Experiments on real world datasets will verify that the pro-posed method is able to generate better prediction results and the model size is successfully shrunk down.

Keywords: Multi-label classification · Random forest · Extreme classification

1 Introduction

Classification is a well-known research in the machine learning field. Data instances with given class labels can be used to train classification models, or called classifiers. After that, these classifiers can be used to predict the labels of unlabeled testing data automatically. Classification is applied in many different areas, such as email spam detection, handwritten recognition, and DNA sequence identification. Various types of classifiers have been proposed, such as Bayesian Classifiers [18, 23], Decision Trees [19, 24], Artificial Neural Networks [21, 25], and Support Vector Machine [22, 26]. However, the design or improvement of classification methods is still a popular research topic.

© Springer International Publishing AG 2017
L. Bellatreche and S. Chakravarthy (Eds.): DaWaK 2017, LNCS 10440, pp. 286–299, 2017.
DOI: 10.1007/978-3-319-64283-3_21

As time goes by, the multi-label classification [10–14] which is a branch of classification techniques comes up. With the single-class classification techniques, the object is assigned to one class. Different from single-label classification, the object of multi-label classification is usually mapped to more than one class. Multi-label classification techniques are also useful in many real applications, such as article categorization, image/video annotation, and e-shopping recommendation. Most of typical classifiers designed for single-label classification can be used in multi-label classification problems in the one-vs-one or one-vs-rest form. In one-vs-one form, a set of classifiers, which are composed of two different labels, are built to predict labels in cascading. One-vs-rest methods predict labels with distinguishing each label from the rest labels. However, single-label classification approaches usually require more calculation and thus cost more time and space for dealing with multi-label classification problems.

In recent years, data volume is getting larger along with the fast development of Internet technologies. Some datasets contain a huge number of labels, dimensions and data points. As a result, some of them cannot be loaded by typical classifiers, and some of them require very long and unacceptable time for execution. Extreme multi-label classification is designed for these issues. For instance, a Wikipedia page can be tagged with a small set of relevant labels which are chosen from a large set consisting of more than a million possible tags in the collection. Google image search engine collects images on the web and automatically assigns tags to them for searching. Extreme multi-label classification differs from traditional multi-label classification in a number of ways including the need for lower execution time, training at an extreme scale with millions of data points, features and labels, etc. Most of extreme multi-label classification approaches are the tree based [1, 2, 20] or embedding based [3, 6, 7]. They use probability or approximation to save time and maintain accuracy. Although single-label classifiers cannot deal with extreme multi-label classification directly because of the huge amount of calculations, distributed systems and cloud computing provide these traditional classifiers a way for dealing with time, space, and calculation problems [4].

However, distributed systems and cloud computing are not available for some environments or users, such as outer space, deep sea, a new company, or students because of the limitation of Internet transmission, hardware restrictions, or limited budget. In these situations, building a large scale distributed system or using a cloud computing system is usually a heavy burden. In order to enhance the practicality, in this paper, we will focus on designing an extreme multi-label classification approach which can be performed on a single personal computer. FastXML [2] is the first approach which emphasizes on being performed on a single computer. PfastreXML [1] is a method modified from FastXML for unbiased label distribution. Our approach is based on the above tree based extreme multi-label classification and further improves the performance on the prediction precision and the storage space.

We propose a two-phase framework named **Reweighting Forest**, abbreviated as ReWF, for dealing with extreme multi-label classification problem. The training data is divided into two parts, the learning part and the pretesting part, to be used in two phases, the reweighting phase and the pretesting phase, respectively. In the reweighting phase, the learning part is taken to generate a forest by the tree based classification approach. The forest is built with many clumps of trees in cascading. Through majority voting to a clump, weights are given to all instances to generate the next clump. The learning of hard-to-classify instances is enhanced by giving larger weights. In the pretesting phase, the pretesting part of the training data is used to evaluate the quality of trees. Trees with lower quality are removed for reducing the consumption of model storage space. With this two-phase mechanism, our method provides higher prediction precision and requires less storage space.

Our contributions are summarized as follows:

1. We propose a reweighting algorithm. The prediction precision ability of hard-to-classify instances is enhanced. In addition, the diversity of the prediction model is improved by reweighting instances for constructing trees repeatedly. As a result, the overall accuracy of the prediction model is improved.
2. The required space and testing time is reduced by shrinking down the model size with the pretesting mechanism.
3. The improvements of our proposed method are verified by experiments on real world datasets.

The rest of this paper is organized as follows. In the next section, related works will be discussed. In Sect. 3, the proposed method and its detailed framework are introduced. In Sect. 4, the experiment environment and results are described, and the conclusion is shared in Sect. 5.

2 Related Work

In this section, we introduce related works of extreme multi-label classification. Extreme multi-label classification approaches can be divided into three types: embedding based [3, 6, 7], tree based [1, 2, 20], and others [4, 5].

Embedding based approaches exploit label correlations and sparsity to compress the number of labels. A low-dimensional embedding of the label space is found typically through a linear projection. Methods mainly differ in the choice of compression and decompression techniques. SLEEC [3] learns low-dimensional embedding which non-linearly capture label correlations by preserving the pairwise distances between only the closest label vectors. RobustXML [6] assumes the tail labels to be outliers and treats them separately in the overall optimization problem. Embedding based approaches usually have higher accuracy and more stable prediction results but spend longer training time.

Tree based approaches have faster predicting and training time because they divide the label or the feature space into small pieces recursively with different type of trees. Owing to the top-down cascading design, the prediction error made at a top-level cannot be corrected at lower levels. As a result, the accuracy of tree based approaches is usually lower. FastXML [2] optimizes an nDCG based ranking loss function. PfastreXML [1] modifies FastXML by providing an unbiased propensity scored loss function. LPSR [20] uses Gini index as a measure of performance. In summary, tree based approaches have faster training time and testing time with the costs of lower accuracy and unstable prediction results.

Finally, other state-of-the-art extreme multi-label classification approaches, which do not belong to the above two types, will be introduced. PD-Sparse [5] uses l_1. regularization along with multi-class loss instead of binary one-vs-rest loss. As a result, it needs to store the intermediary weight vectors during optimization, and hence does not scale to large datasets. DiSMEC [4] is a large-scale distributed framework for learning one-versus-rest linear classifiers coupled with explicit capacity control to control the model size. Distributed frameworks usually can achieve higher accuracy and scale up to huge datasets but require more costs to build the environment.

3 Proposed Method

In this section, we will describe our proposed method, **Reweighting Forest** (ReWF), in detail. This paper aims to allow extreme multi-label classification to be performed on a single personal computer. The problem definition and the proposed framework will be described in Sect. 3.1. Two phases, the reweighting phase and the pretesting phase, will be introduced in Sects. 3.2 and 3.3 respectively.

3.1 Problem Definition and Proposed Framework

In this subsection, the problem definition of extreme multi-label classification will be provided first. After that, the framework of the proposed method will be described (Table. 1).

Table 1. The summary of symbols used in this paper

Symbols	Descriptions	Symbols	Descriptions
D_{Train}	The multi-label training data set	C_j	The j th clump of forest
D_{Test}	The multi-label testing data set	c	The number of clumps
x_i	The i th instance	ω_i	The weight for the i th instance
D_t	The learning part separated from D_{Train}	k	The top k labels of ranking
D_p	The pretesting part separated from D_{Train}	T_{jn}	The t th decision tree in the j th clump
F	The forest composed of decision trees	T_t	The t th tree in prediction model
u	The number of trees in the prediction model	Q	The upper bound of trees in the prediction model
y_i	The label vector assigned to x_i	$\hat{y}_i$	The predicted label vector for x_i
v_i	The voted label vector for x_i	s	The number of trees in a clump

Problem Definition. Suppose all instances have d-dimension features and there are m possible labels. The training data set D_{Train} contains n_{Train} instances. For multi-label classification, it is assumed that training data set D_{Train} can be presented as $\left\{ (x_i, y_i)_{i=1}^{n_{Train}} \right\}$, where x_i is the i th training feature vector and $y_i \in \{0,1\}^m$ is the i th training label vector. For each y_{il} in the vector $y_i = (y_{i1}, y_{i2}, \cdots y_{im})$, $y_{il} = 1$ indicates that the ith instance is assigned the lth label and 0 otherwise. The task of multi-label classification is to learn a prediction model with D_{Train} to predict the unknown label set $\hat{y}_i$ for each instance x_i in testing data set D_{Test}.

Note that extreme multi-label classification usually needs a huge amount of time, memory, and storage space for training due to a large number of data points, labels, and high dimension of features. However, distributed system and cloud computing are not available for everyone to practice. To reduce the burden of implementing an extreme multi-label classification and enhance its practicality, an approach which is able to perform extreme multi-class classification on a single personal computer is proposed in this paper.

As discussed in Sect. 2, among various types of extreme multi-label classification approaches, tree based methods tend to have faster training time and require smaller memory usage but the accuracy of these approaches is usually lower and more unstable. In order to reduce the burden on execution time and storage space, we design our framework from tree based approaches. However, the first challenge is to improve accuracy without increasing execution time and storage space. On the other hand, models of tree based approaches are composed of trees. This characteristic gives us the opportunity to control the model size. Thus, the second challenge is to reduce the model size without sacrificing the quality of classification.

Accordingly, we propose a two-phase framework, ReWF, as shown in Fig. 1, to address these two challenges. At first, we will separate training data D_{Train} into the learning part D_t and the pretesting part D_p. In the reweighting phase, the forest F will be built as the prediction model. The forest F is composed of c clumps and these clumps will be generated in cascading. For each clump C_j, there are s decision trees inside. These trees are generated by picking up data points from D_t. After each clump C_j is built, instances of D_t will be evaluated and given weights to generate the next

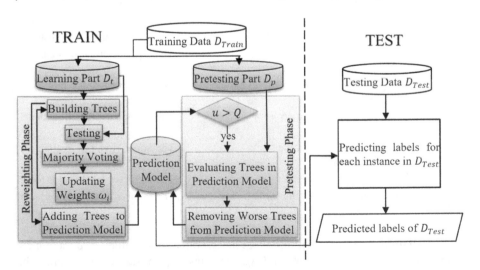

Fig. 1. The framework overview. Note that u is the total number of trees in the prediction model, and Q is a given upper bound of tree quantity.

clump C_{j+1} to enhance the learning of hard-to-classify instances. In the pretesting phase, trees which are generated from the reweighting phase will be added into the prediction model. When the number of trees in the prediction model is larger than a given upper bound, these trees will be tested with the pretesting part D_p and trees with worse quality will be removed until the number of trees is equal to the upper bound. Finally, the remaining forest will be the training model for the testing data.

In the following subsections, we will describe these two phases of our framework in detail individually.

3.2   The Reweighting Phase

As can be observed in real applications, labels usually follow a power law distribution in extreme multi-label learning applications. This phenomenon is illustrated in Fig. 2. Infrequently occurring labels usually have fewer training instance and are generally harder to be predicted than frequently occurring ones. However, sometimes they are more informative and rewarding. In some applications, recommending popular items is more valueless since users know about them already. In contrast, predicting rare items will be more desirable in these cases. These infrequently occurring labels are called tail labels [1, 3, 4, 6, 7]. In previous works, tail labels are handled in different ways. Some of them regard tail labels as outliers or noises [6]. Some of them provide additional learner to treat these labels [1]. Different from previous works, we use an indirect way to enhance the learning of tail labels in this paper.

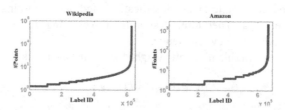

Fig. 2. The number of data points in which each label is present

Inspired by some works about forest weighting rule [8, 9], we devise a reweighting mechanism. [8] decreases weights of attributes which are involving in the already built trees to provide more diversity. [9] increases weights of instances which are hard to be classified by giving these instances more chance to be learned. In our work, we also enhance the learning of hard-to-classify instances. Similar to tail labels, these hard-to-classify instances are more informative than other instances. Besides, if we focus on labels which are hard to be classified, every instance assigned these labels will be enhance. However, it is worthless for easy-to-classify instance to be enhanced. On the other hand, hard-to-classify instances will be neglected if they are assigned label which is easy to be classified. Furthermore, labels which are assigned to the same instance usually are correlative. Focusing on instance will enhance the learning of label correlations. Hence, we design this method to pay more attention to hard-to-classify instances to further improve the precision by successfully classifying more of such instances.

In the reweighting phase, we use the learning part D_t as the input data for generating trees by tree based multi-label classification approaches Other than building the whole forest at the same time, we construct smaller clumps repeatedly. Given an upper bound of trees Q, the forest F will be built with c clumps in cascading. For each clump, some of instances in the training part D_t are picked to build tree by a random variable. Each instance is given a weight to represent the enhancement of increasing the chance for the instance to be picked. Initially, the weights for all instances are the same. For each clump C_j is generated, these trees are sent to the pretesting phase. Simultaneously, the label vectors of all instances in the training part D_t are predicted by each tree T_{jn}. For the nth tree of the jth clump, the predicted label vector of instance x_i is denoted as $\dot{y}_{ijn} \in \{0, 1\}^m$. The voted label vector v_i, which is the result of $\sum_{n=1}^{s} \dot{y}_{ijn}$, is the voting result of predicted labels for each instance x_i. We take the top k voted labels to build up the predicted label vector $\hat{y}_i$. After that, we use the predicted label vector $\hat{y}_i$ to calculate weight ω_i. Weight ω_i for each instance x_i will be updated and these weights will be used to generate the next clump. The reweighting phase, which generate a clump of forest according to the weights obtained from the previous clump, will be repeated until the entire forest F is generated. We follow a state-of-the-art tree based extreme multi-label classification method, PfastreXML [1], for generating trees. Different from PfastreXML, we provide a reweighting mechanism to enhance the learning of hard-to-classify instances. Figure 3(A) illustrate the tree generation and instance reweighting process, and the pseudo-code is shown in Table 3. The idea is inspired by [9], which builds a tiny forest first and then use majority voting to find out which are the hard to be classified instance and gives them weights by error rate. However, [9] is originally designed for single-label classification and its weighting function only accepts single-label data. Therefore, designing a weighting function to deal with multi-label data becomes a new challenge.

To deal with above problem, a new weight for an instance x_i is defined as follows:

$$\omega_i = 1 - P_i@k(v_i, y_i). \tag{1}$$

Here $P_i@k(v_i, y_i)$ is a precision function to measure which instance x_i is hard to be classified by the clump of trees according to the top k predicted labels. The precision function is defined as follows:

$$P_i@k(v_i, y_i) = \frac{1}{k} \sum_{l \in \text{rank}_k(v_i)} y_{il}. \tag{2}$$

Given the ground truth label vector $y_i \in \{0, 1\}^m$ and $\text{rank}_k(v_i)$ returns the indexes of the top k of v_i ranked in descending, the precision at k for a prediction $\hat{y}_i$ can be obtained by Eq. (2).

Example. Given three instances x_1, x_2, x_3 which are assigned the ground truth label vectors $y_1 = (1, 0, 0, 1, 1, 1)$, $y_2 = (1, 1, 1, 0, 0, 1)$, $y_3 = (0, 0, 1, 0, 0, 1)$ and the parameter is set to $k = 5$. After the majority voting, suppose the voted label vectors of x_1, x_2, x_3 are voted for $v_1 = (6, 5, 7, 3, 1, 4)$, $v_2 = (8, 5, 6, 9, 0, 1)$, $v_3 = (5, 6, 0, 1, 7, 2)$ respectively. The indexes of the top ranked 5 labels are sorted as $\text{rank}_5(v_1) = [3, 1, 2, 6, 4]$, $\text{rank}_5(v_2) = [4, 1, 3, 2, 6]$, $\text{rank}_5(v_3) = [5, 2, 1, 6, 4]$. Therefore, the predicted label vectors will be $\hat{y}_1 = (1,1,1,1,0,1)$, $\hat{y}_2 = (1,1,1,1,0,1)$, $\hat{y}_3 = (1,1,0,1,1,1)$.. According to Eq. (2), the precision of x_1, x_2, x_3 is calculated to be: $P_1@5 = \frac{1}{5} \times (0+1+0+1+1) = 0.6$, $P_2@5 = 4/5 = 0.8$, and $P_3@5 = 1/5 = 0.2$. Finally, the instances x_1, x_2, x_3 are assigned the weights $\omega_1 = (1 - 0.6) = 0.4$, $\omega_2 = 0.2$, and $\omega_3 = 0.8$. This example is summarized in Table 2.

For the example above, to data points with lower precision will be given larger

Table 2. Example of the weighting function with $k = 5$

	y_i	v_i	rank$_k(v_i)$	$\hat{y}_i$	$P_i@k(\hat{y}_i, y_i)$	ω_i
x_1	$(1,0,0,1,1,1)$	$(6,5,7,3,1,4)$	$[3,1,2,6,4]$	$(1,1,1,1,0,1)$	0.6	0.4
x_2	$(1,1,1,0,0,1)$	$(8,5,6,9,0,1)$	$[4,1,3,2,6]$	$(1,1,1,1,0,1)$	0.8	0.2
x_3	$(0,0,1,0,0,1)$	$(5,6,0,1,7,2)$	$[5,2,1,6,4]$	$(1,1,0,1,1,1)$	0.2	0.8

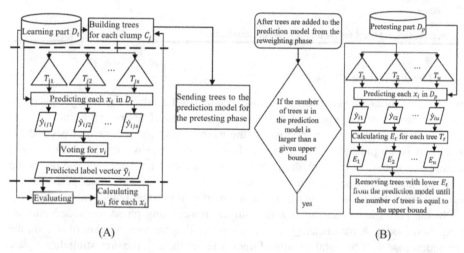

(A)   (B)

Fig. 3. Illustrations of (A) the reweighting phase and (B) the pretesting phase

weights to enhance the learning of these hard-to-classify instances. On the other hand, instances with higher precision, which are more successfully learned in this clump of trees, will be given low probabities to be selected when training the next clump of trees. Furthermore, if the precision of such instances is decreased in the next clump of trees, the weights of them will become higher again and they will have a larger chance to be learned in the following clump. This mechanism not only pays more attention on hard-to-classify instances, but also increases the diversity of the whole forest.

In this subsection, a reweighting mechanism for tree based extreme multi-label classification approaches is proposed. The learning of hard-to-classify instances is enhanced by adjusting weights of instances for each clump of trees. Each clump of trees will be sent to the pretesting phase to generate the prediction model. This method increases the diversity of the forest and the precision of the prediction model as a result.

Table 3. Algorithm of reweighting

Reweighting(c: the number of clumps to be generated, s: the number of trees in a clump, k: for predicting the top k labels, Q: the upper bound of trees)
For each clump C_j, j is from 1 to c
    If $j = 1$
        For each instance $x_i \epsilon D_t$
            Initialize $\omega_i = 1$
        End for
    End if
    For $n = 1$ to s
        $T_{jn} \leftarrow$ Train_trees(ω_i)
    End for
    Add $\{T_{j1}, \cdots, T_{js}\}$ to Prediction Model
    Call Pretesting(Q)
    For each instance $x_i \epsilon D_t$
        $P_i @ k(v_i, y_i) \leftarrow$ Majority_voting($\{T_{j1}, \cdots, T_{js}\}$)
        $\omega_i \leftarrow$ Eq.(1)
    End for
End For
Output: Prediction Model for the testing phase of classification

---

3.3 The Pretesting Phase

In this subsection, we provide a pretesting mechanism to reduce the storage space and keep the quality of the prediction model at the same time. The original training data is divided into two parts, where one part is the learning part for generating trees and the other part is the pretesting part. Different from the previous work [5], which uses the second part to correct the growth direction of its random parameters, we use pretesting part to evaluate the quality of trees. Through removing trees with lower performance, we can shrink down the size of the prediction model and further reduce the testing time.

In the pretesting phase, trees sent from the reweighting phase are added into the prediction model. A parameter Q is given for restricting the total number of trees in the prediction model. If the total amount of trees is larger than Q, the pretesting data D_p are used to check the quality of trees. To evaluate the quality of trees, we define an eliminating index E_t as follows:

$$E_t = \sum\nolimits_{x_i \in D_p} \text{dot}(\dot{y}_{it}, y_i). \tag{3}$$

Here t means the t th tree in the prediction model. $\dot{y}_{it} \in \{0, 1\}^m$ is the predicted label vector of the t th tree for the i th instance in D_p. y_i is the ground truth label vector of the i th instance. $\text{dot}(\dot{y}_{it}, y_i)$ returns the dot product of $\dot{y}_{it}$ and y_i. Accordingly, trees with smaller E_t, which are considered to have relatively lower performance than other trees, can be removed from the prediction model. Each clump of forest generated by the reweighting phase will be added to the prediction model. If the number of tree in the prediction model is larger than a given upper bound, these trees will be evaluated and pruned in the pretesting phase. Finally, the remaining trees will become a prediction model for testing unseen instances. Figure 3(B) illustrate the pretesting process, and the pseudo-code of pretesting is shown in Table 4.

Table 4. Algorithm of pretesting

Pretesting(Q: the upper bound of trees)
    u ←Count trees in Prediction Model
    If $u > Q$
        For each T_t
            E_t ←Test_tree($T_t, x_i \in D_p$)
        End for
        Loop for $u > Q$
            Remove T_t with the smallest E_t
            u--
        End loop
    End if
Output: Prediction Model for the testing phase of classification

---

In summary, a pretesting mechanism is proposed in this subsection. This method preserves trees with good performance as the prediction model for the testing phase of classification and provides comprehensive improvement on storage space, testing time, and prediction precision. These improvements will be verified by real datasets in Sect. 4.

4 Experiments

In this section, we present the experimental results. We will introduce the datasets and the setup used in the experiments first. In Sect. 4.2, the results of experiments will be discussed to evaluate ReWF against other methods.

4.1 Experimental Setup

We use eight datasets in the experiment: Mediamill [10], Bibtex [11], Delicious [12], EURLex-4K [13], AmazonCat-13K [14], Wiki10–31K [15], WikiLSHTC-325K [16], Amazon-670K [14]. All the datasets are publically available and can be downloaded from The Extreme Classification Repository [17]. The statistical information of these datasets is summarized in Table 5. Each dataset was partitioned into the training part and the testing part. For the training part, we further partitioned it into the learning part and the pretesting part.

The methods in this paper were implemented in C++. The experimental environment was Intel i7-6700, 3.40 GHz processors and 16 G of RAM. The number of total trees is set to 50. The clumps of forest C is set to 5. We use *precision@k*, a standard multi-label evaluation criterion, to measure the classification result. The metric *precision@k* is defined as follows:

$$precision@k = \frac{1}{n_{Test}} \sum\nolimits_{x_i \in D_{Test}} \left(\frac{1}{k} \sum\nolimits_{l \in \text{rank}_k(v_i)} y_{il} \right), \tag{4}$$

Table 5. Statistics of the datasets used in experiments. The label cardinality represents the average number of assigned labels for an instance, and the point cardinality represents the average number of points that a label being assigned.

Dataset	#Train	#Test	#Feature	#Label	Label card.	Point card.
Mediamill	30993	12914	120	101	4.38	1902.15
Bibtex	4880	2515	1836	159	2.40	111.71
Delicious	12920	3185	500	983	19.03	311.61
EURLex-4K	15539	3809	5000	3993	5.31	25.73
AmazonCat-13K	1186239	306782	203882	13330	5.04	448.57
Wiki10–31K	14146	6616	101938	30938	18.64	8.52
WikiLSHTC-325K	1778351	587084	1617899	325056	3.19	17.46
Amazon-670K	490449	153025	135909	670091	5.45	3.99

where $\text{rank}_k(v_i)$ returns the indexes the top k of v_i ranked in descending order, and there are n_{Test} instances in the testing data D_{Test}. In the experiments, we take $k = 1$, $k = 3$, $k = 5$ for the evaluation, and *precision@k* is abbreviated as P@k.

4.2   Experimental Results

In this subsection, the experimental results of the proposed method, ReWF, and other competitors will be illustrated. The relationship between the model sizes of tree based methods and the prediction precision will be discussed first, and then the overall performance comparison on various types of the multi-label classification approaches will be presented.

Relationship between Model Size and Precision. The relationship between precision and the model size for tree based extreme multi-label classification approaches is shown in Fig. 4. Two datasets, a smaller one and a larger one, are taken as an example because trends for all datasets are similar. As shown in Fig. 4, compared to other methods, our proposed method ReWF achieves the same quality with a smaller model size. In addition, when given the same model size, the precision of ReWF is higher because we remove trees with worse performance from the prediction model in the pretesting phase.

Discussion on the Classification Performance. Experiment results are summarized in Table 6. As discussed in Sect. 2, we take the state-of-the-art approaches of each type of methods, SLEEC [3], PD-Sparse [5], FastXML [2], PfastreXML [1], as our competitors. Methods requiring distributed systems are excluded because they are incompatible with our goal of using only one personal computer. The bold fonts represent the best results. Because PD-Sparse requires large memory space and it is not able to be performed on our experimental setting for some larger datasets, we use X to mark these entries. In Table 6, it is clearly observed that ReWF outperforms other competitors on most of real datasets, especially among tree based approaches. These results verified that with the enhancement of hard-to-classify instances and the increased diversity of the prediction model, the performance of ReWF is usually good in real world applications.

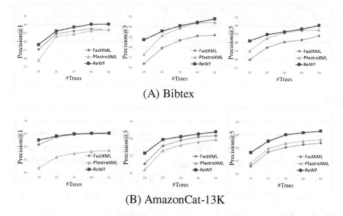

(A) Bibtex

(B) AmazonCat-13K

Fig. 4. The relationship between precision and the model size for tree based methods

Table 6. Results on datasets with *precision@k*

Datasets		Embedding	Other	Tree		
		SLEEC	PD-Sparse	FastXML	PfastreXML	ReWF
Mediamill	P@1	**87.12**	81.86	84.22	83.98	86.72
	P@3	**72.45**	62.52	67.33	67.37	70.23
	P@5	**58.67**	45.11	53.04	53.02	56.66
Bibtex	P@1	**65.08**	61.29	63.42	63.46	64.11
	P@3	39.64	35.82	39.23	39.22	**40.83**
	P@5	28.87	25.74	28.86	29.14	**30.08**
Delicious	P@1	67.59	51.82	69.61	67.13	**70.03**
	P@3	61.38	44.18	64.12	62.33	**64.42**
	P@5	56.56	38.95	59.27	58.62	**59.88**
EURLex-4K	P@1	**79.26**	76.43	71.36	75.45	78.14
	P@3	64.30	60.37	59.90	62.70	**64.36**
	P@5	52.33	49.72	50.39	52.51	**53.71**
AmazonCat-13K	P@1	90.53	X	**93.11**	91.75	93.08
	P@3	76.33		78.20	77.97	**78.42**
	P@5	61.52		63.41	63.68	**64.33**
Wiki10–31K	P@1	**85.88**	X	83.03	83.57	85.06
	P@3	**72.98**		67.47	68.61	70.35
	P@5	**62.70**		57.76	59.10	61.22
WikiLSHTC-325K	P@1	54.83	X	49.75	56.05	**56.54**
	P@3	33.42		33.10	36.79	**37.01**
	P@5	23.85		24.45	27.09	**28.32**
Amazon-670K	P@1	35.05	X	36.99	39.46	**40.10**
	P@3	31.25		33.28	35.81	**36.28**
	P@5	28.56		30.53	33.05	**33.68**

In summary, given only one personal computer, our proposed framework ReWF can generally obtain better results than other competitors. Simultaneously, ReWF can achieve the same quality as others with a smaller model size, and generate better precision with the same model size.

5 Conclusion

In this paper, we aimed to perform extreme multi-label classification methods on a single personal computer in order to enhance practicality. We devised a two-phase framework to improve tree based approaches. In the reweighting phase, prediction precision was increased by assigning larger weights to hard-to-classify instances. In the pretesting phase, trees with lower quality were removed from the prediction model for reducing the model size and increasing the prediction precision. Experiments on real world datasets verified that the proposed method generated better prediction results and the model size was successfully shrunk down.

References

1. Jain, H., Prabhu, Y., Varma, M.: Extreme multi-label loss functions for recommendation, tagging, ranking & other missing label applications. In: Proceedings of the 22nd ACM SIGKDD International Conference on Knowledge Discovery and Data Mining, pp. 935–944. ACM (2016)
2. Prabhu, Y., Varma, M.: FastXML: a fast, accurate and stable tree-classifier for extreme multi-label learning. In: Proceedings of the 20th ACM SIGKDD International Conference on Knowledge Discovery and Data Mining, pp. 263–272. ACM (2014)
3. Bhatia, K., Jain, H., Kar, P., Varma, M., Jain, P.: Sparse local embeddings for extreme multi-label classification. In: Advances in Neural Information Processing Systems, pp. 730–738 (2015)
4. Babbar, R., Shoelkopf, B.: DiSMEC: distributed sparse machines for extreme multi-label classification. In: Proceedings of the Tenth ACM International Conference on Web Search and Data Mining, pp. 721–729. ACM (2017)
5. Yen, I.E., Huang, X., Zhong, K., Ravikumar, P., Dhillon, I.S.: PD-sparse: a primal and dual sparse approach to extreme multiclass and multilabel classification. In: Proceedings of The 33rd International Conference on Machine Learning, pp. 3069–3077. IEEE (2016)
6. Xu, C., Tao, D., Xu, C.: Robust extreme multi-label learning. In: Proceedings of the 22nd ACM SIGKDD International Conference on Knowledge Discovery and Data Mining, pp. 1275–1284. ACM (2016)
7. Yu, H.F., Jain, P., Kar, P., Dhillon, I.S.: Large-scale multi-label learning with missing labels. In: Proceedings of The 31st International Conference on Machine Learning, pp. 593–601. IEEE (2014)
8. Adnan, Md.N., Islam, Md.Z.: Forest CERN: a new decision forest building technique. In: Bailey, J., Khan, L., Washio, T., Dobbie, G., Huang, J.Z., Wang, R. (eds.) PAKDD 2016. LNCS, vol. 9651, pp. 304–315. Springer, Cham (2016). doi:10.1007/978-3-319-31753-3_25

9. Adnan, Md N., Islam, Md Z.: On improving random forest for hard-to-classify records. In: Li, Jinyan, Li, Xue, Wang, Shuliang, Li, Jianxin, Sheng, Quan Z. (eds.) ADMA 2016. LNCS, vol. 10086, pp. 558–566. Springer, Cham (2016). doi:10.1007/978-3-319-49586-6_39

10. Snoek, C.G., Worring, M., Van Gemert, J.C., Geusebroek, J.M., Smeulders, A.W.: The challenge problem for automated detection of 101 semantic concepts in multimedia. In: Proceedings of the 14th Annual ACM International Conference on Multimedia, pp. 421–430. ACM (2006)

11. Katakis, I., Tsoumakas, G., Vlahavas, I: Multilabel text classification for automated tag suggestion. In: ECML/PKDD Discovery Challenge (2008)

12. Tsoumakas, G., Katakis, I., Vlahavas, I.: Effective and efficient multilabel classification in domains with large number of labels. In: ECML/PKDD 2008 Workshop on Mining Multidimensional Data, pp. 30–44 (2008)

13. Mencia, E.L., Fürnkranz, J.: Efficient pairwise multilabel classification for large-scale problems in the legal domain. In: Joint European Conference on Machine Learning and Knowledge Discovery in Databases, pp. 50–65 (2008)

14. McAuley, J., Leskovec, J.: Hidden factors and hidden topics: understanding rating dimensions with review text. In: Proceedings of the 7th ACM conference on Recommender systems, pp. 165–172. ACM (2013)

15. Zubiaga, A.: Enhancing navigation on wikipedia with social tags. Preprint (2012)

16. Partalas, I., Kosmopoulos, A., Baskiotis, N., Artieres, T., Paliouras, G., Gaussier, E., Androutsopoulos, I., Amini, M.-R., Galinari, P.: LSHTC: A benchmark for large-scale text classification. Preprint (2015)

17. The Extreme Classification Repository. http://research.microsoft.com/en-us/um/people/manik/downloads/XC/XMLRepository.html

18. John, G.H., Langley, P.: Estimating continuous distributions in Bayesian classifiers. In: Proceedings of the Eleventh Conference on Uncertainty in Artificial Intelligence, pp. 338–345 (1995)

19. Safavian, S.R., Landgrebe, D.: A survey of decision tree classifier methodology. IEEE Trans. Syst. Man Cybern. 21(3), 660–674 (1991)

20. Weston, J., Makadia, A., Yee, H.: Label partitioning for sublinear ranking. In: Proceeding of ICML, pp. 181–189 (2013)

21. Krizhevsky, A., Sutskever, I., Hinton, G.E.: Imagenet classification with deep convolutional neural networks. In: Advances in Neural Information Processing Systems, pp. 1097–1105 (2012)

22. Suykens, J.A., Vandewalle, J.: Least squares support vector machine classifiers. Neural Process. Lett. 9, 293–300 (1999)

23. Al Bataineh, M., Al-Qudah, Z.: A novel gene identification algorithm with Bayesian classification. In: Biomedical Signal Processing and Control, pp. 6–15 (2017)

24. Goodman, K.E., Lessler, J., Cosgrove, S.E., Harris, A.D., Lautenbach, E., Han, J.H., Tamma, P.D.: A clinical decision tree to predict whether a bacteremic patient is infected with an extended-spectrum β-Lactamase–producing organism. In: Clinical Infectious Diseases, pp. 896–903 (2016)

25. Esteva, A., Kuprel, B., Novoa, R.A., Ko, J., Swetter, S.M., Blau, H.M., Thrun, S.: Dermatologist-level classification of skin cancer with deep neural networks. Nature 542, 115–118 (2017)

26. Geng, Y., Chen, J., Fu, R., Bao, G., Pahlavan, K.: Enlighten wearable physiological monitoring systems: on-body RF characteristics based human motion classification using a support vector machine. IEEE Trans. Mobile Comput. 15, 656–671 (2016)

Social Media and Twitter Analysis

A Relativistic Opinion Mining Approach to Detect Factual or Opinionated News Sources

Erhan Sezerer[1] and Selma Tekir[2]([✉])

[1] Izmir Institute of Technology, 35430 Urla, Izmir, Turkey
erhansezerer@iyte.edu.tr
[2] Izmir Institute of Technology, 35430 Urla, Izmir, Turkey
selmatekir@iyte.edu.tr

Abstract. The credibility of news cannot be isolated from that of its source. Further, it is mainly associated with a news source's trustworthiness and expertise. In an effort to measure the trustworthiness of a news source, the factor of "is factual or opinionated" must be considered among others.

In this work, we propose an unsupervised probabilistic lexicon-based opinion mining approach to describe a news source as "being factual or opinionated". We get words' positive, negative, and objective scores from a sentiment lexicon and normalize these scores through the use of their cumulative distribution. The idea behind the use of such a statistical approach is inspired from the relativism that each word is evaluated with its difference from the average word. In order to test the effectiveness of the approach, three different news sources are chosen. They are editorials, New York Times articles, and Reuters articles, which differ in their characteristic of being opinionated. Thus, the experimental validation is done by the analysis of variance on these different groups of news. The results prove that our technique can distinguish the news articles from these groups with respect to "being factual or opinionated" in a statistically significant way.

1 Introduction

In today's continuous news flow by various news providers the following two questions have gained importance: What makes a news source credible? What are the most trustworthy news sources?

Journalism literature analyze news source credibility through the dimensions of trustworthiness and expertise Gaziano and McGrath (1986). The dimension of trustworthiness is further examined by the factors "is factual or opinionated", "is biased or unbiased", "does or does not separate fact and opinion", and "does or does not tell the whole story" Gaziano and McGrath (1986).

This work covers the factor of "is factual or opinionated" as part of an effort to measure the trustworthiness of news sources. To address this factor, we propose a novel unsupervised probabilistic lexicon-based opinion mining technique. For separating facts from opinions, we take our intuition from the relativism.

© Springer International Publishing AG 2017
L. Bellatreche and S. Chakravarthy (Eds.): DaWaK 2017, LNCS 10440, pp. 303–312, 2017.
DOI: 10.1007/978-3-319-64283-3_22

Our relativistic perspective has two implications; first, it can be said that objectivity is not something that can be computed on its own, rather it is the lack of any positive or negative sentiment. Second, for calculating negativity and positivity we should find how different a word is (how much more negative/positive) from the average word used in English. In order to find this relative strength of the words, we first get the sentiment (positive, negative, and objective) scores from a sentiment lexicon and normalize those scores by their cumulative distribution. Although the proposed technique is classified as a lexicon-based technique, the nature of calculating scores of words relativistically, allows this method to be applied along with any kind of sentiment analysis technique.

We have used three different news sources in our experimental setup: Editorials, New York Times articles, and Reuters articles. Editorials are opinion pieces that are written or evaluated by the editorial staff. The New York Times articles are the regular news articles and Reuters articles are mainly factual wire agency pieces. Therefore, each of these three sources demonstrates different level of opinionatedness. As an experimental hypothesis we claim that editorials are more opinionated than the regular New York Times articles and these regular articles are more opinionated than the articles of the Reuters Agency. Our experimental results prove this claim statistically.

Our contribution comes in two different ways. To our knowledge there isn't any work that examines the credibility of news sources through the use of fact/opinion ratio and second our method is the first method that uses Cumulative Distribution-based normalization in sentiment scores.

In the remaining part of the article, we first refer to existing literature and our contribution in the context. Section 3 describes our experimental setup covering the dataset, knowledge-base and preprocessing, and sentiment analysis parts. In Sect. 4, we present our experimental results. Finally, we conclude our work and state possible future research directions.

2   Related Work

Opinion mining, sentiment analysis, and/or subjectivity analysis deal with the computational treatment of opinion, sentiment, and subjectivity in text Pang and Lee (2012). By opinionated, it means that a document or sentence expresses or implies a positive or negative sentiment Liu (2012). Classifying an opinionated text as either positive or negative is termed as document-level sentiment-polarity classification problem Pang and Lee (2012).

The emerging area of contradiction analysis models and analyzes conflicting opinions. Subjectivity analysis is the general term that covers opinion mining, sentiment analysis, opinion aggregation, and contradiction analysis Tsytsarau and Palpanas (2012).

Our work is an unsupervised probabilistic lexicon-based opinion mining technique that is applied to news texts to determine their level of being factual or opinionated in an effort to measure the credibility of the providing news source.

Our related work is structured as follows: First, the different approaches of opinion mining is described. After that, similar work on news opinion mining is given.

In their survey on subjectivity analysis, Tsytsarau and Palpanas (2012) classifies the different approaches of opinion mining as machine learning and dictionary-based, the latter including the corpus statistics and semantic approaches.

A recent survey on multilingual sentiment analysis Dashtipour et al. (2016) goes over the state-of-the-art sentiment analysis techniques and compares them on two standard datasets both in terms of their accuracy and computational cost. According to the accuracy, their comparison states that Support Vector Machines (SVM) is the best supervised method. In the unsupervised category, on the other hand, semantic orientation method of Singh et al. (2013) outperforms the others.

The machine learning methods assign sentiment scores to documents based on a model learned from document feature vectors and their known sentiments. The document feature vectors are constructed either according to word presence or word frequencies. Pang et al. (2002) in their work "Thumbs up?" predicts the sentence-wise sentiments using the features mainly based on word presence by the classification algorithms of Naive Bayes, Maximum Entropy, and Support Vector Machines (SVM). Their results show that the predicted sentiment scores outperform the random baseline and SVM has the superior performance.

Unsupervised approaches to document-level sentiment analysis are based on determining the semantic orientation (SO) of specific phrases within the document. If the average SO of these phrases is above some predefined threshold the document is classified as positive and otherwise it is deemed negative. There are two main approaches to the selection of phrases: a set of predefined part-of-speech (POS) patterns can be used to select these phrases or a lexicon of sentiment words and phrases can be used Feldman (2013).

The dictionary-based methods take the sentiment of each word from a dictionary and aggregate the sentiment values of words through a weighting formula in order to calculate the sentiment value of a text segment. The recent work in this category uses different document sentiment aggregation formulas that make use of direct polarity scores from dictionaries.

The unsupervised semantic orientation method Singh et al. (2013) is a state-of-the-art lexicon-based method to classify documents (reviews) as either positive or negative. In doing so, they make use of the semantic orientation values of adjectives. They extract adjectives and aggregate the semantic orientation values of them for the document or review. If the value exceeds a given threshold, the review is tagged as positive, otherwise it's evaluated as negative.

The concept of news credibility with its subdimensions is introduced first by Gaziano and McGrath (1986). They define the credibility by trustworthiness and expertise of news source. Moreover, they decompose the dimension of trustworthiness into more concrete factors. Among those factors, "does or does not separate fact and opinion", and "is factual or opinionated" stand out.

Yu and Hatzivassiloglou (2003) uses a Naive Bayes classifier to distinguish articles under News and Business (facts) from articles under Editorial and Letter to the Editor (opinions). Their document-level classifier is evaluated with respect to article type labels (preassigned opinion and fact labels) that are part of the meta-data by the news provider. Our approach differs from this work in three important ways: First, we propose to separate facts from opinions as part of an effort to measure the credibility of news sources. Second, they use a supervised machine learning algorithm to separate facts from opinions at the document level using different features like the average semantic orientation score of the words. We use an unsupervised, statistical technique to make this separation at the corpus level. Last, our experimental design differs. We have three groups of article sets that are known to be different in being factual or opinionated. The experimental hypothesis is our technique is influential in identifying the degree of being factual or opinionated. Analysis of variance is applied to test this hypothesis.

Morinaga et al. (2002) and Bethard et al. (2004) create an opinion-indicator lexicon by looking for terms that tend to be associated more highly with subjective-genre newswire, such as editorials, than with objective-genre newswire Pang and Lee (2012).

3   Experimental Setup

In this section we describe our dataset and the details of our lexicon-based sentiment analysis technique.

3.1   Dataset

Our data consists of articles from three different sources: The New York Times' Editorials, The New York Times' articles Sandhaus (2008) and Reuters' articles Rose et al. (2002). The New York Times (NYT) corpus contains approximately 1.800.000 articles, published in New York Times between January 1st. 1987 and June 19th. 2007. The Reuters corpus has around 800.000 articles, published between July 20th, 1996 and July 19th, 1997.

The reason behind this dataset selection is, they have a distinguishing property that we will try to verify with our tests. While The New York Times is a news source that publishes regular news articles, Reuters is a news agency, and because of this Reuters articles are expected to be briefer and freer from opinion of the authors. On the other hand, The New York Times, although it can be very objective too, is more likely to have opinions. This assumption comes from the fact that, the aim of Reuters and such wire agencies is to provide information to other news sources, and to do so, it should be, and in most of the cases actually is, more fact-based and pure. As Fenby states Fenby (1986), objectivity is the philosophical basis for wire agencies and as these wire services are preselectors and preprocessors of news the emphasis had to be on hard facts rather than comment. Regular news sources, on the other hand, have a particular audience to address, so they tend to have more emotions and opinion on articles to conform with the general view of their audience. Although it is stated that there should be a difference between the two sources in terms of subjectivity, both sources are expected to be objectively written in contrast to the editorials which are written for the purpose of projecting the author's ideas.

By using these three types of data, we aimed to create an environment with three levels of objectivity: editorials, news articles and news agency articles where the objectivity increases from the first to the last.

3.2 Knowledge-Base and Preprocessing

In our work, for calculating words' score we used SentiWordNet Esuli and Sebastiani (2006), which is a sentiment dictionary that relies on the WordNet Miller (1995) dictionary for English words. Since the SentiWordNet scores individual words according to their POS tags, we used maximum-entropy (CMM) part-of-speech tagger of Stanford NLP Group Toutanova et al. (2003).

3.3 Sentiment Analysis

We have built this method on the idea of relativism by making two important analogies to it. First, it can be said that objectivity is not something that can be computed on its own, rather it is the lack of any positive or negative sentiment. Any kind of calculation on objectivity should be built relatively to the negative/positive scores. So, computing the objectivity scores without considering the negative or positive scores will not help us in achieving our goals. Second, calculating negativity and positivity is an important issue in itself. The analogy to relativity helps us particularly here. The notion of negative or positive sentiment comes from the fact that it is more negative/positive than the usual words used by people in ordinary contexts in daily life. There is no such thing as the universal notion of negativity or positivity. So, if we want to determine the negativity of a word, we should find how different it is (how much more negative) than the average word used in English. In order to accomplish that, we had to find the nature of our data and find what the properties of an average word are.

We made use of the Cumulative Distribution Function (CDF) Graphs in order to learn the aforementioned nature in our data. CDF of all words gave us the relative position of a word, which, in turn, gives us an insight of how much that word differs from the average word in dictionary. We extracted the score of every single word on SentiWordNet and using the word's score, we calculated the probability of being positive/negative as the probability of seeing a word with such a positivity/negativity score in a text by finding the underlying area of the curve in CDF. Table 1 shows a sample of outputs from the CDF graphs for each category. For instance, the word "speculation" has SentiWordNet scores of 0.675, 0.125, 0.250 for objectivity, positivity and negativity. Using the CDF based normalization, the scores for objective, positive and negative probabilities become approximately 0.15, 0.86, 0.90 respectively.

Table 1. Example of scores from CDF

SentiWordNet score	Objective prob. (%)	Positive prob. (%)	Negative prob. (%)
0	0	0	0
0.025	0.01	19.36	11.07
0.05	0.14	82.30	71.09
0.1	0.28	85.53	84.66
0.25	1.26	91.44	89.62
0.75	17.58	99.85	99.67
0.9	24.78	99.98	99.86
0.95	58.02	99.98	99.98
0.975	94.58	99.99	99.99
1	100	100	100

4   Experimental Results

We perform the sentimental measurements at the sentence-level, and then combine them to make the corpus-level decision of being factual or opinionated.

Table 2. Summary statistics of positive and negative scores at the corpus levels

	Editorials	NYT	Reuters
Mean of positive scores	0.32783	0.28363	0.22636
Median positive scores	0.32787	0.25194	0.23227
SD of positive scores	0.05323	0.09969	0.10252
IQR of positive scores	0.06816	0.12195	0.12154
Mean of negative scores	0.26938	0.20059	0.17295
Median of negative scores	0.26262	0.20885	0.17648
SD of negative scores	0.04395	0.08377	0.09523
IQR of negative scores	0.04807	0.10634	0.12238

To compute the degree of being factual/opinionated at the corpus levels, we calculated summary statistics of positive and negative scores for all the datasets. Table 2 shows the mean, median, standard deviation, and interquartile range (IQR) of sentences we get from the entire corpus. Mean-standard deviation and median-IQR values are useful pairs to observe the statistical characteristics. By the way, each sentence is taken into account independently without emphasizing which document they belong to.

If we take a look at the positive scores, we will see that the median score is gradually decreasing from 32.78% in editorials to 25.19% in NYT and finally

23.22% in Reuters, as expected. The same thing applies to negativity too. The probability of an average article to be negative is decreasing from 26.26% in editorials to 20.88% in NYT and finally to 17.64% in Reuters articles. There is almost 5% decrease in scores from editorial to NYT.

To test the significance of corpus-level positive and negative score differences, we applied analysis of variance on these different corpora. The analysis of variance (refer to Table 3) generates the F-score of negativity as 2370.54 and the F-score of positivity as 8765.38, and both of them lead to a p-value of 2×10^{-16}. So, we can reject the null hypothesis, stating that editorials, NYT articles and Reuters are equal in objectivity.

Table 3. Results of analysis of variance

	F	p-value
Positive scores	8765.38	2×10^{-16}
Negative scores	2370.54	2×10^{-16}

Furthermore, we applied Scheffé test to see how those three sources can be compared pairwise. By looking at Table 4, we can safely say that all three sources are fairly different from each other with p-scores lower than 0.05.

Table 4. Results of Scheffé tests

Comparison	F	p-value
Negative (Editorials vs. NYT)	7.99	3.38×10^{-4}
Negative (Editorials vs. Reuters)	15.65	1.59×10^{-7}
Negative (NYT vs. Reuters)	2355.56	0
Positive (Editorials vs. NYT)	2.82	5.98×10^{-2}
Positive (Editorials vs. Reuters)	14.85	3.55×10^{-7}
Positive (NYT vs. Reuters)	8751.78	0

As a comparison base, we selected the unsupervised semantic orientation method by (Singh et al. 2013) as according to a recent survey (Dashtipour et al. 2016) it's the best state-of-the-art lexicon-based method in terms of its accuracy.

We performed first the analysis of variance on the semantic orientation scores of editorials, regular New York Times articles, and Reuters articles and then applied the Scheffé test to decide on the pairwise differences on being opinionated. The summary statistics of the semantic orientation scores can be found in Table 5, and the analysis of variance and Scheffé test results can be observed in Table 6. The analysis of variance results show that the unsupervised semantic orientation method can make a separation among groups that are editorials,

Table 5. Summary statistics of semantic orientation scores at the corpus levels

	Editorials	NYT	Reuters
Mean of SO scores	11.06667	11.98192	3.012828
Median of SO scores	11	6	1
SD of SO scores	8.270985	17.17534	5.08865
IQR of SO scores	9.5	17	4

Table 6. Test results of semantic orientation scores

		F	p-value
Analysis of variance	SO scores	60088	2×10^{-16}
Scheffé tests	SO (Editorials vs. NYT)		0.784695
	SO (Editorials vs. Reuters)		0
	SO (NYT vs. Reuters)		0

regular New York Times articles and Reuters articles in a statistically significant way. As for Scheffé test's results, the method fails to separate editorials from regular New York Times articles in a statistically significant way. For this pairwise assessment, our method is better as it does make the separation in a statistically significant way.

Although the aim of this paper is to find the factual/opinionated ratio of news sources, we add an additional experiment to see how it will perform under different circumstances, namely the binary document polarity task using the standard product review dataset (Blitzer et al. 2007). We had an overall accuracy of 52% due to 96% in positive reviews and 10% in negative ones. Our implementation of the unsupervised semantic orientation method, on the other hand, produced an overall accuracy score of 62% attributed to 73% in positive reviews and 53% in the negatives. It seems sufficient to compare against this work as it is the best lexicon-based method for this task and in general the second best after SVM with respect to the accuracy (Dashtipour et al. 2016).

5   Conclusion

We have presented a technique to evaluate news sources' degree of being factual or opinionated in an effort to measure their credibility. The technique assigns positive or negative sentiment scores to words using their cumulative distributions in a sentiment lexicon (SentiWordNet).

As we consider the cumulative distribution of the words in the lexicon, the scores we get are already normalized. Comparisons are done at the news source (corpus) level and each sentence is taken into account independently without emphasizing which document it belongs to. The technique's unsupervised statistical nature allows it to be applied along with any kind of sentiment analysis technique whether it is a lexicon-based model or a machine learning based model.

6   Future Work

Although the results are convincing, any future attempt at this subject should consider the context in which words are written. It is expected to perform much better since the objectivity and sentiment of a word can change drastically depending on the context. Furthermore, finding the context and the subject of the article (finance, sports, politics etc.) is important for any kind of analysis because it is acceptable to assume that in different areas of news, emotions are expressed in different ways. So, in future, topic modeling should be added to this method to calculate the shifts in positivity/negativity and improve the results.

In our project, we focused only one of the factors that leads to the credibility of news sources. However, in order to get the bigger picture it is necessary to investigate other factors, too. In this regard, we are also planning to add a second factor, "biased or unbiased", to our analysis of trustworthiness to cover the credibility of news sources more.

Acknowledgments. This paper is based on work supported by the Scientific and Technological Research Council of Turkey (TÜBİTAK) under contract number 114E784.

References

Bethard, S., Yu, H., Thornton, A., Hatzivassiloglou, V., Jurafsky, D.: Automatic extraction of opinion propositions and their holders. In: Proceedings of the Spring Symposium on Exploring Attitude and Affect in Text, pp. 22–24. AAAI (2004)

Blitzer, J., Dredze, M., Pereira, F.: Domain adaptation for sentiment classification. In: ACL, pp. 187–205 (2007)

Dashtipour, K., Poria, S., Hussain, A., Cambria, E., Hawalah, A.Y.A., Gelbukh, A., Zhou, Q.: Multi-lingual sentiment analysis: state of the art and independent comparison of techniques. Cogn. Comput. 8, 757–771 (2016). 9415[PII] 27563360[pmid] Cognit Comput

Esuli, A., Sebastiani, F.: A publicly available lexical resource for opinion mining. In: Proceedings of the 5th Conference on Language Resources and Evaluation, LREC 2006, pp. 417–422 (2006)

Feldman, R.: Techniques and applications for sentiment analysis. Commun. ACM 56(4), 82–89 (2013)

Fenby, J.: The International News Services. Schocken Books, New York (1986)

Gaziano, C., McGrath, K.: Measuring the concept of credibility. Journal. Q. 63, 451–462 (1986)

Liu, B.: Sentiment analysis and opinion mining. Synth. Lect. Hum. Lang. Technol. 5(1), 1–167 (2012)

Miller, G.A.: WordNet: a lexical database for english. Commun. ACM 38, 39–41 (1995)

Morinaga, S., Yamanishi, K., Tateishi, K., Fukushima, T.: Mining product reputations on the web. In: Proceedings of the Eighth ACM SIGKDD International Conference on Knowledge Discovery and Data Mining, pp. 341–349. ACM, New York (2002)

Pang, B., Lee, L., Vaithyanathan, S.: Thumbs up? Sentiment classification using machine learning techniques. In: Proceedings of the Conference on Empirical Methods in Natural Language Processing, ACL2002, Association for Computational Linguistics, Stroudsburg, PA, pp. 79–86 (2002)

Pang, B., Lee, L., Vaithyanathan, S.: Opinion mining and sentiment analysis. Found Trends Inf. Retr. **2**(1–2), 1–135 (2012)

Rose, T., Stevenson, M., Whitehead, M.: The reuters corpus volume 1 - from yesterday's news to tomorrow's language resources. In: Proceedings of the 3rd International Conference on Language Resources and Evaluation, LREC 2002, pp. 827–832 (2002)

Sandhaus, E.: The new york times annotated corpus overview (2008). https://catalog. ldc.upenn.edu/docs/LDC2008T19/new_york_times_annotated_corpus.pdf. Accessed 16 Sept 2014

Singh, V.K., Piryani, R., Uddin, A., Waila, P., Marisha, R.: Sentiment analysis of textual reviews; Evaluating machine learning, unsupervised and SentiWordNet approaches. In: 2013 5th International Conference Knowledge and Smart Technology (KST), pp. 122–127 (2013)

Toutanova, K., Klei, D., Manning, C., Singer, Y.: Feature-rich part-of-speech tagging with a cyclic dependency network. In: Proceedings of HLT-NAACL, pp. 252–259 (2003)

Tsytsarau, M., Palpanas, T.: Survey on mining subjective data on the web. Data Min. Knowl. Discov. **24**(3), 478–514 (2012)

Yu, H., Hatzivassiloglou, V.: Towards answering opinion questions: separating facts from opinions and identifying the polarity of opinion sentences. In: Proceedings of the Conference on Empirical Methods in Natural Language Processing, pp. 129–136. Association for Computational Linguistics, Stroudsburg, PA (2003)

A Reliability-Based Approach for Influence Maximization Using the Evidence Theory

Siwar Jendoubi[1(✉)] and Arnaud Martin[2]

[1] LARODEC, ISG Tunis, University of Tunis, Avenue de la Liberté,
Cité Bouchoucha, 2000 Le Bardo, Tunisia
jendoubi.siwar@yahoo.fr
[2] DRUID, IRISA, University of Rennes 1, Rue E. Branly, 22300 Lannion, France
Arnaud.Martin@univ-rennes1.fr

Abstract. The influence maximization is the problem of finding a set of social network users, called influencers, that can trigger a large cascade of propagation. Influencers are very beneficial to make a marketing campaign goes viral through social networks for example. In this paper, we propose an influence measure that combines many influence indicators. Besides, we consider the reliability of each influence indicator and we present a distance-based process that allows to estimate the reliability of each indicator. The proposed measure is defined under the framework of the theory of belief functions. Furthermore, the reliability-based influence measure is used with an influence maximization model to select a set of users that are able to maximize the influence in the network. Finally, we present a set of experiments on a dataset collected from Twitter. These experiments show the performance of the proposed solution in detecting social influencers with good quality.

Keywords: Influence measure · Influence maximization · Theory of belief functions · Reliability · Social network

1 Introduction

The influence maximization problem has attracted a great attention in these last years. The main purpose of this problem is to find a set of influence users, S, that can trigger a large cascade of propagation. These users are beneficial in many application domains. A well-known application is the viral marketing. Its purpose is to promote a product or a brand through viral propagation through social networks. Then, several research works were introduced in the literature [1,8,14,21] trying to find an optimal set of influence users in a given social network. However, the quality of the detected influence users stills always an issue that must be resolved.

The problem of identifying influencers was first modeled as a learning problem by Domingos and Richardson [6] in 2001. Furthermore, they defined the customer's network value, *i.e.* "the expected profit from sales to other customers he may influence to buy, the customers those may influence, and so on recursively" [6].

© Springer International Publishing AG 2017
L. Bellatreche and S. Chakravarthy (Eds.): DaWaK 2017, LNCS 10440, pp. 313–326, 2017.
DOI: 10.1007/978-3-319-64283-3_23

Moreover, they considered the market to be a social network of customers. Later in 2003, Kempe *et al.* [14] formulated the influence problem as an optimization problem. Indeed, they introduced two influence maximization models: the *Independent Cascade Model* (ICM) and the *Linear Threshold Model* (LTM). These models estimate the expected propagation size, σ_M, of a given node or set of nodes through propagation simulation models. Besides, [14] proved the NP-Hardness of the maximization of σ_M. Then, they proposed the greedy algorithm to approximate the set of nodes that maximizes σ_M. ICM and LTM just need the network structure to select influencers. However, these solutions are shown in [8] to be inefficient to detect good influencers.

When studying the state of the art of the influence maximization problem, we found that most of existing works use only the structure of the network to select seeds. However, the position of the user in the network is not sufficient to confirm his influence. For example, he may be a user that was active in a period of time, then, he collected many connections, and now he is no longer active. Hence, the user's activity is an interesting parameter that must be considered while looking for influencers. Besides to the user's activity in the network, many other important influence indicators are not considered. Among these indicators, we found the sharing and tagging activities of network users. These activities allow the propagation of social messages from one user to another. Also, the tagging activity is a good indicator of the user's importance in the network. In fact, more he is tagged in others' posts, more he is important for them. Therefore, considering such influence behaviors will be very beneficial to improve the quality of selected seeds.

To resolve the influencers quality issue, many influence indicators must be used together to characterize the influence that exerts one user on another [10]. An influence indicator may be the number of neighbors, the frequency of posting in the wall, the frequency of neighbor's likes or shares, *etc.* Furthermore, a refined influence measure can be obtained through the fusion of two or more indicators. A robust framework of information fusion and conflict management that may be used in such a case is the framework of the theory of belief functions [23]. Indeed, this theory provides many information combination tools that are shown to be efficient [4,24] to combine several pieces of information having different and distinct sources. Other advantages of this theory are about uncertainty, imprecision and conflict management.

In this paper, we tackle the problem of influence maximization in a social network. More specifically, our main purpose is to detect social influencers with a good quality. For this goal, we introduce a new influence measure that combines many influence indicators and considers the reliability of each indicator to characterize the user's influence. The proposed measure is defined through the theory of belief functions. Another important contribution in the paper is that we use the proposed influence measure for influence maximization purposes. This solution allows to detect a set of influencers having a good quality and that can maximize the influence in the social network. Finally, a set of experiments is made on real data collected from Twitter to show the performance of the proposed solution against existing ones and to study the properties of the proposed influence measure.

This paper is organized as follows: related works are reviewed in Sect. 2. Indeed, we present some data-based works and existing evidential influence measures. Section 3 presents some basic concepts about the theory of belief functions. Section 4 is dedicated to explaining the proposed reliability-based influence measure. Section 5 presents a set of experiments showing the efficiency of the proposed influence measure. Finally, the paper is concluded in Sect. 6.

2   Related Works

The influence maximization is a relatively new research problem. Its main purpose is to find a set of k social users that are able to trigger a large cascade of propagation through the word of mouth effect. Since its introduction, many researchers have turned to this problem and several solutions are introduced in the literature [3,8,9,14,15]. In this section, we present some of these works.

2.1   Influence Maximization Models

The work of Kempe *et al.* [14] is the first to define the problem of finding influencers in a social network as a maximization problem. In fact, they defined the influence of a given user or set of users, S, as the expected number of affected nodes, $\sigma_M(S)$, *i.e.* nodes that received the message. Furthermore, they estimated this influence through propagation simulation models which are the *Independent Cascade Model* (ICM) and the *Linear Threshold Model* (LTM). Next, they used a greedy-based solution to approximate the optimal solution. Indeed, they proved the NP-Hardness of the problem.

In the literature, many works were conducted to improve the running time when considering ICM and LTM. Leskovec*et al.* [17] introduced the *Cost Effective Lazy Forward* (CELF) algorithm that is proved to be 700 times faster than the solution of [14]. Kimura and Saito [16] proposed the *Shortest-Path Model (SPM)* which is a special case of the ICM. Bozorgi*et al.* [2] considered the community structure, *i.e.* a community is a set of social network users that are connected more densely to each other than to other users from other communities [22,26], in the influence maximization problem.

The *Credit Distribution model* (CD) [8] is an interesting solution that investigates past propagation to select influence users. Indeed, it uses past propagation to associate to each user in the network an influence credit value. The influence spread function is defined as the total influence credit given to a set of users S from the whole network. The algorithm scans the data (past propagation) to compute the total influence credit of a user v for influencing its neighbor u. In the next step, the CELF algorithm [17] is run to approximate the set of nodes that maximizes the influence spread in the network.

2.2   Influence and Theory of Belief Functions

The theory of belief functions was used to measure the user's influence in social networks. In fact, this theory allows the combination of many influence indicators together. Besides, it is useful to manage uncertainty and imprecision. This section is dedicated to present a brief description of existing works that use the theory of belief functions for measuring or maximizing the influence.

An evidential centrality (EVC) measure was proposed by Wei *et al.* [25] and it was used to estimate the influence in the network. EVC is obtained through the combination of two BBAs defined on the frame {*high, low*}. The first BBA defines the evidential degree centrality and the second one defines the evidential strength centrality of a given node. A second interesting, work was also introduced to measure the evidential influence, it is the work of [7]. They proposed a modified EVC measure. It considers the actual node degree instead of following the uniform distribution. Furthermore, they proposed an extended version of the semi-local centrality measure [3] for weighted networks. Their evidential centrality measure is the combined BBA distribution of the modified semi-local centrality measure and the modified EVC. The works of [7,25] are similar in that, they defined their measures on the same frame of discernment, they used the network structure to define the influence.

Two evidential influence maximization models are recently introduced by Jendoubi *et al.* [10]. They used the theory of belief functions to estimate the influence that exerts one user on his neighbor. Indeed, their measure fuses several influence indicators in Twitter like the user's position in the network, the user's activity, *etc.* This paper is based on our previous work [10]. However, the novelty of this paper is that we not only combine many influence indicators to estimate the user's influence, but also we consider the reliability of each influence indicator in characterizing the influence.

3   Theory of Belief Functions

In this section, we present the *theory of belief functions*, also called *evidence theory* or *Dempster-Shafer theory*. It was first introduced by Dempster [4]. Next, the mathematical framework of this theory was detailed by Shafer in his book *"A mathematical theory of evidence"* [23]. This theory is used in many application domains like pattern clustering [5,19] and classification [11,12,18]. Furthermore, this theory is used for analyzing social networks and measuring the user's influence [7,10,25].

Let us, first, define the *frame of discernment* which is the set of all possible decisions:

$$\Omega = \{d_1, d_2, ..., d_n\} \tag{1}$$

The *mass* function, also called *basic belief assignment* (BBA), m^Ω, defines the source's belief on Ω as follows:

$$2^\Omega \to [0,1]$$
$$A \mapsto m(A) \tag{2}$$

such that $2^\Omega = \{\emptyset, \{d_1\}, \{d_2\}, \{d_1, d_2\}, ..., \{d_1, d_2, ..., d_n\}\}$. The set 2^Ω is called
power set, i.e. the set of all subsets of Ω. The value assigned to the subset
$A \subseteq \Omega$, $m(A)$, is interpreted as the source's support or belief on A. The BBA
distribution, m, must respect the following condition:

$$\sum_{A \subseteq \Omega} m(A) = 1 \qquad (3)$$

We call A focal element of m if we have $m(A) > 0$. The discounting procedure
allows to consider the reliability of the information source. Let $\alpha \in [0, 1]$ be our
reliability on the source of the BBA m, then the discounted BBA m^α is obtained
as follows:

$$\begin{cases} m^\alpha(A) &= \alpha.m(A), \forall A \in 2^\Omega \setminus \{\Omega\} \\ m^\alpha(\Omega) &= 1 - \alpha.(1 - m(\Omega)) \end{cases} \qquad (4)$$

The information fusion is important when we want to fuse many influence
indicators together in order to obtain a refined influence measure. Then, the
theory of belief functions presents several combination rules. The Dempster's
rule of combination [4] is one of these rules. It allows to combine two distinct
BBA distributions. Let m_1 and m_2 be two BBAs defined on Ω, Dempster's rule
is defined as follows:

$$m_{1 \oplus 2}(A) = \begin{cases} \dfrac{\sum_{B \cap C = A} m_1(B)m_2(C)}{1 - \sum_{B \cap C = \emptyset} m_1(B)m_2(C)}, & A \subseteq \Omega \setminus \{\emptyset\} \\ 0 & if\ A = \emptyset \end{cases} \qquad (5)$$

In the next section, we present some relevant existing influence measures and
influence maximization models.

4   Reliability-Based Influence Maximization

In this paper, we propose an influence measure that fuses many influence indicators. Furthermore, we assume that these indicators may do not have the same
reliability in characterizing the influence. Then, some indicators may be more
reliable than the others. In this section, we present the proposed reliability-based
influence measure, the method we use to estimate the reliability of each indicator
and the influence maximization model we use to maximize the influence in the
network.

4.1   Influence Characterization

Let $G = (V, E)$ be a social network, where V is the set of nodes such that
$u, v \in V$ and E is the set of links such that $(u, v) \in E$. To estimate the amount
of influence that exerts one user, u, on his neighbor, v, we start first by defining

a set of influence indicators, $I = \{i_1, i_2, \ldots, i_n\}$ characterizing the influence. These indicators may differ from a social network to another. We note that we are considering quantitative indicators. Let us take Twitter as example, we can define the following three indicators: (1) the number of common neighbors between u and v, (2) the number of times v mentions u in a tweet, (3) the number of times v retweets from u.

In the next step, we compute the value of each defined indicator for each link (u, v) in the network. Then, (u, v) will be associated with a vector of values. In a third step, we need to normalize each computed value to the range $[0, 1]$. This step is important as it puts all influence indicators in the same range.

In this stage, we have a vector of values of the selected influence indicators:

$$W_{(u,v)} = \left(i_{(u,v)_1} = w_1, i_{(u,v)_2} = w_2, \ldots, i_{(u,v)_n} = w_n\right) \tag{6}$$

The elements of $W_{(u,v)}$ are in the range $[0, 1]$, i.e. $w_1, w_2, \ldots w_n \in [0, 1]$, and we define a vector $W_{(u,v)}$ for each link (u, v) in the network. Next, we estimate a BBA for each indicator value and for each link. Then, if we have n influence indicators, we will obtain n BBA to model each of these indicators for a given link. Let us first, define $\Omega = \{I, P\}$ to be the frame of discernment, where I models the influence and P models the passivity of a given user. For a given link (u, v) and a given influence indicator $i_{(u,v)_j} = w_j$, we estimate its BBA on the fame Ω as follows:

$$m_{(u,v)_j}(I) = \frac{w_j - \min_{(u,v) \in E}\left(i_{(u,v)_j}\right)}{\max_{(u,v) \in E}\left(i_{(u,v)_j}\right) - \min_{(u,v) \in E}\left(i_{(u,v)_j}\right)} \tag{7}$$

$$m_{(u,v)_j}(P) = \frac{\max_{(u,v) \in E}\left(i_{(u,v)_j}\right) - w_j}{\max_{(u,v) \in E}\left(i_{(u,v)_j}\right) - \min_{(u,v) \in E}\left(i_{(u,v)_j}\right)} \tag{8}$$

After this step, the influence that exerts a user u on his neighbor v is characterized by a set of n influence BBAs. In the next section, we present the method we use to estimate the reliability of each defined BBA.

4.2   Estimating Reliability

The selected influence indicators may do not have the same reliability in characterizing the user's influence. Then, we estimate the reliability, α_j, of each influence indicator. We assume that *"the farthest from the others the indicator is, the less reliable it is"*. For that purpose, we follow the approach introduced by Martin et al. [20] to estimate reliability. Besides, we note that this operator considers our assumption. In this section we detail the steps of [20] operator we used to estimate the reliability of each influence indicator in this paper.

Let us consider the link (u, v), we have a set of n BBAs to characterize the chosen influence indicators, $\left(m_{(u,v)_1}, m_{(u,v)_2}, \ldots, m_{(u,v)_n}\right)$. Our purpose is to estimate the reliability of each indicator against the others. To estimate the reliability, α_j, of the BBA $m_{(u,v)_j}$, we start by computing the distance between $m_{(u,v)_j}$ and each BBA from the rest of $n-1$ BBAs that characterizes the influence of u on v, i.e. $\left(m_{(u,v)_1}, m_{(u,v)_2}, \ldots, m_{(u,v)_{j-1}}, m_{(u,v)_{j+1}}, \ldots, m_{(u,v)_n} \right)$:

$$\delta_i^j = \delta(m_j, m_i) \tag{9}$$

To estimate these distances, we can use the *Jousselme* distance [13] as follows:

$$\delta(m_j, m_i) = \sqrt{\frac{1}{2}(m_j - m_i)^T \underline{\underline{D}} (m_j - m_i)} \tag{10}$$

such that $\underline{\underline{D}}$ is an $2^N \times 2^N$ matrix, $N = |\Omega|$ and $D(A,B) = \frac{|A \cap B|}{|A \cup B|}$.

Next, we compute the average of all obtained distance values as follows:

$$C_j = \frac{\delta_j^1 + \delta_j^2 + \ldots + \delta_j^n}{n-1} \tag{11}$$

such that $(\delta_1, \delta_2, \ldots, \delta_n)$ are the distance values between $m_{(u,v)_j}$ and $\left(m_{(u,v)_1}, m_{(u,v)_2}, \ldots, m_{(u,v)_n} \right)$, $(\delta(m_j, m_j) = 0)$. We use the average distance C_j to estimate the reliability, α_j, of the jth influence indicator in characterizing the influence of u on v as follows:

$$\alpha_j = f(C_j) \tag{12}$$

where f is a decreasing function. The function f can be defined as [20]:

$$\alpha_j = \left(1 - (C_j)^\lambda \right)^{1/\lambda} \tag{13}$$

where $\lambda > 0$.

After applying all these steps, we obtain the estimated value of the BBA reliability, α_j. To consider this reliability, we apply the discounting procedure described in Eq. (4). Then, we apply these steps for all defined BBAs on every link in the network.

4.3   Influence Estimation

After discounting all BBAs of each link in the network, we use them to estimate the influence that exerts one user on his neighbor. For this purpose, let us consider the link (u,v) and its discounted set of BBAs $\left(m_{(u,v)_1}^{\alpha_1}, m_{(u,v)_2}^{\alpha_2}, \ldots, m_{(u,v)_n}^{\alpha_n} \right)$. We define the global influence BBA that exerts u on v to be the BBA that fuses all discounted BBAs defined on (u,v). For this aim, we use the Dempster's rule of combination (see Eq. (5)) to combine all these BBAs as follows:

$$m_{(u,v)} = m_{(u,v)_1}^{\alpha_1} \oplus m_{(u,v)_2}^{\alpha_2} \oplus \ldots \oplus m_{(u,v)_n}^{\alpha_n} \tag{14}$$

The BBA distribution $m_{(u,v)}$ is the result of this combination.

Consequently, we define the influence that exerts u on v to be the amount of belief given to $\{I\}$ as:

$$Inf(u,v) = m_{(u,v)}(I) \tag{15}$$

The novelty of this evidential influence measure is that it considers several influence indicators in a social network and it takes into account the reliability of each defined indicator against the others. Our evidential influence measure can be considered as a generalization of the evidential influence measure introduced in the work of Jendoubi et al. [10].

To maximize the influence in the network, we need to define the amount of influence that exerts a set of nodes, S, on the hole network. It is the total influence given to S for influencing all users in the network. Then, we estimate the influence of S on a user v as follows [10]:

$$Inf(S,v) = \begin{cases} 1 & if\ v \in S \\ \displaystyle\sum_{u \in S} \sum_{x \in IN(v) \cup v} Inf(u,x).Inf(x,v) & Otherwise \end{cases} \tag{16}$$

where $Inf(v,v) = 1$ and $IN(v)$ is the set of in-neighbors of v, i.e. if (u,v) is a link in the network then u is an in-neighbor of v. Next, we define the influence spread function that computes the amount of influence of S on the network as follows:

$$\sigma(S) = \sum_{v \in V} Inf(S,v) \tag{17}$$

To maximize the influence that exerts a set of users S on the network, we need to maximize $\sigma(S)$, i.e. $\underset{S}{\mathrm{argmax}}\, \sigma(S)$. The influence maximization under the evidential model is demonstrated to be NP-Hard. Furthermore, the function, $\sigma(S)$, is monotone and submodular. Proof details can be found in [10]. Consequently, a greedy-based solution can perform a good approximation of the optimal influence users set S. In such cases, the cost effective lazy-forward algorithm (CELF) [17] is an adaptable maximization algorithm. Besides, it needs only two passes of the network nodes and it is about 700 times faster than the greedy algorithm. More details about CELF-based solution used in this paper can be found in [10].

After the definition of the reliability-based evidential influence measure and the influence spread function, we move to the experiments. Indeed, we made a set of experiments on real data to show the performance of our solution.

5   Results and Discussion

This section is dedicated to the experiments. In fact, we crawled Twitter data for the period between the 08-09-2014 and 03-11-2014. Table 1 presents some statistics of the dataset.

To characterize the influence users on Twitter, we choose the following three influence indicators: (1) the number of common neighbors between u and v, (2)

Table 1. Statistics of the data set [10]

Nbr of users	Nbr of tweets	Nbr of follows	Nbr of retweets	Nbr of mentions
36274	251329	71027	9789	20300

the number of times v mentions u in a tweet, (3) the number of times v retweets from u. Next, we apply the process described above in order to estimate the amount of influence that exerts each user u on his neighbor v in the network.

To evaluate the proposed reliability-based solution, we compare the proposed solution to the evidential model of Jendoubi *et al.* [10]. Furthermore, we choose four comparison criteria to compare the quality of the detected influence users by each experimented model. Those criteria are the following: (1) number of accumulated follow, (2) number of accumulated mention, (3) number of accumulated retweet, (4) number of accumulated tweet. Indeed, we assume that an influence user with a good quality is a highly followed user, mentioned and retweeted several times and active in terms of tweets.

In a first experiment, we compare the behavior of the proposed measure with fixed values of indicator reliability. Figure 1 presents the obtained results for two fixed values of $\alpha_j = \alpha$, which are $\alpha_j = \alpha = 0$ and $\alpha_j = \alpha = 0.2$. In Fig. 1 we have a comparison according to the four criteria, *i.e.* #Follow, #Mention, #Retweet and #Tweet shown in the y-axis of the sub-figures. We compare the experimented measure using the set of selected seed for each value of α_j. Besides, we fixed the size of the set S of selected influence users to 50 influencers, *i.e.* shown in the x-axis of each sub-figure.

According to Fig. 1, we notice that when $\alpha = 0$ (red scatter plots), the proposed reliability-based model does not detect good influencers according to the four comparison criteria. In fact, the red scatter plot ($\alpha = 0$) is very near to the x-axis in the case of the four criteria, which means that the detected influencer are neither followed, nor mentioned, nor retweeted. Besides, they are not very active in terms of tweets. However, we see a significant improvement especially when $\alpha = 0.2$ (blue scatter plots). Indeed, the detected influencers are highly followed as they have about 14k accumulated followers in total. Besides, the model detected some highly mentioned and retweeted influencers, especially starting from the 25th detected influencer. Finally, the influence users selected when $\alpha = 0.2$ are more active in terms of tweets than those selected when $\alpha = 0$.

This first experiment shows the importance of the reliability parameter, α, in detecting influencers with good quality. In fact, we see that when we consider that all indicators are totally reliable in characterizing the influence (the case when $\alpha = 0$), we notice that the proposed model detects influencers with very bad quality. However, when we reduce this reliability ($\alpha = 0.2$) we notice some quality improvement.

In a second experiment, we used the process described in Sect. 4.2 to estimate the reliability of each BBA in the network. Then, each BBA in our network is discounted using its own estimated reliability parameter. We note that the

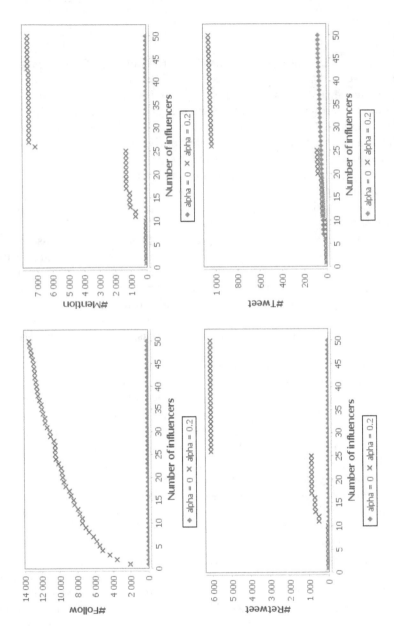

Fig. 1. Comparison of the reliability-based evidential model with three α fixed value (Color figure online)

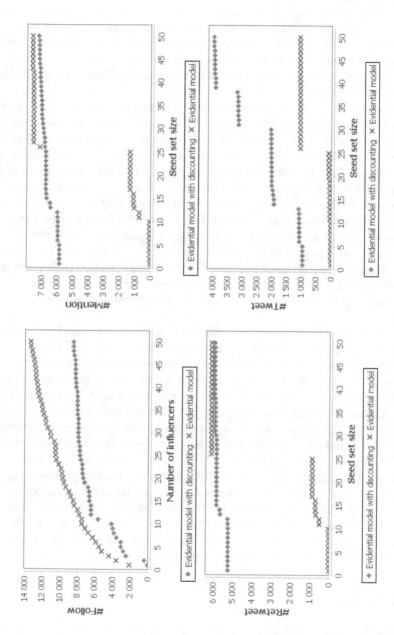

Fig. 2. Comparison between the proposed reliability-based solution and the evidential model of [10]

parameter λ in Eq. (13) was fixed to $\lambda = 5$. To fix this value, we made a set of experiments with different values of λ and the best results are given with $\lambda = 5$. Furthermore, we compare our reliability-based evidential model (also called evidential model with discounting) to the evidential model proposed by [10]. In fact, this last model is the nearest in its principle to the proposed solution in this paper. Besides, we fixed the size of the set S of selected influence users to 50 influencers. Figure 2 presents a comparison between the two experimented models in terms of #Follow, #Mention, #Retweet and #Tweet (shown in the y-axis of the sub-figures).

According to Fig. 2, we note that the two experimented models detect good influencers (shown in the x-axis). However, we see that the best compromise between the four criteria is given by the proposed reliability-based evidential model. In terms of accumulated #Follow, we notice that the most followed influencers are detected by the evidential model, also, our reliability-based model detected highly followed influencers. In terms of #Mention, we see that the evidential model starts detecting some mentioned influencers after detecting about 10 users that are not mentioned. However, the proposed reliability-based model starts detecting highly mentioned users from the first detected influencer. Furthermore, we see a similar behavior in the sub-figure showing the accumulated #Retweet. Indeed, the proposed solution detects highly retweeted influencers from the first user. In the last sub-figure that presents the comparison according to the accumulated #Tweet, the best results are those of the proposed reliability-based model.

This second experiment shows the effectiveness of the proposed reliability-based influence measure against the evidential influence measure of [10]. In fact, the best influence maximization model is always the model that detects the best influencers at first. Indeed, in an influence maximization problem we need generally to minimize the number of selected influencers in order to minimize the cost. For example, this is important in a viral marketing campaign as it helps the marketer to minimize the cost of his campaign and to maximize his benefits. Furthermore, our influence maximization solution gives the best compromise between the four criteria, *i.e.* #Follow, #Mention, #Retweet and #Tweet.

From these experiments we can conclude that the reliability parameter is important if we want to measure the influence in a social network through the consideration of several influence indicators. Indeed, we may have some influence indicators that are more reliable in characterizing the influence of the others. Furthermore, the proposed solution is efficient in detecting influencers with a good compromise between all chosen influence indicators.

6   Conclusion

To conclude, this paper introduces a new reliability-based influence measure. The proposed measure fuses many influence indicators in the social network. Furthermore, it can be adapted for several social networks. Another important contribution of the paper is that we consider the reliability of each chosen influence indicator to characterize the influence. Indeed, we propose to apply a distance-based operator that estimates the reliability of each indicator against to the others and considers the assumption that *"the farthest from the others the indicator is, the less reliable it is"*. Besides, we use the proposed reliability-based measure with an existing influence maximization model. Finally, we present two experiments that show the importance of the reliability parameter and the effectiveness of the reliability-based influence maximization model against the evidential model of Jendoubi *et al.* [10]. Indeed, we obtained a good compromise in the quality of detected influencers and we had good results according to the four influence criteria, *i.e.* #Follow, #Mention, #Retweet and #Tweet.

For future works, we will search to test our influence maximization solution on other social networks. Then, we will collect more data from Facebook and Google Plus and we will prove the performance of the proposed reliability-based influence maximization model. A second important perspective is about the influence maximization within communities. In fact, social networks are generally characterized by a community structure. Then, the main idea is to take profit from this characteristic and to search to select a minimum number of influence users and to minimize the time spent to detect them.

References

1. Aslay, C., Barbieri, N., Bonchi, F., Baeza-Yates, R.: Online topic-aware influence maximization queries. In: Proceedings of the 17th International Conference on Extending Database Technology (EDBT), pp. 24–28, March 2014
2. Bozorgi, A., Haghighi, H., Zahedi, M.S., Rezvani, M.: INCIM: A community-based algorithm for influence maximization problem under the linear threshold model. Inf. Process. Manage. **000**, 1–12 (2016)
3. Chen, D., Lü, L., Shang, M.S., Zhang, Y.C., Zhou, T.: Identifying influential nodes in complex networks. Physica A: Stat. Mech. Appl. **391**(4), 1777–1787 (2012)
4. Dempster, A.P.: Upper and Lower probabilities induced by a multivalued mapping. Ann. Math. Stat. **38**, 325–339 (1967)
5. Denœux, T., Sriboonchitta, S., Kanjanatarakul, O.: Evidential clustering of large dissimilarity data. Knowl.-Based Syst. **106**, 179–195 (2016)
6. Domingos, P., Richardson, M.: Mining the network value of customers. In: Proceedings of KDD 2001, pp. 57–66 (2001)
7. Gao, C., Wei, D., Hu, Y., Mahadevan, S., Deng, Y.: A modified evidential methodology of identifying influential nodes in weighted networks. Physica A **392**(21), 5490–5500 (2013)
8. Goyal, A., Bonchi, F., Lakshmanan, L.V.S.: A data-based approach to social influence maximization. In: Proceedings of VLDB Endowment, pp. 73–84, August 2012
9. Jendoubi, S., Martin, A., Liétard, L., Ben Hadj, H., Ben Yaghlane, B.: Maximizing positive opinion influence using an evidential approach. In: Proceedings of the 12th International FLINS Conference, August 2016
10. Jendoubi, S., Martin, A., Liétard, L., Hadj, H.B., Yaghlane, B.B.: Two evidential data based models for influence maximization in twitter. Knowl.-Based Syst. **121**, 58–70 (2017)

11. Jendoubi, S., Martin, A., Liétard, L., Ben Yaghlane, B.: Classification of message spreading in a heterogeneous social network. In: Laurent, A., Strauss, O., Bouchon-Meunier, B., Yager, R.R. (eds.) IPMU 2014. CCIS, vol. 443, pp. 66–75. Springer, Cham (2014). doi:10.1007/978-3-319-08855-6_8
12. Jendoubi, S., Martin, A., Liétard, L., Ben Yaghlane, B., Ben Hadji, H.: Dynamic time warping distance for message propagation classification in twitter. In: Destercke, S., Denoeux, T. (eds.) ECSQARU 2015. LNCS (LNAI), vol. 9161, pp. 419–428. Springer, Cham (2015). doi:10.1007/978-3-319-20807-7_38
13. Jousselme, A.L., Grenier, D., Bossé, E.: A new distance between two bodies of evidence. Inf. Fusion **2**, 91–101 (2001)
14. Kempe, D., Kleinberg, J., Tardos, E.: Maximizing the spread of influence through a social network. In: Proceedings of KDD 2003, pp. 137–146, August 2003
15. Kempe, D., Kleinberg, J., Tardos, É.: Influential nodes in a diffusion model for social networks. In: Caires, L., Italiano, G.F., Monteiro, L., Palamidessi, C., Yung, M. (eds.) ICALP 2005. LNCS, vol. 3580, pp. 1127–1138. Springer, Heidelberg (2005). doi:10.1007/11523468_91
16. Kimura, M., Saito, K.: Tractable models for information diffusion in social networks. In: Fürnkranz, J., Scheffer, T., Spiliopoulou, M. (eds.) PKDD 2006. LNCS (LNAI), vol. 4213, pp. 259–271. Springer, Heidelberg (2006). doi:10.1007/11871637_27
17. Leskovec, J., Krause, A., Guestrin, C., Faloutsos, C., VanBriesen, J., Glance, N.: Cost-effective outbreak detection in networks. In: Proceedings of KDD 2007, pp. 420–429, August 2007
18. Liu, Z., Pan, Q., Dezert, J., Martin, A.: Adaptive imputation of missing values for incomplete pattern classification. Pattern Recogn. **52**, 85–95 (2016)
19. Liu, Z., Pan, Q., Dezert, J., Mercier, G.: Credal c-means clustering method based on belief functions. Knowl.-Based Syst. **74**, 119–132 (2015)
20. Martin, A., Jousselme, A.L., Osswald, C.: Conflict measure for the discounting operation on belief functions. In: International Conference on Information Fusion, Cologne, Germany, pp. 1003–1010, juillet 2008
21. Mohamadi-Baghmolaei, R., Mozafari, N., Hamzeh, A.: Trust based latency aware influence maximization in social networks. Eng. Appl. Artif. Intell. **41**, 195–206 (2015)
22. Mumu, T.S., Ezeife, C.I.: Discovering community preference influence network by social network opinion posts mining. In: Bellatreche, L., Mohania, M.K. (eds.) DaWaK 2014. LNCS, vol. 8646, pp. 136–145. Springer, Cham (2014). doi:10.1007/978-3-319-10160-6_13
23. Shafer, G.: A Mathematical Theory of Evidence. Princeton University Press, Princeton (1976)
24. Smets, P., Kennes, R.: The transferable belief model. Artif. Intell. **66**, 191–234 (1994)
25. Wei, D., Deng, X., Zhang, X., Deng, Y., Mahadeven, S.: Identifying influential nodes in weighted networks based on evidence theory. Physica A **392**(10), 2564–2575 (2013)
26. Zhou, K., Martin, A., Pan, Q., Liu, Z.: Median evidential c-means algorithm and its application to community detection. Knowl.-Based Syst. **74**, 69–88 (2015)

Sentiment Analysis on Twitter to Improve Time Series Contextual Anomaly Detection for Detecting Stock Market Manipulation

Koosha Golmohammadi[✉] and Osmar R. Zaiane

University of Alberta, Edmonton, AB, Canada
{golmoham,zaiane}@ualberta.ca

Abstract. In this paper, We propose a formalized method to improve the performance of Contextual Anomaly Detection (CAD) for detecting stock market manipulation using Big Data techniques. The method aims to improve the CAD algorithm by capturing the expected behaviour of stocks through sentiment analysis of tweets about stocks. The extracted insights are aggregated per day for each stock and transformed to a time series. The time series is used to eliminate false positives from anomalies that are detected by CAD. We present a case study and explore developing sentiment analysis models to improve anomaly detection in the stock market. The experimental results confirm the proposed method is effective in improving CAD through removing irrelevant anomalies by correctly identifying 28% of false positives.

1 Introduction

Market capitalization exceeded $1.5 trillion in Canada and $25 trillion in USA in 2015[1] (GDP of Canada and USA in 2015 were $1.5 and $17 trillion respectively). Protecting market participants from fraudulent practices and providing a fair and orderly market is a challenging task for regulators. 233 individuals and 117 companies were prosecuted in 2015, resulting in over $138 million in fines, compensation, and disgorgement in Canada. However, the effect of fraudulent activities in securities markets and financial losses caused by such practices is far greater than these numbers suggest as they impact public and market participants trust. Market manipulation and price rigging remain the biggest concerns of investors in today's market, despite fast and strict responses from regulators and exchanges to market participants that pursue such practices. Market manipulation is forbidden in Canada[2] and the United States[3]. We define market manipulation in securities (based on the widely accepted definition in academia and industry) as: *"market manipulation involves intentional attempts to deceive*

---

[1] http://data.worldbank.org/indicator/CM.MKT.LCAP.CD.

[2] Bill C-46 (Criminal Code, RSC 1985, c C-46, s 382, 1985).

[3] Section 9(a)(2) of the Securities Exchange Act (SECURITIES EXCHANGE ACT OF 1934, 2012).

© Springer International Publishing AG 2017
L. Bellatreche and S. Chakravarthy (Eds.): DaWaK 2017, LNCS 10440, pp. 327–342, 2017.
DOI: 10.1007/978-3-319-64283-3_24

investors by affecting or controlling the price of a security or interfering with the fair market to gain profit."

The industry's existing approach for detecting market manipulation is top-down and is based on a set of known patterns and predefined thresholds. Market data such as price and volume of securities (i.e. the number of shares or contracts that are traded in a security) are monitored using a set of rules and red-flags trigger notifications. Then, transactions associated with the detected periods are investigated further, as they might be associated with fraudulent activities. These methods are based on expert knowledge but suffer from two issues: (i) detection of abnormal periods that are not associated with known symptoms (i.e. unknown manipulative schemes), and (ii) adaption to changing market conditions whilst the amount of transactional data is exponentially increasing which makes designing new rules and monitoring the vast data challenging. These issues lead to an increase in false negatives (i.e. there is a significant number of abnormal periods that are left out of the investigation). Data mining methods may be used as a bottom-up approach to detect market manipulation by identifying unusual patterns and data points that merit further investigation, as they are potentially associated with fraudulent activities.

A time series $\{ x_t, t \in T_0 \}$ is the realization of a stochastic process $\{ X_t, t \in T_0 \}$. For our purposes, set T (i.e. the set of time points) is a discrete set and the real-valued observations x_t are recorded on fixed time intervals. Though there has been extensive work on anomaly detection [7], the majority of the techniques look for individual objects that are different from normal objects but do not take the temporal aspect of data into consideration. For example, a conventional anomaly detection approach based on values of data points may not capture anomalous data points in the ECG data where a subsequence with values close to the mean does not follow expected motifs. Therefore, the temporal aspect of data should be considered in addition to the amplitude and magnitude values. Time series anomaly detection methods are successfully applied to different domains including management [24], detecting abnormal conditions in ECG data [15], detecting shape anomalies [25], and credit card fraud detection [11]. Contextual anomalies in time series are data points that are anomalous in a "specific context but not otherwise". For example, Edmonton's average temperature during 2013 was 4.03 degrees Celsius, while the same value during January would be an anomaly (i.e. contextual anomaly). A set of anomalous data points creates an anomalous subsequence (motif). The context is defined both in terms of similarity to the neighbourhood data points of each time series and similarity of time series pattern with respect to the rest of time series in the group. Local anomaly detection methods are particularly useful in non-homogeneous datasets and datasets with changing underlying factors such as financial data. The major motivation for studying local anomaly detection is the development of methods for detecting local anomalies/outliers in complex time series that do not follow a seasonal pattern and are non-parametric, meaning it is difficult to fit a polynomial or deterministic function to the time series data. This is a challenging problem in domains with complex time series such as stock market. Market manipulation periods have been shown to be associated with anomalies in the time series of assets [23], yet the development of effective methods to detect such anomalies remains a challenging problem.

In this paper, we present a formalized method to improve the performance of Contextual Anomaly Detection algorithm that we proposed in our previous work [12]. CAD utilizes an unsupervised learning approach towards detecting anomalies given a set of similar time series. First, a subset of time series is selected based on the window size parameter, Second, a centroid is calculated representing the expected behaviour of time series of the group. Then, the centroid values are used along with the correlation of each time series with the centroid to predict the values of the time series. The proposed method improves recall from 7% to 33% compared to kNN and random walk without compromising precision. The experiments were on S&P industry sectors over 40 years both using daily and weekly data. However, the precision of CAD, kNN and random walk are 0.5% in these experiments (the baseline is less than 0.04% because the number of anomalies in the data is a tiny percentage of samples). We attempt to address this issue by aggregating data from Twitter to reduce the number of false positives that CAD produces.

2   Methods

We adopted big data techniques to improve the performance of the Contextual Anomaly Detection (CAD) method by eliminating false positives. A formalized method is developed to explore the market participants' expectation for each detected datapoint. This information is used to filter out irrelevant items (false positives). Big data techniques are often used to predict consumer behaviour, primarily using social network services such as Twitter, Facebook, Google+ and Amazon reviews. We utilized big data for a novel application in time series anomaly detection, specifically stock market anomalies, by extracting information from Twitter. This information can be integrated into the anomaly detection process to improve the performance of the proposed anomaly detection by eliminating irrelevant anomalies. Although anomalies that are captured using anomaly detection methods represent anomalous data points and periods, some of them may be irrelevant, because there might be a reasonable cause for the anomaly outside time series of market data (for example a news release about a company before the event may explain the abnormal stock return). Using big data techniques to integrate additional information to improve anomaly detection is particularly challenging in securities fraud detection, which are typical challenges in big data problems - velocity, volume, and variability. We are specifically interested in big data techniques to extract information from unstructured data from tweets.

We developed a case study to investigate sentiment analysis on Twitter to improve anomaly detection in the stock market. Figure 1 describes a high-level overview of the process flow in the case study:

(**A.1**) extracting market data for Oil and Gas stocks of S&P 500,
(**A.2**) predicting anomalies in the Oil and Gas stocks,

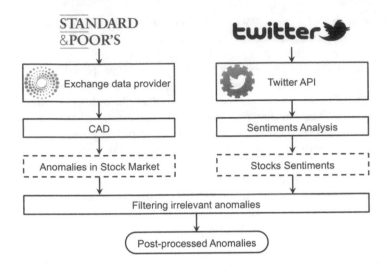

Fig. 1. Utilizing Twitter data to improve anomaly detection in the stock market

(**B.1**) extracting tweets from Twitter for the Oil and Gas stocks in S&P 500,

(**B.2**) preparing a training dataset by extracting tweets for stocks in the Information Technology sector (this data is manually labelled as negative, neutral and positive by an individual who was not involved in developing the methods to preserve fairness of the study),

(**B.3**) building a model for sentiment analysis of tweets that is trained and tested using tweets on stocks (i.e. labelled tweets on the Information Technology sector),

(**B.4**) predicting sentiment of each stock per day using the sentiment analysis model (this produces a time series of sentiments for any given stock returns time series), and,

(**C.1**) filtering irrelevant anomalies based on the respective sentiment on the previous day of every detected anomaly.

2.1 Sentiment Analysis on Twitter

Sentiment analysis is the process of computationally identifying and categorizing people's opinions towards different matters such as products, events, organizations, etc. [4]. Several experiments confirm prediction capabilities of sentiment analysis of social media content such as predicting the size of markets [6] and unemployment rate [1]. Some research works suggest analyzing news and social media such as blogs, micro-blogs, etc. to extract public sentiments could improve predictions in the financial market [22]. Feldman et al. [10] proposed a hybrid approach for stock sentiment analysis based on news articles of companies.

Twitter is the most popular micro-blogging platform. Twitter's technology and popular brand enable millions of people to share their opinions on a variety of topics such as their well-being, politics, products, social events, market conditions and stock market. The flexible architecture and APIs enable researchers and industry to use Twitter for various prediction purposes. Twitter was used to predict movie ticket sales in their opening week with the accuracy of 97.3% [2]. We utilize Twitter in this paper to identify people's opinions about stocks. Sentiment analysis techniques could be used to automatically analyze unstructured data such as tweets in the neighbourhood time period of a detected anomaly. Textual analysis has a long history in the literature [9], however, categorization through sentiments is more recent [20]. The typical approach for representing text for computational processes is based on a the bag-of-words (BOW) [9] where each document is represented by a vector of words. This bag-of-words is called a collection of unigrams. This approach assumes a euclidean space of unigrams that are independent of each other. Thus, documents can be represented as a matrix where each row represents a document. Sentiment analysis methods can be divided into two groups while both use BOW:

1. **lexicon based method** that is an unsupervised approach where a polarity score is assigned to each unigram in the lexicon and the sum of all polarity scores of the text identifies the overall polarity of the text,
2. **machine learning approach** that is a supervised approach where the unigrams or their combinations (i.e. N-grams) are used as features by classifiers.

Social media are increasingly reflecting and influencing the behaviour of other complex systems such as the stock market. Users interactions in social media are generating massive datasets that could explain the "collective behaviour in a previously unimaginable fashion" [14]. We can identify interests, opinions, concerns and intentions of the global population with respect to various social, political, cultural and economic phenomena. Twitter, the most popular micro-blogging platform on internet, is at the forefront of the public commenting about different phenomena. "Twitter data is becoming an increasingly popular choice for financial forecasting" [13]. Researchers have investigated whether the daily number of tweets predicts the S&P 500 stock return [19]. Ruiz et al. used a graph-based view of Twitter data to study the relationship between Twitter activities and the stock market [21]. Some research works utilize textual analysis on Twitter data to find relationships between mood indicators and the Dow Jones Industrial Average (DJIA) [5]. However, the correlation levels between prices and sentiments on Twitter remains low in empirical studies especially when textual analysis is required. More recently, Bartov et al. found aggregated opinions on Twitter can predict quarterly earnings of a given company [3]. These observations suggest a more complicated relationship between sentiments on Twitter and stock returns. Every day, a huge number of messages are generated on Twitter which provides an unprecedented opportunity to deduce the public opinions for a wide range of applications [16]. We intend to use the polarity of tweets to identify the expected behaviour of stocks in the public eyes. Here are some example tweets upon querying the keyword "$xom".

- $XOM flipped green after a lot of relative weakness early keep an eye on that one she's a big tell.
- #OILALERT $XOM »Oil Rises as Exxon Declares Force Majeure on #Nigeria Exports

– Bullish big oil charts. No voice - the charts do the talking. http://ln.is/www.
youtube.com/ODKYG $XLE $XOM $CVX $RDS $SLB @TechnicianApp

The combination of the $ sign along with a company ticker is widely used on
Twitter to refer to the stock of the company. As shown, the retrieved tweets may
be about Exxon Mobil's stock price, contracts and activities. These messages are
often related to people's sentiments about Exxon Mobil Corp., which can reflect
its stock trading. We propose using Twitter data to extract collective sentiments
about stocks to filter false positives from detected anomalies in stocks. We study
the sentiment of stocks at time $t-1$ where t is the timestamp of a detected
anomaly. A sentiment that aligns with the stock return at time t confirms the
return (i.e. aligns with expected behaviour) thus, indicates the detected anomaly is a false positive. We introduce a formalized method to improve anomaly
detection in stock market time series by extracting sentiments from tweets and
present empirical results through a case study on stocks of an industry sector of
S&P 500.

2.2   Data

We use two datasets in this case study: Twitter data and market data. We
extracted tweets on the Oil and Gas industry sector of S&P 500 for 6 weeks
(June 22 to July 27 of 2016) using the Twitter search API. Table 1 shows the
list of 44 Oil and Gas stocks in S&P 500 and the respective number of tweets
constituting 57,806 tweets.

There are two options for collecting tweets from Twitter: the Streaming API
and the Search API. The Streaming API provides a real-time access to tweets
through a query. It requires a connection to the server for a stream of tweets.
The free version of Streaming API and the Search API provide access to a
random sampling of about 1% of all tweets[4]. While the syntax of responses for
the two APIs is very similar, there are some differences such as the limitation on
language specification on queries in Streaming API. We used the Search API to
query recent English tweets for each stock in the Oil and Gas industry sector of
S&P 500 using its cashtag. Twitter unveiled the cashtag feature in 2012 enabling
users to click on a $ followed by a stock ticker to retrieve tweets about the stock.
The feature has been widely adopted by users when tweeting about equities. We
account for the search API rate limits by sending many requests for each stock
with 10-second delays. The batch process runs daily to extract tweets and store
them in a database.

The market data for stocks in Oil and Gas industry sector is extracted from
Thompson Reuters. The stock returns are calculated as $R_t = (P_t - P_{t-1}/P_{t-1})$
where R_t, is the stock return and P_t and P_{t-1} are the stock price on days t and
$t-1$ respectively.

---

[4] The firehose access on Streaming API provides access to all tweets. This is very
expensive and available upon case-by-case requests from Twitter.

Table 1. Tweets about Oil and Gas industry sector in S&P 500

Ticker	Company	Cashtag	Tweets
APC	ANADARKO PETROLEUM	$APC	1052
APA	APACHE	$APA	1062
BHI	BAKER HUGHES	$BHI	1657
COG	CABOT OIL & GAS 'A'	$COG	736
CAM	CAMERON INTERNATIONAL	$CAM	255
CHK	CHESAPEAKE ENERGY	$CHK	4072
CVX	CHEVRON	$CVX	3038
COP	CONOCOPHILLIPS	$COP	1912
CNX	CONSOL EN.	$CNX	1023
DNR	DENBURY RES.	$DNR	1008
DVN	DEVON ENERGY	$DVN	1459
DO	DIAMOND OFFS.DRL.	$DO	1227
ESV	ENSCO CLASS A	$ESV	825
EOG	EOG RES.	$EOG	1149
EQT	EQT	$EQT	669
XOM	EXXON MOBIL	$XOM	5613
FTI	FMC TECHNOLOGIES	$FTI	511
HAL	HALLIBURTON	$HAL	2389
HP	HELMERICH & PAYNE	$HP	838
HES	HESS	$HES	917
KMI	KINDER MORGAN	$KMI	2138
MRO	MARATHON OIL	$MRO	2063
MPC	MARATHON PETROLEUM	$MPC	950
MUR	MURPHY OIL	$MUR	689
NBR	NABORS INDS.	$NBR	384
NOV	NATIONAL OILWELL VARCO	$NOV	827
NFX	NEWFIELD EXPLORATION	$NFX	779
NE	NOBLE	$NE	1102
NBL	NOBLE ENERGY	$NBL	583
OXY	OCCIDENTAL PTL.	$OXY	671
OKE	ONEOK	$OKE	651
BTU	PEABODY ENERGY	$BTU	186
PSX	PHILLIPS 66	$PSX	1205
PXD	PIONEER NTRL.RES.	$PXD	955
QEP	QEP RESOURCES	$QEP	713
RRC	RANGE RES	$RRC	860
RDC	ROWAN COMPANIES CL.A	$RDC	476
SLB	SCHLUMBERGER	$SLB	1962
SWN	SOUTHWESTERN ENERGY	$SWN	1912
SE	SPECTRA ENERGY	$SE	421
TSO	TESORO	$TSO	1086
RIG	TRANSOCEAN	$RIG	1846
VLO	VALERO ENERGY	$VLO	1464
WMB	WILLIAMS COS.	$WMB	2471

2.3 Data Preprocessing

The JSON response for a search query on Twitter APIs (e.g. \$msft) includes several pieces of information such as username, time, location, retweets, etc.[5] For our purposes, we focus on the timestamp and tweet text. We store tweets in a mongoDB database ensuring each unique tweet is recorded once. mongoDB is an open source NoSQL database which greatly simplifies tweet storage, search, and recall eliminating the need of a tweet parser. Tweets often include words and text that are not useful and potentially misleading in sentiment analysis. We remove URLs usernames and irrelevant texts and symbols. Our preprocessing includes three processes:

- **Tokenization** that involves extracting a list of individual words (i.e. bag of words) by splitting the text by spaces. These words are later used as features for the classifier.
- **Removing Twitter Symbols** which involves filtering irrelevant text out such as the immediate word after @ symbol, arrow, exclamation mark, etc.
- **Removing Stopwords** that involves removing words such as "the", "to", "in", "also", etc. by running each word against a dictionary.
- **Recording smiley faces** which involves translating smiley and sad faces to a positive and negative expression in the bag of words.

2.4 Modelling

We adopted three classifiers for determining sentiment of tweets including Naive Bayes, Maximum Entropy and Support Vector Machines. The same features are applied to all classifiers. The anomalous time series of $\{ \eta_k, 0 \leq k \leq n \}$ for the time series $\{ x_1, x_2, \ldots, x_n \}$ where η_k represents an anomaly on day k in the time series (i.e. stock) X. We check sentiment of the stock on day $k - 1$ given η_k. We consider the detected anomaly as a false positive, if the sentiment confirms the change in stock return on day k, however, a sentiment that is in disagreement with the return on the next day implies unexpected stock behaviour, thus anomaly. We study the proposed method for filtering out false positives within detected anomalies by first, running Contextual Anomaly Detection (CAD) method on an anomaly-free dataset, second, removing detected anomalies in the first step that do not conform with their respective sentiment on Twitter. Figure 2 describes an example of stock sentiments on Twitter and anomalies that are detected on XOM (Exxon Mobil). The figure shows 4 anomalies (represented by red circles) that are detected on XOM along with the stock sentiment on Twitter for each day (days with no bars have the neutral sentiment). The data points on June 24 and July 21 are declared irrelevant because the stock's sentiments on the day before these dates confirm the change direction on the next day. However, other two anomalies (July 7 and 15) remain relevant because the sentiments on the day before the anomalies do not confirm the change direction in the stock return.

---

[5] https://dev.twitter.com/rest/reference/get/search/tweets.

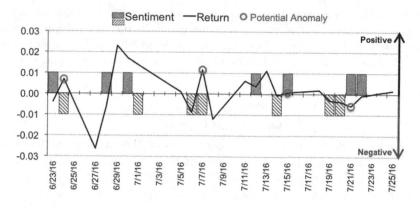

Fig. 2. Identifying false positives in detected anomalies on Exxon Mobil (XOM) (Color figure online)

We found through our preliminary experiments that sentiment analysis using classifiers that are trained on movie reviews or generic tweets that are widely used in literature perform poorly for stock tweets. This is due to different corpus and linguistics that are specific to stock market. We developed a training dataset that is labelled manually to address this issue. This dataset includes 66 stocks in the Information Technology industry sector of S&P 500 and respective tweets constituting over 6,000 tweets. We manually labelled over 2,000 tweets from this dataset. We also used StockTwits[6], a widely popular social media platform that is designed for sharing ideas between investors and traders, to extract messages that are labelled by stock market participants. We developed a tool to query StockTwits for a given stock and extract relevant messages. Then, messages that are labelled by their poster as *Bearish* and *Bullish* are mapped to negative and positive sentiments in our code. The training data is labelled manually with three sentiment labels: negative, neutral and positive. This data is used to train the classifiers that we used. The testing dataset is tweets about the Oil and Gas industry sector of S&P 500. Table 1 shows the list of 44 Oil and Gas stocks in the testing dataset S&P 500 and the respective number of tweets constituting 57,706 tweets in total.

2.5   Feature Selection

Feature selection is a technique that is often used in text analysis to improve performance of results by selecting the most informative features (i.e. words). Features that are common across all classes contribute little information to the classifier. This is particularly important as the number of features grow rapidly with increasing number of documents. The objective is using the words that have the highest information gain. Information gain is defined as the frequency of the

---

[6] http://stocktwits.com/.

word in each class compared to its frequency in other classes. For example, a word that appears in the positive class often but rarely in the neutral and negative classes is a high information word. Chi-square is widely used as a measure of information gain by testing the independence of a word occurrence and a specific class:

$$\frac{N(O_{w_p c_p} * O_{w_n c_n} - O_{w_n c_p} * O_{w_p c_n})^2}{O_{w_p} * O_{w_n} * O_{c_p} * O_{c_n}} \tag{1}$$

Where $O_{w_p c_p}$ is the number of observations of the word w in the class c and $O_{w_p c_n}$ is the number of observations of the word w in other classes (i.e. class negative). This score is calculated for each word (i.e. feature) and used for ranking them. High scores indicate the null hypothesis H_0 of independence should be rejected. In other words, the occurrence of the word w and class c are dependent thus the word (i.e. feature) should be selected for classification. It should be noted that Chi-square feature selection is slightly inaccurate from statistical perspective due to the one degree of freedom. Yates correction could be used to address the issue, however, it would make it difficult to reach statistical significance. This means a small number of features out of the total selected features would be independent of the class. Manning et al. showed these features do not affect the performance of the classifier [18].

2.6   Classifiers

Naive Bayes: A Naive Bayes classifier is a probabilistic classifier based on the Bayes Rule $P(c|\tau) = \dfrac{P(\tau|c)P(c)}{P(\tau)}$ where $P(c|\tau)$ is the probability of class c being negative, neutral or positive given the tweet τ. The best class is the class that maximizes the probability given tweet τ:

$$C_{MAP} = \underset{c \in C}{\operatorname{argmax}} P(\tau|c)P(c) \tag{2}$$

where $P(\tau|c)$ can be calculated using the bag of words as features resulting in

$$C_{MAP} = \underset{c \in C}{\operatorname{argmax}} P(x_1, x_2, \ldots, x_n|c)P(c) \tag{3}$$

$P(c)$ can be calculated based on the relative frequency of each class in the corpus or dataset. There are two simplifying assumption in Naive Bayes which make calculating $P(x_1, x_2, \ldots, x_n|c)$ straightforward, (i) position of the words do not matter, and (ii) the feature probabilities $P(x_i|c_j)$ are independent given the class c:

$$P(x_1, x_2, \ldots, x_n|c) = P(x_1|c) \bullet P(x_2|c) \bullet \cdots \bullet P(x_n|c) \tag{4}$$

in other words, we have the Multinomial Naive Bayes equation as

$$C_{NB} = \underset{c \in C}{\operatorname{argmax}} P(c_j) \prod_{x \in X} P(x|c) \tag{5}$$

Maximum Entropy: MaxEnt eliminates the independence assumptions between features and in some problems outperforms Naive Bayes. MaxEnt is a probabilistic classifier based on the Principle of Maximum Entropy. Each feature corresponds to a constraint in a maximum entropy model. MaxEnt classifier computes the maximum entropy value from all the models that satisfy the constraints of the features for the given training data, and selects the one with the largest entropy. The MaxEnt probability estimation is computed using

$$P(c|f) = \frac{1}{Z(f)} \, exp \left(\sum_i \lambda_{i,c} \, F_{i,c}(f,c) \right) \tag{6}$$

where $Z(f)$ is a normalization function and $F_{i,c}$ is a binary function that takes the input feature f for the class c. λ is a vector of weight parameters that is updated iteratively to satisfy the tweets feature while continuing to maximize the entropy of the model [8]. The iterations eventually converge the model to a maximum entropy for the probability distribution. The binary function $F_{i,c}$ is only triggered when a certain feature exists and the sentiment is hypothesized in a certain class:

$$F_{i,c}(f,c') = \begin{cases} 1 & \text{if } n(f) > 0 \text{ and } c' = c \\ 0 & \text{otherwise} \end{cases} \tag{7}$$

Support Vector Machines: SVM is a linear classification algorithm which tries to find a hyperplane that separates the data in two classes as optimally as possible. The objective is maximizing the number of correctly classified instances by the hyperplane while the margin of the hyperplane is maximized. The hyperplane representing the decision boundary in SVM is calculated by

$$(\boldsymbol{w} \cdot \boldsymbol{x}) + b = \sum_i y_i \alpha_i (\boldsymbol{x_i} \cdot \boldsymbol{x}) + b = 0 \tag{8}$$

where weight vector $\boldsymbol{w} = (w_1, w_2, .., w_n)$ which is the normal vector defining the hyperplane is calculated using the n-dimensional input vector $\boldsymbol{x_i} = (x_{i1}, x_{i2}, .., x_{in})$, outputting the value y_i. α_i terms are the Lagrangian multipliers. Calculating w using the training data gives the hyperplane which can be used to classify the input data instance $\boldsymbol{x_i}$. If $\boldsymbol{w} \cdot \boldsymbol{x_i} + b \geq 0$ then the input data instance is labelled positive (the class we are interested in), otherwise it belongs to the negative class (all of the other classes). It should be noted that although SVM is a linear classifier (as Naive Bayes and Maximum Entropy are) it is a powerful tool to classify text because text documents are typically considered as a linear dataset. It is possible to use Kernel functions for datasets that are not linearly separable. The Kernel is used to map the dataset to a higher dimensional space where the data could be separated by a hyperplane using classical SVM.

There are two approaches for adopting SVM for a classification problem with multiple classes such as sentiment analysis with the classes negative, neutral and positive: (i) one-vs-all where an SVM classifier is built for each class, and

(ii) one-vs-one where an SVM classifier is built for each pair of classes resulting in $M(M-1)/2$ for M classes. We used the latter for classifying sentiments using SVM. In the one-vs-all approach, the classifier labels data instances positive for the class that we are interested in and the rest of instances are labelled negative. A given input data instance is classified with classifier only if it is positive for that class and negative for all other classes. This approach could perform poorly in datasets that are not clustered as many data instances that are predicted positive for more than one class, will be unclassified. The one-vs-one approach is not sensitive to this issue as a data instance is categorized in the class with the most data instances, however, the number of classes can grow rapidly for problems with many classes (i.e. higher numbers of M).

2.7 Classifier Evaluation

We trained classifiers using specifically stock tweets that are carefully labelled manually. We asked a person who has not been involved with training data to label the testing dataset. The testing data includes 1332 stock tweets that are manually labelled. We used 5-fold cross validation for training the classifiers that is sampling the data into 5 folds and using 4 folds for training and 1 fold for testing. This process is repeated 5 times and the performance results are averaged. We used precision and recall for each class in addition to classification accuracy as performance measures to evaluate the classifiers. The precision of a classifier[7] for a given class represents the fraction of the classified tweets that belong to the class, while recall[8] represents the fraction of tweets that belong to the class out of all tweets that belong to the class. The precision for a class measures the exactness or quality, whereas recall measures the completeness or quantity. The classifier with the highest performance is used to predict sentiment of stocks in Oil and Gas industry sector (see Table 1 for the list of stocks in the Oil and Gas sector).

2.8 Calculating Polarity for Each Stock

The Twitter sentiments of stocks are predicted using an SVM classifier that is trained using labelled tweets about stocks. First, the classifier predicts sentiment of each tweet (i.e. negative, neutral and positive). Then, the polarity for each stock is computed using time series of negative, neutral and positive tweets:

- Negative tweets, tw_d^-: the number of negative tweets on day d
- Neutral tweets, tw_d^0: the number of neutral tweets on day d
- Positive tweets, tw_d^+: the number of positive tweets on day d

The polarity for each stock on a given day is the difference between the number of positive and negative tweets as a fraction of non-neutral tweets.

---

[7] $TP/(TP + FP)$.
[8] $TP/(TP + FN)$.

More formally

$$P_{s_d} = \frac{tw_d^+ - tw_d^-}{tw_d^+ + tw_d^-} \tag{9}$$

where P_{s_d} is the polarity of stock s on day d. Figure 3 shows the aggregated polarity of Exxon Mobil. The red dashed lines represent the parameter $sentThreshold$ that we define to control for the minimum magnitude of polarity that is required for declaring a potential anomaly a false positive. For example, the method would not include the polarity of Exxon Mobil on July 14 as an indicator to accept or reject the potential anomaly on July 15 as a false positive because its value is below the threshold. This parameter can be set during preliminary tests by trying a grid on $sentThreshold$ (e.g. 0.2, 0.3, etc.).

3   Results and Discussion

We propose a two-step anomaly detection process. First, the anomalies are predicted on a given set of time series (i.e. stocks in an industry sector) using Contextual Anomaly Detection (CAD). Second, the anomalies are vetted using sentiment analysis by incorporating data in addition to market data. This process gives a list of anomalies that are filtered using data on Twitter. The first step is based on the unsupervised learning algorithm CAD and the second step, relies on state-of-the-art supervised learning algorithms for sentiment analysis on unstructured data on Twitter. We developed a set of experiments for this case study on the Oil and Gas sector of S&P 500 for the period of June 22 to July 27. The correlation of stocks during this 6-week period for the case study is quite high as we expect within an industry sector (70% stocks within the sector have a correlation higher than 0.7)

We studied several other classifiers in addition to the three classifiers that we introduced in Sect. 2.4 (i.e. Multinomial Naive Bayes, MaxEnt, also known as Logistic Regression, and SVM) to build a sentiment analysis model including

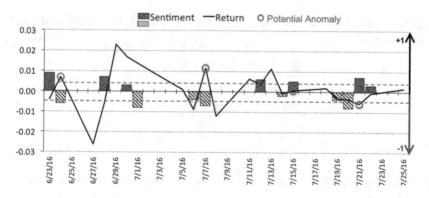

Fig. 3. Polarity of Exxon Mobil stock per day along with potential anomalies that CAD produces (Color figure online)

Bernoulli Naive Bayes, Stochastic Gradient Descent (SGD) and C-Support Vector (SVC). Furthermore, we investigated the performance of sentiment analysis models using different number of features (i.e. 10, 100, 1000 etc. words).

Movie reviews data is typically used for sentiment analysis of short reviews as well as tweets [17]. This dataset includes movie reviews that are collected from IMDB[9]. Our experiments show that sentiment analysis models for stock tweets that are trained using this standard dataset perform poorly (see Fig. 4). The results confirm our hypothesis that training data that is out of context is inappropriate for sentiment analysis of short text samples, particularly on Twitter. We developed a tool to extract labelled data from StockTwits[10] to address this issue (see Sect. 2.3 for more information on data). Figure 4 illustrates that these models outperform models which are trained on movie review data consistently.

We observe that the number of features is an important parameter in the performance of sentiment analysis models. The results show using more features improves the performance results. However, performance of the models decays after hitting a threshold of about 10,000 features. This reiterates our hypothesis on utilizing feature selection to improve sentiment analysis on Twitter.

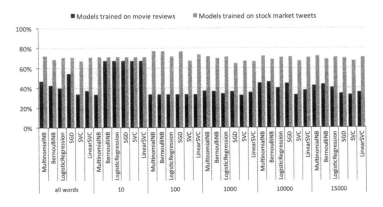

Fig. 4. Accuracy of sentiment analysis models using training datasets in movie reviews and stock market

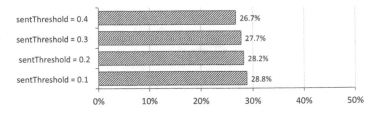

Fig. 5. Filtering irrelevant anomalies using sentiment analysis on Oil and Gas sector

---

[9] http://www.imdb.com/reviews/.
[10] http://stocktwits.com/.

We studied the impact of proposed method in filtering false positives of CAD by first, running CAD on returns of Oil and Gas stocks during June 22 to July 27 of 2016 with no injected anomalies. The predicted anomalies would be false positives because the S&P 500 data is anomaly-free as we explained. Then, using the proposed method we measured how many of false positives are filtered. CAD predicts 261 data points as anomalous given stock market data for the case study (out of 1,092 data points). Figure 5 describes the percentage of irrelevant anomalies that are correctly filtered by sentiment analysis of tweets about stocks. *sentThreshold* is a parameter we use when comparing the aggregated sentiment values for a given stock per day. The results confirm that the proposed method is effective in improving CAD through removing irrelevant anomalies by correctly identifying 28% of false positives.

References

1. Antenucci, D., Cafarella, M., Levenstein, M., Ré, C., Shapiro, M.D.: Using social media to measure labor market flows. Technical report, National Bureau of Economic Research (2014)
2. Asur, S., Huberman, B.A.: Predicting the future with social media. In: 2010 IEEE/WIC/ACM International Conference on Web Intelligence and Intelligent Agent Technology (WI-IAT), vol. 1, pp. 492–499. IEEE (2010)
3. Bartov, E., Faurel, L., Mohanram, P.S.: Can twitter help predict firm-level earnings and stock returns? Available at SSRN 2782236 (2016)
4. Bing, L.: Sentiment Analysis: A Fascinating Problem, pp. 7–143. Morgan and Claypool Publishers (2012)
5. Bollen, J., Mao, H., Pepe, A.: Modeling public mood and emotion: Twitter sentiment and socio-economic phenomena. ICWSM **11**, 450–453 (2011)
6. Bollen, J., Mao, H., Zeng, X.: Twitter mood predicts the stock market. J. Comput. Sci. **2**(1), 1–8 (2011)
7. Chandola, V., Banerjee, A., Kumar, V.: Anomaly detection for discrete sequences: a survey. IEEE Trans. Knowl. Data Eng. **24**(5), 823–839 (2012). http://ieeexplore.ieee.org/lpdocs/epic03/wrapper.htm?arnumber=5645624
8. Daumé III, H.: Notes on CG and LM-BFGS optimization of logistic regression, pp. 1–7 (2004). https://www.umiacs.umd.edu/hal/docs/daume04cg-bfgs.pdf
9. Dillon, M.: Introduction to Modern Information Retrieval: G. Salton and M. Mcgill (1983)
10. Feldman, R., Rosenfeld, B., Bar-Haim, R., Fresko, M.: The stock sonar—sentiment analysis of stocks based on a hybrid approach. In: Twenty-Third IAAI Conference, pp. 1642–1647 (2011)
11. Ferdousi, Z., Maeda, A.: Unsupervised Outlier Detection in Time Series Data, p. 121. IEEE (2006). http://ieeexplore.ieee.org/lpdocs/epic03/wrapper.htm?arnumber=1623916
12. Golmohammadi, K., Zaiane, O.R.: Time series contextual anomaly detection for detecting market manipulation in stock market. In: The 2015 Data Science and Advanced Analytics (DSAA 2015), pp. 1–10. IEEE (2015)
13. Graham, M., Hale, S.A., Gaffney, D.: Where in the world are you? Geolocation and language identification in Twitter. Prof. Geogr. **66**(4), 568–578 (2014)
14. King, G.: Ensuring the data-rich future of the social sciences. Science **331**(6018), 719–721 (2011)

15. Lin, J., Keogh, E., Fu, A., Herle, H.: Approximations to magic: finding unusual medical time series, pp. 329–334. IEEE (2005)
16. Liu, B.: Sentiment analysis and opinion mining. Synth. Lect. Hum. Lang. Tech. **5**(1), 1–167 (2012)
17. Maas, A.L., Daly, R.E., Pham, P.T., Huang, D., Ng, A.Y., Potts, C.: Learning word vectors for sentiment analysis. In: Proceedings of the 49th Annual Meeting of the Association for Computational Linguistics: Human Language Technologies, pp. 142–150. Association for Computational Linguistics, Portland (2011). http://www.aclweb.org/anthology/pp.11-1015
18. Manning, C.D., Raghavan, P., Schutze, H.: Introduction to Information Retrieval, pp. 405–416. Cambridge University Press (2008). Chap. 20
19. Mao, Y., Wei, W., Wang, B., Liu, B.: Correlating S&P 500 stocks with Twitter data. In: Proceedings of the First ACM International Workshop on Hot Topics on Interdisciplinary Social Networks Research, pp. 69–72. ACM (2012)
20. Morinaga, S., Yamanishi, K., Tateishi, K., Fukushima, T.: Mining product reputations on the web. In: Proceedings of the Eighth ACM SIGKDD International Conference on Knowledge Discovery and Data Mining, pp. 341–349. ACM (2002)
21. Ruiz, E.J., Hristidis, V., Castillo, C., Gionis, A., Jaimes, A.: Correlating financial time series with micro-blogging activity. In: Proceedings of the Fifth ACM International Conference on Web Search and Data Mining, pp. 513–522. ACM (2012)
22. Schumaker, R.P., Chen, H.: Textual analysis of stock market prediction using breaking financial news: the Azfin text system. ACM Trans. Inf. Syst. (TOIS) **27**(2), 12 (2009)
23. Song, Y., Cao, L., Wu, X., Wei, G., Ye, W., Ding, W.: Coupled behavior analysis for capturing coupling relationships in group-based market manipulations. In: Proceedings of the 18th ACM SIGKDD International Conference on Knowledge Discovery and Data Mining, pp. 976–984. ACM (2012)
24. Sriastava, A., et al.: Discovering system health anomalies using data mining techniques, pp. 1–7 (2005)
25. Wei, L., Keogh, E., Xi, X.: Sexually explicit images: finding unusual shapes. In: 2006 Proceedings of the Sixth International Conference on Data Mining, ICDM 2006, pp. 711–720. IEEE (2006)

Automatic Segmentation of Big Data of Patent Texts

Mustafa Sofean$^{(\boxtimes)}$

FIZ Karlsruhe, Hermann-von-Helmholtz-Platz 1,
76344 Eggenstein-leopoldshafen, Germany
mustafa.sofean@fiz-karlsruhe.de
http://www.fiz-karlsruhe.de

Abstract. Patent documents are abundant, lengthy and are written in very technical language. Thus, reading and analyzing patent documents can be complex and time consuming. This is where the use of automatic patent segmentation can help. This work attempts to automatically segment the description part of patent texts into semantic sections. Our goal is to develop a robust and scalable segmentation tool for automatic structuring of the patent texts into pre-defined sections that will serve as a pre-processing step to patent text IR(information retrieval) and IE(information extraction) tasks. To do so, an established set of guidelines is exploited for defining the segments in the description part of the patent text. Depending on those guidelines a segmentation tool called PatSeg is developed based on a combination of text mining techniques. A rule-based algorithm is used to identify the headings inside patent text, machine learning technique is used to classify the headings into pre-defined sections, and heuristics are used to identify the sections in patent text that do not contain headings. The performance of our methods achieved up to 94% of accuracy. In addition, we proposed a big data approach based on Hadoop ecosystem modules to apply our methods on the huge amount of patent documents.

Keywords: Patent enrichment · Text mining · Big data

1 Introduction

A patent is an exclusive right granted for an invention, which is a product or a process that provides, in general, a new way of doing something, or offers a new technical solution to a problem [1]. Patents are the largest source of technological information since the text of the patent is a rich source to discover technological progresses, useful to understand the trend and forecast upcoming advances. Additionally, a patent document has much more detailed information about a technology than any other type of scientific or technical publication [2]. The textual part of a patent document contains title, abstract, claims, and Detailed Description of the Invention (DetD). The description is the main body of the detailed content which includes the summary, embodiment, and the description

© Springer International Publishing AG 2017
L. Bellatreche and S. Chakravarthy (Eds.): DaWaK 2017, LNCS 10440, pp. 343–351, 2017.
DOI: 10.1007/978-3-319-64283-3_25

of figures and drawings of the invention. Detailed description texts are abundant, lengthy. Thus, reading and analyzing patent documents can be complex and time consuming. This is where the use of automatic patent segmentation can help. Text segmentation is a process of analyzing the patent text and identifying smaller meaningful segments. The objective of this work is, automatic structuring of patent texts into pre-defined segments and making the content of the individual segments separately searchable. In addition, this process will also serve as a pre-processing step to the IE (information extraction) task. In particular, we identify and recognize the meaningful segments in the description section of patent document. The remainder of the paper is organized as follows. Section 2 describes the related work, Sect. 3 describes the guidelines for patent segmentation, our segmentation methods and evaluation are presented in the Sect. 4, including the workflow of our approach, details on automatic detection of headings, description of heuristic methods which are used to identify segments in free text of patents, approach for segmenting a huge amount of patent documents, and implementation. In the end we summarize our findings and discuss future work.

2   Related Work

In the literature there are many methods of texts segmentation such as [3–6] which aim to identify the topic boundaries in text. Text segmentation has been also used in clinical domain to classify the clinical texts into pre-defined sections [7]. In patent domain, using the traditional segmentation methods such as C99 in [4] and TextTiling in [3] are not significant to segment the patent texts since they use linguistic signs to identify the changes of topics between textual units that are closely related by lexical cohesion and are usually in the same topic. In addition, there is a high similarity between sentences that belong to different segments. Tseng used regular expressions to segment the description part of patents by extracting titles in uppercase [8]. Our work in this paper is based on segmentation guidelines which are provided by patent experts, and our methods are able to extract all segments in description part of patent text regardless of whether there are headlines in the texts or not. Moreover, our technique addresses a need of IP professionals who want to access to any particular segment in the patent text very quickly, as well as the most important segment be used to transform patent information into knowledge that can influence decision-making.

3   Segmentation Guidelines

In the segmentation process of description texts of patents, we want to learn more about patent segments and establish a list of patent segments. Taking into account the characteristics of patent text, it was examined that how well established guidelines that are used to define the patent segments (sections). For this purpose, an established set of guidelines within patent information professional who is working in patent domain for more than 20 years is exploited for defining the sections in the patent text. These guidelines are presented by dividing the description text of patent into different segments. The main segments are presented below:

- **Technical Field** the field to which the invention relates.
- **Background** the background of the invention or prior art.
- **Summary of the Invention** disclosure of the invention, describing the technical problems and their solution.
- **Embodiments and Modes** the embodiments or modes of the invention.
- **Methods and Examples** specific examples of the embodiments or methods of the invention.
- **Advantageous Effect and Industrial Applicability** indication of the advantageous effect, use and industrial applicability of the invention.
- **Drawings and Figures** brief description of figures, drawings, tables, photos, numerals etc.
  Furthermore, the secondary segments that may appear in some patents should be identified if they are found:
- **References** related applications and cited patents.
- **Sequences** disclosure of Nucleotide and/or Amino Acid Sequences.
- **Government interest** this section includes a "statement regarding federally sponsored research" if the invention was made under a government contract, or if federal grant money was used to fund the research.
- **Appendix** the related materials.

4   Methods and Evaluations

According to a dataset of 1,999,147 patents which has been extracted from European patents. 79% of those patents contain at least one header in the description section. We also found that 58% of the dataset contain the summary segment which is located under the headline "Summary of the Invention". Therefore, we use the headlines to signal the start of a respective segment in the description text. The patents with free text or without complete headings on the other hand, they can be processed by using heuristics techniques.

4.1   Workflow

In this work, we attempt to automatically segment the description of the patent text into semantic sections by using a combination of techniques such as machine learning, rule-based algorithm, and heuristics (learning-by-problem-solving pattern). Particularly, our segmentation process was used to classify each patent paragraph in description text of patent into one of several pre-defined semantic sections. Our workflow starts with extracting a description text from the patent document, and checks if the text is structured by headlines. Then the text is divided into paragraphs, a pre-processing step will take place to remove undesired tokens and apply stemming, a rule-based algorithm will be used to identify the headers, and a machine learning model is used to get to pre-defined segment for each header. For the patent that do not have complete structured segments or do not have any heading at all, heuristic methods will be used. The final step is, identify boundaries of each segment and their related text content.

4.2 Headings Identification

In order to discover the headers inside the description text of patent (DetD), we need to get the boundary of the headers. i.e., the header's start and end. We call this operation Header Detection. Then, we identify the text content which is related to each header. The header meaning on the other hand is represented by assigning the header to an appropriate section type (e.g.; summary, example, background, method, etc.). Here, a rule-based approach is more suitable because in the patent domain, there is no sufficient training data for a machine learning algorithm to be successful. To do so, we develop a rule-based algorithm to identify headers and their boundaries. The output consists of all headers and their positions inside the DetD. Our algorithm relies on a collection of headings which are extracted from a dataset of 140,000 patents, and it works as follows: as input we take the DetD as a sequence of paragraphs. Then, we test the following features to decide whether a paragraph is a header or not:

(A) **Length feature1** number of words in the paragraph.
(B) **Length feature2** the length of the paragraph.
(C) **Capitalization feature1** the feature represents whether all letters in the current paragraph are in upper case.
(D) **Capitalization feature2** the feature indicates that all words in the paragraph start with upper case letter.
(E) **Capitalization feature3** the feature respectively represents whether or not the number of words in the line starting with upper case letter more than number of words staring with lower case letter.
(F) **Positive feature** the feature respectively represents whether or not the current paragraph matchs a heading in a heading list which is collected from a collection of patent documents.
(G) **Bullet feature1** the feature value is true if the current paragraph is bullet otherwise false.
(H) **Bullet feature2** the feature value is true if the pervious or next paragraph is a bullet.
(I) **StartsWith feature** determine if the paragraph starts with Fig., Figs., or figure.
(J) **Chemical feature** represents whether or not the current text line contains a simple chemical text.
(K) **Average paragraph length** the average header length in the dataset.
(L) **Average word number** the average word numbers in the dataset.

We use these features on each input paragraph of the DetD to build decision rules for the header detection, and also they could be used as a training set for a machine learning algorithm. Some of the decision rules are listed below:

(i)   C is true and G is false and A$\geq$1 and J is false
(ii) D is true, F is true, A$\geq$1, G is false, and J is false
(iii) G is true, H is false, A$<$L, J is false, B$<$K, and A$\geq$1
(iv) F is true, E is true, J is false, A$\geq$1, and G is false.

If one of the decisions rule is true, the paragraph is a header. We tested our algorithm on a collection of patents, and we found that the average quality reaches an F1-score of 95% for the main segments.

4.3   Meaning of Headings (Semantic of Headings)

In previous sub-section, we described methods for detection of heading and their boundaries in the description text of the patent. But we need to understand the meaning of these headers. Could be method, summary, description of the drawing, and so on. The goal of this work is automatic structuring of description texts of patent into meaningful pre-defined sections. In other words, each header with its content will be assigned into appropriate section. The segmentation task was modeled as a classification task. The task here is to assign each header to one of pre-defined sections, this operation called text categorization (TC). By using our segmentation guidelines, we construct a training dataset by labeling manually each header into related section. Our training set consists of 1495 labeled headers which are distributed for eleven segments.

The training set was first pre-processed by removing undesired tokens like numbers, special symbols, and stop words. Then, we computed weight factor tf *idf (term frequency - inverse document frequency) for all training dataset. In addition, we applied stemming technique on all training set to only convert the plural nouns into singulars. Support Vector Machines (SVMs) are used as a multi-classification technique [9]. The SVMs classifier was trained on the uni-gram features that are extracted from the training set. The eleven sections were combined via one-vs-all classification, and the section of the classifier with the largest output value was selected. The classifier was evaluated by 10-fold cross validation which is a statistical method of evaluating and comparing learning algorithms by dividing data into two segments: first used to learn or train a model and the second used to validate the model. The performance of the categorization achieved up to 90% of accuracy. Table 1 shows all evaluation results for each segment.

4.4   Heuristic Methods

According to European patents dataset, more than 40% of the patents do not have complete headings that cover all categories. So we need to recognize the sections from a free text of patents. For patent text a solution relying on a pre-determined training set is not practical beacuse of the high similarity between the texts of sections in the same patent. Therefore, heuristics are incorporated in

Table 1. Evaluation results of learning with SVMs

Segment	Recall	Precision	F-Score
Embodiment	91%	90%	91%
Background	99%	87%	92%
Summary	82%	85%	84%
Methods and examples	96%	93%	94%
Drawing and figures	80%	84%	82%
Applicability	78%	93%	85%
Technical field	78%	75%	77%
Appendix	50%	80%	62%
Sequences	87%	79%	83%
References	69%	90%	78%
Statements	76%	100%	87%

our segmentation methods to identify the segments from the patents when there are no headlines in the text of the description part. Our strategies are depending on the observation of key terms or phrases that are used to recognize the segments in the patent texts. In cooperation with patent expert, we collected a list of significant term/phrases for each segments from different sources of patent databases such as European, Japanese, Korean, and Chinese patents. In more details, we divided the texts into paragraphs, heuristics are used to assign each paragraph into an appropriate segment by using regular expressions depending on the term and phrases which are previously collected. Moreover, additional heuristic rules are used to recognize the segments. For instance, the first paragraph in the text of DeTD is almost a technical field; the background segment comes almost directly after technical field segment, the summary segment is coming directly after the background segment and it can be obviously observed by the objective, disclosure, or problem and solution of the invention, and finally the drawing segment can be observed by the citation of the drawings and figures (e.g.; figure x shows, or A remark example of the invention is described on the basis the FIGURES 1 to 3. Show..).

4.5 Big Data Approach

Big data technologies are aimed at processing high-volume, high-velocity, and high-variety data and extracting business intelligence (BI) for insight and decision making. Our aim is to apply our segmentation methods on a big data of patents and make the extracted segments available for search and decision making. FIZ-Karlsruhe [13] has more than 20 different databases of patents and can be accessed by the STN Search Engine [10] which is an information service for research and patent information and offers online access to high-quality databases on a neutral platform, with a focus on patent information. Our big data approach is presented as follow: first; we extract patent records from each database and transform them into distributed data repository, Hadoop technology will be used for managing large-scale patent data. Therefore, the patent data are stored on HDFS. To this end, we select Phoenix [14] as a suitable NoSQL database to construct a patent data schema.

Second; we use Hadoop ecosystem to process large data sets in a batch processing manner. Our segmentation methods are implemented as a MapReduce application and run on the patent documents that are stored in distributed repositories based on NoSQL databases. The output will be indexed into Solar index so that the end user can access to any segment directly.

4.6   Implementation

To implement our segmentation tool, we used Apache Maven with Java, and UIMA platform (Unstructured Information Management Architecture) which is software system that analyzes unstructured information to discover, organize, and deliver relevant knowledge to the user [11]. We created four UIMA components (analysis engine); (1) dividing the text into paragraphs and get the positions of each. (2) Identify the heading by using a rule-based algorithm, and get the related segment by using a machine learning model. (3) use heuristics to segment the patent that do not have headlines. (4) Aggregating the segments and get related texts. For the training model, we combined the source code of WEKA toolkit [12] into our java code to build the model and use it for segment prediction. To run our java application on Hadoop cluster, we structured our codes in two phases Map and Reduce. In other words, we converted our code into MapReduce code in order to run it on the Hadoop cluster. We have been using Hadoop modules at FIZ-Karlsruhe for a while for various tasks such as ETL (Extract, Transform and Load) and analysis of patent texts. Our Hadoop cluster consists of two name nodes and 20 data nodes.

4.7   Evaluation

We extracted patent documents for the evaluation by using STN search engine which is developed by FIZ-Karlsruhe [10]. The evaluation was conducted on three different dataset. The first dataset consists of 100 documents which is randomly selected from European patents. 50% of this dataset contains complete segments with headlines inside the texts. The second dataset consists of 50 patents and randomly selected from German patents. The third dataset is randomly extracted from Japanese and Chinese patents, and consists of 50 patents. The datasets 2 and 3 are automatically translated into English and all patents in these datasets do not have any heading inside their texts. All patent documents were annotated by a patent information professional according to the guidelines which were described in the Sect. 3. We run our segmentation tool for each annotated document and Table 2 summarizes the results of our segmentation process. The table presents the accuracy rate for each dataset, number of

Table 2. Evaluation results

Dataset	Accuracy	Hits	Misses	Total number of paragraphs
Dataset 1 with headings	97%	4849	172	5021
Dataset 1 without headings	92%	2159	197	2356
Dataset 2	94%	4168	277	4445
Dataset 3	93%	4051	322	4373
Total	**94%**	**15227**	**968**	**16195**

paragraphs which were correctly identified into related segment (Hits), the number of paragraphs which were assigned into a wrong segment (Misses), and the total number of paragraphs in each dataset. We found that the average accuracy reaches 94% for all datasets.

5   Conclusion

In this paper, we presented an automated approach for automatically segmenting the description part of patent text into structured segments. In particular, we developed a patent segmentation tool PaTSeg which uses a hybrid text mining techniques such as machine learning and rule-based algorithm to extract the semantic sections from the patent text. The performance of our tool achieved up to 94% of accuracy. A big data approach is proposed to segment a huge amount of patent documents and make them available for the search. As future work, we plan to continue improving our methods and ask patents experts to assess the usability and usefulness of our approach.

References

1. World Intellectual Property Organization (WIPO). http://www.wipo.int
2. Thomson Reuter. http://thomsonreuters.com/en.html
3. Hearst, M.: TextTiling: segmenting text into multi-paragraph subtopic passages. Comput. Linguist. **23**, 33–64 (1997)
4. Choi, F.Y.Y.: Advances in domain independent linear text segmentation. In: Proceedings of NAACL 2000, pp. 26–33 (2000)
5. Rojas, L.H., Pagola, J.E.M.: TextLec: a novel method of segmentation by topic using lower windows and lexical cohesion. In: Rueda, L., Mery, D., Kittler, J. (eds.) CIARP 2007. LNCS, vol. 4756, pp. 724–733. Springer, Heidelberg (2007). doi:10.1007/978-3-540-76725-1_75
6. Pérez, R.A., Medina Pagola, J.E.: Text segmentation by clustering cohesion. In: Bloch, I., Cesar, R.M. (eds.) CIARP 2010. LNCS, vol. 6419, pp. 261–268. Springer, Heidelberg (2010). doi:10.1007/978-3-642-16687-7_37
7. Apostolova, E., Channin, D.S., Raicu, D.: Automatic segmentation of clinical texts. In: Conference Proceedings of IEEE Engineering in Medicine and Biology Society (2009)

8. Tseng, Y.-H., Lin, C.J., Lin, Y.I.: Text mining techniques for patent analysis. Inf. Process. Manage. **43**(5), 1216–1247 (2007) ·
9. Cortes, C., Vapnik, V.: Support vector networks. Mach. Learn. **20**, 273–297 (1995)
10. STN International. http://www.stn-international.de
11. Unstructured Information Management applications. https://uima.apache.org/
12. Weka Toolkit. http://www.cs.waikato.ac.nz/ml/weka/
13. FIZ Karlsruhe - Leibniz Institute for Information Infrastructure. https://www. fiz-karlsruhe.de/
14. Apache Phoenix. https://phoenix.apache.org/

Sentiment Analysis and User Influence

Tag Me a Label with Multi-arm: Active Learning for Telugu Sentiment Analysis

Sandeep Sricharan Mukku[1]([✉]), Subba Reddy Oota[2]([✉]), and Radhika Mamidi[1]

[1] LTRC, KCIS, IIIT Hyderabad, Hyderabad, India
sandeep.mukku@research.iiit.ac.in, radhika.mamidi@iiit.ac.in
[2] Teradata, Hyderabad, India
SubbaReddy.Oota@teradata.com

Abstract. Sentiment Analysis is one of the most active research areas in natural language processing and an extensively studied problem in data mining, web mining and text mining for English language. With the proliferation of social media these days, data is widely increasing in regional languages along with English. Telugu is one such regional language with abundant data available in social media, but it's hard to find a labeled training set as human annotation is time-consuming and cost-ineffective. To address this issue, in this paper the practicality of active learning for Telugu sentiment analysis is investigated. We built a hybrid approach by combining different query selection strategy frameworks to increase more accurate training data instances with limited labeled data. Using a set of classifiers like SVM, XGBoost, and Gradient Boosted Trees (GBT), we achieved promising results with minimal error rate.

Keywords: Active learning · SVM · XGBoost · Sentiment analysis · GBT

1 Introduction

Currently, people are commonly found writing comments, reviews, blog posts in social media about trending activities in their regional languages. Unlike English, many regional languages lack resources to analyze these activities. Moreover, English has many datasets available, however, it is not the same with Telugu.

Telugu is a Dravidian language, native to India. It ranks third by the number of native speakers in India and fifteenth in the Ethnologue list[1] of the most spoken languages world-wide[2]. Over the last decade, there has been an increment in movie review sites, newspaper websites, tweets, comments and other blog-posts, etc., written in Telugu. Labeling these reviews with their sentiments would provide a brief summary to the readers.

In this paper, we attempted to perform sentiment analysis in Telugu and classify a sentence with positive or negative polarity. With the dearth of sufficient annotated sentiment data in Telugu language, we needed to increase the

---

[1] https://www.ethnologue.com/statistics/size.
[2] https://en.wikipedia.org/wiki/Telugu_language.

© Springer International Publishing AG 2017
L. Bellatreche and S. Chakravarthy (Eds.): DaWaK 2017, LNCS 10440, pp. 355–367, 2017.
DOI: 10.1007/978-3-319-64283-3_26

existing available labeled datasets. However, annotating abundant unlabeled data manually is very time-consuming, cost-ineffective and resource-intensive. To address this problem, one possible solution is to employ active learning [1]. Active learning algorithms are used in many natural language processing tasks such as sentiment analysis, text categorization etc., where only limited number of instances are actively selected as training data so as to reduce the annotation effort substantially.

In active learning, there are different query selection strategies such as uncertainty sampling [2], random sampling [3], Querying Informative and Representative Examples (QUIRE) [4], Density Weighted Uncertainty Sampling (DWUS) [5] and Query by Committee (QBC) [6]. One may use any of these strategies, however, we observed that using a hybrid of these strategies gives a better performance when compared with individual strategies. In this paper, through active learning, we developed a sentiment analysis model for Telugu language by experimenting with three different classifiers viz., support vector machines (SVM) [7], extreme gradient boosting (XGBoost) [8], gradient boosted trees (GBT) [9]. Our experiments achieved encouraging outputs with minimal error rates.

The contributions of our work are four-fold:

1. We created a word embedding model for Telugu language and examined different regions of the embedding space
2. We proposed a hybrid approach of query selection strategies in active learning
3. We adopted this approach for Telugu language and labeled many unlabeled data instances
4. We built a classification model for Telugu sentiment analysis

This paper is organized as follows: Sect. 2 explains the state-of-the-art systems related to this problem. Section 3 covers dataset generation and feature generation. In Sect. 4, we covered the system architecture, approach and the algorithm. Experimental details and results are covered in Sect. 5. Section 6 briefly discusses future work and finally concludes the paper.

2 Related Work

In this section, we give an overview of related work which is focused on: (i) Analysis of resource poor languages and their labeling techniques, (ii) Active learning and different query selection strategies, (iii) Sentiment analysis for regional languages.

There are many popular approaches like named entity recognition (NER), word sense disambiguation, part-of-speech tagging developed for resource poor languages. [10] developed a part-of speech tagger for Sindhi language. [11] developed a NER system for Arabic language. [12] provided a way of expanding the lexicon for a resource-poor language using a morphological analyzer and a web crawler for small sentences. [13] developed a dialogue system for Telugu, a resource-poor language.

There are many scenarios of active learning algorithms described in the literature: pool-based, stream-based, query synthesis, active class selection and many more discussed in [14]. Two of the most common active learning scenarios are pool- and stream-based active learning.

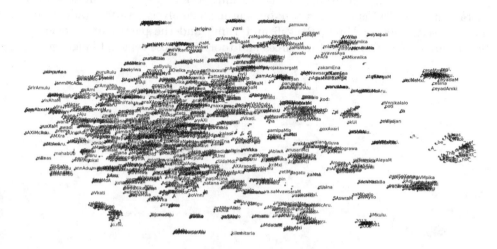

Fig. 1. Word embedding visualization

In stream-based active learning [15–17], the learner is provided with a stream of unlabeled points. On each iteration, learner must decide whether or not to request a label for a new unlabeled point. In our work, we have mainly focused on the pool-based active learning [1], where the learner is presented with both a labeled pool, an unlabeled pool in the beginning, and allowed to access a pool of (unlabeled) instances repeatedly. In pool-based active learning, three independent working groups [18–20] have proposed a similar querying function based on support vector machines.

The most popular existing work on pool-based active learning primarily focused on a reasonable approach for selecting which instance to label. Uncertainty sampling [2] is one such popular approach, which queries the instance that is most uncertain to the classifier. [21,22] studied the combinations of uncertainty sampling. [23] identified issues in uncertainty sampling and proposed a probabilistic approach (PAL) that combines different types of information. Another popular approach is QUIRE [4], which measures the representativeness and informativeness by estimating the possible label-assignments for unlabeled instances. In QBC [6,15], a query sample is chosen according to the principle of maximal disagreement of different classifiers. In DWUS [5,24], the informative instances should not only be those which are uncertain, but also those which are "representative" of the underlying distribution. In Random sampling used by [3], a query sample is chosen randomly from the underlying distribution.

[25] deals with semi-supervised sentiment classification using active deep learning. [26] focused on imbalanced class distribution situations for sentiment

358   S.S. Mukku et al.

analysis and proposed a novel active learning approach. [27] dealt with machine learning experiments with regard to sentiment analysis in blog, review and forum texts and also investigated the role of active learning techniques for reducing the number of examples to be manually annotated. [28] deals with sentiment classification of Telugu text using machine learning techniques with limited labeled data instances. [29] is a shared task on sentiment analysis on three Indian language tweets namely Hindi, Bengali and Tamil and the participants achieved low accuracies due to lack of resources and tools for these regional languages.

Fig. 2. Adjectives   **Fig. 3.** Inflected verbs   **Fig. 4.** Pronouns

3   Dataset Generation

Unlike English, Telugu has neither a large annotated dataset and tools nor any pre-trained models. Telugu data requires indispensable preprocessing to create a word embedding model, and for information extraction and sentiment extraction. For this, we used the wikipedia Telugu dump which is available in Unicode (UTF) format. This data is transliterated [30] to WX notation[3] for the ease of implementing and experimentation, which is then used as raw dataset. See the example given below showing same sentence both in UTF and WX notations:

ఏ తరహ పాత్రలోనైనా జీవించగలిగే ఒక నటుడిగా సూర్య మెప్పించాడు. (UTF Notation)
e warahA pAwralonEnA jIviMcagalige oVka natudigA sUrya meVppiMcAdu.
(WX Notation)

To generate the annotated dataset, we crawled Telugu news websites and collected the data, cleaned and preprocessed. We have given a set of rules to native Telugu speakers and had the data annotated by them. We cross checked the data by using kappa coefficient and obtained an annotation co-efficiency of 0.89. There are a total of around 1000 sentences annotated into positive and negative polarities. Similar to the raw dataset, we transliterated the annotated dataset by using UTF-WX converter [30].

---

[3] https://en.wikipedia.org/wiki/WX_notation.

In our case, we took the annotated data (D) of around 1000 sentences. Initially, we set the test data (D_T) as 200 sentences and in the remaining 800 sentences, we set 10 sentences as labeled data instances (D_L) and 790 sentences as unlabeled data instances (D_U).

3.1   Word Embeddings Generation

We used the word2vec [31] approach for generating the word embedding model. Word vectors are used to contribute to a prediction task about the next word in the sentence [32]. Initially we used Telugu raw dataset (in WX notation) as an input to word2vec for generating the word embedding model. To validate the generated word embeddings, we checked and visualized the nearness of semantically similar words using t-sne [33] shown in Fig. 1.

Using the word embedding model, we generated a 100 dimension feature vector for each sentence of annotated data (D).

Table 1. Adjectives   **Table 2.** Inflected verbs   **Table 3.** Pronouns

WX	UTF	English
wakkuva	తక్కువ	low
eVkkuva	ఎక్కువ	high
cAla	చాల	many
koVMwa	కొంత	some

WX	UTF	English
cesAdu	చేసాడు	he did
ceswAru	చేస్తారు	they do
ceyadaM	చేయడం	to do
ceswU	చేస్తూ	doing

WX	UTF	English
AmeVku	ఆమెకు	to her
Ayanaku	ఆయనకు	to him
imeV	ఇమే	she
iwanu	ఇతను	he

Figures 2, 3 and 4 show how word2vec clustered adjectives, verbs and pronouns on Telugu language data. From Fig. 2, we can observe that all adjectivals are grouped together (examples are shown in Table 1). From Fig. 3, all the inflected forms of the verbs (examples are shown in Table 2), as well as derived words are grouped together. Similarly, pronouns are clustered together depending on their case markers (examples are shown in Table 3). The clustering reflects the morphology of Telugu as shown in the paradigm of nouns, verbs, pronouns, adjectives etc., discussed in [34,35].

3.2   Feature Engineering

For generating features of annotated sentences, we used sentence2vec [36] approach. The input to the sentence2vec is a vocabulary of word vectors generated in the word2vec approach. The average of all the word vectors of the words present in the sentence is calculated. This average vector represents the sentence vector.

4   The Proposed Approach

Our proposed approach is a novel approach called Hybrid query selection approach. The approach solves the task of selecting from a candidate set of algorithms adaptively based on their contributions to the learning performance on a given data set.

4.1   Active Learning

Consider we have a limited number of labeled samples (D_L), abundant unlabeled samples (D_U) and an oracle. In an active learning setting, an active learner may pose queries from unlabeled data instances (D_U) to be labeled by an oracle and then added to labeled data instances (D_L).

4.2   Input to the System

The sentence vectors (D) generated in Sect. 3.1 are given as input to the system. Each sentence vector has a length of 100 dimensions. In the initial setup, we split the entire data (D) into three parts which are unlabeled data instances (D_U), labeled data instances (D_L) and test data (D_T).

4.3   Query Selection Strategies

In active learning, to get the informative samples from unlabeled data instances, there are different query selection strategies proposed in the literature. In uncertainty sampling, an active learner queries the instances which are most uncertain or least confident. We used least confident method in uncertainty sampling, where it queries the instance whose posterior probability of being positive is nearest to 0.5 (for binary classification). In random sampling, an active learner queries the instances randomly. QUIRE provides an organized way of combining both informative and representative sample instances emanating from min-max view of active learning. In DWUS, combining uncertainty with the density of the underlying data helps in reducing the error quickly. We used k-means clustering as a method for underlying data for building a initial training set.

Hybrid Query Selection Approach. Hybrid query selection approach emanated from the multi-arm bandit problem [37]. Multi-arm bandit problem deals with rewarding the arm(s) based on observed feedback. At iteration n, choose an arm a_n, observe the feedback $o_n(a_n)$, and obtain the reward $r_n(a_n)$. In our approach, each arm a_i represents a query selection strategy. If an arm a_i is selected at iteration n, then arm a_i is rewarded or not rewarded based on the previous observed feedback $o_{n-1}(a_i)$ and then the observed feedback $o_n(a_i)$ and reward $r_n(a_i)$ are updated.

As shown in Fig. 5, hybrid query selection approach takes both unlabeled data instances (D_U) and labeled data instances (D_L) as an input. A decision making model is built along with this approach to predict the posterior probability for each instance of D_U. After calculating the sampling query distribution $\phi(D(n))$, based on multi-arm bandit approach a best sample instance $x_i \in D$ is selected for querying. If $x_i \in D_U$, then this selected sample instance (x_i) is labeled with an oracle/labeler as y_i and added to D_L. Now the classifier (C_n) is trained on the updated D_L. This process is repeated until D_U becomes empty. The reward ($r_n(a_i)$) and observation($o_n(a_i)$) is updated by comparing the label y_i given by the oracle/labeler with the classifier ($C_n(x_i)$).

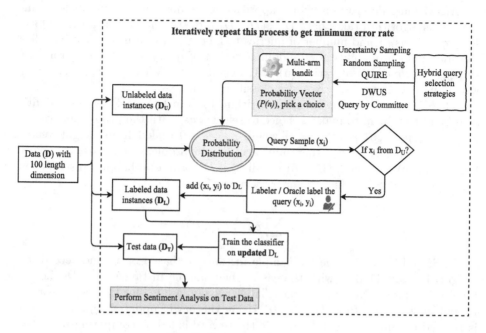

Fig. 5. System architecture

Algorithm 1. Active Learning using Hybrid Query Selection

1: **INPUT** D = {Labeled Data: D_L; Unlabeled Data: D_U}; QS = {$qs_1, qs_2, ..., qs_K$}: query selection strategies; T. Number of queries to label

2: **Initialize** $p_{min} = 1/len(QS)$: minimum probability to select qs_i from QS; n = 1; Train a classifier C_0 on D_L

3: **while** $i = 1 \rightarrow T$ **do**

4:     **Step 1:**Using Multi-arm bandit problem, adopt a weight vector W(n) which

5:     is scaled to probability vector P(n) $\in [p_{min}, 1]$

6:     **Step 2:**Using Multi-arm bandit problem, select qs_i randomly based on P(n)

7:     **Step 3:**For all the samples $x_i \in D$, calculate $q_i(n) = \sum_{l=1}^{K} p_l(n) * \phi_l(x_i)$

8:     **Step 4:**Choose a sample instance x_i based on $q_i(n)$ and record it's weight

9:     $W_n = 1/q_i(n)$

10:     **if** $x_i \in D_U$ **then** (i.e., from unlabeled data)

11:         **a:**label the x_i as y_i and move (x_i, y_i) from $D_U \rightarrow D_L$

12:         **b:**Train the classifier C_n on updated D_L

13:         **c:**$i \rightarrow i + 1$

14:     **end if**

15:     **Step 5:**Caluculate the reward $r = W_n * (C_n(x_i) == y_i)$ and add r to all

16:     query selection strategies which suggest x_i

17:     **Step 6:**$n \rightarrow n + 1$

18: **end while**

Algorithm 1 describes a hybrid approach for annotating the unlabeled samples. This algorithm takes data D = (D_U, D_L), query selection strategies QS = $\{qs_1, qs_2, ..., qs_K\}$ and number of queries to label (T) as an input. We initialize the minimum probability for random selection of query strategy qs_i from QS as $p_{min} = 1/len(QS)$, iteration (n) = 1, weight vector W(n) = $\{w_1, w_2, ..., w_K\}$ where $w_i = 1$; $1 <= i <= K$ and a classifier (C_n).

There are 5 major steps in our algorithm. In Step 1 (Line 4), we used multi-arm bandit problem to make a choice to select a qs_i with the probability vector P(n) = $\{p_1, p_2, ..., p_K\} \in [p_{min}, 1]$ where $p_{min} > 0$ scaled from weight vector W(n). In Step 2 (Line 5) using multi-arm bandit problem, select qs_i randomly based on P(n). In Step 3 (Line 6) for all $x_i \in D$, form a probability distribution ϕ_i using $qs_i \in$ QS and find the query vector q_i using Eq. 1.

$$q_i = \sum_{l=1}^{K} p_l(n) * \phi_l(x_i) \tag{1}$$

In Step 4 (Line 7), select a sample instance x_i based on q_i and record it's weight. Lines 8–11 deal with the cases when sample instance $x_i \in D_U$ i.e. no label exists for x_i. Now using oracle/labeler assign a label for the instance x_i as y_i and move (x_i, y_i) from $D_U \rightarrow D_L$. Using the newly updated D_L, a classifier is trained to build a model. In Step 5, the reward is fed by comparing predicted label $(C_n(x_i))$ and y_i. In Step 6, the iteration number (n) is incremented. This process (i.e., Steps 1–6) is repeated until T.

4.4   Classification Model for Telugu Sentiment Analysis

From the Sect. 4.3, the updated labeled data set (D_L) is used for training the classification model for sentiment analysis. We built the hybrid query selection approach with three models viz., SVM [7], GBT, and XGBoost [8] and picked the best underlying model which gives the lowest error rate on test data (D_T). Using the trained model, we predicted the labels (as positive or negative) for test data (D_T) and found the error rate by comparing true labels with predicted labels. Later using this model, we annotated around 50,000 Telugu sentences and made the sentiment analysis for Telugu facile.

5   Experiments and Results

In order to observe the behavior of error rate with respect to number of queries, we have conducted experiments using uncertainty sampling, random sampling, QUIRE, QBC, DWUS, and Hybrid approach for each classifier namely SVM [7], XGBoost [8], and GBT. Parameters are set for each classifier using cross validation. Since initially there are few test data instances for tuning parameters using cross validation, we used the same hyper parameters obtained based on training and validation dataset samples as shown in Table 4.

Figures 6 show mean error rate over 20 iterations with respect to number of queries and Figs. 7 show median error rate over 20 iterations with respect to

Table 4. Parameters used for each classifier

Classifier	Parameters used
XGB	N_estimators = 100, Learning rate = 1.0, Max_depth = 3
SVM	Kernel = Linear, Regularization, -Parameter(C) = 10
GBT	N_estimators = 100, Learning rate = 1.0, Max_depth = 3

number of queries. From Figs. 6a and 7a, we can observe that for the underlying classifier XGBoost [8], hybrid query selection and uncertainty sampling methods shows a constant error decrease rate as the number of queries are increasing, on the other hand we can observe from the same image that the query selection criteria: DWUS and random sampling are not maintaining a proper reduction in the error rate. Although QUIRE and QBC are showing slight decrement in the error rate on the whole but they are fluctuating too often. However for the SVM classifier as shown in Figs. 6b and 7b, uncertainty sampling and DWUS performed well and shows constant error reduction. The reason is that at each step SVM improves their estimation of uncertainty over a certain training instance changes. The other classifier we used GBT as shown in Figs. 6c and 7c, Hybrid query selection performs well with low error rate.

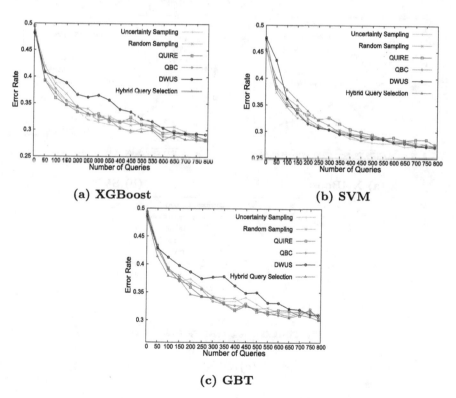

(a) **XGBoost**

(b) **SVM**

(c) **GBT**

Fig. 6. Number of queries vs. mean error rate

It can be observed from the Figs. 6 and 7 that each query selection strategy performs strongly with some classifier but performs badly on the others. Out of all classifiers, XGBoost shows better learning by maintaining reasonable decrement in the error rate. We also observed that XGBoost classifier in combination with Hybrid query selection approach is performing relatively better compared to other query selection approaches. Thus, for the purpose of testing, we used the model obtained from XGBoost in combination with Hybrid query selection approach.

Table 5 shows the average error behavior of each query selection strategy with respect to the classifier. From the Table 5, we observed that the proposed hybrid query selection method's ability of learning semantics of the sentences and accurate prediction producing lower error rate with the two classifiers used being XGBoost and GBT. However, a classifier like SVM produces a lower error rate with both uncertainty and DWUS methods but not with hybrid query selection. We can also observe that, since random and QUIRE query selection methods lack the certainty of selecting queries, their error rate is not a monotonically decreasing graph.

Table 6 shows the precision, recall and F1-scores [38] of test data samples by applying best query selection strategies obtained in the training phase with

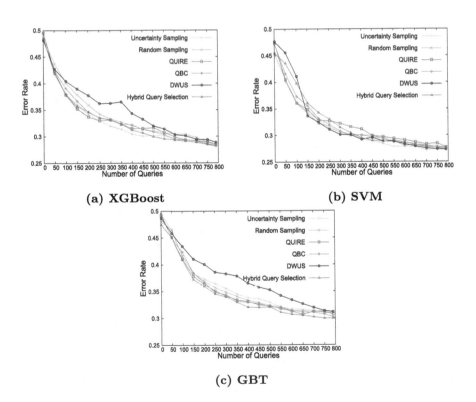

Fig. 7. Number of queries vs. median error rate

Table 5. Minimum average error of each classifier by using query selection strategies

Strategy	XGBoost	SVM	GBT
US	0.310787	**0.305643**	0.349730
RS	0.332918	0.318949	0.346651
QUIRE	0.32543	0.315628	0.331627
QBC	0.327081	0.311489	0.333589
DWUS	0.343632	0.309189	0.366742
HQS	**0.304801**	0.310577	**0.325397**

US: Uncertainty Sampling, RS: Random Sampling
HQS: Hybrid Query Selection

respect to each classifier. Our test data had 200 samples in which 99 samples are positive and 101 samples are negative. From the Table 6, we can observe that the XGBoost classifier is performing slightly better than the other two classifiers. As the results obtained with all three of classifiers are nearly similar, we are planning to use meta-learning approach [39] by combining all three classifiers for annotating all unlabeled samples.

Table 6. Prec., recall, F1-score on test data

Classifier	Strategy	Precision	Recall	F1-score
XGBoost	**HQS**	**0.79**	**0.79**	**0.79**
SVM	US	0.74	0.74	0.74
GBT	HQS	0.73	0.73	0.73

US: Uncertainty Sampling, HQS: Hybrid Query Selection

6   Conclusion

In this paper, we addressed a problem of annotating labels for unlabeled Telugu data using limited labeled data. To handle this issue, we exploited the advantages of active learning and came up with a hybrid approach. In the hybrid approach, we used five different query selection strategies and compared the error rate with each of the individual query selection strategies. We built a sentiment analysis model for Telugu and analyzed with various classifiers.

6.1   Future Work

Using the three classifiers XGBoost, SVM and GBT with their best query selection strategies, we want to perform meta learning on meta-data obtained from the three classifiers. Using meta learning, we will label huge unlabeled data instances and later perform sentiment analysis task using deep learning techniques. We would also like to extend this problem to multi-class classification. We also look to address this for other Dravidian languages as well.

References

1. Settles, B.: Active learning literature survey. Technical report (2010)
2. Lewis, D.D.: A sequential algorithm for training text classifiers: corrigendum and additional data, pp. 13–19 (1995)
3. Kolar Rajagopal, A., Subramanian, R., Ricci, E., Vieriu, R.L., Lanz, O., Kalpathi, R., Sebe, N.: Exploring transfer learning approaches for head pose classification from multi-view surveillance images. Int. J. Comput. Vision **109**, 146–167 (2014)
4. Huang, S.J., Jin, R., Zhou, Z.H.: Active learning by querying informative and representative examples. In: Proceedings of the 23rd International Conference on Neural Information Processing Systems, pp. 892–900 (2010)
5. Settles, B., Craven, M.: An analysis of active learning strategies for sequence labeling tasks. In: EMNLP 2008, pp. 1070–1079. Association for Computational Linguistics (2008)
6. Seung, H.S., Opper, M., Sompolinsky, H.: Query by committee. In: Proceedings of the Fifth Annual Workshop on Computational Learning Theory, pp. 287–294 (1992)
7. Cortes, C., Vapnik, V.: Support-vector networks. Mach. Learn. **20**, 273–297 (1995)
8. Chen, T., Guestrin, C.: XGBoost: a scalable tree boosting system. In: Proceedings of the 22nd ACM SIGKDD International Conference on Knowledge Discovery and Data Mining, pp. 785–794. ACM (2016)
9. Ganjisaffar, Y., Caruana, R., Lopes, C.V.: Bagging gradient-boosted trees for high precision, low variance ranking models. In: Proceedings of the 34th International ACM SIGIR Conference on Research and Development in Information Retrieval, pp. 85–94. ACM (2011)
10. Motlani, R., Lalwani, H., Shrivastava, M., Sharma, D.M.: Developing part-of-speech tagger for a resource poor language: Sindhi
11. Gad-Elrab, M.H., Yosef, M.A., Weikum, G.: Named entity disambiguation for resource-poor languages. In: Proceedings of the Eighth Workshop on Exploiting Semantic Annotations in Information Retrieval, ESAIR 2015, pp. 29–34 (2015)
12. Gasser, M.: Expanding the lexicon for a resource-poor language using a morphological analyzer and a web crawler. In: Proceedings of the International Conference on Language Resources and Evaluation, LREC 2010, Valletta, Malta, 17–23 May 2010
13. Sravanthi, M.C., Prathyusha, K., Mamidi, R.: A Dialogue System for Telugu, a Resource-Poor Language, pp. 364–374 (2015)
14. Settles, B.: Active learning. Synth. Lect. Artif. Intell. Mach. Learn. **6**(1), 1–114 (2012)
15. Freund, Y., Seung, H.S., Shamir, E., Tishby, N.: Selective sampling using the query by committee algorithm. Mach. Learn. **28**, 133–168 (1997)
16. Cohn, D., Atlas, L., Ladner, R.: Improving generalization with active learning. Mach. Learn. **15**, 201–221 (1994)
17. Chu, W., Zinkevich, M., Li, L., Thomas, A., Tseng, B.: Unbiased online active learning in data streams. In: Proceedings of the 17th ACM SIGKDD International Conference on Knowledge Discovery and Data Mining, pp. 195–203 (2011)

18. Tong, S., Koller, D.: Support vector machine active learning with applications to text classification. J. Mach. Learn. Res. **2**, 45–66 (2001)
19. Campbell, C., Cristianini, N., Smola, A., et al.: Query learning with large margin classifiers. In: ICML, pp. 111–118 (2000)
20. Kremer, J., Steenstrup Pedersen, K., Igel, C.: Active learning with support vector machines. Wiley Interdiscip. Rev. Data Min. Knowl. Disc. **4**(4), 313–326 (2014)
21. Fu, Y., Zhu, X., Li, B.: A survey on instance selection for active learning (2012)
22. Reitmaier, T., Sick, B.: Let us know your decision: pool-based active training of a generative classifier with the selection strategy 4DS. Inf. Sci. **230**, 106–131 (2013)
23. Kottke, D., Krempl, G., Spiliopoulou, M.: Probabilistic active learning in datastreams. In: Fromont, E., Bie, T., Leeuwen, M. (eds.) IDA 2015. LNCS, vol. 9385, pp. 145–157. Springer, Cham (2015). doi:10.1007/978-3-319-24465-5_13
24. Settles, B.: Curious machines: active learning with structured instances. ProQuest (2008)
25. Zhou, S., Chen, Q., Wang, X.: Active deep learning method for semi-supervised sentiment classification. Neurocomputing **120**, 536–546 (2013)
26. Li, S., Ju, S., Zhou, G., Li, X.: Active learning for imbalanced sentiment classification. In: Proceedings of the 2012 Joint Conference on Empirical Methods in Natural Language Processing and Computational Natural Language Learning, Association for Computational Linguistics, pp. 139–148 (2012)
27. Boiy, E., Moens, M.F.: A machine learning approach to sentiment analysis in multilingual web texts. Inf. Retrieval **12**(5), 526–558 (2009)
28. Mukku, S.S., Choudhary, N., Mamidi, R.: Enhanced sentiment classification of Telugu text using ML techniques. In: Proceedings of 25th International Joint Conference on Artificial Intelligence, p. 29 (2016)
29. Patra, B.G., Das, D., Das, A., Prasath, R.: Shared task on sentiment analysis in Indian languages (SAIL) tweets - an overview. In: Prasath, R., Vuppala, A.K., Kathirvalavakumar, T. (eds.) MIKE 2015. LNCS, vol. 9468, pp. 650–655. Springer, Cham (2015). doi:10.1007/978-3-319-26832-3_61
30. Gupta, R., Goyal, P., Diwakar, S.: Transliteration among Indian languages using WX notation. In: KONVENS, pp. 147–150 (2010)
31. Mikolov, T., Sutskever, I., Chen, K., Corrado, G.S., Dean, J.: Distributed representations of words and phrases and their compositionality. In: Advances in Neural Information Processing Systems, pp. 3111–3119 (2013)
32. Mikolov, T., Chen, K., Corrado, G., Dean, J.: Efficient estimation of word representations in vector space. arXiv preprint arXiv:1301.3781 (2013)
33. van der Maaten, L., Hinton, G.: Visualizing data using t-SNE. J. Mach. Learn. Res. **9**, 2579–2605 (2008)
34. Krishnamurti, B., Gwynn, J.P.L.: A Grammar of Modern Telugu. Oxford University Press, New York (1985)
35. Krishnamurthi, B.: Telugu verbal bases: a comparative and descriptive study (1961)
36. Le, Q.V., Mikolov, T.: Distributed representations of sentences and documents
37. Mahajan, A., Teneketzis, D.: Multi-armed bandit problems. In: Hero, A.O., Castañón, D.A., Cochran, D., Kastella, K. (eds.) Foundations and Applications of Sensor Management, pp. 121–151. Springer, Boston (2008)
38. Davis, J., Goadrich, M.: The relationship between precision-recall and roc curves. In: Proceedings of the 23rd International Conference on Machine Learning, ICML 2006, pp. 233–240 (2006)
39. Seewald, A.K.: Meta-learning for stacked classification. Audiology **24**(226), 69

Belief Temporal Analysis of Expert Users: Case Study Stack Overflow

Dorra Attiaoui[1,2(✉)], Arnaud Martin[1], and Boutheina Ben Yaghlane[3]

[1] DRIUD IRISA, University of Rennes 1, Rennes, France
dorra.attiaoui@irisa.fr, Arnaud.Martin@univ-rennes1.fr
[2] LARODEC, University of Tunis, Tunis, Tunisia
[3] LARODEC, University of Carthage, Tunis, Tunisia
boutheina.yaghlane@ihec.rnu.tn

Abstract. Question Answering communities have known a large expansion over the last few years. Reliable people sharing their knowledge are not that numerous. Thus, detecting experts since their first contributions can be considered as a challenge. We are interested in studying the activity of these platforms' users during a defined period of time. As the data collected is not always reliable, imperfections can occur. In order to manage these imperfections, we choose to use the mathematical background offered by the theory of belief functions. People say that the more time they spend within a community, the more knowledge they acquire. We investigate this assumption in this paper by studying the behavior of users without taking into consideration the reputation system proposed by Stack Overflow. Experiments with real data from Stack Overflow demonstrate that this model can be applied to any expertise detection problem. Moreover, it allows to identify potential future experts. The analysis allows us to study the behavior of experts and non expert users over time spent in the community. We can see that some users keep on being reliable while others do gain knowledge and improve their expertise measure.

Keywords: Question answering community · Theory of belief functions · Expertise measure · Classification

1 Introduction

With the emergence of Question Answering Communities (Q&A C), several platforms were developed aiming to help people. The main challenge of these websites is to provide helpful, quick and well organized answers for any posted question regarding any specific topic.

One of the most popular platforms is Stack Overflow (SO)[1]. It is the largest online community for programmers. Here, users can post questions, answers them, vote positively or negatively for both answers and questions in order to express their opinion on the quality of the posts.

---

[1] http://stackoverflow.com.

© Springer International Publishing AG 2017
L. Bellatreche and S. Chakravarthy (Eds.): DaWaK 2017, LNCS 10440, pp. 368–382, 2017.
DOI: 10.1007/978-3-319-64283-3_27

Stack Overflow proposes a reputation system to reward active users. Actually, reputation[2] is the summery of users' activity in the web site. It is earned by convincing other users that he/she knows what he/she is talking about. Indeed, reputation reflects how involved a user is in the community and how other people see him/her. If this value is high, it means that a user is able to post fair questions or/and answers and how well he/she can communicate and interact with his/her peers.It also means that we can be in presence of a knowledgeable person. However, we assume this measurement as flawed. Reputation may support competitive gain of points rather than fair contributions in the community. Note that in this paper, we do not consider the reputation because it is a rough measurement of expertise according to only other people's opinion, and not founded on both their activities and opinions [19].

Detecting experts in online communities have been wildly investigated. We can distinguish two different methods: ranking based approaches and attribute based approaches. [15]. On one hand, the ranking based approaches intent to measure a score per user then select the top users [18,19] . On the other hand, attributes based approaches aim to identify a number of features for the users and later apply machine learning techniques in order to classify them as experts and non-expert users [1,12].

However, the literature suffers from few limitations like: (1) the dependence between training data and labels results on supervised machine learning, (2) high time consuming processes, and (3) the proposed approaches consider all the manipulated data as certain and perfect. Thus, this can not be taken into consideration, especially when we are dealing with real world applications. Several theories were proposed to manage uncertainty such as probability theory [14], possibility theory [6] and the theory of belief functions [2,16]. The latter can be presented as a generalization of the other theories. Besides it offers a rich tool able to manage different types of data imperfections. When manipulating uncertainty, information fusion can be an interesting solution to obtain relevant information. Data fusion based on the theory of belief functions has been widely used in classification, image processing [8], clustering [4], etc. and more recently in social networks [10].

In this paper, we propose a new approach of measuring expertise and analyzing the behavior of experts and potential experts during a period of time based on uncertainty theories. The remainder of this paper is organized as follows. In Sect. 2, an overview on experts detection in social networks and in Stack Overflow. In Sect. 3, we present the basic necessary background related to the theory of belief functions. Section 4 details the approach for the representation of the proposed expertise measure and experts detection. Then in Sect. 5, we present the results from experiments on Stack Overflow's data.

---

[2] http://stackoverflow.com/help/whats-reputation.

2   Related Work

Most of the users of Stack Overflow aim to win as much reputation points as possible in order to obtain privileges like creating tags, moderating the forum etc. The reputation is defined according to the system presented in Table 1.

Table 1. Gratification system of Stack Overflow

Action	Reputation
Answer voted up	+10
Question voted up	+5
Accepted answer	+15 (+2 to question asker)
Question voted down	−2
Answer voted down	−2 (−1 to voter)
Spammed answers	−100

Every posted question or answer can be submitted either to positive or negative votes. A positive vote is a reward for the author, while the negative one penalizes him. Each person who posts a question is allowed to choose the best answer that seems to be the most helpful allowing his/her owner to gain reputation points.

Several researches focused on detecting experts in Stack Overflow. For [12], experts are known to provide the best answers in a very short time. They are more reactive and their answers are more useful than usual users. For [9], the authors proposed an analysis of Stack Overflow's reputation system. They focused on the contributors participation model. They consider the reputation as measurement of expertise. Any user with a reputation greater than 2400 points is an expert. However, their approach seems to be strict because it is only based on the value of the reputation gathered during users' activity in the platform. Another approach is proposed in [19] that is not founded on the reputation measure. They propose a metric called "Mean Expertise Contribution" that takes into account two indices: the debate generated by a question and the utility of the provided answers. The first index is related to the number of answers proposed for a given question. The second one is calculated according to the rating of an answer among all the answers provided.

Some other researches were interested in identifying experts and potential future experts in Stack Overflow using temporal analysis. For [11], authors modeled users' behavior based on their early participation in the community and showed that they could use classification as well as ranking algorithms to identify potential experts. They proposed that experts can be effectively identified from their early behavior. In [13], authors considered that expertise is present from the beginning and does not increase with the time spent in the community. Recently, [5] defined early expertise based on the number of best answers given by a user. Besides, they proposed an approach based on the combination of large number of textual, behavioral and time-aware features for detecting early expertise.

In [7], the authors identified three levels of uncertainty in question answering communities. The first level is related to the extraction and integration of the data. The second one deals with information sources, meaning the users of these platforms. The third level covers the uncertainty of the information itself. In the considered case, we are more interested in the evaluation of the sources and the part of uncertainty related to them. The main issue in these communities is that we are dealing with users that we do not usually have an *a priori* knowledge about them. We ignore everything about the sources' reliability, or expertise. In order to deal with this uncertainty, we will use the mathematical background provided by the theory of belief functions. This will help us to consider the problem of early identification of potential experts with an uncertain point of view.

3   Theory of Belief Functions: An Overview

This section recalls the necessary background notions related to the theory of belief functions. This theory has been developed by Dempster in his work on upper and lower probabilities [2]. Afterwards, it was formalized in a mathematical framework by Shafer in [16]. This theory is able to deal and represent imperfect (uncertain, imprecise and /or incomplete) information.

Let us consider a variable taking values in a finite set $\Omega = \{\omega_1, \cdots, \omega_n\}$ called *the frame of discernment*.

A *basic belief assignment* (*bba*) is defined on the set of all subsets of Ω, named power set and noted 2^{Ω}. It affects a real value from $[0, 1]$ to every subset of 2^{Ω} reflecting sources amount of belief on this subset. A bba m verifies:

$$\sum_{X \subseteq \Omega} m(X) = 1. \tag{1}$$

3.1   Particular Belief Functions

Mass function is the common representation of evidential knowledge. Basic belief masses are degrees of support justified by available evidences. This section recalls some particular mass functions.

Categorical Mass Functions. A categorical mass function is a normalized mass function which has a unique focal element X^*. This mass function is noted $m(X)$ and defined as follows:

$$m_{X^*}(X) = \begin{cases} 1 \text{ if } X = X^* \subset \Omega \\ 0 \ \forall \ X \subseteq \ \Omega \text{ and } X \neq X^* \end{cases} \tag{2}$$

We distinguish two particular cases of categorical mass functions: the vacuous mass functions when $X^* = \Omega$ and the contradictory mass functions if $X^* = \emptyset$.

Vacuous Mass Functions. A vacuous mass function is a particular categorical mass function focused on Ω. It means that a vacuous mass function is normalized and has a unique focal element which is Ω. This type of mass functions is defined as follows:

$$m_\Omega(X) = \begin{cases} 1 \text{ if } X = \Omega \\ 0 \text{ otherwise} \end{cases} \tag{3}$$

Vacuous mass function emphasizes the case of total ignorance.

Simple Support Mass Functions. Simple support mass functions are a special type that allow us to model both of the uncertainty and imprecision according the following equation:

$$\begin{cases} m(X) = 1 - \omega, \, X \subset \Omega \\ m(\Omega) = \omega \\ m(Y) = 0, \, Y \neq X \subset \Omega \end{cases} \tag{4}$$

where the mass on $m(\Omega)$ represents the ignorance.

In the theory of belief function, Dempster in [2] proposed the first combination rule. It is defined for two bbas $m_1, m_2, \forall X \in 2^\Omega$ with $X \neq \emptyset$ by:

$$m_{DS}(X) = \frac{1}{1-k} \sum_{A \cap B = X} m_1(A)m_2(B), \tag{5}$$

where k is generally called the inconstancy of the combination, defined by $k = \sum_{A \cap B = \emptyset} m_1(A)m_2(B)$ and $1 - k$ is a normalization constant.

3.2 Discounting

Sometimes, it is possible to quantify the reliability of the body of evidence assessing degrees of support. The reliability of information sources reflects both its degrees of expertise and trust. When handling a mass function, we have to take into account the degree of reliability of its source. the degree of reliability of a source is taken into account by integrating it into all its mass functions. Using discounting operation in belief functions was first introduced in [16].

Discounting a mass function m consists in weighting every mass $m(X)$ by a coefficient $\alpha \in [0, 1]$ called reliability; α is the discount rate. The *bba* is discounted as follows:

$$\begin{cases} {}^\alpha m(X) = \alpha m(X) \qquad \forall, X \subset 2^\Omega \setminus \Omega \\ {}^\alpha m(\Omega) = 1 - (\alpha(1 - m(\Omega)) \end{cases} \tag{6}$$

3.3   Decision Making

In order to make decision within the mathematical background of the theory of belief functions, [17] proposed to transform mass functions into probabilities (called BetP) using the pignistic probability transformation.

To do so, it transforms a *bba* m into a probability measure for all $X \in 2^{\Omega}$:

$$BetP(X) = \sum_{Y \neq \emptyset} \frac{\mid X \cap Y \mid}{\mid Y \mid} \frac{m(Y)}{1 - m(\emptyset)}. \tag{7}$$

where $\mid Y \mid$ is the cardinality of Y.

4   Belief Model of Users in Stack Overflow

In this section, we detail the proposed approach that follows three main phases. In the first, we define the hypothesis describing each category of user and how do they behave in online communities. Then, parameters are estimated and mass functions are constructed. Finally, these latter are integrated in the general combination process. We explain in more details these two big steps in what follows.

First, we present some features related to very user i:

- **Number of votes related to answers** (AV_i): the sum of positive votes collected by posted questions and answers.
- **Number of votes related to questions** (QV_i): the sum of negative votes collected by posted questions and answers.
- **Time activity:** time of the activity of users from their registration to their last connection.
- **Number of posted questions** $(NbQu_i)$: number of questions posted in the dataset during the time activity of a user.
- **Number of posted answers** $(NbAn_i)$: number of answers provided in the dataset during the time activity of a user.
- **Number of posted answers** $(NbAccAn_i)$: number the answers chosen as the best.

4.1   Hypothesis

We applied an ascending hierarchical classification on the dataset allowing us to distinguish between three types of users in online communities. Here, we present these classes and the hypothesis proposed in order to identify each one of them.

- **Occasionals (O):** these users represent the major part of members on the platform. They do not have a lot of knowledge. They occur occasionally only when they need an answer to a specific question that have not been treated before.

- **Apprentices (A):** these users may have some expertise in a given topic. They aim to increase their reputation. To do so, they post a lot of answers that are not always very useful. The quality of their posts is not guaranteed and their answers can be down-voted.
- **Experts (E):** these users are very reliable and recognized by the community. They provide a considerable number of useful answers that are chosen as the best ones. They are very active in the platform and guarantee a high quality content.

According to the previous presentation of the classes of users, we can define the following hypothesis:

Hypothesis 1. *If a user has a high score of answers this might mean that this person is an expert rewarded for the answers provided.*

Hypothesis 2. *If a user has a high score of questions score this might mean that this person is an apprentice seeking for information, and rewarded for posting well asked and interesting questions*

Hypothesis 3. *If a user has a high number of answers posted this can be justified by two facts. First, this person is an expert, providing high quality content. Second, it can be an apprentice trying to become an expert by proving to the community that he/she can be as reliable as an expert.*

Hypothesis 4. *If a user has a high number of questions posted this can represent either an expert or an apprentice. Both of them ask a lot of questions.*

Hypothesis 5. *If a user has a high number of accepted answers this can only represent experts. Experts are frequently chosen as the most helpful answers providers.*

4.2   Definition of Mass Functions

In this section we detail the mathematical model that defines the hypothesis presented bellow. For each hypothesis, we determine how to define the mass functions in order to represent the data relative to each user.

Each user u is characterized by the following features:

- According to the Hypothesis 1, a high score on answers is represented by a mass function on the focal element "**Expert**" (E) and the remainder is given to the ignorance, for a user i:

$$m_1^i(E) = \alpha_1(1 - e^{-\gamma_1 AV_i})$$
$$m_1^i(\Omega) = \alpha_1 e^{-\gamma_1 AV_i}$$

$$(8)$$

- According to the Hypothesis 2, a high score on questions is represented by a mass function on the focal element "**Apprentice**" (A) and the remainder is given to the ignorance, for a user i:

$$m_2^i(A) = \alpha_2(1 - e^{-\gamma_2 QV_i})$$
$$m_2^i(\Omega) = \alpha_2 \, e^{-\gamma_1 QV_i}$$

(9)

- According to the Hypothesis 3, a high number of posted questions is represented by a mass on the union of two classes "**Apprentice ∪ Expert**". Otherwise, when this value is low it is affected to the "**Occasional**" (O) and the reminder to the ignorance. When a mass is on the union, this means that we can not decide which one of these classes is concerned by the mass. For a user i:

$$m_3^i(E \cup A) = \alpha_3\left(1 - e^{-\gamma_3 NbQu_i}\right)$$
$$m_3^i(O) = \alpha_3 e^{-\gamma_3 NbQu_i}$$
$$m_3^i(\Omega) = 1 - \alpha_3$$

(10)

- According to the Hypothesis 4, a high number of answers is represented by a mass on the union of "**Apprentice ∪ Expert**" while on the opposite situation the mass is transferred to the "**Occasional**" and the reminder to the ignorance. For a user i:

$$m_4^i(E \cup A) = \alpha_4(1 - e^{-\gamma_4 NbAn_i})$$
$$m_4^i(O) = \alpha_4 e^{-\gamma_4 NbAn_i}$$
$$m_4^i(\Omega) = 1 - \alpha_4$$

(11)

- According to the Hypothesis 5, a high number of accepted answers is represented by a mass on the focal element "**Expert**" and the reminder to the ignorance, for a user i:

$$m_5^i(E) = \alpha_5(1 - e^{-\gamma_5 NbAccAn_i})$$
$$m_5^i(\Omega) = \alpha_5 \, e^{-\gamma_5 NbAccAn_i}$$

(12)

In the previous equations, we fix $\alpha_1, \alpha_5 = 0.9$, $\alpha_2 = 1$, $\alpha_3 = 0.8$ and $\alpha_4 = 0.5$. The values are fixed after several experimentation's in order to have the best representation of each class of users. These values are used to represent the ingnorance in every mass function as described in [3]. As the apprentices are modeled only one time as focal element in Eq. (10) unlike experts and occasionnals, we choose to affect the value of 1 to α_2. For γ after several experimentation, we decide to keep it as the maximum value of any attribute divided by 100.

Example 1: Let us consider a question posted by a user u_1 in the online community. Two other users u_2 and u_3 will read the question and will try to identify to which class can the asker belong: occasional, apprentice or expert. Thus the frame of discernment Ω is composed by Occasional O, Apprentice A, Expert E, where $\Omega = \{O, A, E\}$.

The corresponding power set $2^\Omega = \{\emptyset, O, A, O \cup A, E, O \cup E, A \cup E, \Omega\}$.

To express their beliefs on the question asker, u_2 will say that this person is an expert at 80% and 20% ignorance (u_2 does not know). User u_3 would say this person could be an expert or an apprentice with a belief of 70% and 20% as an occasional and 10% of ignorance. We obtain the following mass functions:

$$m_{u2}(E) = 0.8, \quad m_{u2}(\Omega) = 0.2 \tag{13}$$

$$m_{u3}(E \cup A) = 0.7, \quad m_{u3}(O) = 0.2, \quad m_{u3}(\Omega) = 0.1 \tag{14}$$

Based on these beliefs, we will explain later how do we proceed to obtain to which class does user u_1 belong.

4.3 Data Aggregation and Decision Making

Coming to the combination of the belief functions for each feature, we adopt the Dempster's combination rule described in Eq. (5) for every time bucket. Finally, the decision process is assured by the pignistic probability (BetP) described in Eq. (7). The final estimated classification label is the one having the higher pignistic probability (Fig. 1).

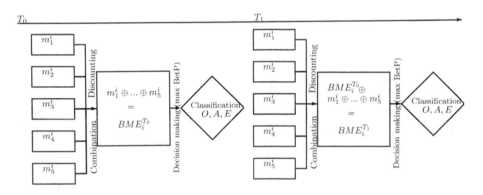

Fig. 1. Flow chart of the Belief Measure of Expertise and decision making

The bucketing system provides us an overview of users' activity in the for a given period of time. For a user, we can then calculate the number of questions, answers and accepted answers posted by a given user during each time snap.

For every 30 days in the data set, we calculate for each user the number of questions asked, answers posted, the scores generated and the number of accepted answers. Each value is transformed into mass functions using Eqs. (8) to (12).

Thus for every period, we obtain for every user 5 features: 5 mass functions for the number of questions, number of answers, score of questions, score of answers and a mass function for the accepted answers.

At t_0 we combine these mass functions using the Demspter's combination rule presented in Eq. (5). Next we apply the pignistic probability and classify the user into Expert, Apprentice or Occasional for this specific time bucket. In t_1, we use the results of the previous period and combine them with the mass functions of this actual period. After, we define the class of belonging. We maintain this combination and classification process for the entire dataset.

The combination process allows us to estimate the actual belief expertise (noted BME) for each user during a period is expressed by the following equation:

$$BME_{t_1}(u_i) = {}^{\alpha^T}BME_{t_0}^i \oplus m_1^i \oplus m_2^i \oplus ... \oplus m_5^i \qquad (15)$$

where α^T is the discounting coefficient related to the time activity of a user. The value $\alpha^T =$ is the inverse of the number of days since the user first connected to the platform. The symbol $\oplus$ represents the operator of combination.

BME will be in the interval [0,1]. This process of combination and classification for every time bucket allows to follow the progress of users monthly during a defined period of time. Furthermore, based on that, we can distinguish clearly the evolution of each user during their time activity within the community. Thus, we can also detect potential experts on the onset of their participation.

Example 2. We keep the same belief functions described in Example 1. We want to determine to which class does user u_1 belong to. Let's assume that the user has been active on the platform for only 30 days. After defining the mass functions previously, during this step, we will first discount the masses m_{u_2} and m_{u_3} based on his time of activity by using $\alpha_{u_1}^T = (1/30)$ and then, combine them using the Dempster's combination rule. We obtain the following results:

$$m_\oplus(O) = 0, \quad m_\oplus(E) = 0.0292, \qquad (16)$$
$$m_\oplus(A \cup E) = 0.0292, \quad m_\oplus(\Omega) = 0.9441$$

After applying the pignistic transformation, we obtain the following probabilities:

$$Bet(O) = 0.3147, \quad Bet(A) = 0.3293, \quad Bet(E) = 0.3560. \qquad (17)$$

We choose the highest probability, thus the user is defined as an Expert.

5   Experimental Evaluation and Analysis

The first step in this analysis of users is to build the temporal series of number of questions, answers and accepted answers given by users during a period of time. To do this, we divide the periods of the dataset into monthly and bi-weekly buckets. The begging of the first bucket is be the time of the earliest question in the dataset, noted t_0, and the end of the first bucket would be $t_0 + 30$ days. We work on data covering 15 months allowing us to have 15 time snaps for monthly buckets.

5.1   Time Analysis of the Data Set

Figure 2 shows the cumulative distribution functions (CDF) related to the mean of the number of questions posted by contributors over a period of time of several months. We notice that apprentices ask more questions than the other users. This is due to the fact that they are seeking for information, and that they lack of knowledge.

We also witness that experts do also questions. This can be justified by the idea that experts can not know everything about anything: they are knowledgeable on some specific topics only. Moreover, they are known to post difficult questions that only other experts can answer.

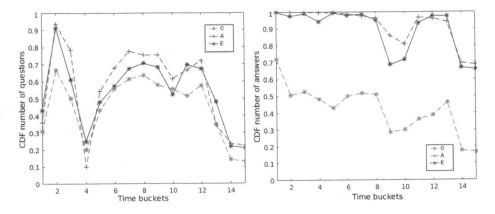

Fig. 2. CDF number of questions   **Fig. 3.** CDF number of answers

The CDF related to the mean of the number of answers is represented in Fig. 3. We notice the same phenomenon described for the CDF of questions. At the beginning both experts and apprentices have almost similar values. However, over time, experts are less and less present within the community. They do not post as much answers as the apprentices. Though, the latter users try to provide a lot of contributions because they are motivated by gaining reputation points in Stack Overflow, sometimes without taking care of the quality of their posts. The fact they anyone posts answers may discourage experts to sharing their knowledge on the platform. This can cause the decrease of their interest on posting helpful answers.

The number of accepted answers is a very important indicator on how to evaluate the expertise of a user in Stack Overflow. Over time buckets, the CDF of the number of best answers provided by each class of users is presented in Fig. 4. We notice that apprentices becoming future experts post a lot of answers that are considered as best. The more time they spend on the community the more expertise they have. However, both experts and apprentices lose interest in the community which is reflected by the decrease of their contributions over time.

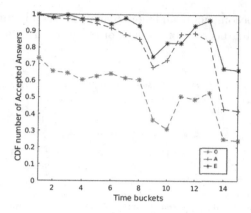

Fig. 4. CDF number of accepted answers

5.2   Analysis of Users' Behavior over Time

In this section we provide an analysis of the activity of the users during the 15 months of the dataset. As described before, we classify users according to the belief expertise measure presented in Eq. (15) for every time bucket. We randomly choose n users from the big dataset and we obtain the results presented in Fig. 5.

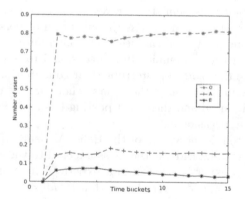

Fig. 5. Evolution of the percentage of Occasionals, Apprentices Experts per time bucket

First of all, we notice that the number of Occasionals is always the highest class of users present in the platform. After that, proportionally to the number of newbies, apprentices are not that numerous. However, we witness that their number changes over the months. Finally, for the experts, we find that their number fluctuates for the period of time described in the dataset. For the last time buckets, they become more and more scarce. The community may risk high-potential users leaving because of the lack of recognition regarding their efforts by other contributors.

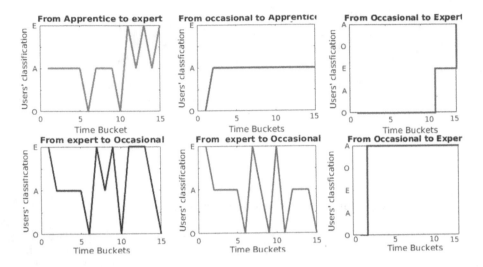

Fig. 6. Evolution of users over time

In Fig. 6, we present the evolution of classification of some users during the hole time buckets. Some of them stay always as experts and some stay always as occasionals. However, we can witness the evolution of persons over the time spent within the community. we can see that contributors may evolve during their time activity in the community from occasional to apprentice to expert. Thus, we may notice that some of interrupt their contribution for some months and then restart posting. For some other users. Therefore, we can find users that can be experts for a period and then start posting less and less until leaving the community becoming occasionals.

In Fig. 7, we present some values of the Belief Measure of Expertise for different classes of users. The BME is the mass affected to the experts. We notice that the value of this expertise measure for experts is high and always close to 1. However, for Occasionals it is very low with a $BME = 0.1$ and decreases over time to 0 if this user does not contribute anymore. Therefore, for the Experts who Apprentices then Occasionals, the value of BME fluctuates over time until reaching 0.

With the value of BME of new users during the first months of their activity, we are able to detect future experts based on their posts and the time spent within the community. The BME being the mass allocated to the experts, potential experts users are detected early based on their behavior. Some of them are identified since the 2 or 3 times buckets like presented in Fig. 7. f where their BME increases from 0.45 to 0.9 in 2 months. However, some of them need a lot more time to acquire knowledge.

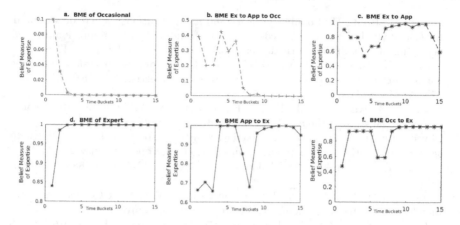

Fig. 7. Belief Measure of Expertise

6   Conclusion

This paper is focused on two major issues: first on identifying three classes of users on Stack Overflow: Occasionals, Apprentices and Experts. Then, detecting potential experts on their early time of activity. The strength of the proposed model is that it could be applied to any topic in the platform. Based on a belief model of the users' behavior, we calculated the general degree of expertise called the BME. This measure takes into account the combination of all the masses that describe a user during a defined period of time of activity on the web site. Once the expertise measure calculated for each time bucket, it allows us to have an overview on the users' behavior. Potential experts can be detected since the early few months of their entrance to the community. In future works, we will search to study the expertise of users based on their topical interests.

References

1. Bouguessa, M., Romdhane, L.B.: Identifying authorities in online communities. ACM TIST **6**(3), 30 (2015)
2. Dempster, A.P.: Upper and lower probabilities induced by a multivalued mapping. In: Yager, R.R., Liu, L. (eds.) Classic Works of the Dempster-Shafer Theory of Belief Functions, vol. 219, pp. 57–72. Springer, Heidelberg (1967)
3. Denoeux, T.: A k-nearest neighbor classification rule based on dempster-shafer theory. IEEE Trans. Syst. Man Cybern. **25**(5), 804–813 (1995)
4. Denœux, T., Kanjanatarakul, O.: Evidential clustering: a review. In: Huynh, V.-N., Inuiguchi, M., Le, B., Le, B.N., Denoeux, T. (eds.) IUKM 2016. LNCS (LNAI), vol. 9978, pp. 24–35. Springer, Cham (2016). doi:10.1007/978-3-319-49046-5_3
5. van Dijk, D., Tsagkias, M., de Rijke, M.: Early detection of topical expertise in community question answering. In: SIGIR, pp. 995–998. ACM (2015)

6. Dubois, D., Prade, H.: Possibility theory and its applications: Where do we stand? In: Kacprzyk, J., Pedrycz, W. (eds.) Springer Handbook of Computational Intelligence, pp. 31–60. Springer, Heidelberg (2015). doi:10.1007/978-3-662-43505-2_3
7. Kasneci, G., Gael, J.V., Stern, D.H., Graepel, T.: Cobayes: bayesian knowledge corroboration with assessors of unknown areas of expertise. In: WSDM, pp. 465–474. ACM (2011)
8. Khaleghi, B., Khamis, A.M., Karray, F., Razavi, S.N.: Multisensor data fusion: a review of the state-of-the-art. Inf. Fusion 14(1), 28–44 (2013)
9. Movshovitz-Attias, D., Movshovitz-Attias, Y., Steenkiste, P., Faloutsos, C.: Analysis of the reputation system and user contributions on a question answering website: Stackoverflow. In: Rokne, J.G., Faloutsos, C. (eds.) ASONAM, pp. 886–893. ACM (2013)
10. Nguyen, V.-D., Huynh, V.-N.: Integrating with social network to enhance recommender system based-on dempster-shafer theory. In: Nguyen, H.T.T., Snasel, V. (eds.) CSoNet 2016. LNCS, vol. 9795, pp. 170–181. Springer, Cham (2016). doi:10.1007/978-3-319-42345-6_15
11. Pal, A., Farzan, R., Konstan, J.A., Kraut, R.E.: Early detection of potential experts in question answering communities. In: Konstan, J.A., Conejo, R., Marzo, J.L., Oliver, N. (eds.) UMAP 2011. LNCS, vol. 6787, pp. 231–242. Springer, Heidelberg (2011). doi:10.1007/978-3-642-22362-4_20
12. Pal, A., Harper, F.M., Konstan, J.A.: Exploring question selection bias to identify experts and potential experts in community question answering. ACM Trans. Inf. Syst. 30(2), 107–134 (2012)
13. Posnett, D., Warburg, E., Devanbu, P.T., Filkov, V.: Mining stack exchange: expertise is evident from initial contributions. In: SocialInformatics, pp. 199–204 (2012)
14. Reyni, A.: Probability Theory. North-Holland, Amsterdam (1962)
15. Sahu, T.P., Nagwani, N.K., Verma, S.: Multivariate beta mixture model for automatic identification of topical authoritative users in community question answering sites. IEEE Access 4, 5343–5355 (2016)
16. Shafer, G.: A Mathematical Theory of Evidence (1976)
17. Smets, P., Hsia, Y.-T., Saffiotti, A., Kennes, R., Xu, H., Umkehren, E.: The transferable belief model. In: Kruse, R., Siegel, P. (eds.) ECSQARU 1991. LNCS, vol. 548, pp. 91–96. Springer, Heidelberg (1991). doi:10.1007/3-540-54659-6_72
18. Tang, X., Yang, C.C.: Ranking user influence in healthcare social media. ACM TIST 3(4), 73 (2012)
19. Yang, J., Tao, K., Bozzon, A., Houben, G.-J.: Sparrows and owls: characterisation of expert behaviour in stackoverflow. In: Dimitrova, V., Kuflik, T., Chin, D., Ricci, F., Dolog, P., Houben, G.-J. (eds.) UMAP 2014. LNCS, vol. 8538, pp. 266–277. Springer, Cham (2014). doi:10.1007/978-3-319-08786-3_23

Leveraging Hierarchy and Community Structure for Determining Influencers in Networks

Sharanjit Kaur[1], Rakhi Saxena[2(✉)], and Vasudha Bhatnagar[3]

[1] Acharya Narendra Dev College, University of Delhi, New Delhi, India
sharanjitkaur@andc.du.ac.in
[2] Deshbandhu College, University of Delhi, New Delhi, India
rsaxena@db.du.ac.in
[3] Department of Computer Science, University of Delhi, New Delhi, India
vbhatnagar@cs.du.ac.in

Abstract. Predicting influencers is an important task in social network analysis. Prerequisite for understanding the spreading dynamics in online social networks, it finds applications in product marketing, promotions of innovative ideas, constraining negative information etc.

The proposed prediction method IPRI (Influence scoring using Position, Reachability and Interaction) leverages prevailing hierarchy, interaction patterns and community structure in the network for identifying influential actors. The proposal is based on the hypothesis that capacity to influence other social actors is an interplay of three facets of an actor viz. (i) position in social hierarchy (ii) reach to diverse homophilic groups in network, and (iii) intensity of interactions with neighbours. Preliminary comparative performance evaluation of IPRI method against classical and state-of-the-art methods finds it effective.

Keywords: k-truss · Hierarchy · Topology · Community · Interaction

1 Introduction

Predicting influential spreaders in Online Social Networks (OSNs) is an important task because of the critical role they play in dissemination of information. The task is also crucial for accelerating the spread of positive vibes and blocking cascade of negative vibes in highly linked contemporary society [1,11].

Early methods for finding influencers in networks were based on classical centrality measures and their variants [2,6,11]. Prediction quality of these methods leaves much to be desired due to limited view of node attributes they take into account and network topology they scrutinize. Taking cues from the real-world, researchers have considered intensity of interactions between individuals for identifying influential nodes [8,9]. Number of links of an actor in diverse communities provides a unique vantage point in aiding spread of information. Method proposed in [16] exploits this idea and uses community structure in addition to weight of links to identify influential nodes. Role of hierarchy in influence spread is admitted and shown to be effective in [6,12].

© Springer International Publishing AG 2017
L. Bellatreche and S. Chakravarthy (Eds.): DaWaK 2017, LNCS 10440, pp. 383–390, 2017.
DOI: 10.1007/978-3-319-64283-3_28

These state-of-the-art methods for finding influencers consider only one facet of the network at a time, and hence overlook the advantage of interplay of three facets mentioned above. In this paper, we address the research gap by exploiting the synergy between community structure, network hierarchy and intensity of interactions with neighbours for spreading influence, and demonstrate improvement over existing methods for prediction of influencers in OSNs.

1.1 Contributions and Organization

In this paper, we introduce a novel scoring method IPRI (Influence scoring using Position, Reachability and Interaction) for identifying influencers by capitalizing on the underlying hierarchy and prevailing homophilic groups in the network. We highlight the contributions of our work below.

i We perform decomposition of the network using k-truss method to capture network hierarchy and to approximate homophilic groups instead of using computationally expensive community detection method (Sect. 3).
ii We capture complex interplay of network hierarchy, prevailing community structure and interaction patterns to differentiate between spreading ability of individuals in social networks (Sect. 4).
iii We evaluate the proposed method (IPRI) using three publicly available networks and compare results against classical and state-of-the art methods (Sect. 5).

2 Related Work

We briefly describe recent approaches that use network topology for identifying influential spreaders in OSNs.

Kitsak et al. have demonstrated that influential spreaders are located in the top hierarchical level of the network where levels are identified using k-core decomposition method [6]. Researchers have extended k-core method to identify better spreaders by incorporating neighbourhood coreness [3] and considering 2-step neighbourhood [10]. Approaches using k-core decomposition are inadequate for fine-grained differentiation since they assign same rank to multiple nodes.

Rossi et al. [12] further refine the set of influential nodes by using k-truss decomposition method to consider position in hierarchy for detecting influential spreaders. It is also shown in [6,13] that influential nodes are not always part of bigger neighbourhood. However, these works do not utilize diversity in neighbourhood contacts to capture node's influence.

Recently, researchers have shown that diverse groups in OSNs affect spreading capability of individuals. Extended Pagerank algorithm for finding influencers incorporates broadness of user's inter-community links to capture diversity of neighbours [16]. However, importance of each identified community is not used in capturing influential spreaders. Liu et al. [9] proposed Trust-Oriented Social Influencers method based on social relationships, trust and similarity preferences between individuals using meta-data.

The proposed method IPRI overcomes limitations of existing methods by incorporating network hierarchy, community structure, and intensity of interactions to discover influential spreaders.

3 Preliminaries

In this section we present the formal notation used in the paper, and briefly describe k-truss decomposition method.

We represent an online social network (OSN) as simple, undirected, unsigned, edge-weighted graph $G = (V, E, W)$ - a triplet formed by (i) finite set of vertices/nodes V, (ii) set of edges $E \in V \times V$, and (iii) an edge weight matrix $W : V \times V \rightarrow \mathbb{R}_{\geq 0}$. Here V models individuals, and $|V|$ $(= n)$ denotes number of individuals in network. Edge $e_{ij} \in E$ models link between individuals v_i and v_j $(v_i, v_j \in V)$, and $|E|$ $(= m)$ denotes number of edges. Weight w_{ij} (in W) of edge e_{ij} quantifies the extent of interaction between v_i and v_j. Degree d_i of v_i is the number of edges incident on it.

Concept of k-truss of graph G was proposed by Cohen [5] as a method to hierarchically decompose G into dense subgraphs with specific properties. We briefly explain k-truss decomposition method and related terminology here.

Definition 1. *A maximal subgraph, $G_k = (V_k, E_k, W_k)$ of $G = (V, E, W)$, induced by set $V_k \subseteq V$ and $E_k = \{e_{ij} | e_{ij} \in V_k \times V_k\}$ is a k-truss, iff each edge in G_k is reinforced by at least (k-2) pairs of edges making a triangle with that edge.*

Informally, a k-truss is a maximal subgraph, in which every edge participates in at least $(k - 2)$ closed triads. The decomposition method produces a nested hierarchy of subgraphs where subgraphs at higher levels represent denser regions of G. Based on decomposition, definition of *trussness* of an edge, adapted from [14], is given below.

Definition 2. *Trussness t_{ij} of $e_{ij} \in E$ has value k, iff $e_{ij} \in G_k \wedge e_{ij} \notin G_{k+1}$.*

The naive k-truss algorithm iteratively removes those edges from G which are not part of $(k-2)$ triangles, until no more edges can be deleted. All the leftover edges in the reduced graph are part of minimum $(k-2)$ triangles and hence, form a k-truss. We use an elegant in-memory k-truss decomposition algorithm proposed by Wang et al. [15]. This algorithm has time complexity $O(m^{1.5})$ and space complexity $O(m + n)$, making the algorithm scalable.

4 Influence Scoring Using Position, Reachability and Interaction

The proposed influence scoring method (IPRI) hypothesises that a node at high position in hierarchy, with strong ties and connectivity to large number of communities has high spreading power. The proposed influence scoring method is detailed in following sections.

4.1   Trussness Based Hierarchical Decomposition

Use of k-truss decomposition method layers out G in hierarchy, thereby exhibiting demarcation among levels in network hierarchy. Trussness t_{ij} of edge e_{ij} indicates number of common neighbours of endpoints of the edge. We define below the trussness τ_i of a node $v_i \in V$.

Definition 3. *The trussness τ_i of node $v_i \in V$ is the maximum trussness of edges incident on it, i.e. $\tau_i = \max\limits_{j}(t_{ij})$.*

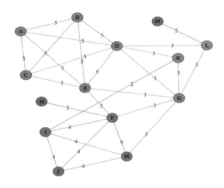

Fig. 1. k-truss decomposition of a toy network with 14 nodes and 28 edges. Vertices and edges with same trussness bear the same colour. Edges are labelled with their trussness. (Color figure online)

High value of trussness indicates occurrence of node in locally dense region of G. Nodes with same trussness actualize a tightly knit group and approximate a homophilic group binding individuals with similar connection patterns. As an example, we show hierarchical structure of homophilic groups obtained by k-truss decomposition of a toy network. Figure 1 shows nodes and edges with same trussness marked with same colour. All 5 nodes coloured green share similar characteristics of being a member of at least 3 triangular associations.

4.2   Positional Index

Network hierarchy reveals positional information of nodes in network. Trussness of a node obtained by hierarchical decomposition of the network proxies for its position. Higher level is indicative of larger neighbourhood span that aids wider spread of information.

Definition 4. *Positional Index of node v_i in G is equal to its trussness τ_i.*

4.3 Reachability Index

Each truss level in G represents a tightly-knit homophilic group and hence can be approximated as a community. A node having connections with more truss levels has higher reachability in terms of information propagation, compared to a node having connections with fewer truss levels [13,16].

We quantify a node's reach to diverse communities as the entropy of the trussness of its neighbours. Entropy is maximum when all neighbours have distinct trussness and minimum when all neighbours have same trussness. Let N_i be the neighbour set of node v_i. We define the probability of an arbitrary neighbour of v_i having trussness k as

$$p_i(k) = \frac{\sum\limits_{v_j \in N_i} I(\tau_j = k)}{|N_i|} \tag{1}$$

where I is an indicator function. The reachability index ρ_i of node v_i quantifies its accessibility to different communities and is formally defined below.

Definition 5. *The reachability index ρ_i of v_i is computed as*

$$\rho_i = \frac{-\sum\limits_{k=2}^{\mathcal{M}} p_i(k) \log_2 p_i(k)}{\log_2 \mathcal{M}} \tag{2}$$

where $\mathcal{M}$ indicates the number of hierarchical levels in G. We normalize the entropy to ensure $0 \leq \rho_i \leq 1$.

4.4 Interaction Index

It is accepted that a node with high degree centrality may not necessarily be efficient in spreading information/influence [6]. Interestingly, propagation of information is governed not only by the strength of interaction with neighbours $(w_{ij}, \forall v_j \in N_i)$, but also by the strength of interaction with 2-steps neighbours $(w_{jk}, \forall v_j \in N_i \wedge v_k \in N_j)$. This 2-steps neighbourhood of node v_i is sufficient for spreading its influence globally [10]. Based on this observation, we use local structure of a node's neighbourhood to determine its ability to spread its influence. The strength ω_j of node v_j is computed as $\omega_j = \sum\limits_{v_q \in N_j} w_{jq}$. The interaction index μ_i of node v_i is formally defined below.

Definition 6. *The interaction index μ_i of v_i is the sum of strength of neighbours scaled by their respective positional index and is computed as*

$$\mu_i = \sum\limits_{v_j \in N_i} \omega_j * \tau_j \tag{3}$$

4.5 Influence Score

Influence score which indicates the ability of an individual to spread information is a real-valued function $\Psi : V \to \mathbb{R}^+$. It is the aggregation of positional index, reachability index and interaction index of a node using a multiplicative function.

$$\Psi = \tau * \rho * \mu \tag{4}$$

The score is indicative of the power to influence other users in the network. Higher the score, more is the influence it exerts on others.

5 Experimental Analysis

The proposed IPRI method is implemented[1] in *Python* (32 bits, v 2.7.3) and is executed on Intel Core i5-3201M CPU @2.50 GHz with 8 GB RAM, running UBUNTU 12.04. Preliminary experiment study is designed to answer the following questions:

- Do influential spreaders predicted by IPRI spread information more widely compared to other measures in simulation tests conducted using SIR epidemic model? (Sect. 5.1)
- Is ranking delivered by IPRI effective in terms of fine grained discrimination? (Sect. 5.2)

5.1 Investigation Using SIR Model

Following previous similar works [3,6,8,12], we perform comparative evaluation of IPRI using SIR epidemic model [4]. SIR model is an artificial stochastic epidemic model in which nodes can be in one of three states: Susceptible (S), Infected (I), or Recovered (R). A small number of nodes are infected initially. At each time step, infected nodes infect their neighbours with probability β (infection rate) and recover with probability γ (recovery rate). Spreading process ceases when no more nodes can be infected. Spreading ability (SA) of the initial set of infected nodes is quantified as the percentage of nodes infected during spreading process.

We report comparison results using three large real-worlds networks [7] shown in Table 1, along with network features. We compare IPRI with a classical measure - degree centrality (DC) and three recent influencer prediction measures - k-core (KC), k-truss (KT), Trust-Oriented Social Influencers (TOSI). Following [6], we set $\gamma = 0.8$ and β as $1/\lambda_1$, where λ_1 is the largest eigenvalue of the adjacency matrix of the network.

For each compared measure, top 20% nodes are taken as initial spreaders and 100 simulations of SIR model are run to capture the average spreading ability (SA) of top-rankers. Figure 2a shows average SA for each measure for three networks. It is clear from the figure that spreading ability of IPRI is higher than competing methods for CollegeMsg and WikiVote networks. For Epinions network it is marginally better.

---

[1] Python code for implemented measures is available on GitHub.

Table 1. Structural properties of networks. $\bar{k}$ - Average degree, k_{max}- Maximum degree, gcc - Global clustering coefficient, L - Number of truss levels, β - Infection rate (as in [6]).

Network	n	m	$\bar{k}$	k_{max}	gcc	L	β
CollegeMsg	1899	59835	63.02	1546	0.05	6	0.0026
WikiVote	7115	103689	29.15	1167	0.12	22	0.0067
Epinions	75879	508837	13.41	3079	0.06	32	0.004

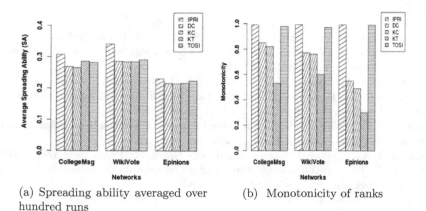

(a) Spreading ability averaged over hundred runs

(b) Monotonicity of ranks

Fig. 2. Results of experimental evaluation of IPRI.

5.2 Monotonicity

In order to capture the uniqueness in ranks assigned by various measures, we quantify fraction of ties in ranks using monotonicity measure defined in [3]. Let R be the vector of ranks assigned to n vertices of G by a measure, then monotonicity $M(R)$ of ranks is defined as below:

$$M(R) = \left[1 - \frac{\sum_r n_r(n_r - 1)}{n(n-1)} \right]^2 \tag{5}$$

where n_r is the number of ties with rank r. If there are no ties in R, monotonicity is 1, and if all ranks are same, then monotonicity is 0. Figure 2b shows monotonicity of all predictive measures on three datasets. It is clearly visible that the proposed method IPRI and comparative measure TOSI are equally good for fine grained discrimination between spreading power of nodes. However, comparatively low spreading ability of TOSI top-rankers (Sect. 5.1) establishes IPRI as relatively better predictor of influencers.

6 Conclusion

The proposed influence scoring method (IPRI) uses position of the actor in network hierarchy, intensity of his interactions with neighbours and extent of his connectivity in different communities to predict influential spreaders. Use of k-truss method confers dual advantage of revealing hierarchy and homophilic groups (approximate communities) in the network, making computation efficient. Preliminary experimentation with publicly available real social networks establishes effectiveness of IPRI scores in terms of wider spread of information and fine grained discrimination as compared to classical and state-of-the-art influencer detection methods.

References

1. Al-Garadi, M.A., Varathan, K.D., Ravana, S.D.: Identification of influential spreaders in online social networks using interaction weighted k-core decomposition method. Phys. A **468**(C), 278–288 (2017)
2. de Arruda, G.F., Barbieri, A.L., Rodríguez, P.M., Rodrigues, F.A., Moreno, Y., da Fontoura Costa, L.: Role of centrality for the identification of influential spreaders in complex networks. Phys. Rev. E **90**(3), 032812 (2014)
3. Bae, J., Kim, S.: Identifying and ranking influential spreaders in complex networks by neighborhood coreness. Phys. A **404**, 549–559 (2014)
4. Barabási, A.L.: Network Science. Cambridge University Press, New York (2016)
5. Cohen, J.: Trusses: cohesive subgraphs for social network analysis (2008). http://www.cslu.ogi.edu/~zak/cs506-pslc/trusses.pdf
6. Kitsak, M., Gallos, L.K., Havlin, S., Liljeros, F., Muchnik, L., Stanley, H.E., Makse, H.A.: Identification of influential spreaders in complex networks. Nat. Phys. **6**(11), 888–893 (2010)
7. Leskovec, J., Krevl, A.: SNAP Datasets: Stanford large network dataset collection, June 2014. http://snap.stanford.edu/data
8. Li, Q., Zhou, T., Lv, L., Chen, D.: Identifying influential spreaders by weighted leaderrank. Phys. A **404**, 47–55 (2014)
9. Liu, G., Zhu, F., Zheng, K., Liu, A., Li, Z., Zhao, L., Zhou, X.: TOSI: a trust-oriented social influence evaluation method in contextual social networks. Neurocomputing **210**, 130–140 (2016)
10. Liu, Y., Tang, M., Zhou, T., Do, Y.: Identify influential spreaders in complex networks, the role of neighborhood. CoRR abs/1511.00441 (2015)
11. Pei, S., Muchnik, L., Andrade, J.S., Zheng, H., Makse, H.A.: Searching for super-spreaders of information in real-world social media. Sci. Rep. **4**, 5547 (2014)
12. Rossi, M.E.G., Malliaros, F.D., Vazirgiannis, M.: Spread it good, spread it fast: identification of influential nodes in social networks. In: Proceedings of the 24th International Conference on World Wide Web, pp. 101–102. ACM (2015)
13. Ugander, J., Backstrom, L., Marlow, C., Kleinberg, J.: Structural diversity in social contagion. PNAS **109**, 5962–5966 (2012)
14. Wang, J., Cheng, J.: Truss decomposition in massive networks. Proc. VLDB Endow. **5**(9), 812–823 (2012)
15. Wang, M., Wang, C., Yu, J.X., Zhang, J.: Community detection in social networks: an in-depth benchmarking study with a procedure-oriented framework. Proc. VLDB Endow. **8**(10), 998–1009 (2015)
16. Wang, S., Wang, F., Chen, Y., Liu, C., Li, Z., Zhang, X.: Exploiting social circle broadness for influential spreaders identification in social networks. World Wide Web **18**(3), 681–705 (2015)

Using Social Media for Word-of-Mouth Marketing

Nagendra Kumar[✉], Yash Chandarana, Konjengbam Anand,
and Manish Singh

Indian Institute of Technology Hyderabad, Kandi 502285, India
{cs14resch11005,cs14btech11040,cs14resch11004,msingh}@iith.ac.in

Abstract. Nowadays online social networks are used extensively for personal and commercial purposes. This widespread popularity makes them an ideal platform for advertisements. Social media can be used for both direct and word-of-mouth (WoM) marketing. Although WoM marketing is considered more effective and it requires less advertisement cost, it is currently being under-utilized. To do WoM marketing, we need to identify a set of people who can use their authoritative position in social network to promote a given product. In this paper, we show how to do WoM marketing in Facebook group, which is a question answer type of social network. We also present concept of reinforced WoM marketing, where multiple authorities can together promote a product to increase the effectiveness of marketing. We perform our experiments on Facebook group dataset consisting of 0.3 million messages and 10 million user reactions.

1 Introduction

Marketing is a process by which products and services are introduced and promoted to potential customers. Marketing leads to increase in sales, build the reputation of company and maintain healthy competition. To do effective marketing, one has to identify the best customers, understand their needs and implement the most effective marketing method. There are many marketing methods such as direct marketing, field marketing, account-based marketing, B2P marketing, online marketing, word-of-mouth marketing, etc. With the emergence of Internet, online marketing has become one of the biggest sources of marketing. In online marketing, advertisements provide range from basic text descriptions with links to rich graphics with slideshows. However, the problem with most of these advertisement strategies is the lack of trust that users have on these information sources. People are being bombarded with so many online advertisements that they have grown immune to online advertisements. Word-of-Mouth (WoM) marketing has the advantage that the advertisement is done by people who are trusted by the person whom we try to market. According to Whitler[1], 64% of marketing executives

---

[1] http://www.forbes.com/sites/kimberlywhitler/2014/07/17/why-word-of-mouth-marketing-is-the-most-important-social-media/7762b8f07a77.

© Springer International Publishing AG 2017
L. Bellatreche and S. Chakravarthy (Eds.): DaWaK 2017, LNCS 10440, pp. 391–406, 2017.
DOI: 10.1007/978-3-319-64283-3_29

indicate that they believe WoM marketing is the most effective form of marketing. Incite [10] also stated that 91% of B2B (business-to-business) buyers are influenced by WoM marketers when making their buying decisions. In WoM marketing the information is passed from person to person through the WoM communication. People believe on the words of people whom they know such as friends, family and closely known authorities. If we do WoM marketing through only friends and family then the marketing will be quite restricted. Since people use multiple social media to access different types of information, we propose the use of social media to do widespread WoM marketing.

There are different social network models, such as friend-to-friend, follower-following, question-answer, etc. In this paper, we use question-answer (QA) type of network to do WoM marketing. QA network can lead to more widespread marketing compared to other types of networks because the influential users in such networks are known by more number of users compared to other types of networks. For example, in a friend-to-friend network a user may just have few hundred friends and the user may not even have an authoritative status amongst his peers. There are many QA networks, such as Quora, Stack Overflow, Facebook groups, etc. In this paper, we use online social groups (OSGs) such as Facebook groups. The members of a Facebook group have more focused interest compared to a generic friends or follower-following networks. For example, Java, Java for developers, C/C++ programming, etc., are some very popular and focused public groups having more than 20,000 group members.

Since Facebook groups are focused on specific topic and have large number of members, one can use the prominent and reliable members of such groups to do marketing. Prominent members, whom we call influential members, post lots of important and relevant information. In Facebook groups, members can ask questions from other members of the group to get solution to their problems. Influential members help other members by posting useful information in the form of posts and comments, and in return they get publicity in the form of reactions, such as likes, comments, shares, from other members. Organizations can use these trusted influential users to market their products by giving them incentives. Since the recommendations are made by one of their trusted peers, with influential position, members of group pay more attention to such recommendations. Let's consider an example task to better appreciate our problem.

Example 1. A book publisher wants to advertise a book, say a DBMS book, in Facebook group. It has a limited advertisement budget. It would like to find few influential users who can promote the book to a large audience. The publisher can attract such influential users by giving some free sample copy or discount. Following are some questions that would be of interest to the publisher:

1. For a given Facebook group, who are the top-k influential users?
2. What fraction of the group would be influenced by a selected set of top influential users?
3. How to do reinforced marketing so that each topic is marketed jointly by at least k influential users?
4. What is the best time to start promotion in the group?

In this paper, we answer all the above questions. Our key contributions are as follows:

- We use Facebook groups for product marketing.
- We analyze the characteristics of Facebook groups.
- We propose different marketing strategies and give solution using social network analysis.
- We present a topical relevance method to create social interaction graph from the users' activities in the group.
- We find important characteristics of influential users in Facebook groups and examine the dynamics of user influence across topics and time.
- We evaluate our algorithms on a large dataset containing 0.3 million posts and 10 million reactions.

2 Related Work

The subject of social media and network marketing has attracted significant research attention [5,18]. Trusov et al. [18] stated that social networking firms earn from either showing advertisements to site visitors or being paid for each click/action taken by site visitors in response to an advertisement. Domingos et al. [5] studied the mining of the network value of customers. They have shown that network marketing which exploits the network value of customers, can be extremely effective. According to Ogilvy Cannes[2] 74% of consumers identify WoM marketer as a key influencer in their purchasing decisions.

According to MarketShare [6], WoM has been shown to improve marketing effectiveness up to 54%. It has been shown by Wu et al. [22] that less than 1% of the social network users produce 50% percent of its content, while the others have much less influence and completely different social behavior [21]. However, in case of Facebook groups we observe that 6.5% users generate 85% content of the group and less than 2% of these users are able to influence 80% population of the group. These statistics show the users' behavior of Facebook groups and it is quite different from other social networks such as friend network in Facebook, Twitter, etc.

To discover top authorities in social network, we need to understand topological structure of social network, flow and diffusion of information in social network [4,9,15]. Many researchers in this domain have studied the structure of social network to study the similar problem of influence maximization [3,12]. The goal of Influence maximization is to maximize product penetration, while minimizing the promotion cost by selecting the subset of users which are also called influential users. Vogiatzis et al. [19] stated that influential users could spread the news of the product or service may reach up to maximum possible level. However, our approach is complementary to the existing approaches of

---

[2] http://www.adweek.com/prnewser/ogilvy-cannes-study-behold-the-power-of-word-of-mouth/95190?red=pr.

finding influential users. These works did not focus on finding topical influential users based on interaction activities. Our findings show that influence of a user varies across the topics and influential users may not be interested in chosen advertising product. Moreover, users' friend network is small and most of them do not mention their interests. We find out the users' interests from their activities in social network groups which are focused communities and choose top users who are authoritative users as well as interested in the advertising the product.

3   Problem Definition

We define the problem of finding the influential users for WoM marketing in terms of the following sequence of sub-problems:

Problem 1 (Create social interaction graph): *Given a topic T, a Facebook group F and the activities A in the group F, create a social interaction graph $G(V, E)$, where the vertices V represent the members of the group F and edges E represent the interaction between group members.*

Problem 1 is to create a social interaction graph for a given Facebook group. The activities in a group include creation of posts and reactions to posts, such as likes, comments, likes on comment, and shares. The members of the group represent vertices and interaction among members (users) represent edges of the graph. It is a topic sensitive graph, where the edge weights are dynamically computed based on the given topic T. We assign a weight to the edge based on the given topic T, type of reaction that a user had done to the post or comment created by an other user.

Problem 2 (Finding influential users): *Given a social interaction graph $G(V, E)$ and a topic T, find the top-K influential users I from the graph $G(V, E)$, who can give maximum visibility to the topic T in the corresponding Facebook group of the given interaction graph $G(V, E)$.*

People form Facebook groups to explore about certain topic. Naturally, in such groups some members with more knowledge become authorities, whose words have great influence on the other group members. In this problem our goal is to find the influential users for a given marketing topic.

Problem 3 (Reinforced marketing): *Given a social interaction graph $G(V, E)$, a topic T and reinforcement parameter r, find the set of influential users I_R such that the marketing of each influential user from I_R can be reinforced by at least $(r - 1)$ other influential users.*

The social position of a user has important effect on the marketing. If someone who is not an authority markets a product, the marketing will hardly have any impact. If one authority markets the product, the marketing will be more effective. The marketing will be even more effective if multiple authorities can collectively market the product. When people hear the same message reinforced by multiple authorities that they trust, it is more likely they will consider buying the product. Thus, we need to find authorities in such a way that if one authority markets the product, there are at least $(r-1)$ other authorities in the set I_R, who can support the marketing. These $(r-1)$ other authorities should be closely related with other members whom the above mentioned one authority will market.

4   Analysis of Online Social Groups

In this section, we present structure of OSGs to get insight into the users' activities. We use bow tie structure [1] to analyze the general structure of OSGs. It has five distinct components namely *core, in, out, tendrils* and *tube.* In bow tie structure, *core* is a strongly connected component and contains users who often help each other. The *in* component contains users who only react to the posts. The *out* consists of users who only post the contents. *Tendrils* and *tubes* contain the users who connect to either *in* or *out* or both but not to the *core*. *Tendrils* users only react to the posts created by *out* users or whose posts are only reacted by *in* users. *Tubes* users connect to both *in* and *out.*

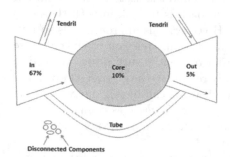

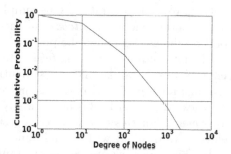

Fig. 1. Bow tie structure of the OSGs

Fig. 2. Degree distribution in groups

We present the bow tie structure of OSGs in Fig. 1. We observe that there are 10% *core* users, 67% *in* users, 5% *out* users, 15% *tendril* users, 0.2% *tube* users and 2.8% *disconnected* users. OSGs have much bigger *in* component compared to the *out* and *core*. We find that in OSGs about 10% of the users do both, post contents and react to the posts of each other. Most of the users (67%) only react to the posts, and 5% of the users only post the contents. These results indicate that OSGs are the information seeking communities where most of the users consume the information and very few people generate the information. Most of the members join the groups to keep themselves updated by getting the information related to the topics of shared interest.

Next, we perform degree analysis to get more insight into the users' connectivity in the groups. Degree is a general way to illustrate users' relative connectedness in a large complex network. Degree distribution reports the number of users (cumulative probability of users) in the network with a given degree. As we can see in Fig. 2 that the degree distribution appears to follow a power law. The most of the users have very less degree which signifies that these users are connected to just a few other users; however, there are very few users who are connected to a large number of users. As these few users have large number of connections, they can easily spread the information to a large audience.

5   Social Interaction Graph

In this section, we give a solution to Problem 1. We generate the topic sensitive social interaction graph based on the group activities. We consider the topical relevance of group members to create the graph.

5.1   Measuring Topical Relevance

We measure the topical relevance of users by analyzing the content of their posts. For a given topic T, we find the users who are interested in topic T. For example, to market a *database* book, we need to find the users who have posted contents related to *database*. One simple approach is to look into all the posts which contain the word *database* in them. However, this approach fails to give good results as there might be posts which are actually relevant to the database but do not contain the word *database*. In order to identify such posts, we need to generate a list of words which are semantically related to the given seed word. For example, for the word *database*, some related words can be *sql, query, schema,* etc. Clearly, the word *sql* is more closely related to *database* as compared to the word *query*. For this task we need a system which, given a word, gives a list of relevant words along with it's relevance score.

In this paper, we use Semantic Link[3] system, which gives a list of words which are semantically related to the seed word. It uses the fact that some words occur frequently together. For example, the words *database* and *sql* often occur together. These are semantically related words, meaning that their co-occurrence is not due to chance but rather due to some non-trivial relationship. Semantic Link attempts to find such relationships between words and uses these relationships to find the related words. Semantic Link analyzes the text of the English Wikipedia and attempts to find all pairs of words which are semantically related. It uses a statistical measure called Mutual Information (MI), which is a measure of the mutual dependence between two topics. Higher the MI score for a given pair of topics, higher the chance that they are related. MI score is defined as follows:

$$MI(x,y) = \sum_{y \in Y} \sum_{x \in X} p(x,y) \log \frac{p(x,y)}{p(x)p(y)} \qquad (1)$$

---

[3] http://semantic-link.com/.

Where, X and Y are two random set of topics. $p(x, y)$ is the joint probability distribution function of X and Y. p(x) and p(y) are the marginal probability density functions of X and Y respectively.

After getting a list of related words, we find posts relevant to these words. One approach is to filter out the posts which do not contain any of the related words. However, this approach has a limitation that the users who don't have relevant posts (related words in their posts) will have no in-links. Such users will get low ranks while applying authority measures on the graph and thus their out-links will not contribute much to the rank of the relevant users. In such a case, only the popularity of the relevant users will matter while determining the ranks. This approach ignores the relationship of relevant users with non-relevant users. A better approach is to give higher weight to the relevant posts and their interactions. The weight (boosted relevance) is calculated on the basis of the presence of relevant words in the posts. First, we calculate *relevance* score for every post which is the sum of the MI score of all the related words that are present in the post. We then compute boosted relevance (*bRelevance*) based on the *relevance* score.

Algorithm 1. Algorithm for Computing Boost to Interactions

Input: T: set of topic words
            P: Post
Output: *bRelevance*: boost of post P
Method:
1: *relevance* $\leftarrow 0$
2: *postWords* $\leftarrow P.getWords()$
3: **for all** *tWord* $\in T$ **do**
4:      **for all** *pWord* $\in postWords$ **do**
5:          *relevance*+ $= Similarity(tWord, pWord)$
6:      **end for**
7: **end for**
8: *bRelevance* $= 1 + \alpha * ln(1 + relevance)$
9: **return** *bRelevance*

Algorithm 1 shows the method that we use to compute the *bRelevance* of different posts. Lines 3 to 7 show how to compute the semantic similarity (*relevance*) between topic word and post. Line 8 reveals the equation to compute the boosted relevance (*bRelevance*). This equation indicates that if a person tries to spam the system with too many words related to the product, the logarithmic function *bRelevance* is not increased too much (in our experiment, we set the value of constant factor α to 20). Similarly, we compute *bRelevance* for textual comments based on the relevant words present in it and assign higher weight to the relevant comments and their interactions. We drive the graph structure of the group by representing each user of the group as a vertex of the graph and each user interaction such as 'like on comment', 'like', 'comment', 'share' as an edge of the graph. We assign the weights 1, 2, 4 and 8 for like on comment, like, comment, and share respectively [2]. We create an edge from user u_i to u_j, if the user u_i has reacted to any post or comment that is created by the user u_j. The weights of the edges are determined by the product of the weight corresponding to the type of interaction with the boosted relevance.

6   Finding Influential Users in OSG

In this section, we describe how to find influential users in OSGs. This is a solution to Problem 2.

We use PageRank [16] algorithm to find the influential users. One of the reasons to use PageRank that it considers the importance of each user while finding the influential user unlike other authority measures [7,8,23]. PageRank was originally developed to rank the web pages for search results. Web pages are connected together by hyperlinks. Similarly, we have topic sensitive social interaction graph where users are connected through social interactions. So, we can apply PageRank algorithm on the social interaction graph to find influential users. We rank the users based on decreasing order of PageRank score, and select the top-k users to be the potential WoM marketers.

Example: Consider an example graph in Fig. 3. We apply PageRank on both weighted and unweighted version of the graph to investigate the effect of weighted edges in the computation of users' ranking. In this graph, we consider vertices as users and edges as reactions. Weights of each type of reaction are assigned based on type of interaction as described in Sect. 5. We do not consider *brelevance* in this example for ease of understanding. We apply PageRank on unweighted version of this graph, we get rank of each node as x1 = 0.7210, x2 = 0.2403, x3 = 0.5408, x4 = 0.3605, whereas we get x1=0.7328, x2=0.1466, x3=0.5374, x4=0.3908 for weighted version of the graph. We can see that assigning weights to the edges alters the ranks from the ranks calculated without weights. We can see that x2 has reduced from 0.2403 to 0.1466 because it has only one like on its comment which carries less weight. Here, we can also see that x1 has higher rank compare to x3 so it is not only the number reactions determine the authority, it also depends on the importance of users who react on the content. When we apply the PageRank on larger graph, users' rank change significantly as well as their ranking orders. We also compare the effectiveness of PageRank algorithm with other authority measure algorithms such as HITS [13], Z-score [23], Eigen vector [17], Betweenness [8] and Closeness [7].

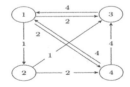

Fig. 3. Social interaction graph

7 Reinforced Marketing

In this section, we give a solution to Problem 3. In OSGs users interact with other users having similar topics of interests. This type of user interaction leads to formation of sub-groups. Since user interactions may get confined to sub-groups, it is important to find multiple influential users from each sub-group so that they can collectively promote the product, which will be more effective in giving trust to users about the product. We need to do this for all the important sub-groups.

We find sub-groups by finding weakly connected components in the graph. A weakly connected component is a maximal sub-graph of a directed graph such that for every pair of vertices in the sub-graph, there is an undirected path. So each member of a weakly connected component may have reacted on someone's post in the group or would have received a reaction from someone else in the group. We choose to target a sub-group only if it contains enough users. If users in the sub-group are less than threshold th, we do not select that sub-group for marketing.

We apply the best authority measure algorithm (described in Sect. 8) in the topic sensitive social interaction graph to find the top-k topical influential users of the group. For each of the sub-group, we select top-r ($r<k$) users from the set of k users such that these r users also belong to the sub-group. These r influential users can support each other by advertising the same product to their sub-group(s).

8 Evaluations

In this section, we describe our dataset and evaluation metrics. We also compare the performance of various authority measure algorithms and show the characteristics of influential users through some anecdotal examples.

8.1 Experimental Setup

We use dataset of Facebook groups for the experiment as these are focused groups with large number of audience. Facebook groups are community of people where they share their common interests in the form of posts and comments. Members of groups can react to the posts/comments created by each other. Reactions consist of a textual comment and a unary rating score in the form of likes and shares. We use Facebook Graph API to collect the dataset. The dataset contains 100 of Facebook groups having at least 20,000 members. It includes 0.3 million posts and 10 million of reactions that were created in 5 years (from 2011 to 2015). We perform various text pre-processing tasks on text content of dataset such as stop words removal, stemming and lemmatization.

8.2   Evaluation Metrics

We show the effectiveness of algorithms by using three metrics namely correlation, precision, and influence.

Correlation: We use correlation metric to measure the strength of association between two ranks. We use Pearson correlation [14] to evaluate the effectiveness of authority finding algorithms. We find the correlation between rank assigned to users by authority measure algorithms and the baseline influence metrics (described later in this section).

Precision and Normalized Discounted Cumulative Gain: We use these measures to check the quality of authority finding algorithms by measuring the relevancy of top-k influential users generated by these algorithms. Precision is fraction of retrieved instances that are relevant. Normalized Discounted Cumulative Gain is computed based on the discounted cumulative gain [11] which includes the position of users in the consideration of their importance.

Influence: We use two influence metrics as baselines to evaluate the user's authority position in the group namely, centrality and popularity. Degree is a centrality measure that evaluates the user's connectivity whereas votes and topical votes are popularity measures that evaluate the user's prestige. For each user, we compute votes by taking the weighted sum of all the audience reactions received by the user over all his posts, comments. However, we compute topical votes by taking the weighted sum of audience reactions over all his posts, comments that contain the advertisement topic itself or the topics semantically related to the advertisement topic.

8.3   Effectiveness of Algorithms

It is important to understand the effectiveness of authority finding algorithms in OSGs. The algorithm which is appropriate for one network may not be appropriate for other because of user behavior dynamics. To measure the effectiveness of authority finding algorithms in OSGs, we find the correlation of top-200 influential users generated by authority finding algorithms with the votes received by these users. As all of these groups are technical groups having similar number of users, we compute the overall correlation by averaging the correlation across all the groups.

Figure 4 shows the correlation of the top-200 influential users ranked by the various authority finding algorithms with the votes. HITS performs better than other algorithms for top-10 users whereas PageRank outperforms for top-50 or more users. One of the reasons is that PageRank is a global measure and it does not trap in local neighborhood however HITS suffers from topic drift. Betweenness tends to produce slightly better results than most of the other algorithms. One of the reasons is that nodes having high Betweenness are the bridges of two parts of the graph (sub-graph) and have the potential to disconnect graph if removed. If a user having high Betweenness posts an update, there is high chance that it will spread rapidly across the sub-graphs.

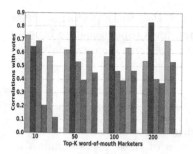
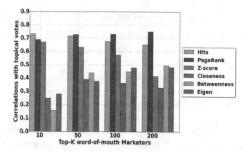

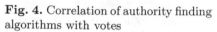

Fig. 4. Correlation of authority finding algorithms with votes

Fig. 5. Correlation of authority finding algorithms with topical votes

Figure 5 shows the correlation of the top-200 influential users ranked by the various authority finding algorithms with the topical votes. HITS performs better for top-10 users whereas PageRank performs better than HITS for top-50 or more users. In conclusion, PageRank can be utilized for finding influential users for general marketing as it shows high correlation with both votes and topical votes.

8.4 Precision Analysis

We evaluate the correctness of authority finding algorithms by using Mean Average Precision (MAP) and Normalized Discounted Cumulative Gain (NDCG), which are standard measures to evaluate the effectiveness of web page ranking algorithms. We consider top-50 influential users of the groups generated by algorithms. We ask five students of our research lab to join these technical groups and manually judge whether a user is influential or not from their viewpoints for a given topic T. We also ask to rank these users for a given topic. We provide all the posts and reactions of influential users to the students. These students label the data independently, without influencing each other. The average percentage of agreement among the students was 92%. We use this label data as a ground truth for finding the MAP and NDCG of algorithms. We compute the overall MAP, NDCG by averaging the MAP, NDCG across all the groups respectively.

Table 1. MAP and NDCG of authority finding algorithms

Authority measures	MAP	NDCG
PageRank	0.91	0.83
HITS	0.87	0.75
Z-score	0.70	0.65
Eigen	0.72	0.69
Betweenness	0.76	0.70
Closeness	0.73	0.67

As can be observed in Table 1, for a given topic T PageRank performs better than other authority finding algorithms. PageRank finds topic sensitive influential users with the highest accuracy. So, we use PageRank for our analysis in rest of the paper.

8.5   Marketing Across Topics

In this section, we analyze behaviour of influential users across different topics and investigate how widely the rank correlation of these users changes by changing the topics.

Top influential users (top users) for all the query topics are not different. Top users tend to express their opinions on many popular topics of the group. To examine dynamic behavior of top users across different topics, we compare the relative order of their ranks across topics. We ignore the least popular topics and focus on the set of relatively popular topics. We apply Topical N-Grams [20] on the posts to find popular topics of the Java For Developers[4] group (Java group). Web, Servlet, and Constructor are some popular topics in the group, so we choose these topics to measure the variation in top users ranking across these topics. We use correlation to compare the ranking patterns of top users for pairs of topics.

We observe in Table 2 that correlation is high for the top-20 users which implies that these users post over a wide range of topics. Among topic pairs, {Web, Servlet} shows the highest correlation for the top-20 users. This is because these two topics are closely related in Java. Servlets are used in Web programming. This analysis indicates that top users hold significant influence over a range of topics and could be used to spread the information about variety of topics.

To get more insight into variation in correlation of top users across topics, we perform the experiment on wide range topics in Java group. We select 20 topics from each of popular topics, less popular topics and unpopular topics. We compute Mutual Information (MI) score for all these topics with respect to group topic (shared group interest). We derive top-200 topic sensitive influential

Table 2. Correlation in top users ranking for popular topics

Topics	Top-20 users	Top-200 users
Web vs. Servlet	0.79	0.56
Web vs. Constructor	0.53	0.46
Constructor vs. Servlet	0.49	0.39

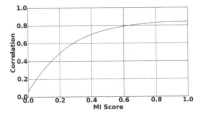

Fig. 6. Correlation of top users across variety of topics

***

[4] https://www.facebook.com/groups/java4developers/.

users by using these topics and measure the correlation of these users with top-ical votes. As can be observed in Fig. 6 that correlation decreases as MI score decreases. If a chosen topic has very less dependency with the group topic, then authority measure algorithms show very less correlation. It indicates that quality of top users also depends on the topic. If a query topic is less related to shared group interest then it is not possible to get prominent topical users who can influence the whole group as the quality of these users decreases. Therefore, it is recommended that advertising business should select a query topic which is highly related to shared group interest to do effective marketing in OSGs.

8.6   Empirical Evaluation

In order to investigate influential users' characteristics and behavior dynam-ics, we find the connectivity of influential users and their structural position in OSGs. First, we find indegree connectivity of top-k influential users in the Java For Developers group (Java group) having 35,000 members at the time of exper-iment. We observe that average indegree of top-20 users is 1604 whereas average indegree of the whole group is 8. The reason for this is that authority finding algorithm strongly correlate with the indegree of the top users. Moreover, we observe that 6.5% users post the 85% content of the group content and less than 2% of them are able to influence 80% users of the group.

To get more insight into the structural position of influential users in the group, we present the network structure of influential users which is a undirected network constructed in a similar way as mentioned in Sect. 5. We take a small instance of Java group with 707 nodes, 1187 edges and visualize the network structure of the Group. The users of the network can be divided into two types: top users and ordinary users. The green color nodes represent the top-20 users, and the red color nodes represent ordinary users of the group.

Figure 7 shows that top users are strongly connected with the large number of members of group. Statistics reveal that average degree of the group is 3.35 whereas average degree of top-20 influential users is 72. Moreover, average num-ber of reactions received by a user of the group is 5.2 whereas average reaction received by top-20 influential users is 98.

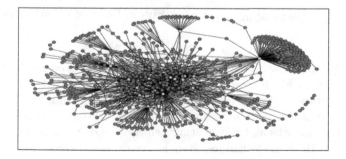

Fig. 7. Structure of Java group

Furthermore, we also analyze the reactions received by users in the Java group to examine the popularity of influential users in the group. Top users receive large number of reactions as these are the prestigious users of the group. As rank of the user increases, reaction received by the user decreases exponentially. This difference indicates that it is more beneficial to target popular users for marketing than to employ a massive number of non-popular users.

8.7 Temporal Dynamics

We analyze the action (posting) and reaction behavior of influential users over a period of time and find the right time to start promotion in the group to maximize content visibility. Our results are based on 5 years of temporal data.

In order to examine the influential users' posting behavior, we pick the top 1000 influential users based on their ranks from the Java group. We divide top users into three groups based on their ranks such as top 200 users, top 201–500 users and top 501–1000 users. Our aim is to analyze the differences in posting behavior of these users. We compute the probability of posting a post for all these three groups in each month of the year. Figure 8 shows the time evolution of the posts of the influential users (top users).

Our findings about the posting behavior of top users reveal two interesting observations. First, top users post significant updates over a period of time. Top 200 users post lots of information compared to top 500 and top 1000 users. Second, lots of posts are posted during the month of March, April, and October. This is perhaps due to various competitive and semester exams in India during these months, which motivates the top users to post a lot of information about various topics. So, it is better to choose these periods of the year for marketing.

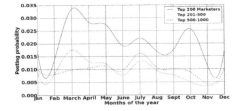

Fig. 8. Posting behavior of top users   **Fig. 9.** Reaction behavior of top users

We also perform the similar experiment on reactions received by top users. As can be seen in Fig. 9, reaction pattern follows the same trend as posting pattern, i.e., more number of audience reactions in the month of March, April, and October. It is due to a large number of posts created by top users during these periods of months and this posting behavior leads to increase the number of audience reactions. As lots of users are active during these periods of months, advertising companies can target more number of top users to promote products during these periods.

9   Conclusion

In the paper, we propose methods to use OSGs for WoM marketing. We present an algorithm to create topic sensitive social interaction graph from the activities of the group. We apply authority finding algorithm on social interaction graph to find topic specific influential users. Organizations can promote the product through these influential users by giving them incentives. We propose the concept of reinforced marketing to perform effective marketing where multiple influential users collectively market a product. We also analyze the important characteristics of influential users such as these users post most of the content of the group and able to influence most of the population of the group. We find that influential users post over a wide range of topics and receive lots of audience reactions. Finally, we show the best time of the year to start marketing in Facebook groups to improve the effectiveness of marketing.

References

1. Broder, A., Kumar, R., Maghoul, F., Raghavan, P., Rajagopalan, S., Stata, R., Tomkins, A., Wiener, J.: Graph structure in the web. Comput. Netw. **33**(6), 309–320 (2000)
2. Bucher, T.: Want to be on the top? algorithmic power and the threat of invisibility on facebook. New Media Soc. **14**(7), 1164–1180 (2012)
3. Chen, W., Wang, Y., Yang, S.: Efficient influence maximization in social networks. In: SIGKDD. ACM (2009)
4. Cheng, J., Adamic, L., Dow, P.A., Kleinberg, J.M., Leskovec, J.: Can cascades be predicted?. In: WWW. ACM (2014)
5. Domingos, P., Richardson, M.: Mining the network value of customers. In: SIGKDD. ACM (2001)
6. Forbes: What are they saying about your brand? (2013). http://www.forbes.com/sites/pauljankowski/2013/02/27/quick-what-are-they-saying-about-your-brand/#ee5ff7371a8d
7. Freeman, L.C.: Centrality in social networks conceptual clarification. Soc. Netw. **1**(3), 215–239 (1978)
8. Freeman, L.C.: A set of measures of centrality based on betweenness. Sociometry **40**(1), 35–41 (1977)
9. Guille, A., Hacid, H., Favre, C., Zighed, D.A.: Information diffusion in online social networks: A survey. SIGMOD **42**(2), 17–28 (2013)
10. Incite: How social media amplifies the power of word-of-mouth (2014). http://www.incite-group.com/brand-management/how-social-media-amplifies-power-word-mouth
11. Järvelin, K., Kekäläinen, J.: Cumulated gain-based evaluation of IR techniques. ACM Trans. Inf. Syst. (TOIS) **20**(4), 422–446 (2002)
12. Kempe, D., Kleinberg, J., Tardos, É.: Maximizing the spread of influence through a social network. In: SIGKDD. ACM (2003)
13. Kleinberg, J.M., Kumar, R., Raghavan, P., Rajagopalan, S., Tomkins, A.S.: The web as a graph: measurements, models, and methods. In: Asano, T., Imai, H., Lee, D.T., Nakano, S., Tokuyama, T. (eds.) COCOON 1999. LNCS, vol. 1627, pp. 1–17. Springer, Heidelberg (1999). doi:10.1007/3-540-48686-0_1

14. Lawrence, I., Lin, K.: A concordance correlation coefficient to evaluate repro-
    ducibility. Biometrics **45**(1), 255–268 (1989)
15. Leskovec, J., McGlohon, M., Faloutsos, C., Glance, N.S., Hurst, M.: Patterns of
    cascading behavior in large blog graphs. In: SDM. SIAM (2007)
16. Page, L., Brin, S., Motwani, R., Winograd, T.: The pagerank citation ranking:
    bringing order to the web (1999)
17. Ruhnau, B.: Eigenvector-centrality—a node-centrality? Soc. Netw. **22**(4), 357–365
    (2000)
18. Trusov, M., Bodapati, A.V., Bucklin, R.E.: Determining influential users in internet
    social networks. J. Mark. Res. **47**(4), 643–658 (2010)
19. Vogiatzis, D.: Influential users in social networks. In: Anagnostopoulos, I.,
    Bieliková, M., Mylonas, P., Tsapatsoulis, N. (eds.) Semantic Hyper/Multimedia
    Adaptation. Springer, Heidelberg (2013)
20. Wang, X., McCallum, A., Wei, X.: Topical n-grams: Phrase and topic discovery,
    with an application to information retrieval. In: Data Mining, ICDM (2007)
21. Weng, J., Lim, E.P., Jiang, J., He, Q.: Twitterrank: finding topic-sensitive influ-
    ential twitterers. In: WSDM. ACM (2010)
22. Wu, S., Hofman, J.M., Mason, W.A., Watts, D.J.: Who says what to whom on
    twitter. In: WWW. ACM (2011)
23. Zhang, J., Ackerman, M.S., Adamic, L.: Expertise networks in online communities:
    structure and algorithms. In: WWW. ACM (2007)

Knowledge Discovery

Knowledge Discovery of Complex Data Using Gaussian Mixture Models

Linfei Zhou[1], Wei Ye[1], Claudia Plant[2], and Christian Böhm[1(✉)]

[1] Ludwig-Maximilians-Universität München, Munich, Germany
{zhou,ye,boehm}@dbs.ifi.lmu.de
[2] University of Vienna, Vienna, Austria
claudia.plant@univie.ac.at

Abstract. With the explosive growth of data quantity and variety, the representation and analysis of complex data becomes a more and more challenging task in many modern applications. As a general class of probabilistic distribution functions, Gaussian Mixture Models have the ability to approximate arbitrary distributions in a concise way, making them very suitable for the representation of complex data. To facilitate efficient queries and following analysis, we generalize Euclidean distance to Gaussian Mixture Models and derive the closed-form expression called Infinite Euclidean Distance. Our metric enables efficient and accurate similarity calculations. For the analysis of complex data, we model two real-world data sets, NBA player statistic and the weather data of airports, into Gaussian Mixture Models, and we compare the performance of Infinite Euclidean Distance to previous similarity measures on both classification and clustering tasks. Experimental evaluations demonstrate the efficiency and effectiveness of Infinite Euclidean Distance and Gaussian Mixture Models on the analysis of complex data.

1 Introduction

With the increase of generated and stored data quantity and variety, the analysis of complex data faces great challenges. One of the most important aspect is how to represent and retrieve data in an efficient way. On one hand, modern applications like speaker recognition [1,2], content-based image and video retrieval [3,4], biometric identification and stock market analysis can benefit from the retrieval and analysis of complex data, on the other hand, they also limit these applications. Take player statistics for example, far more than field goal made, rebounds and etc., SportVU utilizes six cameras to track the real-time positions of NBA players and the ball 25 times per second [5]. Comprehensive and sophisticated data generated by SportVU provides a possibility to make the best game strategy or to achieve the most effective team building, but it increases the difficulty of following modeling and analysis as well.

Many representation methods for complex data have been proposed, ranging from feature vectors to complicated models [6–10]. As a general class of Probability Density Function (PDF), Gaussian Mixture Models (GMMs) consist of

© Springer International Publishing AG 2017
L. Bellatreche and S. Chakravarthy (Eds.): DaWaK 2017, LNCS 10440, pp. 409–423, 2017.
DOI: 10.1007/978-3-319-64283-3_30

a weighted sum of univariate or multivariate Gaussian distributions, allowing a
concise but exact representation of data distributions. Storing complex data as
GMMs will dramatically reduce the resource consumption and guarantees the
accuracies of retrieval operations. GMMs are capable of representing large classes
of distributions, and another advantage of representing data as GMMs is that
the complexity of the model is constant with the variable number of instances.
As shown in Fig. 1(a), there are some records of Munich Airport weather statistics are missing, which is very common for real-world data. When we regard the
statistics of each day as an instance, the effect of missing data for modeling the
distribution is insignificant, as illustrated in Fig. 1(b).

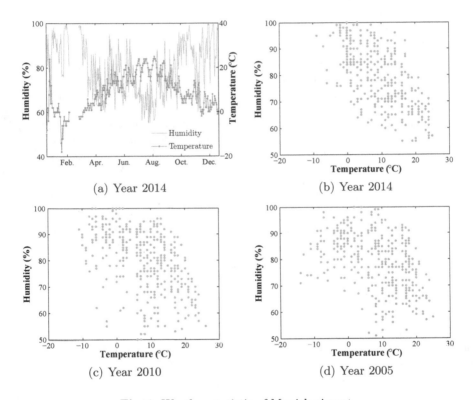

(a) Year 2014

(b) Year 2014

(c) Year 2010

(d) Year 2005

Fig. 1. Weather statistic of Munich airport.

Comparing the distributions of instances in Fig. 1(b), (c) and (d), we can tell
that the numbers of cold days in Munich tend to decrease from 2005 to 2014.
However, quantitative indicators are needed to get a more accurate description of weather changes. The design of similarity measures aims at facilitating
indexes and further analysis. A closed-form expression is essential to efficient
calculations, otherwise, approximate methods like Monte-Carlo sampling are
needed, which results in time consuming and/or inaccurate. What is more, since

GMMs might have different numbers of Gaussian components in them, traditional indexes designed for fixed-length vectors can not be applied here. For those indexes that are based on similarity measures, for instance, M-tree [11] and VP-tree [12], the properties of metric (e.g., triangle inequality) are required for the similarity measure to guarantee the effectiveness and efficiency queries. As we will demonstrate, the main contributions of this paper are:

- We generalize Euclidean distance to Infinite Euclidean Distance (IED) on PDFs, prove its metric properties and derive the closed-form expression for GMMs.
- Our experimental evaluations on both synthetic and real-world data sets demonstrate the effectiveness and efficiency of IED and the better performances than previous similarity measures for GMMs.

The rest of this paper is organized as follows. In Sect. 2, we survey the previous work. Section 3 gives the basic definition of GMMs, and introduces IED, a metric with closed-form expression for GMMs. Section 4 shows the experimental studies for verifying the efficiency and effectiveness of the proposed similarity measure. Section 5 summarizes the paper.

2   Related Work

This section gives a survey and discussion of previous work on Multiple-Instance Learning, similarity measures and indexes for GMMs.

2.1   Data Representations

For objects with inherent structures, Multiple-Instance (MI) is a natural way to describe them. First motivated by the problem of drug activity predictions, Multiple-Instance Learning (MIL) deals with MI objects that are sets (or bags) of instances [13]. MIL algorithms can be grouped into three categories, the instance space based paradigm, the bag space based paradigm and the embedded space based paradigm [14], the last of which maps MI objects into new feature spaces.

Many mapping methods have been proposed to represent the data of instances, including feature vectors and complex models. The vocabulary based mapping clusters all instances from all bags into k clusters (vocabularies), and then uses the histogram information, i.e., the counts of instances that belong to these clusters, to obtain a k-dimensional feature vector for each bag [7,14,15]. The instance based mapping models instances as a feature. For example, DD-SVM (Diverse Density SVM) [16] uses instance prototypes that obtained according to DD measure, while MILES [17] chooses one of instances as the feature. The model based mapping trains each bag to a model, for instance, each bag is represented by a k-component mixture model (EM-clustering [8], PPMM [18], miFV [9]), a Gaussian distribution [19], a graph in miGraph [10], a joint optimization concept [20] and so on.

Having the ability to approximate arbitrary distributions, GMMs can achieve a more accurate representation of data than the feature vectors and other models, and it is a concise model as well [21].

2.2 Similarity Measures

Similarity measures for GMMs can be grouped into two categories, having closed-form expressions for GMMs or not. For measures that have no closed-form expressions, Monte Carlo sampling or other approximation approaches are applied, which may be time consuming or imprecise.

Kullback-Leibler (KL) divergence [22] is a common way to measure the distance between two PDFs. It has a closed-form expression for Gaussian distributions, but no such expression for GMMs exists.

To compute the distance between GMMs by KL divergence, several approximation methods have been proposed. For two GMMs, a commonly used approximation for KL divergence between them is Gaussian approximation. It replaces two GMMs with two Gaussian distributions, whose means and covariance matrices depend on those of GMMs. Another popular way is to use the minimum KL divergence of Gaussian components that are included in two GMMs. Moreover, Hershey et al. [23] have proposed the product of Gaussian approximation and the variation approximation, but the former tends to greatly underestimate the KL divergence between GMMs while the latter does not satisfy the positivity property. Besides, Goldberger et al. [24] have proposed the matching based KL divergence (KLm) and the unscented transformation based KL divergence (KLt). KLm works well when the Gaussian elements are far apart, but it cannot handle the overlapping situations which are very common in real-world data sets. KLt solves the overlapping problem based on a non-linear transformation. Cui et al. [25] have compared the six approximation methods for KL divergence with Monte Carlo sampling, where the variation approximation achieves the best result quality, while KLm give a comparable result with a much faster speed.

Besides the approximation similarity methods for GMMs, several methods with closed-form expressions have been proposed. Helén et al. [26] have described a squared Euclidean distance (ES), which integrates the squared differences over the whole feature space. Sfikas et al. [27] have presented a KL divergence based distance C2 for GMMs. Jensen et al. [28] used a normalized L2 (NL2) distance to measure the similarity of GMMs in mel-frequency cepstral coefficients from songs. Beecks et al. have proposed Gaussian Quadratic Form Distance (GQFD) for modeling image similarity in image databases [29]. However, only GQFD fulfills the properties of metric on condition that a proper setting of parameters is given.

2.3 Indexes

For the indexes of GMMs, there are several techniques available, including universal index structures designed for uncertain data and GMM-specific methods.

U-tree provides a probability threshold retrieval on general multi-dimensional uncertain data [30]. It pre-computes a finite number of Probabilistically Constrained Regions (PCRs) which are possible appearance regions with fixed probabilities, and uses them to prune unqualified objects. Although U-tree works well with single PDFs, its effectiveness deteriorates for mixture models such as GMMs. The reason behind this is that it is difficult for PCRs to represent mixture models, especially when the component numbers increase.

Rougui et al. [31] have designed a bottom-up hierarchical tree and an iterative grouping tree for GMM-modeled speaker retrieval systems. Both approaches provide only two index levels, and are lack of a convenient insertion and deletion strategy. Furthermore, they can not guarantee reliable query results.

Instead of indexing curves as spatial objects in feature spaces, Probabilistic Ranking Query (PRQ) technique [32] and Gaussian Component based Index [33] search the parameter space of the means and variances of GMMs. However, PRQ can not guarantee the query accuracy since it assumes that all the Gaussian components of candidates have relatively high matching probabilities with query objects, which is not common in general cases. For both indexes, their prune strategies are highly effected by the distributions of Gaussian components.

Similarity measures that have the properties of metric can easily be supported by metric trees like M-tree [11] and VP-tree [12]. Otherwise, special designed structures are needed to guarantee efficient queries.

3   Methods

In this section, firstly we summarize the formal notations for GMMs, then we introduce IED for distributions and give the proof of its metric properties. Finally we derive the closed-form expression of IED for GMMs.

3.1   Gaussian Mixture Models

A GMM is a probabilistic model that represents the probability distribution of observations. The definition is shown as follows.

Definition 1 *(Gaussian Mixture Model). Let $\mathbf{x} \in \mathbb{R}^D$ be a variable in a D-dimensional space, $\mathbf{x} = (x_1, x_2, ..., x_D)$. A Gaussian Mixture Model $\mathcal{G}$ is the weighted sum of m Gaussian functions, defined as:*

$$\mathcal{G}(\mathbf{x}) = \sum_{1 \leq i \leq m} w_i \cdot \mathcal{N}_i(\mathbf{x}) \tag{1}$$

where $\sum_{1 \leq i \leq m} w_i = 1$, $\forall i \in [1, m], w_i \geq 0$, and Gaussian component $\mathcal{N}_i(\mathbf{x})$ is the density of a Gaussian distribution with a covariance matrix Σ_i:

$$\mathcal{N}_i(\mathbf{x}) = \frac{1}{\sqrt{(2\pi)^D |\Sigma_i|}} \exp\left(-\frac{1}{2}(\mathbf{x} - \mu_i)^T \Sigma_i^{-1}(\mathbf{x} - \mu_i)\right)$$

As we can see in Definition 1, a GMM can be represented by a set of m components, and each of them is composed of a mean vector $\mu \in \mathbb{R}^D$ and a covariance matrix $\Sigma \in \mathbb{R}^{D \times D}$. Modelling complex data into GMMs will dramatically reduce the resource consumption. What is more, with the increase of components number m, GMMs provide more and more precious representations of the original data.

3.2   Infinite Euclidean Distance for Distributions

Euclidean distance is the basic distance function for feature vectors in Euclidean space. Here we generalize Euclidean distance into IED, a distance measure for PDFs. We determine square differences between the values of the corresponding PDFs and sum them up by integration. The definition of IED is shown as follows.

Definition 2 *(Infinite Euclidean Distance). Given two PDFs $f(\mathbf{x})$ and $g(\mathbf{x})$ in a D-dimensional space, Infinite Euclidean Distance between them is defined as:*

$$d_{IED}(f, g) = \left(\int_{\mathbb{R}^D} |f(\mathbf{x}) - g(\mathbf{x})|^2 \mathrm{d}\mathbf{x} \right)^{\frac{1}{2}} \tag{2}$$

Metric Properties. The metric properties of similarity measures facilitate the applications of metric trees for efficient queries, while for similarity measures without metric properties, special structures are needed to guarantee the accuracy and efficiency of queries. What is more, some analysis techniques like DBSCAN [34] also require the properties of a metric. A metric, e.g. Euclidean distance, is a distance function that fulfills three metric properties. Next we give the proof that IED is a metric.

Lemma 1. *IED is a metric.*

Proof. (M_1) **Positive Definiteness:** As a PDF, the integrated function is everywhere greater or equal to zero. If and only if f_1 and f_2 are exactly equal, $(f_1(\mathbf{x}) - f_2(\mathbf{x}))^2 = 0$ for all $\mathbf{x}$, thus $d_{\mathrm{IED}}(f_1, f_2) = 0$. If in some positions, f_1 and f_2 are not equal, then it will have a positive influence on the integral. In that case, $d_{\mathrm{IED}}(f_1, f_2) > 0$.

(M_2) **Symmetry:** Obviously, $d_{\mathrm{IED}}(f_1, f_2) = d_{\mathrm{IED}}(f_2, f_1)$, because of the absolute value in $|f_1(\mathbf{x}) - f_2(\mathbf{x})|^2$.

(M_3) **Triangle Inequality:** The triangle inequality of IED states that for any PDF, the following inequality always holds.

$$d_{\mathrm{IED}}(f_1, f_2) + d_{\mathrm{IED}}(f_2, f_3) \geq d_{\mathrm{IED}}(f_1, f_3)$$

Since for $A, B, C \geq 0$, inequality $A + B \geq C$ is equivalent to $(A + B)^2 \geq C^2$. The inequality can be transformed into:

$$(d_{\mathrm{IED}}(f_1, f_2) + d_{\mathrm{IED}}(f_2, f_3))^2 \geq (d_{\mathrm{IED}}(f_1, f_3))^2$$

To prove this inequality, we substitute IED into the objective function Obj as shown below.

$$Obj = (d_{\text{IED}}(f_1, f_2) + d_{\text{IED}}(f_2, f_3))^2 - (d_{\text{IED}}(f_1, f_3))^2$$

$$= 2\int_{\mathbb{R}^D} (f_2 - f_3)(f_2 - f_1)\mathrm{dx} + 2\sqrt{\int_{\mathbb{R}^D} (f_1 - f_2)^2\mathrm{dx} \int_{\mathbb{R}^D} (f_2 - f_3)^2\mathrm{dx}}$$

$$\geq 2\int_{\mathbb{R}^D} (f_2 - f_3)(f_2 - f_1)\mathrm{dx} + 2\left|\int_{\mathbb{R}^D} (f_2 - f_3)(f_2 - f_1)\mathrm{dx}\right| \geq 0$$

Thus we obtain $d_{\text{IED}}(f_1, f_2) + d_{\text{IED}}(f_2, f_3) \geq d_{\text{IED}}(f_1, f_3)$.

Closed-form Expression. Firstly we derive the closed-form expression for the inner product of GMMs. Let $\mathcal{G}_1$ and $\mathcal{G}_2$ be two GMMs with diagonal covariance matrices, and they have m_1 and m_2 Gaussian components, respectively. Let $\mathbf{x}$ be a feature vector in the space $\mathbb{R}^D$. The inner product of $\mathcal{G}_1$ and $\mathcal{G}_2$ can be derived as:

$$\langle \mathcal{G}_1, \mathcal{G}_2 \rangle = \int_{\mathbb{R}^D} \sum_{i=1}^{m_1} w_{1,i} \cdot \mathcal{N}(\mu_{1,i}, \sigma_{1,i}^2) \sum_{j=1}^{m_2} w_{2,j} \cdot \mathcal{N}(\mu_{2,j}, \sigma_{2,j}^2)\mathrm{dx}$$

$$= \sum_{i=1}^{m_1}\sum_{j=1}^{m_2} w_{1,i}w_{2,j} \prod_{l=1}^{D} \frac{1}{2\pi\sqrt{\sigma_{1,i,l}^2\sigma_{2,j,l}^2}} \int e^{-\frac{(x-\mu_{1,i,l})^2}{2\sigma_{1,i,l}^2} - \frac{(x-\mu_{2,j,l})^2}{2\sigma_{2,j,l}^2}}\mathrm{dx} \qquad (3)$$

$$- \sum_{i=1}^{m_1}\sum_{j=1}^{m_2} w_{1,i}w_{2,j} \prod_{l=1}^{D} \frac{e^{-\frac{(\mu_{1,i,l}-\mu_{2,j,l})^2}{2(\sigma_{1,i,l}^2+\sigma_{2,j,l}^2)}}}{\sqrt{2\pi(\sigma_{1,i,l}^2 + \sigma_{2,j,l}^2)}}$$

where $\sigma_{1,i,l}$ and $\sigma_{2,j,l}$ are the l-th diagonal elements of $\Sigma_{1,i}$ and $\Sigma_{2,j}$, respectively.

The relation of the inner product to IED is:

$$d_{\text{IED}}(\mathcal{G}_1, \mathcal{G}_2) = \sqrt{\langle \mathcal{G}_1, \mathcal{G}_1 \rangle + \langle \mathcal{G}_2, \mathcal{G}_2 \rangle - 2\langle \mathcal{G}_1, \mathcal{G}_2 \rangle} \qquad (4)$$

A closed-form expression is intrinsically valuable for computations. It saves extra efforts to get a good approximation by avoiding simulation methods, like Monte-Carlo sampling, which may cause a significant increase in computation time and the loss of precision. Therefore, closed-form expressions are well received in many applications, especially in real-time applications. It is worth noting that only for GMMs that have diagonal covariance matrices, IED for GMMs has a closed-form expression, so are the other similarity measures to the best of our knowledge.

4   Experimental Evaluations

In this section, we provide experimental evaluations on both synthetic and real-world data sets to show the efficiency and effectiveness of complex data analysis using GMMs and the proposed similarity measure.

For KL divergence based similarity measures, only KLm is included in the comparison since it is one of the best-performing approximations [25]. We set the parameter α of GQFD as 10E-5 following the original paper [29]. As for Hausdorff distance, we use the following equations to calculate the distance between GMM $\mathcal{G}_1$ and $\mathcal{G}_2$.

$$
d_{\text{Hausdorff}}(\mathcal{G}_1, \mathcal{G}_2) = \max\{\sup_{\mathcal{G}_1} \inf_{\mathcal{G}_2} \frac{e(g_{1i}, g_{2j})}{w_{1i}w_{2j}}, \sup_{\mathcal{G}_2} \inf_{\mathcal{G}_1} \frac{e(g_{1i}, g_{2j})}{w_{1i}w_{2j}}\}
$$

$$
e(g_1, g_2) = \sqrt{\sum_{i=1}^{2D}(v_{1,i} - v_{2,i})^2} \tag{5}
$$

where $v = \{\mu_l, \sigma_l^2\}_{l=1}^{D}$ is the parameter vector of the Gaussian distribution in a D-dimensional space.

All the experiments are implemented with Java 1.7, and executed on a regular workstation PC with 3.4 GHz dual core CPU equipped with 32 GB RAM. For all the experiments, we use the 10-fold cross validation and report the average results over 100 runs.

4.1   Data Sets

The data sets used in this paper consists of a synthetic data set[1] and two real-world data sets[2].

We collect 3,769 NBA players statistic data that includes 1,023,731 match logs until 2014. Seventeen statistics (WL, MIN, FGM, FGA, FG3M, FG3A, FTM, FTA, OREB, DREB, REB, AST, STL, BLK, TOV, PF and PTS[3]) are used as features for each player, and we estimate GMMs with ten components from statistic data using the Expectation-Maximization (EM) algorithm. To tune the number of GMMs into its optimum, Bayesian Information Criterion can be applied.

For another real-world data, we use daily weather data of 2,946 airports in the whole year 2014. Because of missing data, there are 961,308 pieces of records in the data set. The selected features are temperature and humidity, and only the average values of each day are used. For each airport, a ten-component GMM is estimated by EM algorithm based on the whole year weather data.

---

[1] https://drive.google.com/open?id=0B3LRCuPdnX1BSTU3UjBCVDJSLWs.
[2] https://drive.google.com/open?id=0B3LRCuPdnX1BUW5TbzNSdDBoaVk.
[3] http://stats.nba.com/help/glossary/.

4.2 Query Performance

We study the performance of the only two metrics, IED and GQFD[4], when using VP-tree to facilitate efficient queries. The query results on synthetic data are reported in Fig. 2. As shown in Fig. 2(a), the query time acceleration ratio (comparing with linear scan) of IED increases with the number of stored objects while that of GQFD almost remains unchanged. As for the query time (Fig. 2(b)), IED costs more run-time than GQFD at the beginning, then achieves a much better results than GQFD with the increase of the number of objects.

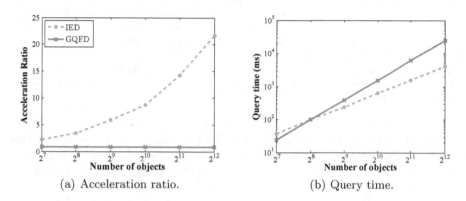

(a) Acceleration ratio.   (b) Query time.

Fig. 2. 1-Nearest Neighbour query results of IED and GQFD on synthetic data using VP-tree. The capacity of nodes in the VP-tree is set to 32.

4.3 Classification on NBA Data

Since there is no label information for NBA data, we evaluate classification results by comparing them to subjective opinions.

There is a famous question about the NBA players: who plays most like M. Jordan. To answer the question with the support of data, we model the statistics into GMMs and then apply k-Nearest Neighbour (k-NN) algorithm, setting the query object as the GMM of M. Jordan. Most of the similarity measures, except GQFD, successfully pick Jordan as the'most like' player to himself. The other results are shown in Table 1 in the form of ranking lists, and opinions from two experts J. Kiang[5] and F. Ewere[6] are also included. To have a quantified criteria, we define an accuracy function shown as follows to evaluate the rankings.

$$Accuracy = \frac{|QR \bigcap (EO_1 \bigcup EO_2)|}{k}$$

---

[4] With the given parameter, the query accuracies of GQFD using VP-tree is guaranteed for the synthetic data.

[5] http://bleacherreport.com/articles/537852-michael-jordan-and-his-nba-heirs-the-10-most-like-mike-players-in-the-league.

[6] http://www.rantsports.com/nba/2015/07/12/10-current-nba-players-who-emulate-michael-jordans-competitiveness/.

where QR is the query results of k-NN, and EO_1 and EO_2 are expert opinions. As shown in Table 1, the highest accuracy is obtained by IED and SE, and four out of eight candidates picked by them meet the opinions of experts.

Table 1. Eight NBA players that play most 'like' Jordan

	1	2	3	4	5	6	7	8	Accu
Kiang	Bryant	D. Rose	Wade	Durant	James	Westbrook	Anthony	Ellis	–
Ewere	Bryant	Westbrook	Wade	C. Paul	Garnett	P. Pierce	Ginobli	Durant	–
IED	Barkley	Iverson	Wade	P. Pierce	Durant	Nowitzki	Powell	Bryant	4/8
SE	Barkley	Iverson	Wade	P. Pierce	Durant	Nowitzki	Powell	Bryant	4/8
C2	Wade	Iverson	Robinson	Mashburn	Rose	Malone	Bryant	Nowitzki	3/8
NL2	Wade	Worthy	D. Rose	Robinson	Iverson	R. Gay	Mashburn	Westbrook	3/8
KLm	Drexler	Bird	Wade	Aguirre	Bryant	Carter	Wilkins	Anthony	3/8
GQFD	Hanson	Nickerson	Johnson*	Johnson†	Werdann	Lewis	Stokes	Claxton	0/8
Hausdf	R. White	T. Tyler	Nimphius	McDaniel	Sobers	Lucas	Silas	Churchwell	0/8

*Darryl Johnson †DeMarco Johnson

Because of their definitions, IED and SE provide the same results on this query task. Picking three levels of NBA players (See Table 2) from their ranking lists, we demonstrate the multidimensional scaling (MDS) plot of IED and SE. A good similarity measure should be able to put the candidates of the same level closer than the other levels in the MDS plot. As shown in Fig. 3, IED not only assigns NBA players from the same level closer than ES, but also achieves a more distinguishable layout between levels than ES.

Table 2. Sub-dataset: three levels of NBA players

Level					
A	M. Jordan	K. Bryant	D. Wade	A. Iverson	P. Pierce
B	W. Burton	T. Chambers	A. Peeler	M. Fizer	R. Pack
C	C. Laettner	B. Miller	W. Person	R. Seikaly	B. Gordon

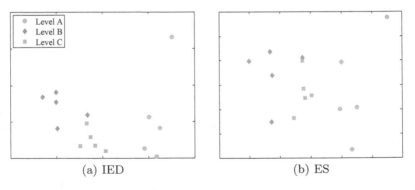

(a) IED   (b) ES

Fig. 3. Multidimensional scaling plot of 15 players using different similarity measures. Each marker (dot, diamond and square) represents a NBA player.

4.4   Clustering on Weather Data

For Weather data, we perform clustering experiments to compare the usability of the proposed similarity measure for unsupervised data mining. Instead of k-means algorithm, the k-medoids is used since it works with arbitrary similarity measures, making it more suitable here. We evaluate the clustering results using two widely used criteria, Purity and Normalized Mutual Information (NMI).

According to Peel et al. [35], the world climate can be divided into a total of 29 categories using Köppen climate classification, which is based on average annual and monthly temperature and precipitation, as well as the seasonality of precipitation. These features are highly relevant to Weather data, thus we assign each airport to one of the categories according to its location and get 25 classes of climate types (Cwc, Dsd, Dwd and EF are not included) in total.

Table 3. Two clustering criteria on weather data

	Purity	NMI
IED	**0.363 ± 0.010**	**0.245 ± 0.000**
SE	0.342 ± 0.014	0.241 ± 0.000
C2	0.347 ± 0.010	0.231 ± 0.010
NL2	0.357 ± 0.010	0.237 ± 0.000
KLm	0.337 ± 0.014	0.224 ± 0.010
GQFD	0.198 ± 0.024	0.143 ± 0.083
Hausdorff	0.219 ± 0.014	0.085 ± 0.010

Table 3 shows the clustering results of Weather data. We can see that IED achieves the highest Purity and NMI among all the similarity measures. To get an visualized impression of the best two results from IED and NL2, we mark the airports of different clusters as dots with different colors in Fig. 4.

Figure 4(a) shows the ground truth of Köppen climate classification of all the world. Figure 4(b) and (c) demonstrate the clustering results of IED and NL2, respectively. To compare these two results, we focus on the clusters in areas like Africa, North America and Southeast Asian Islands. Climate type BWh locates in North Africa and the most part of Australia. The result of IED indicates the same trend with the ground truth. NL2 clusters airports that locate in North Africa and Australia in the same group, however, it also includes airports that locate in the south part of Africa. For airports in North America and Southeast Asian Islands, IED outperforms NL2 with a more clear categories.

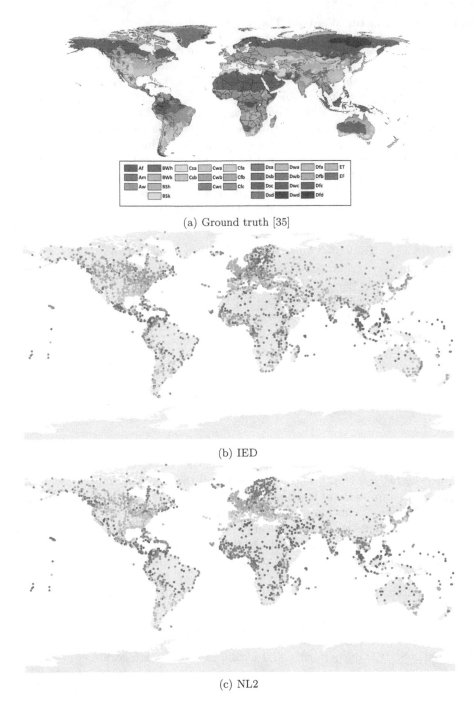

(a) Ground truth [35]

(b) IED

(c) NL2

Fig. 4. Clustering results of Weather data. It is worth noting that dots with same color on different figures may indicate different clusters.

5   Conclusions and Future Work

In this paper, we generalize Euclidean distance to Infinite Euclidean Distance for probability distribution functions, and derive its closed-form expression for Gaussian Mixture Models. The metric properties of the proposed similarity measure enable the usage of metric trees for indexing Gaussian Mixture Models. Representing complex data that have inherent structures as Gaussian Mixture Models, we apply classification and clustering analysis on real-world data with different similarity measures. Experimental evaluations demonstrate the efficiency and effectiveness of the proposed similarity measure and better performances than its comparisons.

For the future work, a Gaussian Mixture Models-specific index structure that uses Infinite Euclidean Distance as the similarity measure is a perspective to outperform general metric trees.

References

1. Campbell, W.M., Sturim, D.E., Reynolds, D.A.: Support vector machines using GMM supervectors for speaker verification. IEEE Signal Process. Lett. **13**(5), 308–311 (2006)
2. Reynolds, D.A., Quatieri, T.F., Dunn, R.B.: Speaker verification using adapted gaussian mixture models. Digit. Signal Proc. **10**(1–3), 19–41 (2000)
3. KaewTraKulPong, P., Bowden, R.: An improved adaptive background mixture model for real-time tracking with shadow detection. In: Remagnino, P., Jones, G.A., Paragios, N., Regazzoni, C.S. (eds.) Video-Based Surveillance Systems, pp. 135–144. Springer, Boston (2002)
4. Zivkovic, Z.: Improved adaptive gaussian mixture model for background subtraction. In: ICPR, pp. 28–31 (2004)
5. STATS description. https://www.stats.com/sportvu-basketball-media/. Accessed 25 Feb 2017
6. Cheplygina, V., Tax, D.M.J., Loog, M.: Dissimilarity-based ensembles for multiple instance learning. IEEE Trans. Neural Netw. Learn. Syst. **27**(6), 1379–1391 (2016)
7. Sivic, J., Zisserman, A.: Video google: a text retrieval approach to object matching in videos. In: ICCV, pp. 1470–1477 (2003)
8. Kriegel, H.-P., Pryakhin, A., Schubert, M.: An EM-approach for clustering multi-instance objects. In: Ng, W.-K., Kitsuregawa, M., Li, J., Chang, K. (eds.) PAKDD 2006. LNCS, vol. 3918, pp. 139–148. Springer, Heidelberg (2006). doi:10.1007/11731139_18
9. Wei, X., Wu, J., Zhou, Z.: Scalable multi-instance learning. In: ICDM, pp. 1037–1042 (2014)
10. Zhou, Z., Sun, Y., Li, Y.: Multi-instance learning by treating instances as non-I.I.D. samples. In: ICML, pp. 1249–1256 (2009)
11. Ciaccia, P., Patella, M., Zezula, P.: M-tree: an efficient access method for similarity search in metric spaces. In: VLDB, pp. 426–435 (1997)
12. Yianilos, P.N.: Data structures and algorithms for nearest neighbor search in general metric spaces. In: ACM/SIGACT-SIAM SODA, pp. 311–321 (1993)
13. Dietterich, T.G., Lathrop, R.H., Lozano-Pérez, T.: Solving the multiple instance problem with axis-parallel rectangles. Artif. Intell. **89**(1–2), 31–71 (1997)

422   L. Zhou et al.

14. Amores, J.: Multiple instance classification: Review, taxonomy and comparative study. Artif. Intell. **201**, 81–105 (2013)
15. Weidmann, N., Frank, E., Pfahringer, B.: A two-level learning method for generalized multi-instance problems. In: Lavrač, N., Gamberger, D., Blockeel, H., Todorovski, L. (eds.) ECML 2003. LNCS, vol. 2837, pp. 468–479. Springer, Heidelberg (2003). doi:10.1007/978-3-540-39857-8_42
16. Chen, Y., Wang, J.Z.: Image categorization by learning and reasoning with regions. J. Mach. Learn. Res. **5**, 913–939 (2004)
17. Chen, Y., Bi, J., Wang, J.Z.: MILES: multiple-instance learning via embedded instance selection. Pattern Anal. Mach. Intell. **28**(12), 1931–1947 (2006)
18. Wang, H., Yang, Q., Zha, H.: Adaptive p-posterior mixture-model kernels for multiple instance learning. In: ICML, pp. 1136–1143 (2008)
19. Vatsavai, R.R.: Gaussian multiple instance learning approach for mapping the slums of the world using very high resolution imagery. In: SIGKDD, pp. 1419–1426 (2013)
20. Sikka, K., Giri, R., Bartlett, M.S.: Joint clustering and classification for multiple instance learning. In: BMVC, p. 71.1–71.12 (2015)
21. Reynolds, D.: Gaussian mixture models. In: Li, S.Z., Jain, A. (eds.) Encyclopedia of Biometrics, pp. 827–832. Springer, New York (2015)
22. Kullback, S.: Information Theory and Statistics. Courier Dover Publications, Mineola (2012)
23. Hershey, J.R., Olsen, P.A.: Approximating the kullback leibler divergence between gaussian mixture models. In: ICASSP, pp. 317–320 (2007)
24. Goldberger, J., Gordon, S., Greenspan, H.: An efficient image similarity measure based on approximations of KL-divergence between two gaussian mixtures. In: ICCV, pp. 487–493 (2003)
25. Cui, S., Datcu, M.: Comparison of kullback-leibler divergence approximation methods between gaussian mixture models for satellite image retrieval. In: IGARSS, pp. 3719–3722 (2015)
26. Helén, M.L., Virtanen, T.: Query by example of audio signals using euclidean distance between gaussian mixture models. In: ICASSP, vol. 1, pp. 225–228 (2007)
27. Sfikas, G., Constantinopoulos, C., Likas, A., Galatsanos, N.P.: An analytic distance metric for gaussian mixture models with application in image retrieval. In: Duch, W., Kacprzyk, J., Oja, E., Zadrożny, S. (eds.) ICANN 2005. LNCS, vol. 3697, pp. 835–840. Springer, Heidelberg (2005). doi:10.1007/11550907_132
28. Jensen, J.H., Ellis, D.P.W., Christensen, M.G., Jensen, S.H.: Evaluation of distance measures between gaussian mixture models of MFCCs. In: ISMIR, pp. 107–108 (2007)
29. Beecks, C., Ivanescu, A.M., Kirchhoff, S., Seidl, T.: Modeling image similarity by gaussian mixture models and the signature quadratic form distance. In: ICCV, pp. 1754–1761 (2011)
30. Tao, Y., Cheng, R., Xiao, X., Ngai, W.K., Kao, B., Prabhakar, S.: Indexing multidimensional uncertain data with arbitrary probability density functions. In: VLDB, pp. 922–933 (2005)
31. Rougui, J.E., Gelgon, M., Aboutajdine, D., Mouaddib, N., Rziza, M.: Organizing Gaussian mixture models into a tree for scaling up speaker retrieval. Pattern Recogn. Lett. **28**(11), 1314–1319 (2007)
32. Böhm, C., Kunath, P., Pryakhin, A., Schubert, M.: Querying objects modeled by arbitrary probability distributions. In: Papadias, D., Zhang, D., Kollios, G. (eds.) SSTD 2007. LNCS, vol. 4605, pp. 294–311. Springer, Heidelberg (2007). doi:10.1007/978-3-540-73540-3_17

33. Zhou, L., Wackersreuther, B., Fiedler, F., Plant, C., Böhm, C.: Gaussian component based index for GMMs. In: ICDM, pp. 1365–1370 (2016)
34. Ester, M., Kriegel, H., Sander, J., Xu, X.: A density-based algorithm for discovering clusters in large spatial databases with noise. In: SIGKDD, pp. 226–231 (1996)
35. Peel, M.C., Finlayson, B.L., McMahon, T.A.: Updated world map of the köppen-geiger climate classification. Hydrol. Earth Syst. Sci. Discuss. 4(2), 439–473 (2007)

Optimized Mining of Potential Positive and Negative Association Rules

Parfait Bemarisika[1,2](✉) and André Totohasina[1]

[1] Laboratoire de Mathématiques et d'Informatique, ENSET,
Université d'Antsiranana, Antsiranana, Madagascar
bemarisikap7@yahoo.fr, andre.totohasina@gmail.com
[2] Laboratoire d'Informatique et de Mathématiques EA2525,
Université de La Réunion, Saint-Denis, France

Abstract. The negative association rules are less explored compared to the positive rules. The existing models are limited to the structure of binary data requiring of the repetitive accesses to the context, and the traditional couple support-confiance which is not effective in the presence of the dense data. For that, we propose a new model of optimization by using a new structure of data, noted MATRICESUPPORT, and a new more selective couple, support-M_{GK}.

Keywords: Association rules · Optimized extraction · Support-M_{GK}

1 Introduction and Motivations

The extraction of the positive and negative association rules [6–8,13] is a major challenge in the community of the data mining. Although the negative rules have obvious advantages [3,4,11], particularly in the extraction of knowledge (production of bases rules, or the construction of classifiers), they remain little explored compared to positive rules. We think that one of the significant disadvantages of these negative rules resides in their difficult extraction, they increase four times more than the whole together. The limited effectiveness of the existing algorithms rises mainly from the handling of the structures of the data and the indices of quality used. In fact, the immense majority of these existing use the structure of binary data of Apriori [1] requiring repetitive accesses to the context, and are limited to the use of the traditional couple support-confiance [1]. However, this couple is not effective in the presence of the strongly correlated data and/or for the thresholds of the weak support: the discovery of the potentially relevant association rules is even more delicate since many rules prove to be uninteresting and redundant. In order to exceed these limits, we propose a new method of optimization by using a new structure of data, noted MATRICESUPPORT, and a new couple more selective support-M_{GK}. The remainder of this paper is organized as follows. Section 2 introduced the formal concepts used into the continuation of paper. Section 3 details the approach suggested, OM2PNR-Optimized Mining of Potential Positive and Negative Rules association. Section 4 synthesizes the experiments. A conclusion is given in Sect. 5.

© Springer International Publishing AG 2017
L. Bellatreche and S. Chakravarthy (Eds.): DaWaK 2017, LNCS 10440, pp. 424–432, 2017.
DOI: 10.1007/978-3-319-64283-3_31

2   Preliminary Concepts

A transactional context is a triplet $\mathcal{B} = (\mathcal{T},\mathcal{I},\mathcal{R})$, where $\mathcal{T},\mathcal{I},\mathcal{R}$ are finite and not empty sets. An element of $\mathcal{I}$ is called item (or pattern), an element of $\mathcal{T}$ is a transaction (or object) represented by an identifier TID, and $\mathcal{R}$ a binary relation between $\mathcal{T}$ and $\mathcal{I}$. The Table 1 represents an example of it.

Table 1. Database $\mathcal{B}$, $\mathcal{T} = \{1,2,3,4,5,6\}$ and $\mathcal{I} = \{A,B,C,D,E\}$

TID	Items
1	ACD
2	BCE
3	ABCE
4	BE
5	ABCE
6	BCE

For is any X of $\mathcal{I}$, $\overline{X} = \mathcal{I}\backslash X$ is called the logical negation of X. A k-itemset is a itemset of size k. The connection of Galois [9] is a pair of functions (f,g), where $g(\mathcal{I}) = \{t \in \mathcal{T}|\forall i \in \mathcal{I}, i\mathcal{R}t\}$ and $f(\mathcal{T}) = \{i \in \mathcal{I}|\forall t \in \mathcal{T}, i\mathcal{R}t\}$. $g(\mathcal{I})$. $g(\mathcal{I})$ is the transactions set of $\mathcal{T}$ having commonly all the item for $\mathcal{I}$ and f is the dual function of g. The applications $\gamma = fog$ and $\gamma' = gof$ indicate the operators of closures Galois, where o indicates the composition of functions. Caution! The calculation of a closure is very exponential, because it requires site of the correspondences f and g. The calculation of a single correspondence requires to itself to traverse the context. For that, we developed a new technique (see Sect. 3), without calculating the closures. For all X and Y of $\mathcal{I}$, the couple (X,Y) noted $X \rightarrow Y$, such as $X \cap Y = \emptyset$, is called an association rules, where X is the premise and Y the conclusion of the rule. In the same way, one will speak about the negative rules if one at least of two items is negative, being able to be type of $X \rightarrow \overline{Y}$, or $\overline{X} \rightarrow Y$, or $\overline{X} \rightarrow \overline{Y}$. The relevance of an association rules is measured by quality measures. With that, we use two measure, such as support [1] and M_{GK} [10,12]. For all X of $\mathcal{I}$, the support of X is defined by: $supp(X) = \frac{|\{t\in\mathcal{T}|X\subseteq t\}|}{|\mathcal{T}|}$ and $supp(\overline{X}) = 1 - supp(X)$, where $|\mathcal{A}|$ indicate the cardinality of $\mathcal{A}$. Let a $minsupp$, threshold of support. X is said as frequent if $supp(X) \geq minsupp$. For all X and Y of $\mathcal{I}$, the support of rule $X \rightarrow Y$ is defined by $supp(X \rightarrow Y) = supp(X \cup Y) = \frac{|\{t\in\mathcal{T}|X\cup Y\subseteq t\}|}{|\mathcal{T}|}$. For all X and Y of $\mathcal{I}$, such as $X \cap Y = \emptyset$, we define M_{GK} of rule $X \rightarrow Y$,

$$M_{GK}(X \rightarrow Y) = \begin{cases} \frac{P(Y|X)-P(Y)}{1-P(Y)}, \text{ si } X \text{ favours } Y, \ P(Y) \neq 1 \\ \frac{P(Y|X)-P(Y)}{P(Y)}, \text{ si } X \text{ disfavours } Y, \ P(Y) \neq 0 \end{cases} \quad (1)$$

where X favours (resp. disfavours) Y means $P(Y|X) > P(Y)$ (resp. $P(Y|X) < P(Y)$). Given a minimum support, called $minsupp$, and a threshold $\alpha \in [0,1]$ fixed, an association rule $X \rightarrow Y$ is then said as valid according to our approach support-M_{GK} if $supp(X \cup Y) \geq minsupp$ et $M_{GK}(X \rightarrow Y) \geq vc_\alpha$, where $vc_\alpha = \sqrt{\frac{1}{n}\frac{n-n_X}{n_X}\frac{n_Y}{n-n_Y}\chi^2(\alpha)}$ is the critical value of the rule $X \rightarrow Y$ according to M_{GK}, $\chi^2(\alpha)$ is one of the chi-deux with a degree of liberty to the threshold α and $(n = |\mathcal{T}|,\ n_X = |X|,\ n_Y = |Y|)$. The criticalvalue vc_α is used to establish the statistical conclusion of an obtained association rule.

3 OM2PNR Algorithm

Firstly, we present the data structure MATRICESUPPORT and the mathematical results justifying the model. A MATRICESUPPORT is a projection of the database $\mathcal{B}$ with respect to its attributes. The idea is to gather progressively the data with the structure and to store them. The Table 2 below illustrate its formalism on a database $\mathcal{B}$. On each attribute corresponds to one cell of this one in which the frequency is associated, noted v_{ij}, which represents the number of times that the item v_j appears with the item v_i, where i (resp. j) indicates the i-th line (resp. j-th column) of this projected base. This frequency is then used to calculate the supports: the 1-item will be recovered by diagonal, the 2-item of the superior part, via a Lemma 1.

Table 2. Formalism of MATRICESUPPORT on a basis $\mathcal{B}$

Database $\mathcal{B}$

TID	Attributs
1	ACD
2	BCE
3	ABCE
4	BE
5	ABCE
6	BCE

MATRICESUPPORT formalism

MATRICESUPPORT

i/j	A	B	C	D	E
A	3	2	3	1	2
B	-	5	4	0	5
C	-	-	5	1	4
D	-	-	-	1	0
E	-	-	-	-	5

Lemma 1. *Given a set of transactions $\mathcal{B} = (\mathcal{T}, \mathcal{I}, \mathcal{R})$ and 1 or 2-item X of this data $\mathcal{B}$, the support of X is defined by $supp(X) = \frac{v_{ij}}{|\mathcal{B}|}$.*

For example, $supp(A) = v_{11}/6 = 3/6$ and $supp(BC) = v_{23}/6 = 4/6$. The MATRICESUPPORT is limited particularly to a itemset for size lower than 3, which limits also the range of Lemma 1. Thus, we exploited the concept of the minimal generators. An item X is known as generator if it is minimal (in sense of inclusion) in its class of equivalence. The class of equivalence of X, according to [2], defined by $[X] = \{X_1 \subseteq \mathcal{I} | \gamma(X_1) = \gamma(X)\}$. The elements of $[X]$ share the same support. For example, from Table 2, the itemset AB, ABC, ABE and $ABCE$ have the same support, therefore they have the same class of equivalence, AB is then minimal generator. The follow proposition present a new technic for itemset not generator.

Proposition 1. *For all X not generator, $supp(X) = min\{supp(X')|X' \subset X\}$*

In addition to the property 2 [1], we introduce proposal 3 for pruning. For lack of place, the reader can refer to [3] for the demonstration.

Proposition 2. *Any subset of a frequent pattern is frequent. Any superset of a not frequent pattern is not frequent.*

Proposition 3. *Any subset of a generator pattern is also generator. Any superset of a not generator pattern is also not generator.*

3.1 Optimization of the Research the Frequent Patterns

The model proposed remains on our Algorithm 1 [5]. The step is to calculate all the items candidates and to prune the not frequent ones by minimizing the costs of calculations, particularly the number of access to the database $\mathcal{B}$. It is done in two stapes. The first stape is the generation of the 1 and 2-items in only one scanning. We have then used the data structure MATRICESUPPORT. The second stape generates the itemset for higher size. For that, we used proposals 2 and 3. These results are central, one does not calculate the support of a candidate as soon as it is not generator, because it can be derived from its subsets. For example, AC and BE of the Table 2 (see. MATRICESUPPORT) are not generators, because $supp(AC) = supp(A) = 3/6$ and $supp(BE) = supp(B) = supp(E) = 5/6$. No scanning is done for the supersets ABC, ABE, ACE and BCE, inevitably not generators. Their support is obtained by applying the Proposition 1. If it is generator, an access to the context will permit to recover its support, by using the classical support Apriori [1]. But, it is in general of low size, that is to say that on a certain level, all the candidates can be not generators. Consequently we do not have to access to the context any more. Algorithm 1 synthesizes these different optimizations. It takes as input a context $\mathcal{B}$, a minimum support $minsupp$, and gives at output a set of the frequent patterns, where C_k indicates the set of the k-itemsets candidates and $\mathcal{CGM}_k$ that of the k-candidates generators. The database $\mathcal{B}$ is built by the function ConstruireBase (line 1). $\mathcal{F}_1$ and $\mathcal{F}_2$

Algorithm 1. EOMF-Extraction Optimised of the motifs frequents

Require: A database $\mathcal{B}$, a list of items $\mathcal{I}$, a minimum support $minsupp$.
Ensure: A set of frequents $\mathcal{F}_k$.
```
1:  MATRICESUPPORT ← BuildBase(B, I); //Build the database B
2:  F₁ ← {c₁ ∈ MATRICESUPPORT|supp(c₁) ≥ minsupp}; //Generate the 1-Itemsets
3:  F₂ ← {c₂ ∈ MATRICESUPPORT|supp(c₂) ≥ minsupp}; //Generate the 2-itemsets
4:  for (k = 3; F_{k-1} ≠ ∅; k + +) do
5:      C_k ← EOMF-Gen(F_{k-1}); //Generate candidates patterns
6:      for all (candidate c ∈ C_k) do
7:          if (c ∈ CGM_k) then
8:              for all (transaction t ∈ B) do
9:                  C_t ← subset(C_k, t); //Select candidates
10:                 for all (candidate c ∈ C_t) do
11:                     supp(c) + +;
12:                 end for
13:             end for
14:         else
15:             supp(c) ← min{supp(c')|c' ⊂ c};
16:         end if
17:         F_k ← F_k ∪ {c}; //Generate frequents patterns
18:     end for
19: end for
20: return ∪_k F_k;
```

---

are then generated on the only pass (lines 2 and 3), instead of 2 passes for the esistant algorithms. EOMF-Gen is called to generate the candidates (line 5). For each c of C_k, EOMF-Gen traverses two stapes. If c is generator (lines 7 to 13), an access to the database will permit to recover the candidates in the set of $\mathcal{CGM}_k$

(lines 10 to 12). Otherwise, c is not generator (lines 14 to 16), its support is equals to the minimum supports of its subsets (line 15). Thus, no access to the database is done. After the execution of these two phases, the algorithm returns the frequent patterns sets (line 20).

The procedure EOMF-Gen (Algorithm 2) functions like Apriori-Gen [1]. It takes in entry a frequent pattern set $\mathcal{F}_{k-1}$, and returns Ck set of the candidates. It proceeds in two phases. The first phase consists of mining the itemsets

---

Algorithm 2. EOMF-Gen procedure

Require: A set $\mathcal{F}_{k-1}$ of the $(k-1)$-itemsets fréquents
Ensure: A set C_k of the k-itemsets candidates
1: **for all** itemset $p \in \mathcal{F}_{k-1}$ **do**
2:    **for all** itemset $q \in \mathcal{F}_{k-1}$ **do**
3:       **if** $(p[1] = q[1], \ldots, p[k-2] = q[k-2], p[k-1] < q[k-1])$ **then**
4:          $c \leftarrow p \cup q(k-1)$; //Generate candidates
5:          **for all** (itemset candidate $c \in C_k$) **do**
6:             **for all** $((k-1)$-subset s de c) **do**
7:                **if** $(s \notin \mathcal{F}_{k-1})$ **then**
8:                   Delete c from C_k;
9:                **else**
10:                   $supp(c) \leftarrow min\{supp(c), supp(s)|s \subset c\}$;
11:                   **if** $(s \notin C\mathcal{GM}_k)$ **then**
12:                      candidate=fals;
13:                   **end if**
14:                **end if**
15:             **end for**
16:             **if** $(c \notin C\mathcal{GM}_k)$ **then**
17:                $supp(c) \leftarrow min\{supp(c')|c' \subset c\}$;
18:             **end if**
19:          **end for**
20:       **end if**
21:    **end for**
22: **end for**
23: **return** $\bigcup_k C_k$;

---

candidates. Firstly, the itemset candidates with the iteration k are created from the frequent patterns of the stape $(k-1)$. The generation of the k-itemsets candidates is done by auto-joint of $(k-1)$-itemsets of the previous iteration. For example, the joint of the itemset ABC and ABD give $ABCD$, by contrast, the joint of ABC and CDE does not give anything, because there is no $(k-2)$ itemset in common. The second stape consists of pruning the not frequent itemset. For that, it is necessary to test for each item for the joint if all the $(k-1)$-subset items are frequent. We test for each candidate c of C_k if it's generator or not (lines 5 to 21). In that case, the procedure gets in 2 under-stapes. The first consists, for each subset s of c, to test that if s is frequent or not (lines 7 to 14). If s is not frequent ($s \notin \mathcal{F}_{k-1}$), c is then eliminated (lines 7 to 9), if not it could recover via a minimum support of its subsets (line 10). At the same time, we test if s it is generator or not (lines 11 to 13). At the time of the second sub-operation, we test for each element c of C_k if it is generator or not (lines 16 to 18). EOMF-Gen returns the C_k set of the patterns candidates (line 23).

3.2   Optimization of the Course of Research the Potential Rules

We show that it is useless to study the set of the entire candidates, only half is sufficient. This approach is to generate the potential valid rules of association deriving from the family $\mathcal{F}$ of the frequent patterns by sharing this set of the candidates in two classes, *attraction classe* and *repulsion classes*, according to the membership of rule $X \to Y$. For that, we introduced the following proposals. For lack of space, the evidence of these proposals is not detailed in this paper. The reader can refer to [4].

Proposition 4. *If $P(Y|X) > P(Y)$, then $M_{GK}(X \to Y) = M_{GK}(\overline{Y} \to \overline{X})$, and $M_{GK}(Y \to X) = M_{GK}(\overline{X} \to \overline{Y})$, and $M_{GK}(X \to Y) < M_{GK}(Y \to X)$, and $M_{GK}(X \to Y) < M_{GK}(\overline{X} \to \overline{Y})$.*

The proposal 4 means that M_{GK} is implicative. This implicative character is advantageous for pruning: only one rule $X \to Y$ will permit to deduce the three others. What permits to optimize notably the space of research.

Proposition 5. *If $P(Y|X) < P(Y)$, then $M_{GK}(X \to \overline{Y}) = M_{GK}(Y \to \overline{X})$, and $M_{GK}(\overline{X} \to Y) = M_{GK}(\overline{Y} \to X)$, and $M_{GK}(X \to \overline{Y}) < M_{GK}(\overline{X} \to Y)$*

The proposal 5 guarantees the interest of index M_{GK} to the extraction of the unilateral negative rules. If $X \to \overline{Y}$ is valid, then $\overline{X} \to Y$, $Y \to \overline{X}$ and $\overline{Y} \to X$ will be it also. Thus, only one rule, $X \to \overline{Y}$, will permit to infer the three others. What also reduces the space of research significantly.

These results gives a powerful strategy of pruning. The Algorithm 3 synthesizes some. It takes in entry a family $\mathcal{F}$, a minimum support *minsupp* and a threshold α, and returns a set $\mathcal{E}_{PNR}$ of the potential positive and negative rules.

Algorithm 3. Optimized mining of potential association rules to support-M_{GK}

Require: A set $\mathcal{F}$ of frequents patterns, a *minsupp* and threshold α.
Ensure: A set $\mathcal{E}_{PNR}$ of potential positive and negative association rules.
1: $\mathcal{E}_{PNR} = \emptyset$;
2: **for all** $(X \in \mathcal{F})$ **do**
3:     **for all** $(Y \in \mathcal{F})$ **do**
4:         **if** $(P(Y|X) > P(Y))$ **then**
5:             calculate $supp(X \to Y)$; $M_{GK}(X \to Y)$; vc_α;
6:             **if** $(supp(X \to Y) \geq minsupp \wedge M_{GK}(X \to Y) \geq vc_\alpha)$ **then**
7:                 $\mathcal{E}_{PNR} \leftarrow \mathcal{E}_{PNR} \cup \{X \to Y, Y \to X, \overline{Y} \to \overline{X}, \overline{X} \to \overline{Y}\}$;
8:             **end if**
9:         **else**
10:            calculate $supp(X \to \overline{Y})$; $M_{GK}(X \to \overline{Y})$; vc_α;
11:            **if** $(supp(X \to \overline{Y}) \geq minsupp \wedge M_{GK}(X \to \overline{Y}) \geq vc_\alpha)$ **then**
12:               $\mathcal{E}_{PNR} \leftarrow \mathcal{E}_{PNR} \cup \{X \to \overline{Y}, Y \to \overline{X}, \overline{Y} \to \overline{X}, \overline{X} \to Y\}$;
13:            **end if**
14:         **end if**
15:     **end for**
16: **end for**
17: **return** $\mathcal{E}_{PNR}$;

The algorithm starts by initializing the set $\mathcal{E}_{PNR}$ (line 1). It proceeds in two recursive phases. The first stape (lines 4 to 9) consists of mining the rules of attraction class (i.e. $P(Y|X) > P(Y)$). In that case, it traverses only one rule $X \rightarrow Y$ (line 6), i.e. if $supp(X \rightarrow Y) \geq minsupp$ and $M_{GK}(X \rightarrow Y) \geq vc_\alpha$, the set $\mathcal{E}_{PNR}$ is updated by adding $X \rightarrow Y$, $Y \rightarrow X$, $\overline{Y} \rightarrow \overline{X}$ and $\overline{X} \rightarrow \overline{Y}$ (line 7). The second stape (lines 9 to 14) consists of mining the rules of repulsion class (i.e. $P(Y|X) < P(Y)$). So it traverses only one rule $X \rightarrow \overline{Y}$ (line 11), i.e. if $supp(X \rightarrow \overline{Y}) \geq minsupp$ and $M_{GK}(X \rightarrow \overline{Y}) \geq vc_\alpha$, the set $\mathcal{E}_{PNR}$ is also updated by adding $X \rightarrow \overline{Y}$, $Y \rightarrow \overline{X}$, $\overline{Y} \rightarrow X$ et $\overline{X} \rightarrow Y$ (line 12). Finally, it returns the set $\mathcal{E}_{PNR}$ of the positive and negative association rules (line 17).

4   Experimental Resultants

Our approach, OM2PNR, is implemented in R and experimented on a PC of 4 revolving Go of RAM under Windows system. We will compare the set results on 4 databases of the UCI, with those of RAPN [11] and Wu [13] of the interesting approaches to which OM2PNR seems very similar. The first three columns of Table 3 give the characteristics of the data files, the six last indicate the number of the rules and the execution times for each algorithm. For lack of space, we limit ourselves only on one rough number $|\mathcal{E}_{PNR}|$, we could not present numbers of any type nor the graphs which defer the pace of these execution times.

Table 3. Characteristics of the data of test and results for the extraction

Data	Object	Item(s)	minsupp	α	Wu et al. 2004		RAPN		OM2PNR	
					$|\mathcal{E}_{PNR}|$	time(s)	$|\mathcal{E}_{PNR}|$	time(s)	$|\mathcal{E}_{PNR}|$	time(s)
Adult	48842	115	1%	30%	100581	300	89378	280	28784	40
			2%	20%	57786	200	55120	185	26500	30
			3%	10%	39817	150	22588	150	18925	25
German	1000	71	1%	30%	53637	200	42613	195	27421	30
			2%	20%	41235	175	39598	165	19379	20
			3%	10%	10845	125	19132	120	15392	14
Income	6876	50	1%	30%	3808	50	3151	48	1834	15
			2%	20%	2867	40	2783	37	1616	11
			3%	10%	1785	30	1478	28	1005	8
Iris	150	15	1%	30%	2952	40	2048	37	1939	15
			2%	20%	2424	30	1407	37	1347	10
			3%	10%	1463	15	1111	17	1019	7

The number of the rules certainly increases when the minimum minsupp decreases. Except on the dense data (**Adult** and **German**), RAPN and Wu produce very prohibitory numbers, that is to say respectively 3.10 and 3.50 more than those of OM2PNR, in the minimum $minsupp = 1\%$. The OM2PNR thus showed the greatest concision in front of RAPNR and Wu. As the number of the rules, the time execution also increases when the support decreases. The execution times of RAPN and Wu clearly exceed those of OM2PNR, that is to say (280 or 300) against 40 s. The OM2PNR thus gained 7 times more of best execution speed. For the slightly correlated data (`Income` and `Iris`), the three algorithms give of a number of the rules and the very reasonable response times.

These different performances can be explained by the fact that RAPN and Wu are limited to the traditional structure of data of Apriori requiring of the repetitive accesses to the context, given that the data are dense and voluminous. Moreover, they use mainly more the traditional couple support-confiance of Apriori which easily produces uninteresting and redundant rules. Thanks to the set optimizations, OM2PNR, will permit to extract the frequent set on a single pass to the context, and not to interest but in the half of the association rules candidates.

5   Conclusion

This model permit to generate all the frequent items in only one pass with the context, and to extract only half of the candidates, where the existing algorithms undergo some difficulty. This results validate the idea of using the MATRICESUP-PORT and concept of minimum support for the frequents patterns. Moreover, they show that the implicative character of the measure M_{GK} notably reduce the cost of the extraction. A study on the conjunction of the positive and negative items was not initially developed in this paper, which constitutes a track to explore much of methodological point of view than algorithmic one.

References

1. Agrawal, R., Srikant, R.: Fast algorithms for mining association rules. In: Proceedings of 20th VLDB Conference, Santiago, Chile, pp. 487–499 (1994)
2. Bastide, Y., Taouil, R., Pasquier, N., Stumme, G., Lakhal, L.: Mining frequent patterns with counting inference. SIGKDD Explor. 2(2), 66–75 (2000)
3. Bemarisika, P.: Extraction de régles d'association selon le couple support-M_{GK}: Graphes implicatifs et Application en didactique des mathématiques. Université d'Antsiranana, Madagascar (2016)
4. Bemarisika, P., Totohasina, A.: Apport des régles négatives l'extraction des régles d'association. SFC 2014, CNRST-Raba-Maroc, pp. 99–104 (2014)
5. Bemarisika, P., Totohasina, A.: Eomf: Un algorithme d'extraction optimiseée des motifs fréquents. In: Proceedings of AAFD & SFC, Marrakech Maroc, pp. 198–203 (2016)
6. Boulicaut, J.-F., Bykowski, A., Jeudy, B.: Towards the tractable discovery of association rules with negations. In: Larsen, H.L., Andreasen, T., Christiansen, H., Kacprzyk, J., Zadrożny, S. (eds.) Conference on FQAS 2000. Advances in Soft Computing, vol. 7, pp. 425–434. Springer, Heidelberg (2000)
7. Brin, S., Motwani, R., Silverstein, C.: Bayond market baskets: Generalizing association rules to correlation. In: Proceedings of the ACM SIGMOD, pp. 265–276 (1997)

8. Cornelis, C., Yan, P., Zhang, X., Chen, G.: Mining Positive and Negative Association Rules from Large Databases. Proceedings of the IEEE, pp. 613–618 (2006)
9. Ganter, B., Wille, R.: Formal Concept Analysis: Mathematical Foundations. Springer, New York (1999)
10. Guillaume, S.: Traitement de données volumineuses. Mesure et algorithmes d'extraction de régles d'association et régles ordinales. Université de Nantes (2000)
11. Guillaume, S., Papon, P.-A.: Extraction optimiséé de Régles d'Association Positives et Négatives (RAPN). Clermont Université, LIMOS (2013)
12. Totohasina, A., Ralambondrainy, H.: ION: A pertinent new measure for mining information from many types of data. IEEE, SITIS, pp. 202–2007 (2005)
13. Wu, X., Zhang, C., Zhang, S.: Efficient mining of both positive and negative association rules. ACM Trans. Inf. Syst. **3**, 381–405 (2004)

Extracting Non-redundant Correlated Purchase Behaviors by Utility Measure

Wensheng Gan[1], Jerry Chun-Wei Lin[1(✉)], Philippe Fournier-Viger[2], and Han-Chieh Chao[1,3]

[1] School of Computer Science and Technology,
Harbin Institute of Technology (Shenzhen), Shenzhen, China
`wsgan001@gmail.com, jerrylin@ieee.org`
[2] School of Natural Sciences and Humanities,
Harbin Institute of Technology (Shenzhen), Shenzhen, China
`philfv@hitsz.edu.cn`
[3] Department of Computer Science and Information Engineering,
National Dong Hwa University, Hualien, Taiwan
`hcc@ndhu.edu.tw`

Abstract. In the high-utility itemset mining (HUIM) model, the low-utility patterns sometimes with a very high-utility pattern will be considered as a valuable pattern even if this behavior may be not highly correlated. A more intelligent system that provides non-redundant and correlated behavior based on utility measure is desired. In this paper, we first present a novel method, called extracting non-redundant correlated purchase behaviors by utility measure, to determine the high qualified patterns, which can lead to higher recall and better precision. In the proposed projection-based approach, efficient projection mechanism and a sorted downward closure property are developed to reduce the database size. Two pruning strategies are further developed to efficiently and effectively discover the desired patterns. An extensive experimental study showed that the proposed algorithm considerably outperforms the existing HUIM algorithms.

Keywords: Purchase behavior · Utility · Correlation · Projection

1 Introduction

The tasks of frequent pattern mining (FPM) [9] and association rule mining (ARM) [2,3] are highly important and have been extensively studied since they have numerous real-world applications. However, FPM [2,3,9] only measures the interestingness of patterns based on the co-occurrence frequencies of items or itemsets in transactional databases. Other implicit factors such as the weight, interest, risk or profit of itemsets are not considered. Besides, all items are considered to be equally important in traditional FPM, and hence the items or itemsets that are really important to users may not be found by FPM algorithms.

© Springer International Publishing AG 2017
L. Bellatreche and S. Chakravarthy (Eds.): DaWaK 2017, LNCS 10440, pp. 433–446, 2017.
DOI: 10.1007/978-3-319-64283-3_32

Purchase behavior patterns have been utilized successfully in many areas, previous studies have, however, several limitations in real-life situations such as: (1) each user may perform the same action/event multiple times; (2) multiple accessed events product different profit; (3) the overall utility of traditional pattern is limited since they rarely consider the inherent correlation. In recent years, a new mining and searching framework of high-utility itemset/pattern mining (HUIM) [5,7,16,21] incorporates both quantity and profit values of an item/set to identify those items or itemsets which can bring valuable profits for the retailers or managers. HUIM [5,12,14–16,19] has a critical role in data analysis and has been widely utilized to discover knowledge and mine valuable information in recent decades. Many studies have been carried to develop the efficient algorithms in HUIM, such as Two-Phase [16], IHUP [5], UP-Growth [18], UP-Growth+ [19], HUI-Miner [15], and FHM [10], HUP-Miner [11], HUI-MMU [14], etc. Among them, most algorithms were developed to efficiently discover HUIs by adopting the transaction-weighted downward closure (TWDC) property of the Two-Phase algorithm.

However, these algorithms are designed under a user-specified minimum utility threshold, and ignored the inherent correlation of items inside of patterns. Thus, an important limitation of current algorithms in HUIM is that they may find a huge amount of itemsets containing items that are weakly correlated. For example, it is common that retail stores sell some products/items at a loss to stimulate the sale of other related products/items. Since the item "diamond" has a very high utility, any itemset containing "diamond" would be regarded as a high-utility pattern (group of items) in traditional HUIM, but most of discovered information is meaningless, redundant or non-discriminative (have happened by chance). The patterns "diamond & beer" and "diamond & diaper" may be found but do not represent the interesting correlation. Hence, it is a critical issue to design the efficient algorithms to discover correlated and high-utility purchase behaviors w.r.t. HUIs based on the utility and correlation measures.

In the field of pattern-based mining framework, some correlation measures such as the support and confidence [8], all-confidence [17], frequency affinity [6], and coherence [17], etc. have been studied. Up to now, only the works of HUIPM [6] and FDHUP [13] have been applied in high-utility pattern mining by employing the measure of frequency affinity. However, they only consider the co-occur frequency in each transaction as the correlation factor. In order to enrich the efficiency and effectiveness of HUIM, an efficient projection-based algorithm named non-redundant correlated high-utility itemset mining from transaction databases with strong positive correlation (CoHUIM) is developed in this paper. Major contributions are summarized as follows:

– A new knowledge called non-redundant correlated high-utility itemsets (CoHUIs), is introduced to consider both the utility and positive correlation measures. To the best of our knowledge, it is the first work that considers the correlation of items in HUIM, which is more applicable and realistic in real-life environment.

- A projection-based approach named CoHUIM is developed to mine CoHUIs by spanning the sub-projected databases of candidates. Thus, the search space and memory usage can be gradually reduced.
- The developed *global sorted downward closure (SDC)* property guarantees the global anti-monotonicity to discover the complete set of CoHUIs. Thus, the CoHUIM algorithm can easily prune a huge number of unpromising candidates and speed up mining performance.
- An extensive experimental study showed that fewer but more interesting CoHUIs can be discovered and the proposed CoHUIM algorithm considerably outperforms the existing HUIM algorithms.

The rest of this paper is organized as follows. In Sect. 2, we present the preliminaries and then define the problem. Section 3 introduces the properties of the CoHUIM, and the proposed *sorted downward closure (SDC)* property. Section 4 evaluates the performance with experimental results. Finally, Sect. 5 concludes this paper.

2 Preliminaries and Problem Statement

Let $I = \{i_1, i_2, \ldots, i_m\}$ be a finite set of m distinct items in a transactional database $D = \{T_1, T_2, \ldots, T_n\}$, where each quantitative transaction $T_q = \{q(i_1, T_q), q(i_2, T_q), \ldots, q(i_j, T_q)\}$ is a subset of I, and has an unique identifier (*TID*). Note that the $q(i_j, T_q)$ in each T_q is the different purchase quantity of each item. An unique profit $pr(i_j)$ is assigned to each item $i_j \in I$, which represents its importance (e.g., profit, interest, risk), and they are stored in a profit-table $ptable = \{pr(i_1), pr(i_2), \ldots, pr(i_m)\}$. An itemset with k distinct items $\{i_1, i_2, \ldots, i_k\}$ is called a k-itemset. An itemset X is said to be contained in a transaction T_q if $X \subseteq T_q$. As a running example, Table 1 shows a transactional database containing 10 transactions. Assume that the *ptable* is defined as $\{pr(a){:}6,\ pr(b){:}1,\ pr(c){:}7,\ pr(d){:}3,\ pr(e){:}12\}$.

Definition 1. The utility of an item i_j in each T_q is denoted as $u(i_j, T_q)$, and defined as: $u(i_j, T_q) = q(i_j, T_q) \times pr(i_j)$, where $q(i_j)$ is the quantity of i_j in T_q.

For example, the utility of item (d) in transaction T_1 is calculated as $u(d, T_1) = q(d, T_1) \times pr(d) = 3 \times 3 = 9$.

Definition 2. The utility of an itemset X in transaction T_q is denoted as $u(X, T_q)$, and defined as: $u(X, T_q) = \sum_{i_j \in X \wedge X \subseteq T_q} u(i_j, T_q)$.

For example, the utility of the itemset (ad) in T_1 is calculated as $u(ad, T_1) = u(a, T_1) + u(d, T_1) = q(a, T_1) \times pr(a) + q(d, T_1) \times pr(d) = 3 \times 6 + 3 \times 3 = 27$.

Definition 3. The utility of an itemset X in a database D is denoted as $u(X)$, and defined as: $u(X) = \sum_{X \subseteq T_q \wedge T_q \in D} u(X, T_q)$.

For example, the utility of itemset (ad) is calculated as $u(ad) = u(ad, T_1) + u(ad, T_2) + u(ad, T_{10}) = 27 + 15 + 15 = 57$.

Table 1. An example database

TID	Transaction (item, quantity)
T_1	a:3, c:2, d:3
T_2	a:2, d:1, e:2
T_3	b:3, c:5
T_4	a:1, b:2, c:2, e:4
T_5	b:1, d:3, e:2
T_6	b:2, d:2
T_7	a:3, c:2
T_8	a:2, b:1, c:4
T_9	c:3, d:2, e:1
T_{10}	a:2, c:2, d:1, e:3

Table 2. Derived HUIs and CoHUIs

Itemset	$Kulc(X)$	$u(X)$
(c)	1.000	140
(e)	1.000	144
(ac)	0.774	150
(ae)	0.550	138
(bc)	0.514	83
(ce)	0.514	145
(de)	0.733	117
$(ace)*$	0.340	130
$(ade)*$	0.356	90
$(cde)*$	0.340	92

Definition 4. The transaction utility of T_q is denoted as $tu(T_q)$, and defined as: $tu(T_q) = \sum_{i_j \in T_q} u(i_j, T_q)$, in which j is the number of items in T_q.

Definition 5. The total utility in D is the sum of all transaction utilities in D and denoted as TU, which can be defined as: $TU = \sum_{T_q \in D} tu(T_q)$.

For example, in Table 1, $tu(T_1) = u(a, T_1) + u(c, T_1) + u(d, T_1) = 18 + 14 + 9 = 41$. Thus, the total utility in D is calculated as: $TU = tu(T_1) + tu(T_2) + tu(T_3) + tu(T_4) + tu(T_5) + tu(T_6) + tu(T_7) + tu(T_8) + tu(T_9) + tu(T_{10}) = 41 + 39 + 38 + 70 + 34 + 8 + 32 + 41 + 39 + 65 = 407$.

Definition 6. An itemset X in a database is a HUI iff its utility is no less than the minimum utility threshold ($minUtil$) multiplied by the TU as: $HUI \leftarrow \{X | u(X) \geq minUtil \times TU\}$.

As stated in the introduction, an important limitation of current algorithms in HUIM is that they may find a huge amount of itemsets containing items that are weakly correlated. In the study [6], a new measure called frequency affinity was proposed to discover high-utility interesting patterns having a strong frequency affinity. Only the minimum purchase quantity among purchase quantities of items in an itemset is regarded as the affinitive frequency. It cannot discover the real inherent correlation of among items inside the desired patterns. In this paper, we evaluate the inherent correlation of a generalized k-itemset by means of the Kulczynsky (abbreviated as $Kulc$) correlation measure [17,20], and the new knowledge namely correlated high-utility itemset (CoHUI) is defined as follows:

Definition 7. Correlation among an itemset measures the strength of the correlation between its items. In general, there are (1) *positive correlation*, (2) *non-correlation*, and (3) *negative correlation* among the items of an itemset.

Definition 8. The *Kulc* value is defined as: $Kulc(X) = \frac{1}{k} \sum_{i_j \in X} \frac{sup(X)}{sup(i_j)}$ [20], where i_j is the j-item in X which contains k items. It is used as an interesting measure to evaluate the correlation between items in an itemset X.

Thus, the range of *Kulc* is $[0, 1]$. Unlike other traditional itemset correlation measures, *Kulc* has the **null** (transaction)-invariant property, which implies that the correlation measure is independent of the dataset size.

Definition 9. An itemset X in a database D is defined as a CoHUI if it satisfies the following two conditions: (1) $u(X) \geq minUtil \times TU$; (2) $Kulc(X) \geq minCor$. The *minUtil* is the minimum utility threshold and *minCor* is the minimum positively correlation threshold; both of them can be specified by users' preferences.

Example 1. In the given example, when *minCor* and *minUtil* are respectively set at 0.4 and 20%; the itemset (ace) is a HUI since its utility is $u(ace)$ ($= 130 >$ $minUtil \times TU = 81.4$), but it is a not CoHUI since its *Kulc* value is $Kulc(ace)$ ($= 0.340 < 0.4$). For the running example, the complete set of CoHUIs with strong positively correlation is shown in Table 2 except three itemsets $((ace),$ (ade) and $(cde))$ which were only considered as the HUIs and remarked with notation *.

Problem statement: Given a quantitative transactional database (D), a profit table $(ptable)$, a user-specified minimum positive correlation threshold $(minCor)$ and a minimum utility threshold $(minUtil)$, the purpose of non-redundant correlated high-utility itemset mining (CoHUIM) is to efficiently find the complete set of CoHUIs while considering both the positive correlation and the utility constraints of the itemsets.

3   Proposed CoHUIM Algorithm for Mining CoHUIs

3.1   Properties of the CoHUI

According to the previous studies, the utility measure does not hold monotonic or anti-monotonic property. In other words, an itemset may have lower, equal or higher utility than any of its subsets. For example in Table 2, the itemset (d) is not a HUI, but its supersets $\{(de): 117\}$, $\{(ade): 90\}$, and $\{(cde): 92\}$ are the HUIs. Therefore, the strategies used in FPM and ARM to prune the search space based on the anti-monotonicity cannot be directly applied to discover HUIs as well as the desired CoHUIs. Moreover, in the literature, it has been pointed out that the *Kulc* measure does, however, not hold the anti-monotonicity [20], thus the search space is hard to be efficiently reduced in the mining process. To seep up mining performance, we introduce the projection mechanism to effectively reduce the search space, and the TWU *downward closure* property [16] is extended in the designed algorithm while exploring the CoHUIs.

Property 1 (non-anti-monotonicity of CoHUI). Based on the designed CoHUI, the CoHUI does not hold anti-monotonicity. In other words, a CoHUI may have lower, equal or higher utility or Kulc value than any of its subsets.

Definition 10 (transaction-weighted utilization, TWU). The *transaction-weighted utilization* (TWU) of an itemset X is defined as the sum of the transaction utility of transactions containing X, i.e., $TWU(X) = \sum_{X \subseteq T_q \wedge T_q \in D} tu(T_q)$ [16].

3.2 Reducing Database Size Using Projection Mechanism

Definition 11 (total order $\prec$ on items). Assume that the total order $\prec$ on items in the addressed CoHUI mining framework is one specified order on items (i.e., lexicographic order, support ascending/descending order).

Definition 12 (projected transaction). Given an itemset X and a transaction T. The projection of T using X is the collection of postfixes of items in T w.r.t. prefix X, denoted as $T_{|X}$ and defined as $T_{|X} = \{i|i \in T \wedge i \in X\}$.

Definition 13 (projected database). Given an itemset X and a transaction database D. The projection database from D using X is called X-projected database, denoted as $D_{|X}$ and defined as $D_{|X} = \{T_{|X}|T_{|X} \in D \wedge T_{|X} \neq \emptyset\}$.

Consider our running example in Table 1 and $X = b$. The projected database $D_{|b}$ contains five transactions: $D_{|b} = \{T_{3|b} = \{c{:}5\}; T_{4|b} = \{c{:}2, e{:}4\}; T_{5|b} = \{d{:}3, e{:}2\}; T_{6|b} = \{d{:}2\}; T_{8|b} = \{c{:}4\}\}$. The projection mechanism of database can greatly reduce the cost of database scans since the size of transactions will become smaller as more itemsets are explored. Moreover, we can quickly calculate the necessary information (i.e., $u(X)$, $TWU(X)$ and $sup(X)$, etc.) from the projected databases only without rescanning the whole database. However, an important issue is how to efficiently implement database projection. A naive and inefficient approach is to make physical copies of transactions for projection. An efficient approach used in CoHUIM is to first sort the transactions in the original database according to the $\prec$ total order. Then, a projection mechanism is performed as a pseudo-projection, that is each projected transaction is represented by an offset pointer of the corresponding original transaction.

3.3 Proposed Sorted Downward Closure Property

As mentioned above, the *Kulc* measure does, however, not hold anti-monotonicity, it is extremely hard to efficiently prune the search space in the mining process.

Lemma 1 (sorted downward closure property of *Kulc*). *Assume the items in transactions of database are sorted by their support-ascending order. The Kulc measure has the sorted downward closure property as:* $Kulc(a_1...a_k a_{k+1}) \leq Kulc(a_1...a_k)$.

Proof. Since the items in the set $\{a_1, a_2, ..., a_k, a_{k+1}\}$ are sorted in support-ascending order, i.e., $sup(a_1) \leq sup(a_2) \leq ... \leq sup(a_k) \leq sup(a_{k+1})$, then:

$$Kulc(a_1...a_k a_{k+1}) = \frac{sup(a_1...a_k a_{k+1})}{k+1} \times \{\frac{1}{sup(a_1)} + ... + \frac{1}{sup(a_k)} + \frac{1}{sup(a_{k+1})}\}$$

$$= \frac{sup(a_1...a_k a_{k+1})}{k+1} \times \{\frac{1}{sup(a_1)} + \frac{1}{ksup(a_{k+1})} + ... + \frac{1}{sup(a_k)} + \frac{1}{ksup(a_{k+1})}\}$$

$$\leq \frac{sup(a_1...a_k a_{k+1})}{k+1} \times \{\frac{1}{sup(a_1)} + \frac{1}{ksup(a_1)} + ... + \frac{1}{sup(a_k)} + \frac{1}{ksup(a_k)}\}$$

$$= \frac{sup(a_1...a_k a_{k+1})}{k+1} \times \{\frac{k+1}{ksup(a_1)} + ... + \frac{k+1}{ksup(a_k)}\}$$

$$= \frac{sup(a_1...a_k a_{k+1})}{k} \times \{\frac{1}{sup(a_1)} + ... + \frac{1}{sup(a_k)}\}$$

$$\leq \frac{sup(a_1...a_k)}{k} \times \{\frac{1}{sup(a_1)} + ... + \frac{1}{sup(a_k)}\}$$

$$= Kulc(a_1...a_k).$$

Thus, the *Kulc* measure holds anti-monotonicity if the processed items are sorted in support-ascending order.

In the running example, assume that the items are sorted as $\{e \prec b \prec a \prec d \prec c\}$. Since *Kulc* of the itemsets (ba) and (ead) are respectively 0.367 and 0.356; none of the supersets of (ba) and (ead) would be regarded as a CoHUI. We called this anti-monotonic of *Kulc* as *sorted downward closure (SDC)* property. Consequently, if *Kulc* of an $(k$-1$)$-itemset is below the minimum positive correlation threshold, it can be discarded in subsequent iterations, i.e., if an itemset is not a strongly correlated itemset (also not a CoHUI), any superset (by adding items with higher support count) cannot be a strongly correlated itemset (also not a CoHUI).

Definition 14. An itemset X is a correlated high utility upper-bound itemset (CHUUBI) if its *Kulc* value is no less than the minimum positive correlation threshold, and its *TWU* value is no less than the minimum utility value. Thus:

$$CHUUBI \leftarrow \{X|TWU(X) \geq minUti \times TU \wedge Kulc(i) \geq minCor\}. \quad (1)$$

For example, $Kulc(cd) = 0.464$, and $TWU(cd) = tu(T_1) + tu(T_9) + tu(T_{10}) = 41 + 39 + 65 = 145$; itemset (cd) is a CHUUBI.

Theorem 1 (The sorted downward closure property). *Let X^k be an itemset such that X^{k-1} is a subset of X^k. The SDC of CoHUI states that $Kulc(X^k) \leq Kulc(X^{k-1})$ and $TWU(X^k) \leq TWU(X^{k-1})$. Thus, if an itemset X is not as a CHUUBI, all its supersets are not CHUUBIs nor CoHUIs.*

Proof. The TWU model [16] indicates that $TWU(X^k) \leq TWU(X^{k-1})$, and the Lemma 1 shows that $Kulc(X^k) \leq Kulc(X^{k-1})$, this theorem holds.

Theorem 2 (CoHUIs $\subseteq$ CHUUBIs). *If the Kulc and TWU values of a 1-itemset donot meet the criteria of CHUUBI, it is thus discarded based on the* *sorted downward closure (SDC) property. We have that* $CoHUIs \subseteq CHUUBIs$, *which indicates that if an itemset is not a CHUUBI, it is not a CoHUI as well as all its supersets.*

Proof. Let X^k be a k-itemset such that X^{k-1} is its subset. We have that $u(X) = \sum_{X \subseteq T_q \wedge T_q \in D} u(X, T_q) \leq \sum_{X \subseteq T_q \wedge T_q \in D} tu(T_q) = TWU(X)$. From Theorem 1, it can be obtained that $Kulc(X^k) \leq Kulc(X^{k-1})$ and $TWU(X^k) \leq TWU(X^{k-1})$. Thus, if X^{k-1} is not a CHUUBI, none of its supersets are CoHUIs.

Lemma 2. *The discovered set of CoHUIs by the proposed CoHUIM algorithm is correct and complete.*

Proof. According to Lemma 1, Theorems 1 and 2, the CoHUIM algorithm can ensure that any unpromising itemset will be discarded (**completeness**) and the related information can be exactly obtained from the sub-projected database (**correctness**).

3.4   Procedure of the Projection-Based CoHUIM Algorithm

Based on the above lemmas and theorems, two efficient pruning strategies are designed in the developed model to early prune unpromising candidates. Thus, a more compressed search space can be obtained to reduce the computation. By utilizing the *SDC* property, we only need to initially construct the projected databases for those promising itemsets w.r.t. the $CHUUBI^1$ as the input for later recursive process. Furthermore, based on the pruning strategy 2, the designed CoHUIM algorithm can early prune the itemsets with lower *Kulc* and utility count, without constructing their projected database of extensions.

Strategy 1. *After the first database scan, we can obtain the Kulc and TWU value of each 1-item in the database. If the TWU and Kulc of a 1-item i_j ($TWU(i_j)$ and $Kulc(i_j)$) do not satisfy the two conditions of CHUUBI, this item can be directly pruned, and none of its supersets is concerned as CoHUI.*

Strategy 2. *Assume that the total order $\prec$ on items adopted the support-ascending order of items, when traversing the sub-projected database of an itemset (a prefix) based on a projected strategy, if an itemset is not a CHUUBI, then none of the extension nodes of this prefix is concerned as CoHUI, stop the projection with this prefix.*

Based on the above properties and pruning strategies, the pseudo-code of the proposed projection-based CoHUIM algorithm is described in Algorithm 1.

The projection-based $CoHUIM$ algorithm takes as input (1) D, a transaction database with quantity values; (2) *ptable*, the user-specified profit table; (3) *minCor*, the minimum positive correlation threshold; and (4) *minUtil*, the minimum utility threshold. It mainly consists of two phases. In the first phase, it first scans D to calculate the $sup(i_j)$ and $TWU(i_j)$ of each item $i_j \in I$ in D (Line 1), then find the set of $CHUUBI^1$ and $CoHUI^1$ (Lines 2 to 7).

Algorithm 1. Projection-based CoHUIM algorithm

Input: D; $ptable$; $minCor$; $minUtil$.

Output: The set of complete CoHUIs.

1 scan D to calculate the $sup(i_j)$ and $TWU(i_j)$ of each item $i_j \in I$ in D;

2 **for** *each 1-item $i_j \in D$* **do**

3 | **if** $(TWU(i_j) \geq minUtil \times TU) \wedge (Kulc(i_j) \geq minCor)$ **then**

4 | | $CHUUBI^1 \leftarrow CHUUBI^1 \cup i_j$;

5 | | calculate $u(i_j) := pr(i_j) \times sup(i_j)$;

6 | | **if** $(u(i_j) \geq minUtil \times TU) \wedge (Kulc(i_j) \geq minCor)$ **then**

7 | | | $CoHUI^1 \leftarrow CoHUI^1 \cup i_j$;

8 sort $CHUUBI^1$ in support-ascending order;

9 **for** *each 1-item $i_j \in CHUUBI^1$* **do**

10 | scan D to project all transactions within i_j into the sub-database $db_{|i_j}$ of i_j;

11 | set $k = 1$, **call Project-CHUUBI**$(i_j, db_{|i_j}, k)$;

12 scan D to calculate actual utilities for all candidates in $CHUUBIs$ by using the TID-index;

13 **for** *each itemset $X \in CHUUBIs$* **do**

14 | **if** $u(X) \geq minUtil \times TU$ **then**

15 | | $CoHUIs \leftarrow CoHUIs \bigcup X$;

16 **return** $CoHUIs$

Algorithm 2. Project-CHUUBI$(X, db_{|X}, k)$ procedure

Input: X, a prefix itemset; $db_{|X}$, the projected database in which each transaction contains an itemset X; k, the length of an itemset X.

Output: The set of CHUUBIs with a prefix X.

1 generate the $(k+1)$-itemset X' consisting of the prefix itemset X and each i_j in $CHUUBI^1$;

2 put X' into the set of potential $(k+1)$-candidates PC^{k+1};

3 **for** *each $(k+1)$-itemset $X' \in PC^{k+1}$* **do**

4 | scan $db_{|X}$ to calculate the projected sub-database $db_{|X'}$ of X' and $sup(X')$;

5 | calculate the $TWU(X')$ and $Kulc(X')$ value;

6 | **if** $(TWU(X') \geq minUtil \times TU) \wedge (Kulc(X') \geq minCor)$ **then**

7 | | $CHUUBI^{k+1} \leftarrow CHUUBI^{k+1} \cup X'$;

8 | | **call Project-CHUUBI**$(X', db_{|X'}, k+1)$;

9 | $CHUUBIs \leftarrow \bigcup CHUUBIs^{k+1}$;

10 **return** $CHUUBIs$ with the prefix X

After that, it sorts $CHUUBI^1$ in support-ascending order as the initially set (Line 8). For each item in $CHUUBI^1$, it scans D to project all transactions within i_j into the sub-database $db_{|i_j}$ of i_j (Line 10). The algorithm then sets k as 1, and recursively performs using the **Project-CHUUBI** procedure and the candidate set of $CHUUBIs$ are returned (Line 11). In the second phase, it scans D again to quickly calculate actual utilities for all candidates in *CHUU-BIs* by using the TID-index (Line 12) and outputs the final set of the desired *CoHUIs* (Lines 12 to 16). As shown in **Project-CHUUBI** (cf. Algorithm 2), each itemset X' with a prefix X is recursively determined to directly produce the complete set of *CHUUBIs* (Lines 1 to 10). Details are performed as: it first generates the $(k+1)$-itemset X' consisting of the prefix itemset X and each i_j in $CHUUBI^1$ (Line 1). After that, put them into the set of PC^{k+1} (Line 2). Each itemset in PC^{k+1} is then processed and the $db_{|X'}$ is scanned to calculate the projected sub-database as $db_{|X'}$ of X', $TWU(X')$ and $sup(X')$ (Line 4). Two constraints are then applied to further determine whether its extensions should be explored for the later projection search (Lines 6 to 8). If one itemset is regarded as a *CHUUBI*, the **Project-CHUUBI** procedure is continuously executed (Line 8). Notice that each projection of X' is a 1-extension of itemset X; all of them should be put into the set of *CHUUBIs* (Line 9). At last, the **Project-CHUUBI** procedure returns *CHUUBIs* with a prefix X (Line 10).

4   Experimental Results

Substantial experiments were conducted to verify the effectiveness and efficiency of the proposed CoHUIM algorithm. Since the HUIPM [6] and FDHUP [13] algorithms are performed to discover the high-utility itemsets based on frequency affinity constraint, we use the faster FDHUP as a benchmark to evaluate the discovered CoHUIs by CoHUIM and the improvement achieved by CoHUIM's pruning strategies. The traditional HUIM algorithm, FHM [10], was executed to derive traditional HUIs, which can provide a benchmark to verify the efficiency of the proposed algorithm.

4.1   Dataset and Experimental Setup

The experiments are conducted on two real-life datasets (*foodmart* and *retail* [1]) and two synthetic datasets (*T10I4D100K* [4] and *T5I2N2KD100K* [4]). A standard benchmark simulation model, which has been widely used in the field of HUIM [15,16,19], was developed to generate the quantity and profit value of items on the test datasets except for the foodmart dataset since it already obtains the quantity and profit value of items. A log-normal distribution was used to randomly assign quantities in the [1,5] interval, and item profit values in the [1,1000] interval. These tested datasets having varied characteristics.

- *foodmart* was acquired from Microsoft foodmart SQL Server 2000. It contains 21,556 transactions and 1,559 distinct items. It is a very sparse dataset. The average number of items per transaction is 4 and the largest transaction contains 11 items.
- *retail* is a sparse dataset which contains 88,162 transactions with 16,470 distinct items and an average transaction length of 10.30 items.
- *T10I4D100K* is a synthetic dataset that contains 870 distinct items, 100,000 transactions, and has an average length of 10.1 items.

– *T5I2N2KD100K* dataset contains total 100,000 transactions which are generated by the IBM Quest Synthetic Data Generator [4].

All the algorithms in the experiments were implemented in Java language and performed on a personal computer with an Intel Core i5-3460 dual-core processor and 4 GB of RAM and running on the 32-bit Microsoft Windows 7 operating system. It is important to notice that both the FHM and FDHUP algorithms are varied by one parameter $minUtil$, while the CoHUIM algorithm discovers the CoHUIs by using both correlation and utility constraints. Therefore, experiments are conducted on datasets with varying $minUtil$. In addition, the $minCor$ is adjusted with five times on each dataset to verify the efficiency of the proposed CoHUIM algorithm and respectively denoted them as $CoHUIM_{minCor1}$, $CoHUIM_{minCor2}$, $CoHUIM_{minCor3}$, $CoHUIM_{minCor4}$, and $CoHUIM_{minCor5}$.

4.2 Pattern Analysis

With the problem parameters set above, the derived patterns of HUIs (discovered by FHM), DHUIs(discovered by FDHUP) and CoHUIs (discovered by the proposed CoHUIM algorithm) are evaluated to show the acceptable of the knowledge representation of CoHUI and proposed CoHUIM approach. Results under various parameters are shown in Fig. 1.

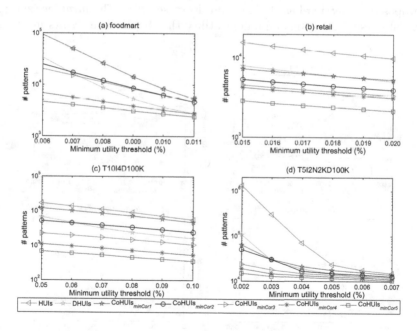

Fig. 1. Number of patterns under various parameters.

In general, the number of DHUIs and CoHUIs are always smaller than that of HUIs. In addition, the number of discovered patterns of the designed CoHUIM algorithm under five $minCor$ is always: $CoHUIs_{minCor1} \geq CoHUIs_{minCor2} \geq CoHUIs_{minCor3} \geq CoHUIs_{minCor4} \geq CoHUIs_{minCor5}$. The number of discovered patterns by the CoHUIM is fewer but more valuable than those found by FHM for two reasons. First, the correlated HUIs indicates that the discovered information is more interesting to users. Numerous patterns are redundant and generated by the traditional algorithm in HUIM, and only fewer patterns have strong correlation of the items. This situation can be found when the minimum utility threshold is set low. On the other hand, the patterns of HUIs and DHUIs do not hold high correlation. Hence, those HUIs and DHUIs may not the interesting patterns for real-life applications such as market-basket analysis, where a manager or retailer wants to make effective decisions. The discovered CoHUIs can be viewed as patterns that are more valuable than that of HUIs and DHUIs since both the utility and correlation aspects are considered. The designed CoHUI is more feasible and realistic in real-life situations. It can also be observed that the number of discovered patterns decreases when $minUtil$ increases for all compared algorithms on all datasets.

4.3   Runtime Analysis

With the same settings of parameters, the runtime of the compared algorithms under various $minUtil$ are shown in Fig. 2. It can be observed that the runtime of the proposed projection-based algorithm decreases when the $minCor$ increases on all datasets with various minimum utility thresholds. The results under five

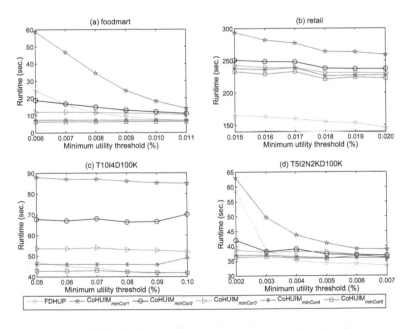

Fig. 2. Runtime under various parameters.

$minCor$ is always $CoHUIM_{minCor1} \leq CoHUIM_{minCor2} \leq CoHUIM_{minCor3} \leq CoHUIM_{minCor4} \leq CoHUIM_{minCor5}$. The reason is that fewer patterns satisfy the higher $minCor$ with the fixed $minUtil$, and fewer candidates are generated and the search space can be accordingly reduced. It also indicates that the minimum positive correlation threshold has significant influence on the runtime.

In general, CoHUIM with a higher $minCor$ is faster than the state-of-the-art FDHUP algorithm. This is reasonable since both the utility and correlation constraints are considered in the CoHUIM to discover the complete set of CoHUIs while only the utility constraint is concerned in the traditional utility-based algorithms to find HUIs. With more constraints, the search space can be further reduced; fewer but more interesting patterns are discovered. Based on the designed projection mechanism, *SDC property* and pruning strategies, the runtime of the proposed CoHUIM algorithm can be greatly reduced.

5   Conclusions

On web search and data mining, previous studies have several limitations in real-life situations. The derived patterns by traditional approaches is limited since they rarely consider the inherent utility factor or the correlation. In this paper, we incorporate two measures, both the utility measure and inherent correlation among patterns, into purchase behavior pattern mining for high quality pattern. An efficient mining algorithm named CoHUIM is proposed to both consider traditional utility measure and the inherent correlation of items inside the patterns. A projection-based approach is first developed to mine CoHUIs by spanning the sub-projected databases of candidates; the search space and memory usage can thus be reduced gradually. To guarantee the global anti-monotonicity of CoHUIs and prune the search space effectively, a novel *sorted downward closure* property is further proposed. An extensive experimental study shows that the proposed algorithm outperforms the state-of-the-art FDHUP algorithm in terms of generated patterns and execution time. Moreover, the number of derived CoHUIs is much smaller but more valuable and interesting than the that of HUIs; the large amount of redundant and meaningless information can thus be avoided.

Acknowledgments. This research was partially supported by the National Natural Science Foundation of China (NSFC) under grant No. 61503092 and by the Tencent Project under grant CCF-Tencent IAGR20160115.

References

1. Frequent itemset mining dataset repository. http://fimi.ua.ac.be/data/
2. Agrawal, R., Imielinski, T., Swami, A.: Database mining: a performance perspective. IEEE Trans. Knowl. Data Eng. **5**, 914–925 (1993)
3. Agrawal, R., Srikant, R.: Fast algorithms for mining association rules in large databases. In: The International Conference on Very Large Data Bases, pp. 487–499 (1994)

4. Agrawal, R., Srikant, R.: Quest synthetic data generator. http://www.Almaden. ibm.com/cs/quest/syndata.html
5. Ahmed, C.F., Tanbeer, S.K., Jeong, B.S., Le, Y.K.: Efficient tree structures for high utility pattern mining in incremental databases. IEEE Trans. Knowl. Data Eng. **21**(12), 1708–1721 (2009)
6. Ahmed, C.F., Tanbeer, S.K., Jeong, B.S., Choi, Y.K.: A framework for mining interesting high utility patterns with a strong frequency affinity. Inf. Sci. **181**(21), 4878–4894 (2011)
7. Chan, R., Yang, Q., Shen, Y.D.: Mining high utility itemsets. In: The International Conference on Data Mining, pp. 19–26 (2003)
8. Geng, L., Hamilton, H.J.: Interestingness measures for data mining: a survey. ACM Comput. Surv. 38(3), Article 9 (2006)
9. Han, J., Pei, J., Yin, Y., Mao, R.: Mining frequent patterns without candidate generation: a frequent-pattern tree approach. Data Min. Knowl. Discov. **8**(1), 53–87 (2004)
10. Fournier-Viger, P., Wu, C.W., Zida, S., Tseng, V.S.: FHM: faster high-utility itemset mining using estimated utility co-occurrence pruning. Found. Intell. Syst. **8502**, 83–92 (2014)
11. Krishnamoorthy, S.: Pruning strategies for mining high utility itemsets. Expert Syst. Appl. **42**(5), 2371–2381 (2015)
12. Lin, J.C.W., Gan, W., Hong, T.P., Tseng, V.S.: Efficient algorithms for mining up-to-date high-utility patterns. Adv. Eng. Inform. **29**(3), 648–661 (2015)
13. Lin, J.C.-W., Gan, W., Fournier-Viger, P., Hong, T.-P.: Mining discriminative high utility patterns. In: Nguyen, N.T., Trawiński, B., Fujita, H., Hong, T.-P. (eds.) ACIIDS 2016. LNCS, vol. 9622, pp. 219–229. Springer, Heidelberg (2016). doi:10. 1007/978-3-662-49390-8_21
14. Lin, J.C.W., Gan, W., Fournier-Viger, P., Hong, T.P.: Mining high-utility itemsets with multiple minimum utility thresholds. In: ACM International Conference on Computer Science & Software Engineering, pp. 9–17 (2015)
15. Liu, M., Qu, J.: Mining high utility itemsets without candidate generation. In: ACM International Conference on Information and Knowledge Management, pp. 55–64 (2012)
16. Liu, Y., Liao, W., Choudhary, A.: A two-phase algorithm for fast discovery of high utility itemsets. In: Ho, T.B., Cheung, D., Liu, H. (eds.) PAKDD 2005. LNCS, vol. 3518, pp. 689–695. Springer, Heidelberg (2005). doi:10.1007/11430919_79
17. Omiecinski, E.R.: Alternative interest measures for mining associations in databases. IEEE Trans. Knowl. Data Eng. **15**(1), 57–69 (2003)
18. Tseng, V.S., Wu, C.W., Shie, B.E., Yu, P.S.: UP-growth: an efficient algorithm for high utility itemset mining. In: ACM SIGKDD International Conference on Knowledge Discovery and Data Mining, pp. 253–262 (2010)
19. Tseng, V.S., Shie, B.E., Wu, C.W., Yu, P.S.: Efficient algorithms for mining high utility itemsets from transactional databases. IEEE Trans. Knowl. Data Eng. **25**(8), 1772–1786 (2013)
20. Wu, T., Chen, Y., Han, J.: Re-examination of interestingness measures in pattern mining: a unified framework. Data Min. Knowl. Discov. **21**(3), 371–397 (2010)
21. Yao, H., Hamilton, J., Butz, C.J.: A foundational approach to mining itemset utilities from databases. In: SIAM International Conference on Data Mining, pp. 211–225 (2004)

Data Flow Management and
Optimization

Detecting Feature Interactions in Agricultural Trade Data Using a Deep Neural Network

Jim O'Donoghue$^{(\boxtimes)}$, Mark Roantree, and Andrew McCarren

School of Computing, Insight Centre for Data Analytics,
DCU, Collins Avenue, Dublin 9, Ireland
jim.odonoghue@insight-centre.org,
{mark.roantree, andrew.mccarren}@dcu.ie

Abstract. Agri-analytics is an emerging sector which uses data mining to inform decision making in the agricultural sector. Machine learning is used to accomplish data mining tasks such as prediction, known as *predictive analytics* in the commercial context. Similar to other domains, hidden trends and events in agri-data can be difficult to detect with traditional machine learning approaches. Deep learning uses architectures made up of many levels of non-linear operations to construct a more holistic model for learning. In this work, we use deep learning for *unsupervised* modelling of commodity price data in agri-datasets. Specifically, we detect how appropriate input signals contribute to, and interact in, complex deep architectures. To achieve this, we provide a novel extension to a method which determines the contribution of each input feature to *shallow, supervised* neural networks. Our generalisation allows us to examine *deep* supervised *and* unsupervised neural networks.

1 Introduction and Motivation

Agri-analytics is an emerging sector which uses data mining and analytics to improve decision making in a market contributing €24 billion to Ireland's economy. In an industry of this size, decision makers in the agricultural (agri) sector require appropriate tools and up-to-date information to make appropriate predictions across a range of products and areas. Despite a high volume of both free-to-use online datasets and highly processed pay-to-use datasets, this is still very difficult. For example, the prediction of future prices for many agri products. Determining how complex input data signals interact and contribute to agri-datasets, would enable us to better *understand* this data and leverage it for prediction. Deep learning refers to a recent breakthrough in machine learning, where *deep architectures*, made up of many levels or *layers* of non-linear operations are used to model data. A central premise behind deep learning is that, like the brain, these algorithms can learn high-level, abstract features from data [1]. These high-level features better represent the outcome or dataset being modelled

---

Research funded by Science Foundation Ireland under grant number SFI/12/RC/2289.

L. Bellatreche and S. Chakravarthy (Eds.): DaWaK 2017, LNCS 10440, pp. 449–458, 2017.
DOI: 10.1007/978-3-319-64283-3_33

and correspond to *latent* variables in the dataset. The lower layers correspond to localised, specific features and as the data progresses through the architecture, it is transformed into more abstract, higher level representations.

Motivation. An important task in data mining and machine learning is the detection of how variables interact when predicting an outcome or describing data. These interactions enable us to extract the complex relationships between inputs, a highly important task in argi-analytics. Neural networks, especially deep neural networks have great potential in this regard as they can successfully learn multiple layers of highly complex feature interactions or *feature representations*, but they are notoriously hard to interpret and lack a methodology to interpret these learnt features and thus, are often seen as "black box" solutions.

Contribution. The connection weight method [14] was designed for a single hidden layer, single output *supervised* classification *shallow* neural network. Our contribution is an extension to this. We present the *deep Connection Weight*, or dCW method. The dCW approach gives us a *generic* means to interpret supervised *or* unsupervised *deep* networks and the multiple layers of highly complex features therein. Furthermore, we use the dCW and a Deep Belief Network (DBN) to extract highly complex interactions between the multiple input signals in agri trade data.

Paper Structure. The paper is structured as follows: in Sect. 2 we discuss related research; in Sect. 3, we present detailed description of the methods employed in this paper, in terms of the algorithmic optimisation and configuration; in Sect. 4 we present our novel dCW method for interpreting the feature representation of a deep network; Sect. 5, describes our dataset, transformations, experiment setup as well as the results and analysis; and finally in Sect. 6, we present our conclusions.

2   Related Research

Deep and shallow neural networks are often seen as uninterpretable "black boxes". However, there have been studies [6,14] to compare and examine various techniques to interpret *shallow, supervised* networks. In [6,14] they discuss: Garson's algorithm [5]; the connection weight (CW) method; partial derivatives; perturbing inputs; sensitivity analyses; and various forms of stepwise addition, elimination and selection. In [6] they compare many of these methods on a dataset of ecological information. The review of [14] employs a synthetic dataset with known properties. Therefore, the comparisons performed in [14] are more accurate and can be generalised to other datasets. Garson's algorithm [5] was one of the first popular methods for summarising neural networks. However, it only accounts for the cumulative *magnitude* of the weight. The CW method [14] examines the cumulative magnitude *and* sign (positive or negative) of weights. The CW method was also shown to outperform all others [14]. Thus, it was selected as the basis of our approach. The CW method can successfully interpret single classification, single hidden layer *shallow* neural networks but does not generalise to *deep* networks. The research presented here extends this method to encompass these more complicated networks.

Table 1. Hyper-parameter search space

Hyper-parameter	Description	Bounds (low, high)
b_size	Samples in a mini batch	(856, 8560)
l_hidden	Number of hidden layers	(1, 3)
o_nodes	Number of hidden layer nodes	(1, 204)
learning_rate α	Co-efficient for weight updates	(0.00001, 0.1)
Patience	Minimum epochs to iterate	(10, 150)
max_epochs	Max possible epochs to iterate	(10, 1000)

3   Deep Belief Network Components

In this section we describe our approach to construct and optimise a Deep Belief Network (DBN). Section 3.1 describes the two parameter-types involved and the procedures we adopt for their initialisation and optimisation. Section 3.2 describes how DBNs are configured and relate to the strategies in Sect. 3.1.

3.1   Parameter Initialisation and Optimisation

Parameter optimisation enables the discovery of the best learner model for pattern extraction. There are two types and thus two levels of optimisation in an experiment, which we define. *Hyper*-parameters are optimised at the *Trial* level and influence algorithm architecture and training. *Model* parameters, optimised at the *Run* level, are used for the predictive or descriptive learning task.

Model Parameter Initialisation and Optimisation. The procedure for *model parameter* initialisation was adopted from the literature [7]. This encompasses initialising all bias terms to zero and for logistic activation units, randomly initialising layer weights between the bounds given in Eq. (1). The number of inputs to a layer is given by n_{in} and the number of nodes in a layer by o_{nodes}. Model parameters were optimised with *mini-batch stochastic gradient descent* (MSGD) [4] and *early-stopping* [15].

$$[-4 \sqrt{\frac{6}{n_{in} + o_{nodes}}}, 4 \sqrt{\frac{6}{n_{in} + o_{nodes}}}] \tag{1}$$

Hyper-parameter Initialisation and Optimisation. Table 1 shows hyper-parameter initialisation bounds. We discuss the upper and lower bound for each. Batch size (b_size) - number of samples in an MSGD data batch - was bounded at roughly 1% and 10% of total dataset samples. Layer counts (L_hidden) were bounded at 1 (shallow) and 3 (deep). Node counts in each layer (o_nodes) were searched between 1 and $2 \cdot n_{inputs} + n_{inputs}$. Learning rate bounds are typical of the literature [2], with a reduced upper bound to avoid vanishing gradients. Finally, early stopping parameters $patience$ - minimum number of dataset iterations (epochs) - and max_epochs - maximum number - had a lower bound of 10 and an upper bound of 150 and 1000, respectively. Random search was used to optimise hyper-parameters [2].

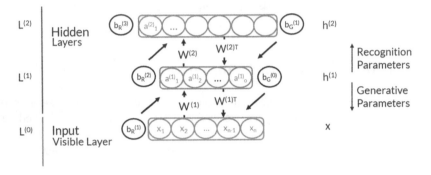

Fig. 1. Unsupervised DBN configuration

3.2 Architectural Configuration

A Deep Belief Network (DBN) is an algorithm architecture characterised by an unsupervised pre-training phase and an *optional* supervised or unsupervised fine-tuning stage [1]. A DBN is trained in a greedy, unsupervised, layer-wise fashion where each layer is trained as a Restricted Boltzmann Machine (RBM). The first RBM is completely optimised before its hidden values are treated as the visible inputs for the next layer, and so on, until all layers are combined as presented in Fig. 1. Values can propagate from $L^{(0)}$ through $L^{(1)}$, to the final hidden layer $L^{(|L|)}$ via relevant parameters and functions and can propagate down to form an accurate *reconstruction* of inputs. When the DBN can accurately reconstruct the data from its hidden layers, it has learnt an abstract representation of the inputs. Thus, a method to examine these abstract representations would allow us to determine how input features interact. To train RBMs we used CD_1, one-step *Contrastive Divergence*. This amounts to 1 step of **Gibbs sampling** at each iteration of MSGD [8]. We refer the reader to [13] for (hidden) activation, energy, cost and derivative functions.

4   An Approach to Interpreting Deep Representations

The CW method was designed for a *single* hidden layer and *single* output *supervised* feed-forward network. Its inputs are: the first hidden layer weight matrix $W^{(1)}$ and the output weight vector $W^{(2)}$. These respectively contain connection weights between: every input to every hidden node; and every hidden node to the classification. Consider the first row in $W^{(1)}$, which maps the first input to each hidden node. Each value in this row is multiplied by each output weight in $W^{(2)}$. This gives a number of values equal to the number of hidden nodes and the contribution of the first input to each input1-hidden-out pathway. Summing these values gives the contribution of this input to the model overall. This process repeats for every row in $W^{(1)}$, to give the contribution of each input. Equation (2) shows the CW method formulated as the dot product between the input and output tensors. This formulation facilitates our generalisation and the method that follows.

$$CW_{score} = W^{(1)} \cdot W^{(2)} \tag{2}$$

Equation (3) shows the calculation of $\varphi \in \mathbb{R}^{n \times H}$, where n is the number of input features and H is the number of nodes $|N^{(|L|)}|$, in the final layer of a neural network with a set of L layers. We call this the feature *score matrix*. It is the first part of our deep connection weight (dCW) method and generalises the CW approach to networks with *arbitrary numbers* of hidden layers and nodes in the final layer. The *score matrix* is the result of a cumulative dot product of all weight matrices $W^{(1)} \cdot W^{(2)} \ldots W^{(|L|)}$ in the network being examined. The score matrix φ, calculates a single value for the *importance* of each input feature to each node in the top layer of a neural network. This allows us to rank input feature contributions to each node in the final hidden layer nodes and thus, on further analysis examine how they *interact*, which is very useful.

$$\varphi = \prod_{i=0}^{|L|} W^{(i)} \tag{3}$$

In some cases, it is necessary to determine the *rank* of each input feature's importance and its contribution to the model as a whole, *or* how much each of the hidden nodes contribute to the overall model. The next steps summarise the score matrix φ and allow us to do this. Equation (4) shows how to determine the overall contribution of each input feature to the learner model. For each feature row vector $f_1, f_2, \ldots f_n$ in the score matrix φ, we sum the value in each column $j \in H$ to obtain the *contribution score* for each *input* feature. Summing over the opposite axis gives φ_N, the contribution of each *node* in the final hidden layer to the entire learner model.

$$\varphi_{f_i} = \sum_{j=1}^{H} \varphi_{ij} \tag{4}$$

5   Experiments

In this section, we present the results of the contributions of inputs and the interactions discovered in the agri dataset. We first describe our experimental setup and the dataset used, before discussing and analysing our results.

5.1   Setup

Experimental Environment. Experiments were run on Ubuntu 14.04.4 LTS and a NVIDIA GeForce GT 620 810 MhZ Graphical Processing Unit. Code was developed using 64-bit Python 2.7.6 in PyCharm 2016.2.3. IDE. Experiments use the CDN and Deep NoSQL Toolkit [12,13], Theano 0.7.0 [3], Pandas 0.18.0 [10], PyMongo 2.6.3 [11], Mongo 3.2.11. For each trial of 128, bootstrap sampling and nested cross-validation was used.

Dataset. The dataset was generated from 5 fact tables in an agri data-warehouse [9]. European Union trade data 2010–2014 (inclusive) was used to eliminate missing values, resulting in 85,602 samples. Table 2 shows a summary of the 68 input features and data transformations performed on each. For description we have divided them into 11 categories. *Category* provides a label for each category, *Num. Vars.* is the number of variables each contains and *Transformation* describes how the data was processed. Variables in categories 1–3 were one-hot encoded. Other variables were normalised between 0 and 1. The *time* category relates to month and year. *Countries* relates to the country exporting (27 possible) and importing (14 possible). *Product* describes the *type* of pork cut being exported. *Primary trade info.* relates to the volume of product being exported in kilograms, and its value in Euro. *Secondary trade info* (weight and value) relates to the trade volumes in kgs and value in euro from other major exporting

Table 2. Input feature summary

Cat.Num.	Category	Num.Vars.	Transformation
1	Time	2	One-hot
2	Countries	2	One-hot
3	Product	1	One-hot
4	Primary trade info.	2	Normalised
5	Weight secondary trade info.	6	Normalised
6	Value secondary trade info.	7	Normalised
7	Feed price	10	Normalised
8	Live price	10	Normalised
9	Slaughter info.	17	Normalised
10	Meat price	3	Normalised
11	Currency info.	8	Normalised

regions in the world in the same month, as well as the entire EU. The value, but not the weight of Russian trade was available as secondary trade information, hence the 6:7 discrepancy. *Live price* contains the price per kg of live animal for countries where this was available. *Feed price* contains all available information for the price of pig feed in the 5 regions studied. *Slaughter info.* contains available information on how many pigs were slaughtered in each region (supply of product). *Meat price* relates to the price per kg of pork product in Europe, the US and Canada. Finally, *currency info.* contains exchange rates for a wide range of currencies.

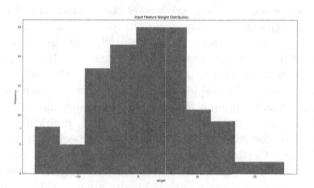

Fig. 2. Histogram of input node contributions

5.2   Results and Analysis

We first explore top architectural hyper-parameters and associated error rates. These provide insight into interaction analyses that follow, but are not central to the paper. Due to space restrictions, we omit training hyper-parameters and in-depth analysis of these results. Table 3 shows the top 5 training (t_score), validation (v_score) and test (tst_score) reconstruction cross entropy scores and associated hyper-parameters of the 128 trials. Error scores closer to zero are best and are ranked according to validation performance. Trial 52 obtained the smallest validation *and* test generalisation scores at -0.0018 and -0.001 respectively. Trial 126 obtained the lowest training score of -0.005. All configurations

Table 3. Top 5 error scores and architecture hyper-parameters

Rank	trial_id	l_hidden	o_nodes1	o_nodes2	o_nodes3	t_scores	v_scores	tst_scores
1	52	**2**	**5**	**130**	na	-0.0163	**-0.0018**	**-0.001**
2	74	3	169	63	60	74	0.0978	-0.0025
3	126	2	29	79	Na	**-0.005**	-0.0035	-0.0197
4	31	2	149	27	Na	0.0318	0.0082	0.0661
5	94	2	118	46	Na	0.0255	0.0087	-0.0697

are deep - four have 2 hidden layers and one has 3. The top performing - Trial 52 - has 5 nodes in its first hidden layer and 130 in its second. The next best has 169 nodes in the first, 63 in the second and 60 in the third hidden layer. For all 5, there is roughly a ratio of at least 3:1 nodes between input-hidden or hidden-input layers suggesting a combination of data compression or summary and then expansion or separation of signals, best extracts the interactions between inputs.

Feature Representation and Interaction Results. For analysis of features and interactions we focus on φ_f and top ranked contributors therein. Figure 2 shows the distribution of cumulative weight scores in φ_f. The x-axis shows the cumulative weight score and the y-axis shows the number of features with that score. It highlights that a small proportion of inputs has the largest contribution to the model and that contributions are normally distributed. Figure 2 shows roughly: 50 to 60 features have scores in the range 0 to 10; fifteen have scores in the range 10 to 30; forty are in the range 0 to -10; and about 12 or 13 have scores lower than -10. Table 4 shows these scores at a greater granularity. IrishKill and NIKill (animals slaughtered in Ireland and Northern Ireland) account for the largest weight contribution at 5.47%. FranMill (price of millet in France) accounts for 2.05%; partner1 (country exporting to Australia) contributes 2.03%; and GerRap (price of rapeseed meal in Germany) contributes 1.99% to the model. The final contributors are: the weight of monthly exports from Brazil (BrazilKg); February (month1); the amount the reporter exported in kg (ReporterKg); whether a country is exporting to Mexico; and 2014 (year4). For interactions, cumulative weight is used, similar to regression co-efficients.

Table 4. Top 10 contributing features

Rank	Feature	Cumulative weight	Contribution %
1	IrishKill	25.03578609	2.94
2	NIKill	21.53403839	2.53
3	FranMill	17.48791526	2.05
4	partner1	-17.31823953	2.03
5	GerRap	16.97575512	1.99
6	BrazilKg	16.40203897	1.93
7	month1	-16.19787237	1.90
8	ReporterKg	16.14323915	1.90
9	partner8	16.13083378	1.89
10	year4	-15.68426263	1.84

Feature Representation and Interaction Analysis. The above findings would suggest that *in our dataset*, that the supply coming from Ireland has the largest impact on trade. Perhaps this is the case in Europe, or perhaps we have richer data from Ireland. Further investigation is necessary. The price of feed in Germany and France stands to reason as having a large effect, as these are some of the largest countries in Europe. Finally, the amount in Kg the reporter is exporting having a large impact also stands to reason as this is what the model is developed to describe. More unusual findings were a particular year and month having a large impact. Seasonality effects trade but further investigations will have to be made into these time-points. Interactions also require further investigation as they are somewhat unintuitive, describing impact on the final hidden representation.

6 Conclusions and Future Work

Agri-analytics plays a significant role in decision making for today's agri sector. Determining multi-variate, unsupervised, feature interactions is extremely difficult, but can greatly aid understanding of agri data. Deep neural networks can extract complex multi-variate interactions via learned representations, but no generic method exists to interpret these representations. We presented the dCW method for summarisation of complex *deep* representations. Furthermore, we successfully applied our method to complex, *unsupervised* agri trade data and determined the most relevant input signals and interactions. Future work will extend dCW to discern each input's effect on individual variable reconstructions.

References

1. Bengio, Y., Lamblin, P., Popovici, D., Larochelle, H., et al.: Greedy layer-wise training of deep networks. Adv. Neural Inf. Process. Syst. **19**, 153–160 (2007)
2. Bergstra, J., Bengio, Y.: Random search for hyper-parameter optimization. J. Mach. Learn. Res. **13**, 281–305 (2012)
3. Bergstra, J., Breuleux, O., Bastien, F., Lamblin, P., Pascanu, R., Desjardins, G., Turian, J., Warde-Farley, D., Bengio, Y.: Theano: a CPU and GPU math expression compiler. In: Proceedings of the Python for Scientific Computing Conference (SciPy), June 2010. Oral Presentation
4. Bottou, L.: Stochastic gradient learning in neural networks. Proc. Neuro-Nimes **91**(8) (1991)
5. Garson, G.D.: Interpreting neural-network connection weights. AI Expert **6**(4), 46–51 (1991). http://dl.acm.org/citation.cfm?id=129449.129452
6. Gevrey, M., Dimopoulos, I., Lek, S.: Review and comparison of methods to study the contribution of variables in artificial neural network models. Ecol. Model. **160**(3), 249–264 (2003)
7. Glorot, X., Bengio, Y.: Understanding the difficulty of training deep feedforward neural networks. In: International Conference on Artificial Intelligence and Statistics, pp. 249–256 (2010)
8. Hinton, G.: A practical guide to training restricted Boltzmann machines. Momentum **9**(1), 926 (2010)
9. McCarren, A., McCarthy, S., Sullivan, C.O., Roantree, M.: Anomaly detection in agri warehouse construction. In: Proceedings of the Australasian Computer Science Week Multiconference, ACSW 2017, pp. 17:1–17:10. ACM (2017)

10. McKinney, W.: Data structures for statistical computing in python. In: van der Walt, S., Millman, J. (eds.) Proceedings of the 9th Python in Science Conference, pp. 51–56 (2010)
11. MongoDB: Pymongo (2016). Accessed 12 Dec 2016
12. O'Donoghue, J., Roantree, M.: A toolkit for analysis of deep learning experiments. In: Boström, H., Knobbe, A., Soares, C., Papapetrou, P. (eds.) IDA 2016. LNCS, vol. 9897, pp. 134–145. Springer, Cham (2016). doi:10.1007/978-3-319-46349-0_12
13. O'Donoghue, J., Roantree, M., Boxtel, M.V.: A configurable deep network for high-dimensional clinical trial data. In: 2014 International Joint Conference on Neural Networks (IJCNN). IEEE, July 2015
14. Olden, J.D., Joy, M.K., Death, R.G.: An accurate comparison of methods for quantifying variable importance in artificial neural networks using simulated data. Ecol. Model. **178**(3), 389–397 (2004)
15. Prechelt, L.: Early stopping - but when? In: Orr, G.B., Müller, K.-R. (eds.) Neural Networks: Tricks of the Trade. LNCS, vol. 1524, pp. 55–69. Springer, Heidelberg (1998). doi:10.1007/3-540-49430-8_3

Air Quality Monitoring System and Benchmarking

Xiufeng Liu$^{(\boxtimes)}$ and Per Sieverts Nielsen

Technical University of Denmark, Kongens Lyngby, Denmark
{xiuli,pernn}@dtu.dk

Abstract. Air quality monitoring has become an integral part of smart city solutions. This paper presents an air quality monitoring system based on Internet of Things (IoT) technologies, and establishes a cloud-based platform to address the challenges related to IoT data management and processing capabilities, including data collection, storage, analysis, and visualization. In addition, this paper also benchmarks four state-of-the-art database systems to investigate the appropriate technologies for managing large-scale IoT datasets.

Keywords: IoT-based · Dashboard · Cloud computing · Benchmarking

1 Introduction

With the development of urbanization, cities are facing an increasing challenge to reduce the climate impact. Cities are the large energy consumers accounting for 80% of the total carbon emissions [10]. Carbon emissions are mainly responsible for the greenhouse effect, and close monitoring of carbon emissions is an effective way to reduce the climate impact. Today more than 1,400 cities worldwide regularly report on their greenhouse gas (GHG) emissions through the Carbon Climate Register and the Governors Convention initiative [1]. However, emissions monitoring at the city level is often costly and time-consuming because they relate to a high degree of uncertainty. Most cities in Europe do not currently possess the capacity to measure the actual emissions within their urban space. On the other hand, carbon reduction has become a city development strategy. For example, the European Union (EU) aims to cut its primary energy consumption by 27% by 2030. In Denmark, the government has set the goal of reducing GHG emission by 40% by 2020, and becoming a fossil-fuel free country by 2050. This requires innovating approaches to reporting air quality for politician and citizens to make quick and effective decision makings. In this context, IoT technologies can be used to address the challenge of real-time monitoring of air quality, such as the detection of pollutant concentration levels and trends.

In this paper, we present an IoT-based air quality monitoring system developed under our *Carbon Track and Trace (CTT)* project [3]. The system can track the real-time greenhouse gas emissions at the urban-street level. This work makes the following contributions: First, we present an IoT-based solution for air

L. Bellatreche and S. Chakravarthy (Eds.): DaWaK 2017, LNCS 10440, pp. 459–470, 2017.
DOI: 10.1007/978-3-319-64283-3_34

quality monitoring. Second, we propose a cloud-based platform for managing air quality and other IoT related datasets, which provides the necessary functionality for adding high-frequency sensor data and accessing data from the database. Third, we develop a dashboard for presenting the insight of the data. Fourth, we investigate and benchmark alternative database technologies for managing IoT datasets, including specialized time-series database, in-memory columnar store, key-value store and relational database.

The rest of this paper is organized as follows. Section 2 describes the system design and implementation. Section 3 benchmarks the state-of-the-art database systems for managing IoT data. Section 4 surveys the related works. Section 5 concludes the paper and presents the future works.

2   System Design and Implementation

The system is designed with two purposes: one is to enable city officers, decision makers, citizens, and other stakeholders to visualize the emission measurements of the whole city for decision making or monitoring purpose, while the other is to establish a scalable data management system that can manage other IoT data used to study the impact on the air quality, such as weather data and traffic data. The data management system is required with a high scalability for supporting the large-scale deployments of sensors within the cities in the future. Besides, the platform should ease the integration of new data sources, and provide standard access service for enabling the data to be used by other applications or users. The system should make use of existing open source technologies, including the IoT network, data management, and visualization systems to reduce cost.

We develop the system based on the existing technologies, including sensors, sensor platforms, and IoT network. Air quality data (including CO_2 and NO_x) are collected by low-cost sensors deployed in outdoor environments. All sensor nodes are also equipped with climate sensors for collecting weather conditions (e.g., ambient temperature, pressure, wind speed and humidity) and particulate matter (PM) sensors for measuring dust particle size (e.g., PM1, 2.5 and 10). The sensors collect these measurements every five minutes and send them to the server on the cloud over Internet of Things network (TTN) via a low-power LoRaWAN gateway. The server aggregates the data from the TTN and saves them in the cloud database from which the dashboard reads the data to generate real-time views. Figure 1 shows the system architecture. In order to achieve high scalability, the architecture employs the distributed time series database, *OpenTSDB* [16], as the cloud database for managing air quality data, as well as weather and traffic data for correlation analysis. The reason for choosing OpenTSDB is that it is a NoSQL database that can handle massive amounts of data which is anticipated. As air quality data and their associated data (such as traffic and weather data) are captured, they will never be updated or changed. We expect thousands of sensors to be deployed in the city in the future. Therefore, it is preferable to select a distributed database to meet the city-scale IoT data management. OpenTSDB provides REST-based services for adding and accessing data, and supporting advanced queries for complex data analytics.

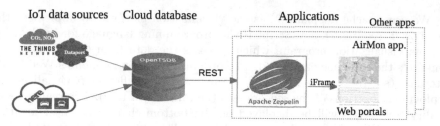

Fig. 1. Overview of the system architecture.

The time series of saving into OpenTSDB are given a unique name, called *metric*. To discriminate different data sources, we add a prefix to the name of metric, e.g., AQ_- for air quality data, TF_- for traffic data and WT_- for weather data. In addition, each time series is labeled with multiple *tags* for its dimensions or features. A tag is a key/value pair, and multiple tags can be combined to for doing complex queries. List 1 is an example of a data point with a CO_2 metric tagged with the device id (dev_eui) and the sensor location (longitude and latitude).

List 1: A single data point written to Opentsdb

```
  {
1   "metric": "AQ_GP_CO2",
2   "timestamp": 1488207776,
3   "value": 11.487,
4   "tags": {
5       "dev_eui": "00000000902FBDD2",
6       "longitude": 9.5324,
7       "latitude": 55.70805
8   }
  }
```

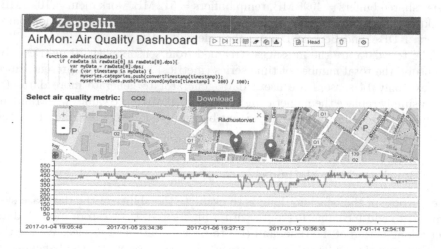

Fig. 2. Data analysis and visualization on Zeppelin.

We use Apache Zeppelin [2] as the visualization platform to implement the dashboard (see Fig. 2). JavaScript is the programming language for implementing the visualization program, which accesses the data from OpenTSDB, and generates the real-time dashboard. The program is run on Zeppelin's Web-based interface (see the top in Fig. 2). The open source Leaflet map visualizes the deployment locations of the sensors (see the middle in Fig. 2). The real-time air quality time series will be displayed on the bottom chart if a sensor location marker is clicked (see the bottom in Fig. 2). The chart is implemented using Highchart JavaScript library. The chart is updated dynamically when a new reading has been received. The chart can also be exported as an *iframe* to be embedded in other web pages, e.g., on a city government portal.

3   Benchmarking

3.1   Experimental Settings

In this section, we will benchmark four representative database systems for managing IoT data, including OpenTSDB [16], BerkeleyDB [15], PostgreSQL [20] and KDB+ [11]. They are the state-of-the-art database technologies in the following four categories: distributed time series database (OpenTSDB), key-value store (BerkelyeDB), relational database (PostgreSQL), and column store (KDB+). BerkelyeDB and KDB+ are main memory based. KDB+ is the commercial database system, and we are permitted to publish its benchmarking data. Only OpenTSDB is an distributed database system. To align the settings, we evaluate OpenTSDB on a single server environment.

We benchmark the database systems in the private Cloud (*SciCloud*, www.science-cloud.dk). A virtual machine instance is created, with 8 cores and 16 GB RAM, running 64-bit Ubuntu 16.04. PostgreSQL 9.6 is used, with the settings "shared buffers = 4096 MB, temp buffers = 512 MB, work mem = 1024 MB, checkpoint segments = 64" and default values for the other configuration parameters. KDB+ and BerkeleyDB use their default settings.

The test data are the messages streamed from sensors with 10 metrics. For n sensors, the total number of time series (metrics) is $10 \times n$. At present, since we have only 14 sensors, we use a data generator to generate more datasets, which simply replace the numeric and string typed attribute values with random numbers. These values are not important because we are only benchmarking the performance of the database systems.

3.2   Benchmarking Methods

The purpose of the benchmarking is to evaluate the ability of the database systems (i) to add incoming IoT data with a high speed, (ii) to store big IoT data over a long period of the time, and (iii) to provide data as a service. We use the appropriate data model supported by each database system for the benchmarking. This means that OpenTSDB uses its default data model; BerkelyDB uses

Algorithm 1. Benchmark data loading

```
1: function LOAD( records, insertNumOfRecords)
2:      T ← {}                                        ▷ Initialize the set of throughput
3:      n ← 0                                         ▷ Inserted data point counter
4:      st ← Get the wall clock time
5:      for all r ∈ records do
6:          deviceId, metricId, timestamp, reading, tags ← Get attribute values from r
7:          addRecordToDB(deviceId, metricId, timestamp, reading, tags)
8:          n ← n + 1
9:          if i mod 10000 == 0 then
10:             ed ← Get the wall clock time
11:             t ← 10000/(ed − st)
12:             T ← T ⋃{t}
13:             st ← ed
14:         if n == insertNumOfRecords then
15:             return
```

Algorithm 2. Benchmark data query

```
1: function QUERY(deviceId, metricId, queryNumOfRecords)
2:      T ← {}                                        ▷ Initialize the set of throughput
3:      n ← 0                                         ▷ Queried record counter
4:      st ← Get the wall clock time
5:      for deviceId ← 0...maxDeviceId do
6:          for metricId ← 0...maxMetricId do
7:              cur ← fetchRecordsOrderByTimestamp(deviceId, metricId)
8:              while hasNext(cur) do
9:                  deviceId, metricId, timestamp, reading, tags ← next(cur)
10:                 n ← n + 1
11:                 if i mod 10000 == 0 then
12:                     ed ← Get the wall clock time
13:                     t ← 10000/(ed − st)
14:                     T ← T ⋃{t}
15:                     st ← ed
16:                 if n == queryNumOfRecords then
17:                     return
```

key-value pair data model, where device id and metric id are saved as the key, bundled with the timestamp as a key object, and the other attributes are saved as the value; PostgreSQL uses a relational data model, with the table layout of *(device id, metric, metric value, timestamp, tags)* with a composite index created on the device id and metric attribute; KDB+ uses a columnar data model with its default settings.

The pseudocode for loading and querying is shown in Algorithms 1 and 2, respectively. The target table is empty before loading. We insert the required number of data points into the table, measure the elapsed time and calculate the throughput for every 10,000 data points added. This approach allows us to investigate the impact of the amount of data on the subsequent load performance. The target table is padded with the data before the query is benchmarked. We are interested in evaluating the queries that need to scan the table, such as aggregations of min, max, sum, or avg. Aggregation can be performed easily in OpenTSDB, PostgreSQL, and KDB+, but BerkelyeDB is difficult because it is a key-value store which does not provide built-in aggregation functions. Therefore, we use Algorithm 2 to benchmark the query for individual time series and compare their performance.

3.3   Benchmarking Results

We now evaluate the loading and query performance of the four database systems using the air quality IoT datasets. We generate two datasets for our testing: (1) *big data set* contains 1 million sensors (with 10 metrics for each) and (2) *small data set* contains 10 thousand sensors. The reading interval is ten minutes, and the duration of the monitoring is one day (a total of 144 data points). Thus, the total number of data points for the big data set is 1.44 billion (i.e., $1000000 \times 10 \times 144$) and its size is 198.8 GB. In the following, we will use the generated data to benchmark the performance of loading, time series queries, and analysis.

Load Performance. When loading, we use a bulk with the size of 10,000 for OpenTSDB. For PostgreSQL, we use batch insertion with the same bulk size. For KDB+ and BerkeleyDB, we use their default insert or add method. We measure the performance of loading the big and small datasets, and show the results in Figs. 3 and 4, respectively. The loading test uses Algorithm 1, which measures the elapsed time per 10,000 operations and calculates the corresponding throughput. The results are shown as box plots where the red line represents the median of the throughput values, and the height represents the variability. According to the results, KDB+ has the highest throughput, up to 38,000 operations per second. PostgreSQL shows a significant difference in loading different sizes of datasets. The throughput of the small data set is 4.3 times the throughput of the big data set. For the investigation, we use the Linux command `vmstat` to check IO and find that for the big data set, the write latency is 5–15% longer than the small data set. In addition, the I/O speed is quite variable, the read speed of 2–15MB/s and the write speed of 0–10MB/s. This may be due to the construction of the index, which involves reading the b-tree pages from disk to memory to update the index. According to the height of box plots, KDB+ has the smallest variability.

We now compare the database sizes after the loading (see Table 1). PostgreSQL is larger than the others because it saves the data in a de-normalized table, and uses extra space for the index. In contrast, the tags in OpenTSDB are saved once in row keys, which saves space. Although it has an additional index table, the table does not seem to cause a significant increase in space usage. KDB+ ranks the second place of using the least space which is due to the data compression in columns. BerkeleyDB hasn't shown space efficiency with its default settings.

Table 1. Database size after the loading the big data set

Raw data	OpenTSDB	PostgreSQL	KDB+	BerkeleyDB
198.8 GB	164.8 GB	344.5 GB	176.4 GB	352.8 GB

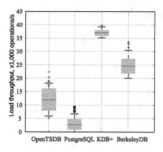

Fig. 3. Throughput of loading big data set.

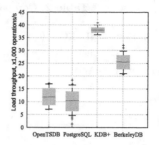

Fig. 4. Throughput of loading small data set.

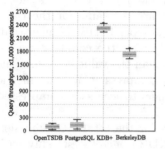

Fig. 5. Throughput of querying big data set.

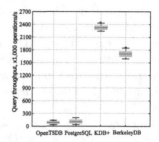

Fig. 6. Throughput of querying small data set.

Time Series Query Performance. After the loading experiment, the next step is to benchmark query performance. The query time series is done by using Algorithm 2, which scans each time series with specific metric and device ID. Figures 5 and 6 show the results for the big and small datasets, respectively. The results indicate that the in-memory-based solutions, KDB+ and BerkeleyDB, have better performance for time series query.

Analysis Query Performance. Analysis query benchmarking is performed only on OpenTSDB, PostgreSQL and KDB+ since Berkeley DB does not provide the operators for analysis queries. We compare three queries on CO_2 emissions, which are shown in Table 2. Different to PostgreSQL and KDB+ which support SQL statements, OpenTSDB queries data through the RESTful interface, and a query is serialized as a JSON object sent over HTTP to the query endpoint (see [16] for more details).

(i) *Total number of CO_2 readings:* Fig. 7 shows the performance of counting the total number of CO_2 readings in the database with the load of the big data set. There are different ways for counting the number of the readings. In this case, we test using "SELECT COUNT(*)" statements without and with specifying the attributes (on metric or value attribute), respectively. The motivation for carrying out this test is to quantify the implication on query

Table 2. Comparison of analysis queries

Analytic Queries	SELECT			WHERE			ORDER BY		
	metric	dev_eui	value	metric	dev_eui	value	metric	dev_eui	value
Total number of readings	ANY			✓					
Total CO_2 emission			✓	✓					
Top CO_2 emission			✓	✓					✓

performance for different encoding methods used in different columns, e.g., in KDB+. But, for OpenTSDB, it doesn't matter which attribute is chosen for the counting. Therefore, the query times are the same for the three tests, which are around 1.4 s. The query time of OpenTSDB is longer than the other two systems. It is interesting to see that the counting on the metric column in KDB+ takes a longer time than the value column. This illustrates the trade-off for achieving better compression for the metric column and the dataset overall, but it takes longer to access the compressed data.

(ii) *Total CO_2 emission:* Fig. 8 shows the results for the total CO_2 emission query. The times used by OpenTSDB and KDB+ increase steadily with more sensor data added (indicated as the number of sensors). The performance of KDB+ far outperforms the other two systems, due to its memory-based technology.

(iii) *Top CO_2 emission:* Fig. 9 shows the results for the analysis of top CO_2 emission. OpenTSDB and KDB+ show better performance than PostgreSQL, and the times increase slightly with more sensor data. In contrast, the time by PostgreSQL increases faster and almost linearly to the data size. In addition, PostgreSQL uses more time when querying top CO_2 emission, as compared to the time of querying total CO_2 emission in Fig. 8, because additional time is required for ordering.

Discussion. Based on the results of the above experiments, we can observe that the memory-based column store, KDB+, indicates the best load and query performance for managing IoT data. KDB+ is a non-distributed database solution suitable for real-time analysis of high-frequency time series data, e.g., IoT and trading data. Key-value stores (e.g., BerkeleyDB), also displays good performance, but lack analysis operators, while these operators are often needed for providing data as a service (DaaS) to other applications. IoT data are generated by many sensors, with the characteristics of fine granularity, high frequency, high volume, and high dimensionality. Comparatively, it is challenging to use traditional databases, e.g., PostgreSQL, to manage this type of datasets. The distributed open source time series database, OpenTSDB, has shown good performance in both loading and query. Furthermore, OpenTSDB offers built-in analytic operators for manipulating time series, which well suit our needs in the

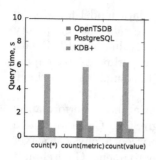

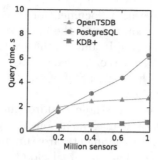

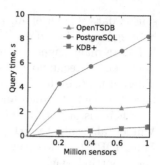

Fig. 7. CO_2 reading number.

Fig. 8. Total CO_2 emission.

Fig. 9. Top CO_2 emission.

implementation of our monitoring system, i.e., open source, scalable, and good performance.

There are some limitations on this benchmarking work. Among them, OpenTSDB is a distributed database system. Technically, it is unfair to compare with the others on the single-server setting. Second, all database systems are evaluated under their most basic settings, i.e., without advanced optimizations. There is still much room for optimizing the performance for each system. For example, in PostgreSQL, the applicable optimizations can be tuning different configuration settings, table partitioning, clustered indexed, and using a normalized table, etc. For KDB+, it can use a partitioned table; and for OpenTSDB, it can also tune different database settings, and ideally be tested on a cluster. In addition, parallelism is worthwhile for the test, as well as testing more advanced analysis queries (e.g., time series or pattern similarity). However, through this work, we have obtained the initial results of choosing a cloud database for managing IoT data. Obviously, there are a lot of variables that can be considered to obtain comprehensive benchmarking results, for space reason we would like to leave this as our future work.

4 Related Work

IoT-based air quality monitoring has received intensive research effort in recent years, e.g., [4,7,9,17,21]. These efforts focus on the implementation of the IoT, whilst rarely introduce how to manage IoT data. In contrast, we detail IoT data management, as well as benchmark the alternative technologies. Several solutions have been proposed to manage IoT data. Li et al. present a NoSQL-based storage system IOTMDB to manage scalable IoT data and discuss how to store large-scale data efficiently [12]. However, it is unclear how the data is visualized and accessed. Pintus et al. present a social web-based system [19] of using MongoDB to store IoT data expressed in form of key-value pairs. Ding et al. conversely use the relational data management system PostgreSQL to store sensor data [6]. Di et al. present a document-oriented data model for storing heterogeneous IoT data [5]. Contrary to these works, we take the air quality IoT data management as an example, but emphasize the evaluation of IoT data management technologies. We have conducted extensive experiments to compare different types of database technologies: SQL and NoSQL, column and relational based, memory and non-memory.

There is a large number of studies addressing database performance. Van et al. evaluate IoT data management on PostgreSQL, Cassandra, and MongoDB [23]. However, the experiments are run between a physical server and a virtual machine, rather than in a real cloud environment. Goldschmidt et al. evaluate the scalability and robustness of three open source time series databases (OpenTSDB, KairosDB and Databus) in a cloud environment, and the results indicate that KairosDB is the best to meet the initial assumptions about scalability and robustness [8]. Sanaboyina et al. evaluate the time series databases OpenTSDB and InfluxDB in terms of energy consumption when database read/write operations are carried out [22]. Phan et al. compare the performance and complexity of NoSQL databases with SQL databases (including MySQL, MongoDB, CouchDB and Redis) [18]. IoT data (sensor readings) and multimedia data are used for their evaluation. The comparison mainly focuses on MongoDB and MySQL, and shows that MongoDB has better scalability. They conclude that scalability is a key point that could potentially make NoSQL better than SQL databases. In our previous work [13,14], we evaluate the database systems and techniques for managing energy time series data, including traditional, columnar store, and distributed databases. This work focuses primarily on the ability to analyze the well-formatted energy consumption data. IoT data are usually more complicated, for example, with different types, formats, and complex metadata; and the data are often used by IoT applications for different purposes. In addition, there is still a lack of technical work on the evaluation of memory-based technologies for IoT data management. This paper bridges this gap by evaluating the performance, and compares it to traditional and distributed database technologies.

5   Conclusions and Future Work

Air quality monitoring has received increasing attention in recent years. Smart cities necessitate effective tools for air quality monitoring and data management. In this paper, we have presented an air quality monitoring system for smart cities, and established an open source IoT-based platform including software and hardware. We have proposed a scalable IoT data management solution for data collection, analytics and visualization. For the benchmarking, we have evaluated four state-of-the-art database technologies for managing large-scale IoT datasets, and conducted a comprehensive comparison by experimenting on the server. According to the results, the memory-based column store outperforms the others on a dedicated server environment, while the specialized time-series database can also achieve high-throughput writes and reads.

For the future work, we will further improve the system by adding data analytic capabilities, for example, to show the correlation between air quality and other factors such as traffic and weather conditions. We will conduct a more comprehensive benchmarking of database technologies for managing IoT data, and evaluate more advanced analytic queries.

Acknowledgements. This research is supported by the CTT project funded by Local Governments for Sustainability (LoCaL), and the CITIES project funded by Danish Innovation Fund (1035-0027B).

References

1. Ahlers, D., Driscoll, P.A., Kraemer, F.A., Anthonisen, F.V., Krogstie, J.: A measurement-driven approach to understand urban greenhouse gas emissions in Nordic Cities. In: NIK (2016)
2. Apache Zeppelin. http://zeppelin.apache.org/. Accessed 1 June 2017
3. Carbon track and trace. http://www.carbontrackandtrace.com. Accessed 1 June 2017
4. Chen, X., Liu, X., Xu, P.: IoT-based air pollution monitoring and forecasting system. In: Proceedings of ICCCS, pp. 257–260 (2015)
5. Di, F.M., Li, N., Raj, M., Das, S.K.: A storage infrastructure for heterogeneous and multimedia data in the internet of things. In: Proceedings of GreenCom, pp. 26–33 (2012)
6. Ding, Z., Xu, J., Yang, Q.: SeaCloudDM: a database cluster framework for managing and querying massive heterogeneous sensor sampling data. J. Supercomput. **66**(3), 1260–1284 (2013)
7. Fioccola, G.B., Sommese, R., Tufano, I., Canonico, R., Ventre, G.: Polluino: an efficient cloud-based management of IoT devices for air quality monitoring. In: Proceedings of Research and Technologies for Society and Industry Leveraging a better tomorrow (RTSI), pp. 1–6 (2016)
8. Goldschmidt, T., Jansen, A., Koziolek, H., Doppelhamer, J., Breivold, H.P.: Scalability and robustness of time-series databases for cloud-native monitoring of industrial processes. In: 7th International Conference on Cloud Computing, pp. 602–609 (2014)
9. Jadhav, D.A., Patane, S.A., Nandarge, S.S., Shimage, V.V., Vanjari, A.A.: Air pollution monitoring system using Zigbee and GPS module. Int. J. Emerg. Technol. Adv. Eng. **3**(9), 533–536 (2013)
10. Kamal-Chaoui, L., Robert, A.: Competitive cities and climate change. OECD Regional Development Working Papers, 2009(2), 1 (2009)
11. KDB+. https://kx.com/. Accessed 1 June 2017
12. Li, T.L., Liu, Y., Tian, Y., Shen, S., Mao, W.: A storage solution for massive IoT data based on NoSQL. In: Proceedings of IEEE International Conference on Green Computing and Communications, pp. 50–57 (2012)
13. Liu, X., Golab, L., Golab, W., Ilyas, I.F., Jin, S.: Smart meter data analytics: systems, algorithms and benchmarking. ACM Trans. Database Syst. (TODS) **42**(1), 1–39 (2016)
14. Liu, X., Golab, L., Golab, W., Ilyas, I.F.: Benchmarking smart meter data analytics. In: Proceedings of EDBT, pp. 385–396 (2015)

15. Olson, M.A., Bostic, K., Seltzer, M.I.: Berkeley DB. In: Proceedings of USENIX Annual Technical Conference, FREENIX Track, pp. 183–191 (1999)
16. OpenTSDB. http://OpenTSDB.net/. Accessed 1 June 2017
17. Pavani, M., Rao, P.T.: Real time pollution monitoring using Wireless Sensor Networks. In: Proceedings of IEMCON, pp. 1–6 (2016)
18. Phan, T.A.M., Nurminen, J.K., Di Francesco, M.: Cloud databases for Internet-of-things data. In: Proceedings of GreenCom, pp. 117–124 (2014)
19. Pintus, A., Carboni, D., Piras, A.: Paraimpu: a platform for a social web of things. In: Proceedings of the 21st International Conference on World Wide Web, pp. 401–404 (2012)
20. PostgreSQL. https://www.postgresql.org/. Accessed 1 June 2017
21. Prasad, R.V., Baig, M.Z., Mishra, R.K., Desai, U.B., Merchant, S.N.: Real time wireless air pollution monitoring system. ICTACT J. Commun. Technol. $2(2)$, 370–375 (2011)
22. Sanaboyina, T.P.: Performance Evaluation of Time series Databases based on Energy Consumption. Master thesis, Blekinge Institute of Technology (2016)
23. Van der Veen, J.S., Van der Waaij, B., Meijer, R.J.: Sensor data storage performance: SQL or NoSQL, physical or virtual. In: Proceedings of Cloud Computing, pp. 431–438 (2012)

Electric Vehicle Charging Station Deployment for Minimizing Construction Cost

Kai Li$^{(\boxtimes)}$ and Shuai Wang

Department of Computer Science and Technology,
Harbin Institute of Technology, Harbin, China
likai.1991.cs@gmail.com, wangshuaipower@hit.edu.cn

Abstract. Compared with traditional Gasoline Vehicles, Electric Vehicles have a huge advantage in green energy. But lagging construction of charging stations cannot keep up with development of the electric vehicle. The expensive charging stations require reasonable deployment plan aiming at construction cost. The construction cost of charging stations mainly consists of two items: charging equipment cost and charging station building cost depending on the capacity of charging station and the location of charging station respectively. However, the existing charging stations deployment researches focus on minimizing the time or the distance of charging-trip, instead of construction cost. In this paper, we propose CCOA to optimize location and capacity of charging stations in order to optimize construction cost. To our best knowledge, this is the first charging station deployment study with consideration of both equipment cost and building cost. Evaluation results show that 99.98% charging demand can be meet and 24% cost can be saved by CCOA. Through the comparison with other baselines, our approach can achieve more reliable deployment plan decision with less construction cost.

Keywords: Electric vehicle · Charging station deployment · Charging demand denoising · Grids clustering · Minimizing construction cost

1 Introduction

Compared with traditional Gasoline Vehicles, the Electric Vehicles have a huge advantage in green energy. The annual emissions of carbon dioxide-equivalent per Electric Vehicle are 4587 pounds, less than a third of the Gasoline Vehicles' [1]. Since there is no tail pipe air pollutants and noise pollution, Electric Vehicle (EV) is becoming increasingly popular while the shortage of charging stations is still a severe problem. In addition, lagging construction of charging stations cannot keep up with the development of electric vehicle [8]. Due to high construction cost, Ontario State even has poured $20M for only 5 quick public charging stations [7].

Therefore, it is very critical to optimize the construction cost of charging station deployment plan. The construction cost of charging stations mainly consists of two items: charging equipment cost and charging station building cost [6].

© Springer International Publishing AG 2017
L. Bellatreche and S. Chakravarthy (Eds.): DaWaK 2017, LNCS 10440, pp. 471–485, 2017.
DOI: 10.1007/978-3-319-64283-3_35

The charging equipment cost depends on the capacity of charging station (number of charging piles). In addition to the capacity, the building cost depends on the location of charging station and the corresponding estate price. Even through the unit charging equipment costs are assumed same across the city, the building costs vary remarkably in different areas [2].

Although charging station deployment has became an important part of intelligent transportation system (ITS), none of the existing researches aim at the construction cost in consideration of both charging stations capacity and location. Some existing work [10,14,16,18] focus on the average time or distance to the charging station, thus minimizing the drivers' discomfort. Some related researches [12,13,15] minimize the number of charging stations, but they do not consider the capacity allocation and spatial distribution of building cost (estate price).

In this paper, we present **CCOA** (Construction Cost Optimization Approach) to optimize location and capacity of charging stations in order to minimize construction cost. To our best knowledge, this is the first charging station deployment study in consideration of both equipment cost and building cost. We formulate this problem into an optimization problem where the objective is cost and constraints contain charging demands, limitation of grids and number of charging stations (L0-norm formulated constraint). Theoretically, this formulation is a non-convex form. We loose this non-convex problem into a L1-norm problem which can be solved by convex solver directly. As a result, a well-perform approximation solution: sparse and efficient charging station deployment plan could be gotten.

More specifically, building cost and the spatial distributions of charging demand are extracted from on-line estate price website and the trajectories of taxis in Shenzhen city which is a representative metropolis in China. Intuitively, the more target region's charging demands are, the more charging capacity should be allocated while the higher target region's estate price is, the less charging capacity should be allocated. The framework of our design is shown by Fig. 1.

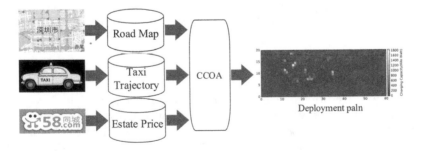

Fig. 1. Framework of our deployment decision system

Our main contributions are summarized as follows:

- We conduct the first work to make charging station deployment decision for minimizing construction cost with consideration of both equipment cost and building cost. And Our deployment approach can take the pre-existing charging stations into account to further reduce cost.

– The experimental results show that CCOA can achieve about 24% cost saving on construction cost compared with other baseline approaches. And 99.98% charging demand can be met by our approach.

The rest of the paper is organized as follows. How to extract useful information from multiple data sources are discussed in Sect. 2. The cost optimization model is presented in Sect. 3. Section 4 presents detail deployment optimization and analyses the results. The related works are reviewed in Sect. 5. Finally, we conclude this paper in Sect. 6.

2 Information Extraction

Charging demand and estate price should be collected before we introduce our main design. In our paper, we utilize trajectories of taxis and housing information (http://sz.58.com) in Shenzhen city to estimate the charging demand and estate price respectively. Due to privacy protection, it is hard to obtain the private cars' trajectory data, and our paper dose not consider the private cars' travel pattern. The voluntary private cars' full-time trajectory data can be added with the existing taxis' idle-trip trajectory data as the charging demand model.

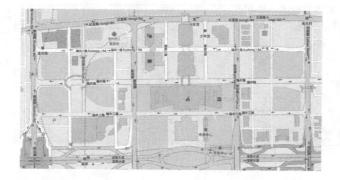

Fig. 2. Taxis' trajectory to charging station

2.1 Idle Trip

According to [14], most taxi drivers would like to go to charging station only when there is no passenger in taxis (idle-state). For example, Fig. 2 illustrates two taxi trajectories and their charging events. The GPS of each taxi is recorded and sent to data center via cellular network. As shown in Table 1, Record 1 (R_1) contains taxi ID current GPS, status, speed, and timestamp. The status of a taxi is a binary value while '0' indicates no passenger and '1' indicates there are some passengers in the taxi. Idle trip is defined as consecutive records where the status of a specific taxi are all '0's.

Table 1. Raw data format

Record	TaxiID	Time	Lat	Lon	Status	Speed
...	...	...	...	...	...	...
R_1	2252	19:00:00	22.528	113.969	1	32
R_2	2252	19:01:07	22.531	113.972	0	38
R_3	2252	19:02:11	22.538	113.968	0	21
R_4	2252	19:02:57	22.533	113.971	0	42
R_5	2252	19:03:39	22.536	113.972	1	35
...	...	...	...	...	...	...

2.2 Charging Demand Model

We assume that the urban area is partitioned into a number of non-overlapping square cells: $CP = \{C_1, C_2, \cdots, C_n\}$ where C_i represents the ith area in the city regardless of noise. The corresponding index set: $I = \{1, 2, \cdots, n\}$. In our experiment, the urban area of Shenzhen city is partitioned into 20×60 cells.

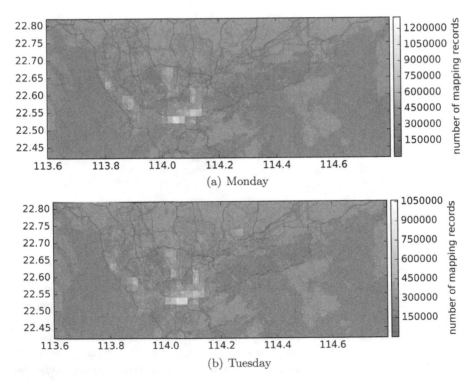

(a) Monday

(b) Tuesday

Fig. 3. Distribution of idle-trips mapping records number

The idle-trips pattern presents a certain and strong periodicity regardless of different days. We map idle-trip records into the corresponding cell based on GPS. The Fig. 3(a) and (b) show the different day's distribution of amount of mapping records. The slight difference caused by noise can be eliminated by our de-noise algorithm later on. Then we define the idle-trips pattern matrix $M(d \times n)$ where d represents how many days our data set cover. For instance, the element $m[k, j]$ of M is the number of kth day's idle-trips records mapping into cell C_j. The Principal Component Analysis (PCA) of M is shown by Fig. 4. Apparently, idle-trips pattern matrix is low-rank, and the periodicity is further verified. The rare fluctuation and the emergency should be regarded as the noise for charging demand model.

In order to avoid the noise's influence, we de-noise the charging demand by singular value decomposition (SVD) method:

$$M(d \times n') = U \times S \times V \tag{1}$$

where singular value matrix S can be gotten. The diagonal elements of S indicating the weight of each components are already sorted. Since the last (small) 5% diagonal elements of S is the result of noise, we enforce these small values to zero and get the de-noised singular value matrix SP. And the de-noised idle-trips pattern matrix:

$$MP(d \times n) = U \times SP \times V \tag{2}$$

The corresponding charging demand of CP:

$$D = \{D_1, D_2, \cdots, D_n \mid D_j = \frac{1}{n} \sum_{x=1}^{n} mp[x, j]\} \tag{3}$$

The $mp[x, j]$ is the corresponding element of matrix $MP(d \times n)$. In our experiment, the result is showed by Fig. 5 indicating that the demand of different cell is not even.

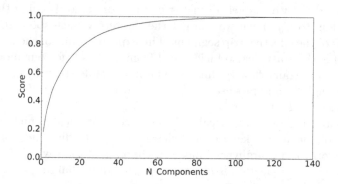

Fig. 4. PCA with $M(d \times n')$

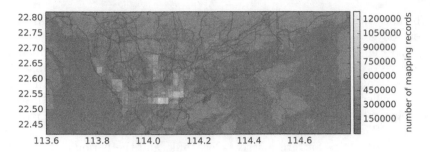

Fig. 5. The charging demand of CP: D

2.3 Impact of Traffic Condition

The charging-trip's traffic condition has a great effect on driver's charging decision. It's obvious that: the greater the spatial distance of trip between electric vehicle and charging station is, the less able or willing will driver be to charge. Similar to the trip spatial distance, the product of the statistical vehicle speed and the traffic flow volume also characterize the drivers' charging willing: In general, the high product result means good traffic condition. So the larger product result between electric vehicle and charging station is, the more willing to charge drivers would be.

Hence, we take both trip spatial distance and the product mentioned above into account when defining the charging willing. Before we define the charging willing, first we will define the following concept: statistical trip distance, statistical vehicle speed and traffic flow volume.

Definition 1: statistical trip distance. For cell C_x and cell C_y, we can get all idle-trip trajectory records passing by C_x and approaching to C_y. Then, the statistical trip distance between C_x and C_y (denoted by STDM(x,y)) is defined as median value of number of different cells from C_x to C_y in all trajectories.

Definition 2: statistical vehicle speed: Similar to statistical trip distance, for cell C_x and cell C_y, we can get all idle-trip trajectory records going through C_x and proceeding to C_y. Then we can get the discrete vehicle speed distribution from records of the effective trip segment. The expected value of the distribution: $SVSE(x,y)$ is the statistical vehicle speed from C_x to C_y for our model.

Definition 3: traffic flow volume: The number of idle-trip trajectory records, which go through C_x and proceed to C_y, is the corresponding traffic flow volume. It is formalized as $TFV(x,y)$.

As mentioned above, statistical trip distance, statistical vehicle speed and traffic flow volume are the key items of charging-trip's traffic condition. And the three items form the charging willing. For charging station which is in cell C_y and the driver whose location is C_x, the corresponding charging willing is:

$$CW(x,y) = \frac{SVSE(x,y) \times TFV(x,y)}{STDM(x,y)} \tag{4}$$

2.4   Estate Price Model

The construction costs of charging station varies over the location and capacity significantly. The urban areas with low estate price should be deployed more charging piles than the high price areas. We use the average estate price (sourced by http://sz.58.com/) in each cell as the unit building cost indicator. The CP's corresponding average on-line estate price is:

$$EP = \{EP_1, EP_2, \cdots, EP_n\} \tag{5}$$

Figure 6 illustrates the spatial distribution of on-line estate price across Shenzhen City. It's obvious that the Fig. 6 is similar with the Fig. 5, and it means that the deployment decision without consideration of estate price's distribution may lead to the much higher construction cost. Only considering the distribution of charging demand while ignoring estate price, can not be the optimal deployment plan for construction cost.

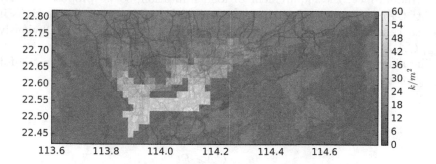

Fig. 6. The spatial distribution of estate price

3   Construction Cost Optimization

We aim to minimize construction cost of charging station deployment. Our paper formulate the deployment optimization problem as an capacity allocation problem. In detail, let the allocation plan $CA = \{CA_1, CA_2, \cdots, CA_n\}$ denotes the deployment configuration, with each CA_{ij} representing the charging capacity should be deployed in cell CP_j. If CA_j is 0, it means that there is no charging station construction for cell CP_j. Our paper's optimization model is shown below:

$$\underset{CA}{\arg\min} \quad \sum_{k \in I} (\lambda \times CA_k + \beta \times CA_k \times EP_k)$$

$$s.t. \quad \sum_{x \in I} CA_x \times CW(k, x) \geqslant D_k, \quad \forall k \in I$$

$$CA_k \geqslant EC_k, \quad \forall k \in I$$

$$CA_k \leqslant L_k, \quad \forall k \in I$$

$$\sum_{x \in I} CA_x{}^0 - \sum_{x \in I} EC_x{}^0 \leqslant K$$

We use λ to denote the equipment cost of unit charging capacity, and $\sum_{k \in I} \lambda \times CA_k$ represents the equipment cost of total charging capacity. The β is the spatial space of unit capacity equipment and the EP_k is the corresponding estate price of cell CP_k, so the $\sum_{k \in I} \beta \times CA_k \times EP_k$ is the building cost of all charging stations. As mentioned in introduction, the sum of the equipment cost and the building cost: $\sum_{k \in I} (\lambda \times CA_k + \beta \times CA_k \times EP_k)$ is the construction cost.

The constraint: $\sum_{x \in I} CA_x \times CW_k(x) \geqslant D_k, \forall k \in I$ guarantees that the different cell's charging demand can be met by the deployment. The EC_k is the cell C_k's existing charging station capacity. $EC_k = 0$ indicates that initially no charging station is deployed in cell C_k, otherwise the C_k has a charging station with EC_k volume capacity. The constraint: $CA_k \geqslant EC_k, \forall k \in I$ makes our approach flexible. The existing charging stations can be sufficiently utilized by our approach. The load limitation of power network is another important constraint for charging station deployment. For cell C_k, the L_k is the maximum limitation of power load for charging station. The capacity deployment configuration of C_k should meet the requirement of load limitation, so we propose the constraint: $CA_k \leqslant L_k, \forall k \in I$. Since there exist $\sum_{x \in I} EC_x{}^0$ charging stations in different cells of CP, the number of charging stations to deploy is K. Obviously, $\sum_{x \in I} CA_x{}^0 - \sum_{x \in I} EC_x{}^0 \leqslant K$ should hold.

The above optimization problem may be conveniently expressed in the following vector and matrix form:

$$\underset{\overrightarrow{CA}}{\arg\min} \quad \overrightarrow{\lambda}^{\top} \times \overrightarrow{CA} + \beta \times \overrightarrow{CA}^{\top} \times \overrightarrow{EP}$$

$$s.t. \quad \mathbf{CW} \times \overrightarrow{CA} \geqslant \overrightarrow{D}$$
$$\overrightarrow{CA} \geqslant \overrightarrow{EC}$$
$$\overrightarrow{CA} \leqslant \overrightarrow{L}$$
$$\|\overrightarrow{CA}\|_0 - \|\overrightarrow{EC}\|_0 \leqslant K$$

Table 2 provides notations in above formulation. Because L0-norm is nondifferentiable and non-convex, it is hard to directly solve the optimization problem above. The constraint: $\|\overrightarrow{CA}\|_0 - \|\overrightarrow{EC}\|_0 \leqslant K$ is to limit the number of charging station, thus minimizing the number of charging stations to deploy. In general, L1-norm is the tightest convex relaxation of L0-norm due to the intrinsic sparsity of L1-norms [9,11,17]. So we try to guarantee the sparsity of $\overrightarrow{CA}$ without changing optimization object. We have the new L1-norm based optimization problem without L0-norm constraint:

$$\underset{\overrightarrow{CA}}{\arg\min} \quad \lambda \times \|\overrightarrow{CA}\|_1 + \beta \times \overrightarrow{CA}^{t} \times \overrightarrow{EP}$$

$$s.t. \quad \mathbf{CW} \times \overrightarrow{CA} \geqslant \overrightarrow{D}$$
$$\overrightarrow{CA} \geqslant \overrightarrow{EC}$$
$$\overrightarrow{CA} \leqslant \overrightarrow{L}$$

It is obvious that the above L1-norm based optimization problem is convex: the object is the sum of L1-norm and linear programming; the constraint is linear constraint. So, some mature tools for convex optimization can be applied directly in our deployment approach. Our experiments also take advantage of the state-of-art convex optimization tools, which will be described in the following section. It is important to note that the optimization object $\lambda \times ||\overrightarrow{CA}||_1$ is equal to $\overrightarrow{\lambda}^\top \times \overrightarrow{CA}$ because of the non-negativity constraint: $\overrightarrow{CA} \geqslant \overrightarrow{EC}$.

Table 2. Key notations

Notations	Description
$\overrightarrow{\lambda}$	$\overrightarrow{\lambda} \in R(n \times 1)$, and $\forall x \in \{1, 2, \cdots, n\}$, $\overrightarrow{\lambda}[x, 1] = \lambda$
$\overrightarrow{CA}$	$\overrightarrow{CA} \in R(n \times 1)$, and $\forall x \in \{1, 2, \cdots, n\}$, $\overrightarrow{CA}[x, 1] = CA_x$
$\overrightarrow{EP}$	$\overrightarrow{EP} \in R(n \times 1)$, and $\forall x \in \{1, 2, \cdots, n\}$, $\overrightarrow{EP}[x, 1] = EP_x$
$\overrightarrow{D}$	$\overrightarrow{D} \in R(n \times 1)$, and $\forall x \in \{1, 2, \cdots, n\}$, $\overrightarrow{D}[x, 1] = D_x$
$\overrightarrow{EC}$	$\overrightarrow{EC} \in R(n \times 1)$, and $\forall x \in \{1, 2, \cdots, n\}$, $\overrightarrow{EC}[x, 1] = EC_x$
$\overrightarrow{L}$	$\overrightarrow{L} \in R(n \times 1)$, and $\forall x \in \{1, 2, \cdots, n\}$, $\overrightarrow{L}[x, 1] = L_x$
CW	$\mathbf{CW} \in R(n \times n)$, and $\forall x, y \in \{1, 2, \cdots, n\}$, $\mathbf{CW}[x, y] = CW(x, y)$

4   Evaluation

We perform a series of experiments based on real data to evaluate the performance of CCOA. All experiments are ran on the same computer with INTEL core i5-6400 CPU at 2.7 GHz and 16 GB RAM, and conducted in PYTHON 2.7 and INTEL Math Kernel Library [4] environment.

4.1   Data Set

The taxis' trajectory data set is collected from about 14000 taxis in Shenzhen City [19] in different days. The average trajectory sampling rate is about 60 seconds. The estate price data set is sourced from http://sz.58.com/. The raw unit estate price is in terms of Renminbi yuans. The average number of estate price sample is about 20 in each cell.

4.2   Baselines

Three baselines, denoted by **Rand-PA**, **GPDA**, **OCSD** [14] and **OSS** [12], are compared against the proposed optimization approach, CCOA.

- **Rand-PA** Random Placement and Assignment: This baseline approach chooses the $K - ||\overrightarrow{EC}||_0$ locations randomly and uniformly. Then random capacity is assigned.

- **GPDA** Greedy-Coverage Placement and Demand-based Assignment: In detail, for the target area with n cells, we can get $L = \lfloor n/K \rfloor - 1$. The target cells can be divided into two sets: *not-in-coverage* and *in-coverage* (initialized as empty set), the *not-in-coverage* $\cup$ *in-coverage* = {target cells} and *not-in-coverage* $\cap$ *in-coverage* = $\emptyset$. We always choose the cell $C_{x'}$ from *not-in-coverage* as the station-located cell, and $C_{x'}$ must meet the following requirements: $\{C_{x_1}, \cdots, C_{x_L}\}$ is the set consisted of neighbors of $C_{x'}$ in *not-in-coverage*; the corresponding charging willing to $C_{x'}$ is $\{CW(x_1, x'), \cdots, CW(x_L, x')\}$; there is no other $C_{y'} \in$ *not-in-coverage* to make $CW(x_1, x') + ... + CW(x_L, x') < CW(y_1, y') + ...+ CW(y_L, y')$. Then $C_{x'}$ and the top(L) neighbor cells are deleted from *not-in-coverage* and added into *in-coverage*. The above process need to repeat for K times to pick up K cells for locations of charging stations. At last, the capacity allocation depends on the proportion of each L-cells group in all cells.
- **OCSD** Optimal Charging Station Deployment: The OCSD also takes the taxi trajectory data, road map data and existing charging station information as input. The OCSD includes following two main parts: Optimal Charging Station Placement (OCSP) and Optimal Charging Point Assignment (OCPA). Without consideration of construction cost, the OCSD aims to minimize the average time to travel to charging station, and the average waiting time for an available charging point, respectively. As a baseline, the OCSD does not need the estate price as input.
- **OSS**: The cost of OSS consists of the charging station construction costs and charging costs. OSS aims to optimize the number, and locations of charging stations to minimize the overall cost.

4.3   Evaluation Metrics

The following definitions are used to evaluate the efficiency of the different approaches:

- **cc construction cost**: cc represents the construction cost of charging station deployment plan CA. As discussed in Sect. 4, the cc is:

$$cc = \lambda \times ||\overrightarrow{CA}||_1 + \beta \times \overrightarrow{EC}^\top \times \overrightarrow{EP}$$

  The value of parameters λ, β will be discussed in the following subsection.
- **wdsr weighted demand satisfaction rate**: $wdsr$ represents the sum of charging demand satisfaction rate for different weighted cells. In other words, $wdsr$ is the sum ratio of different weighted cells' support to demand. The weight of different cells depend on the corresponding charging demand.

$$wdsr = \sum_{k \in I}(DW_k \times \min(\frac{DS_k}{D_k}, 1))$$

For cell C_k, DS_k is the corresponding charging service support: $DS_k = \sum_{x \in I} CA_x \times CW(k, x)$; the charging demand: D_i is defined by above section.

Because of the charging demand difference between cells, the different cells should have different importance when making decision about charging stations' location and capacity. And the charging demand weight: $DW_k = D_k/\sum_{x\in I} D_x$. For cell C_x, $\forall x \in I$, the corresponding demand satisfaction rate should be less than 1.

- **wsrr weighted support redundancy rate**: $wsrr$ represents the sum of charging support redundancy rate for different weighted cells. The practical deployment plan should consider the redundancy capacity for unexpected demand. In detail, $wsrr$ is the sum ratio of different cells' redundancy support to demand. The $wsrr$'s weight is similar to the $wdsr$'s.

$$wdrr = \sum_{k\in I}(DW_k \times \max(\frac{DS_k - D_k}{D_k},0))$$

The notation in $wdrr$ is same with $wdsr$'s. But the need to pay attention to is that if the charging support DS_k is less than the corresponding charging demand D_k, the support redundancy rate of cell C_k should be zero.

4.4   Experiment Settings and Evaluation Results

As mentioned in above section, λ is the equipment cost of unit charging capacity, and β is the spatial space of unit charging capacity. According to the on-line introduction of charging pile, the $\lambda = \$1.2k$ and $\beta = 0.0185\,\mathrm{m}^2$ when the unit of charging capacity is kilo watt (kW).

Our convex optimization problem can be solve by mature convex optimization tools directly. Because of the good adaptability to large-scale data, Splitting Conic Solvers (SCS) [5] is our solver with cvxpy [3].

Table 3 shows the comparison on cc, when applying our construction cost optimization based approach (CCOA) and other baselines. Due to the consideration of optimization for construction cost, our CCOA has advantage on the aspect of cc.

Table 3. Construction cost

Approach	Construction cost
CCOA	$204M
OSS	$267M
Rand-PA	$385M
GPDA	$279M
OCSD	$329M

Figure 7 shows the comparison on $wdsr$ and $wsrr$, when applying our construction cost optimization based approach (CCOA) and other baselines. The experiment result indicates that our CCOA performs better than other baselines on $wdsr$ because of the support-demand constraint: $\sum_{x\in I} CA_x \times CW(k,x) \geqslant$

$D_k, \forall k \in I$. Our CCOA's $wdsr(99.98\%)$ is slightly lower than 100%, which mainly results from the complex solution space of global optimization. In practical application, there are several ways to avoid less-than-100% $wdsr$ solution, such as the clustering of cells. And this is our future work. Although $GPDA$ and $OCSD$ perform slightly better than our CCOA on $wsrr$, the construction cost of $CCOA$ is lower. As the most foundation constraint of deployment plan, $wdsr$ is much more important than $wsrr$. Overall, our CCOA has obvious competitive advantage on construction cost.

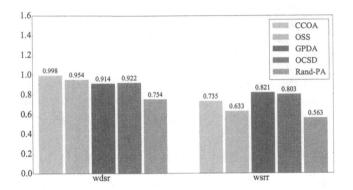

Fig. 7. The experiment results on $wdsr$ and $wsrr$.

But GPDA should be concerned because of its higher efficiency of $wsrr$, which may be of great value for incremental deployment in the future. The deployment plan by our CCOA approach is shown by Fig. 8. Another interesting phenomenon is that our solution for the higher demand area is: deploying more charging stations with less capacity. It is similar with the deployment strategy of gasoline station designed by expert, the smaller gasoline stations are dense deployment in downtown, the larger stations are sparse in suburban.

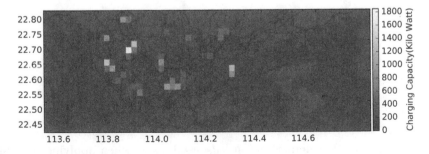

Fig. 8. The deployment plan by CCOA.

5 Related Work

The charging station research is an important part of ITS research now. The existing charging station deployment researches can be divided into two major categories: *to minimize discomfort of EV drivers* and *to minimize cost of charging stations*.

To Minimize Discomfort of EV Drivers: [1] aim to minimize the social cost (the all EVs time cost) of charging-trips, and the time cost consist of charging-trips travel time cost and charging queuing time at station. [18] also considers EV drivers strategic behaviors to minimize their time cost. [14] propose the Optimal Charging Station Deployment (OCSD) framework. The taxi trajectory data, road map data and existing charging station information is the input of the framework. Similar to [18], OCSD aims to minimize the average time for the drivers charging-trip travel time and queuing time. But different with [18], OCSD performs Optimal Charging Station Placement (OCSP) and Optimal Charging Point Assignment (OCPA). The location of charging stations is decided by OCSP first, and then do capacity decision by (OCPA). [10] try to predict when and where vehicles are likely to be parked. It is assumed that the distribution of charging demand can be represented as the pattern of parking demand. The deployment plan is decided by minimizing total system travel distances to the closest charging station. [16] aims to minimize the overall number of charging stations required and the aggregate distance drivers need to travel to reach the closet charging station simultaneously. But the NP-hard problem make that it is difficult to solve directly. [16] propose a near-optimal solution based on greedy and genetic algorithms.

[14, 18] only constrain the number of charging stations, without consideration of relationship between construction cost and charging stations' configuration: location and capacity. And [10, 14, 16, 18] assume that the drivers only choose the closest charging station, which is obviously unreasonable.

To Minimize Cost of Charging Stations: In [13], the road network is formulated as an undirected graph, where the vertices denote intersection of road and edges denote the road. The vertices are the location candidate of charging station deployment. Given the each vertex's charging station construction cost, [13] aim to minimize the total construction cost under constrains that (i) for any vertex, there is a charging station whose distance from the vertex is less than threshold value (ii) the demand of each vertex must be satisfied by the total capacity of the surrounding charging stations. In [15], the optimization objective is to minimize the cost of charging stations, which includes the annual average construction cost of charging stations, the annual operating cost and the power supply operation cost. It is assumed that the distribution of charging demand can be represented as the distribution of traffic flow. Similar to [12, 13] builds the charging stations within road networks. The optimization object of [12] is to minimize the integrated cost of investment and operation. The optimization problem is formulated as a linear optimization problem.

[12,13] don't take into account the difference of capacity between charging stations, without consideration of relationship between construction cost and charging stations' capacity. For [12,13], the charging station candidate of deployment plan can be only the intersections of roads, which is obviously unreasonable. The optimization problems of [12,13,15] are linear optimization problem without the constraint: the number of charging station, which leads to the dense deployment.

6 Conclusion

While charging station deployment has been an hot topic on smart city area, the existing work aim at charging time or distance, and do not take construction cost into account. Motivated by this, we propose a novel approach: CCOA, which can deploy charging stations to minimize the construction cost subject to the constraints of charging QoS. Our experiment results are presented from three aspects, (1) construction cost, (2) weighted demand satisfaction ratio and (3) weighted support redundancy ratio. We compare our deployment results with baselines based on large-scale data experiment and present their differences in detail. Our CCOA approach can provide 24% cost saving on construction cost than the baselines.

References

1. Emissions from Hybrid and Plug-In Electric Vehicles. http://www.afdc.energy.gov/vehicles/electric_emissions.php
2. EV Charging Station Infrastructure Costs. https://cleantechnica.com/2014/05/03/ev-charging-station-infrastructure-costs/
3. Introduction of CVXPY. http://www.cvxpy.org/en/latest/
4. Introduction of Intel Math Kernel Library. https://software.intel.com/en-us/intel-mkl
5. Introduction of Splitting Conic Solver. https://github.com/cvxgrp/scs
6. Investors 'Pile' into China's Public EV Charging Industry Despite Lack of Profitability. http://www.renewableenergyworld.com/articles/2016/01/investors-pile-into-china-s-public-ev-charging-industry-despite-lack-of-profitability.html
7. Ontario spends $20M to build electric vehicle charging stations. http://www.cbc.ca/news/canada/toronto/ontario-electric-vehicle-charging-stations-1.3355595
8. Skeptical motorists, shortage of charging stations the biggest obstacles in Ontario's electric car plan. http://www.cbc.ca/news/technology/electric-vehicles-climate-change-plan-ontario-1.3624289
9. Candes, E.J., Tao, T.: Near-optimal signal recovery from random projections: universal encoding strategies? IEEE Trans. Inf. Theory 52(12), 5406–5425 (2006)
10. Chen, T.D., Kockelman, K.M., Khan, M., et al.: The electric vehicle charging station location problem: a parking-based assignment method for seattle. In: Transportation Research Board 92nd Annual Meeting, vol. 340, pp. 13–1254 (2013)
11. Donoho, D.L., Elad, M.: Optimally sparse representation in general (nonorthogonal) dictionaries via minimization. Proc. Natl. Acad. Sci. 100(5), 2197–2202 (2003)

12. Jia, L., Hu, Z., Song, Y., Luo, Z.: Optimal siting and sizing of electric vehicle charging stations. In: 2012 IEEE International Electric Vehicle Conference (IEVC), pp. 1–6. IEEE (2012)
13. Lam, A.Y., Leung, Y.-W., Chu, X.: Electric vehicle charging station placement. In: 2013 IEEE International Conference on Smart Grid Communications (SmartGridComm), pp. 510–515. IEEE (2013)
14. Li, Y., Luo, J., Chow, C.-Y., Chan, K.-L., Ding, Y., Zhang, F.: Growing the charging station network for electric vehicles with trajectory data analytics. In: 2015 IEEE 31st International Conference on Data Engineering (ICDE), pp. 1376–1387. IEEE (2015)
15. Liu, Z., Zhang, W., Ji, X., Li, K.: Optimal planning of charging station for electric vehicle based on particle swarm optimization. In: 2012 IEEE Innovative Smart Grid Technologies-Asia (ISGT Asia), pp. 1–5. IEEE (2012)
16. Vazifeh, M.M., Zhang, H., Santi, P., Ratti, C.: Optimizing the deployment of electric vehicle charging stations using pervasive mobility data. arXiv preprint arXiv:1511.00615 (2015)
17. Wang, S., Li, Z., Jiang, S.: Distributed energy-efficient power control algorithm of delay constrained traffic over multi fading channels. In: 2014 IEEE 12th International Conference on Dependable, Autonomic and Secure Computing (DASC), pp. 344–349. IEEE (2014)
18. Xiong, Y., Gan, J., An, B., Miao, C., Bazzan, A.L.: Optimal electric vehicle charging station placement. In: IJCAI, pp. 2662–2668 (2015)
19. Zhang, D., Zhao, J., Zhang, F., He, T.: Urbancps: a cyber-physical system based on multi-source big infrastructure data for heterogeneous model integration. In: Proceedings of the ACM/IEEE Sixth International Conference on Cyber-Physical Systems, pp. 238–247. ACM (2015)

Author Index

Printed in the United States
By Bookmasters